NSTAR 2001

Proceedings of the Workshop on
The Physics of Excited Nucleons

NSTAR 2001

Proceedings of the Workshop on The Physics of Excited Nucleons

Mainz, Germany 7–10 March 2001

Editors

D. Drechsel & L. Tiator

Institut für Kernphysik, Universität Mainz, Germany

World Scientific

New Jersey • London • Singapore • Hong Kong

Published by

World Scientific Publishing Co. Pte. Ltd.
P O Box 128, Farrer Road, Singapore 912805
USA office: Suite 1B, 1060 Main Street, River Edge, NJ 07661
UK office: 57 Shelton Street, Covent Garden, London WC2H 9HE

British Library Cataloguing-in-Publication Data
A catalogue record for this book is available from the British Library.

NSTAR 2001
Proceedings of the Workshop on The Physics of Excited Nucleons

ISBN 981-02-4760-5

Printed in Singapore.

NSTAR 2001

Workshop on
THE PHYSICS OF EXCITED NUCLEONS
Universität Mainz, Germany
March 7–10, 2001

Organization

Conference Chairmen
V. Burkert (Jefferson Lab), D. Drechsel, and Th. Walcher (Mainz)

Organizing Committee
R. Beck (Mainz), V. Burkert (Jefferson Lab), D. Drechsel (Mainz), E. Klempt (Bonn), M. Ripani (Genova), B. Saghai (Saclay), H. Schmieden (Mainz), P. Stoler (RPI), L. Tiator (Mainz), Th. Walcher (Mainz)

Advisory Committee
G. Anton (Erlangen), C. Bennhold (GWU), L. Cardman (Jefferson Lab), C. E. Carlson (Jefferson Lab), E. de Sanctis (Roma), S. Dytman (Pittsburgh), A. Faessler (Tübingen), R. Frascaria (Orsay), N. Isgur (Jefferson Lab), H. Lee (Argonne), V. Metag (Gießen), R. Milner (MIT), T. Nakano (Osaka), B. Nefkens (UCLA), E. Oset (Valencia), C. Papanicolas (Athen), W. Plessas (Graz), A. Radyushkin (Jefferson Lab), D. O. Riska (Helsinki), A. Sandorfi (Brookhaven), B. Schoch (Bonn), S. Simula (Roma), H. Ströher (Jülich), W. Weise (München), R. Workman (GWU), S.-N. Yang (NTU Taipei), B.-S. Zou (Beijing)

Institutional Sponsors
Deutsche Forschungsgemeinschaft
Jefferson Laboratory
Johannes Gutenberg-Universität Mainz
State Government of Rhineland-Palatinate

Foreword

This Volume contains both the invited lectures and the contributions to the Workshop on *The Physics of Excited Nucleons (NSTAR 2001)*, which was held at the Johannes Gutenberg-Universität Mainz, March 7–10, 2001. The origin of this workshop series goes back to the conference on *Excited Baryons*, which was initiated and organized by our late friend Nimai Mukhopadhyay and his colleagues at the Rensselaer Polytechnic Institute in 1988. More recent workshops of the series on N* physics took place at the Florida State University (1994), at CEBAF (1995), at the Institute for Nuclear Theory in Seattle (1996), at the George Washington University (1997), at the ECT* in Trento (1998), and at the Thomas Jefferson Laboratory in February 2000. Immediately before NSTAR 2001, the *Baryon Resonance Analysis Group (BRAG)* had its working group meetings on *resonance extraction and interpretation, partial wave analysis, and database issues.*

It was the aim of the Workshop to present the recent experimental and theoretical results in the field of nucleon resonance physics. A wealth of new high precision data was presented from facilities around the world, such as BES, BNL, ELSA, GRAAL, JLab, MAMI, MIT/Bates, SPring8, and Yerevan. Particular emphasis was laid on polarization degrees of freedom and large acceptance detectors as precision tools to study small but important transition amplitudes, and the helicity (spin) structure of the nucleon. On the theory side, the impact was two-fold. First, there were new results describing the nucleon resonance structure on the basis of Quantum Chromodynamics, either directly in terms of quarks and gluons by means of lattice gauge theory, or in terms of hadrons in the framework of chiral field theories. A status report on duality showed the surprising connections between the physics of the low-energy nucleon resonance region and the realm of quark structure functions in deep inelastic scattering. Second, three sessions and the BRAG workshop were devoted to an improved understanding of resonance structure in terms of quark and other dynamical models and to a detailed data analysis by partial wave expansions and coupled channels calculations, with the aim to establish resonance properties in a unique and practically model-independent way.

The Workshop was attended by about 130 scientists from more than 50 universities and laboratories from 20 countries. We thank all the participants for their valuable contribution to make the Workshop a success.

We wish to thank the institutional sponsors for their support: Deutsche Forschungsgemeinschaft, Thomas Jefferson National Accelerator Facility, Johannes Gutenberg-Universität at Mainz, and State Government of Rhineland-

Palatinate. We gratefully acknowledge the advice of the members of the International Advisory Committee and the help of the members of the Local Organizing Committee. Our special thanks goes to all the speakers who provided the basis for a successful conference.

We are very grateful to all the students who helped to organize the Symposium and to Monika Baumbusch, Sabine Bua, Roswitha Drescher and Felicia Ohl for handling the numerous administrative details. Finally, a special note of thanks is due to Felicia Ohl for her expert assistance in preparing these proceedings.

Mainz, June 2001

V. Burkert, D. Drechsel, L. Tiator, and Th. Walcher

CONTENTS

Appendix

NUCLEON RESONANCES IN THE QUARK MODEL

S. CAPSTICK
Department of Physics,
Florida State University
Tallahassee, FL, 32306-4350, USA
E-mail: capstick@csit.fsu.edu

Recent developments in the description of nucleon resonances using quark models are surveyed, with an emphasis on how such models can be made more consistent, both internally and with expectations from QCD.

1 Introduction

The study of nucleon resonances in the quark model can be used to identify the effective degrees in baryons, and their properties. These effective degrees of freedom undergo a confining interaction, and in most models they also have a 'residual' interaction between them at short distances. Inclusive models are described which attempt to describe all baryon states and a wide range of their properties, and which differ mainly in their description of this residual interaction between the quarks. Implementation of constraints due to chiral symmetry remains to be achieved in inclusive models. Recent results from lattice QCD calculations are outlined which may constrain models.

An important goal is to identify hybrid baryons (baryon states with the confining glue in an excited state), and models such as the flux-tube model can provide their quantum numbers and estimates of their masses and decay properties.

Couplings to baryon-meson intermediate states can have important effects on the baryon spectrum and should not be overlooked. Similarly, in order to make models of nucleon resonances more realistic, their predictions for masses and basic amplitudes must be combined with models of reaction dynamics in order to describe observables in scattering reactions.

2 Effective Degrees of Freedom

In potential models of baryon structure the effective degrees of freedom are taken to be constituent quarks, with effective light quark masses in models with a relativistic kinetic energy of roughly 220 MeV, and in non-relativistic models roughly 330 MeV. Strange quarks have masses from 420-550 MeV. The constituent quarks are not point-like, but have electromagnetic and strong form factors. The strong form factors are usually taken to be Gaussian or monopole, with heavier quarks being more point like. These form factors necessarily make finite the contact interactions between the quarks, which are otherwise proportional to $\delta^3(\mathbf{r}_i - \mathbf{r}_j)$. Electromagnetic form factors for the quarks have proven necessary to fit the nucleon form factors, even in relativistic calculations based on light-cone dynamics [1,2]. A typical choice is a monopole form for the F_1^i form factor of quark flavor i, and a dipole form for F_2^i, proportional to an assumed anomalous magnetic moment κ_i. It has been shown in a lattice QCD calculation [3] (similar results are obtained by considering baryon-meson intermediate state contributions to baryon structure) that the κ_i can not be considered universal, but are dependent on the baryon state in which the quarks reside.

Baryons have been described as a quark plus a diquark cluster. The simplest of these models uses a tightly-bound scalar-isoscalar diquark, a channel in which the short-range forces between quarks are known to be attractive. Such models have fewer degrees of freedom at energies too low to break up the diquark, which implies fewer low-lying excited states. The identification of several positive-parity baryon states in the 1.7-2.1 GeV range which are predicted by symmetric quark models, but currently 'missing' from analyses, would rule out such models. A model based on a spectrum-generating algebra [4] describes excited states in terms of collective excitations of string-like objects which carry the quark quantum numbers, with radial excitations arising from rotations and vibrations of these strings. As the degrees of freedom are extended, there are more excited states, some of which would be described in other models as excitations of the confining glue.

The constituent quark model in its standard form does not possess the chiral symmetry present in QCD. The incorporation of chiral symmetry into a model which describes the structure of low-lying baryons was accomplished using the cloudy-bag model [5], which puts together light quarks in a bag with pions coupled to the surface. Here the effective degrees of freedom are current quarks moving relativistically inside a (spherical) bag surface, and pions. Although successful at describing the electromagnetic properties of predominantly spherical ground states, the generalization of such a model to excited states of the quarks likely to deform the bag away from a spherical shape is technically involved. Calculations of nucleon and Δ properties using the Schwinger-Dyson Bethe-Salpeter approach [6] are in progress. These difficult calculations are covariant, and exhibit exact chiral symmetry with dynamically generated running quark masses. Work is also underway to build a chiral quark model based on the Hamiltonian approach [7] to QCD.

3 Confining Interaction

A modern picture of the confining interaction between the quarks is that it is provided by flux tubes of lengths l_i emanating from the quarks which meet at a junction. In the adiabatic approximation this gives a confining interaction from the minimum length of these tubes [8] for a given set of quark positions $V(r) = \sigma \sum_i l_i = \sigma L_{\rm min}$, where σ is the meson string tension, which is obviously linear for large quark-junction separations. This picture implies the existence of new excited states based on excitations of the flux tubes.

Confirmation of this picture for heavy-quark systems has been provided by a quenched lattice measurement of the QQQ potential using a Wilson loop for three static quarks [9]. The resulting potential is fit to a form

$$V_{3Q} = -A_{3Q} \sum_{i<j} 1/|\mathbf{r}_i - \mathbf{r}_j| + \sigma_{3Q} L_{\rm min} + C_{3Q}. \tag{1}$$

The $Q\bar{Q}$ potential is also measured and fit to a constant plus (attractive) Coulomb plus linear form, in order to compare the strengths of the linear and Coulomb terms. The results show that the string tensions are similar in heavy-quark mesons and baryons, and that the coefficients of the baryon and meson Coulomb terms are in the ratio 1/2 expected from one-gluon-exchange (OGE) or a similar $\lambda_i \cdot \lambda_j$ potential. By examining many sets of quark positions it is also shown that the confining potential is not fit well by a constant times the sum of the inter-quark separations (a 'Δ' potential), which reinforces the flux-tube picture described above, at least for heavy quarks.

4 Residual Interactions

The spectrum of the flavor octet and decuplet ground-state baryons shows flavor dependent short-range (or contact) interactions between the quarks. A good fit to the masses of these states can be made assuming that its source is a OGE color-magnetic dipole-dipole interaction of the form

$$M = \sum_{i=1}^{3} m_i + \frac{2\alpha_s}{3}\frac{8\pi}{3}\langle\delta^3(\mathbf{r})\rangle \sum_{i<j=1}^{3} \frac{\mathbf{S}_i \cdot \mathbf{S}_j}{m_i m_j}. \tag{2}$$

This yields, for example, the relation $(M_\Sigma - M_\Lambda)/(M_\Delta - M_N) = (2/3)(1 - m_{u,d}/m_s)$ which is capable of explaining the $\Sigma - \Lambda$ mass difference with a ratio of light to strange quark masses which fits other physics of these states, such as the magnetic moments. This interaction also has the advantage that it explains regularities in the meson spectrum, such as the evolution of the vector-pseudoscalar mass splitting with quark mass. Note that this is a non relativistic form and so it is not clear why it should work for light quark systems.

The adoption of OGE for the contact interaction implies the presence of tensor and spin-orbit interactions between the quarks, with strengths commensurate with that of the contact interaction. While there is no strong evidence for the tensor interaction from fits to masses which come out of partial wave analyses, there is some evidence from the strong and electromagnetic decays of these states that the mixings between states caused by a one-gluon (or other vector) exchange tensor interaction are present. A good example is the large $N\eta$ decay width for the lighter of the two S_{11} states at 1535 MeV and the small $N\eta$ width for the heavier state at 1650 MeV, which can be explained as due to mixing between the quark-spin-1/2 and 3/2 states caused by the tensor interaction of the strength implied by the contact interaction [10].

Although there are also spin-orbit interactions arising from the Thomas precession of the quarks in the confining potential, and there is a partial cancellation of these with the OGE spin-orbit interactions, the spin-orbit interactions which accompany the contact and tensor interactions in the non-relativistic model cause splittings which are unphysically large.

A variational calculation in a large basis allows the use of a relativistic kinetic energy and parametrized relativistic corrections to the potentials of the kind expected from examining the OGE T-matrix element away from the non relativistic limit [11]. Using a string model for confinement plus the associated spin-orbit interactions, as well as all of the interactions expected from OGE, a reasonable fit to the spectrum of all baryons can be made. Spin-orbit interactions are smaller, partly due to the use of a smaller α_s once the contact interactions are smeared over the constituent quark size and evaluated without resort to wave function perturbation theory, and partly by choosing the relativistic parameters to make them so.

Another possibility for the residual interactions are caused by the exchange between light quarks of pions [12,13], or an octet of pseudo-Goldstone-boson pseudoscalars [14]. This gives a contact interaction with explicit SU(3) flavor dependence in addition to the dependence on quark masses. Using this model it is possible to arrange for the lightest radial excitations, like $N1/2^+(1440)$, $\Delta 3/2^+(1600)$, and $\Lambda 1/2^+(1600)$ with masses similar to, or in the case of the Roper resonance, lighter than those of low-lying negative-parity states of the same flavor, by fitting the radial matrix elements of the contact potential to the spectrum. More sophisticated calculations in large variational bases exist [15] which calculate these matrix elements along with the associated tensor interactions, a relativistic kinetic energy and string confinement, with additional potentials due to the exchange of a nonet of vector mesons and a scalar expected from exchange of two pions. This model can also provide a reasonable fit to the low-lying states in the spectrum.

A second explicitly flavor-dependent possibility is that the residual interactions are instanton induced, which can cause an short-range (contact) attractive interaction between two quarks in an S-wave, $I = 0$, $S = 0$ state. The result is a model [16] with few parameters which has been applied to the ground and excited states with reasonable success for the ground states and non-strange orbitally excited states, but with splittings in the orbitally excited Σ states and with positive-parity states generally too heavy by about 250 MeV.

5 Lattice Results

Quenched lattice results are available for the spectrum of ground state baryons based on improved actions which require mild continuum extrapolations [17], with agreement within 10% with experiment. These calculations require (chiral) extrapolations to reach physically meaningful quark masses, which may not be linear as is usually assumed. The known structure of the chiral limit can and should be incorporated [18].

A description of a successful calculation of the mass of the lightest $N1/2^-$ baryon mass using domain-wall fermions to incorporate chiral symmetry is elsewhere in these proceedings [19]. Interestingly, this quenched calculation finds the lightest $N1/2^+$ resonance heavier than the lightest $N1/2^-$ state, in agreement with other lattice determinations [20,21]. This suggests that effects not present in the flavor-independent OGE quark model or in quenched lattice QCD, like threshold effects or coupling to baryon-meson intermediate states, may be responsible for the low actual mass of the Roper resonance. It has been shown that $N\pi \to N\pi\pi$ reaction dynamics can generate a pole in the region of the Roper resonance with no need for a qqq excitation of this mass [22]. Similarly, these lattice calculations find the lightest $\Lambda 1/2^-$ [corresponding to $\Lambda(1405)$] roughly degenerate with the lightest $N1/2^-$ [corresponding to $N(1535)$], hinting that the $\Lambda 3/2^-(1520)$ – $\Lambda 1/2^-(1405)$ splitting may also have the same source.

6 Hybrid Baryons

Identification of hybrid baryons is complicated by the fact that they have the same quantum numbers as conventional three-quark excitations, and so should mix with conventional excitations. Their identification is also somewhat model dependent, as it relies on the separation of the gluon and quark degrees of freedom inherent in some models. In the flux-tube model these are described as baryon states built on excited flux tubes, so that in an adiabatic approximation three quarks move in a confining potential generated by excited states of the flux tubes. It can be shown [23] that the excitation energy of these tubes, for a given configuration of the quarks, is approximately that of a moving junction with a calculable effective mass. Hybrid baryon masses can then be found by solving for qqq energies in the usual manner, but using a modified confining potential which includes this excitation energy, calculated variationally, for all possible quark positions.

In this model the junction motion adds $l^{\pi} = 1^{+}$ to the orbital angular momentum of the quarks, with the result that the lightest states have $L^P = 1^+$. With consideration of the quark spin and excited flux tube exchange symmetry, and with the usual spin-spin interaction between the quarks, the lightest hybrids are nucleons with $J^P = 1/2^+$ and $3/2^+$ at 1870±100 MeV, or Δ states with $J^P = 1/2^+$, $3/2^+$, or $5/2^+$ at 2075±100 MeV, significantly heavier than the roughly 1500 MeV predicted for states containing 'constituent' gluons in the bag model [24]. Such states are in the middle of the region of positive-parity baryons predicted by symmetric quark models but missing from analyses of scattering data.

7 Effects of Decay Channel Couplings

It is reasonable to expect that baryon self energies due to the presence of baryon-meson or $qqq(q\bar{q})$ intermediate states, into which baryon states can decay, should be comparable to their widths. If this is the case, the differences between most inclusive calculations of the baryon spectrum become irrelevant if they ignore the mass splittings induced by adding many such self energies. The best calculations of these splittings [25,26,27] calculate the self energies of ground and orbitally excited non strange and Λ and Σ states due to intermediate states made up of ground state baryons and the pseudoscalar octet and vector nonet of mesons. Mesons are coupled to baryons as elementary particles coupled directly to the quarks (elementary-meson emission, EME) or using a pair-creation ($^3P_0$) model. Self energies are calculated by using time-ordered perturbation theory [26,27], or by using dispersion relations [25] to evaluate shifts in the mass squared.

The effects on baryon mass splittings are found to be substantial, of the order of 50-100 MeV, making it possible to coarsely fit many aspects of the spectrum entirely without residual interactions between the quarks. Some of these baryon-meson intermediate state splittings resemble spin-orbit effects [27].

However, such calculations lack a self-consistent treatment of external and intermediate states (lacking orbitally excited intermediate-state baryons, for example) and so, by analogy to a similar non relativistic calculation using a pair-creation model in mesons [28], the sum over intermediate states may not have converged. Strong decay amplitudes calculated in this fashion may not be realistic far off shell, although the loop integrals required to calculate self energies require knowledge of them there. This convergence has been shown to be faster in a covariant model of meson self energies [29], where these off-shell amplitudes can be realistically modeled.

8 Describing Reactions Using the Quark Model

The description of scattering observables requires a model of the reaction dynamics to which 'bare' quark model masses and momentum-dependent decay amplitudes can be input. Such models exist based on both Hamiltonian [30] and relativistic [31] approaches. The dynamical input required from quark models to describe meson photo production, for example, are the momentum-dependent helicity amplitudes required to describe the resonance electromagnetic couplings, and the momentum-dependent strong-decay amplitudes for the decay of intermediate baryons to the final-state hadrons.

A calculation which puts quark model input together with a Hamiltonian approach to the reaction dynamics for ω photoproduction is underway [32], and a joint collaboration has been proposed [33] which would allow direct comparison of various inclusive quark-model predictions for meson photoproduction observables. This will ultimately require the dynamical input above for all model states in a given mass range, but as a first step the first two states in a few partial waves can be utilized. What is interesting about this approach is that it will require renormalization of the 'bare' quark model parameters (quark masses, string tension, quark-quark interaction parameters, etc.) while fitting directly to the data.

9 Summary

The study of nucleon resonances using quark models is more than simply stamp collecting: we are exploring the consequences of QCD for the spectrum and decay of baryons; identifying the important degrees of freedom for the description of the majority of hadrons; and discovering the nature of the properties and interactions of those degrees of freedom. Progress hinges on the solution of a unique theoretical challenge, which is the resolution of overlapping broad resonances in a strongly-coupled multi-channel system.

Acknowledgments

The author wishes to acknowledge the kind support of the organizers of N*2001. This work was supported by the U.S. Department of Energy under Contract DE-FG02-86ER40273.

References

1. P. L. Chung and F. Coester, *Phys. Rev.* D **44**, 229 (1991).
2. F. Cardarelli, E. Pace, G. Salme, and S. Simula, *Phys. Lett.* B **357**, 267 (1995).
3. D. B. Leinweber, R. M. Woloshyn, and T. Draper, *Phys. Rev.* D **43**, 1659 (1991).
4. R. Bijker, F. Iachello, and A. Leviatan, *Phys. Rev.* **D55**, 2862 (1997).
5. S. Theberge, A. W. Thomas, and G. A. Miller, *Phys. Rev.* D **22**, 2838 (1980).
6. M. Oettel, M. Pichowsky, and L. von Smekal, *Eur. Phys. J.* A **8**, 251 (2000); J. C. Bloch, C. D. Roberts, and S. M. Schmidt, *Phys. Rev.* C **61**, 065207 (2000); M. Oettel, R. Alkofer, and L. von Smekal, *Eur. Phys. J.* A **8**, 553 (2000).
7. A. P. Szczepaniak and E. S. Swanson, hep-ph/0006306.
8. J. Carlson, J. Kogut, and V. R. Pandharipande, *Phys. Rev.* D **27**, 233 (1983).
9. T. T. Takahashi, H. Matsufuru, Y. Nemoto, and H. Suganuma, *Phys. Rev. Lett.* **86**, 18 (2001).
10. N. Isgur and G. Karl, *Phys. Rev.* D **18**, 4187 (1978).
11. S. Capstick and N. Isgur, *Phys. Rev.* D **34**, 2809 (1986).
12. D. Robson, *Proceedings of the Topical Conference on Nuclear Chromodynamics*, Argonne National Laboratory (1988), Eds. J. Qiu and D. Sivers (World Scientific), p. 174.
13. G. Wagner, A. J. Buchmann, and A. Faessler, *Phys. Lett.* B **359**, 288 (1995).
14. L. Y. Glozman and D. O. Riska, *Phys. Rept.* **268**, 263 (1996).
15. L. Y. Glozman, W. Plessas, L. Theussl, R. F. Wagenbrunn, and K. Varga, *PiN Newslett.* **14**, 99 (1998).
16. W. H. Blask, U. Bohn, M. G. Huber, B. C. Metsch, and H. R. Petry, *Z. Phys.* A **337**, 327 (1990).
17. UKQCD Colloboration, *Phys. Rev.* D **62**, 054506 (2000).
18. D. B. Leinweber, A. W. Thomas, K. Tsushima, and S. V. Wright, *Nucl. Phys. Proc. Suppl.* **83**, 179 (2000) [hep-lat/9909109].

19. S. Sasaki, *these proceedings.*
20. Frank X. Lee, *these proceedings.*
21. D. G. Richards [UKQCD Collaboration], *Nucl. Phys. Proc. Suppl.* **94**, 269 (2001) [hep-lat/0011025].
22. J. Speth, O. Krehl, S. Krewald, and C. Hanhart, *Nucl. Phys.* A **680**, 328 (2000).
23. S. Capstick and P. R. Page, *Phys. Rev.* D **60**, 111501 (1999).
24. T. Barnes and F. E. Close, *Phys. Lett.* B **123**, 89 (1983); E. Golowich, E. Haqq, and G. Karl, *Phys. Rev.* D **28**, 160 (1983).
25. P. Zenczykowski, *Ann. Phys. (NY)* **169**, 453 (1986).
26. W. Blask, M. G. Huber, and B. Metsch, *Z. Phys.* A **326**, 413 (1987).
27. B. Silvestre- Brac and C. Gignoux, *Phys. Rev.* D **43**, 3699 (1991).
28. P. Geiger and N. Isgur, *Phys. Rev.* D **47**, 5050 (1993).
29. M. A. Pichowsky, S. Walawalkar, and S. Capstick, *Phys. Rev.* D **60**, 054030 (1999).
30. T. Sato and T. S. Lee, *Phys. Rev.* C **54**, 2660 (1996).
31. F. X. Lee, C. Bennhold, S. S. Kamalov, and L. E. Wright, *Phys. Rev.* C **60**, 034605 (1999).
32. Y. Oh, *these proceedings.*
33. Proposal to Baryon Resonance Analysis Group (BRAG) members, October 2000.

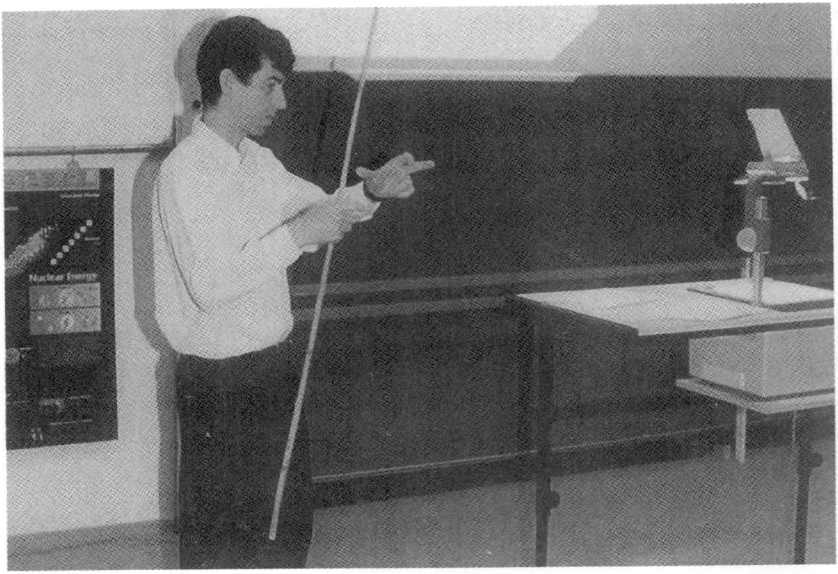

Simon Capstick

Costas Papanicolas

QUADRUPOLE STRENGTH IN THE $N \to \Delta$ TRANSITION

C. N. PAPANICOLAS
Institute of Accelerating Systems and Applications and
University of Athens, Athens, GREECE
E-mail: cnp@cc.uoa.gr

The detailed investigation of the $N \to \Delta$ transition offers one of the best avenues to understand the structure of hadrons and the intricate dynamics of their constituents. It is the subject of investigations at every medium energy electron accelerator. A brief overview of the field is presented, followed by a presentation of the Bates $N \to \Delta$ program.

1 Introduction

The possibility of nucleon deformation raised more than 20 years ago [1] still remains an open one. The presence of resonant quadrupole amplitudes in the transition to the only isolated excited state of the nucleon is regarded as the definitive signature of deviation from the simplistic spherical models of the nucleon and/or the delta. Their origin differs in the various nucleon models. For instance, in "QCD-inspired" constituent quark models it arises from the intra-quark effective color-magnetic tensor forces [2] while in chiral bag models [3] most of the deformation can be attributed to the asymmetric coupling of the meson cloud to the spin of the nucleon.

In the spherical quark model of the nucleon, the $N \to \Delta$ excitation is a pure $M1$ ($M_{1+}^{3/2}$) transition. The resonant quadrupole multipoles $E2$ ($E_{1+}^{3/2}$) and $C2$ ($S_{1+}^{3/2}$) contain the pertinent information. It has become standard practice to quote experimental and theoretical results in terms of the Electric- and Scalar(Coulomb)-to-Magnetic-Ratios of amplitudes defined as $R_{EM} = \text{Re}(E_{1+}/M_{1+})$ and $R_{SM} = \text{Re}(S_{1+}/M_{1+})$ respectively. QCD inspired models predict values of R_{SM} in the range of -1% to -4%, at low momentum transfers, $Q^2 \leq 1.0$ (GeV/c)^2. However, the isolation and interpretation of R_{EM} and R_{SM} is complicated by the presence of the nonresonant "background processes" which are coherent with the resonant excitation of the $\Delta(1232)$. These interfering processes (such as the pion pole, Born terms, tails of higher resonances) need to be constrained in order to isolate the resonant contributions to R_{EM} and R_{SM} which contain the physics of interest. As a result these quantities are invariably extracted with substantial model error which is poorly known and rarely quoted. Fig. 1 offers a recent compilation of the current status of R_{EM} and R_{SM} as a function of Q^2. The progress achieved

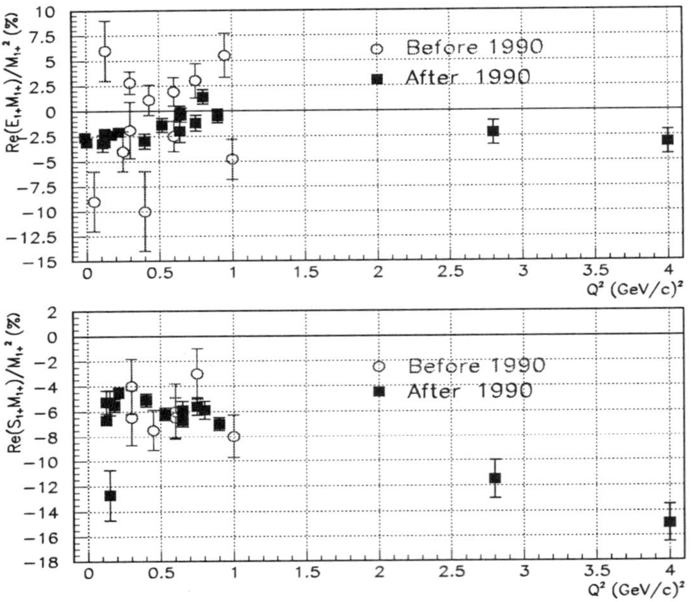

Figure 1. Experimentally derived R_{EM} and R_{CM} values. Filled symbols denote results from recent experiments. In most cases the error shown is purely statistical. Rarely a model error is given.

in recent years is obvious.

R_{SM} and R_{EM} are extracted through one of the following two approaches: a) Most or all of the background multipoles are neglected (e.g. see [4,5,6]) assuming that at resonance only the resonant terms contribute significantly. b) A phenomenological reaction framework with adjustable quadrupole amplitudes is used to perform a model extraction (e.g. see [6,7]). It is assumed that the reaction is controlled at the level of precision required for the disentanglement of the background from the resonance. Neither of these approaches has been adequately tested for consistency.

Fig. 2 shows performed and programmed experiments at several laboratories [8,5,6,7,10,4,11]. The high Q^2 range can only be reached by Jefferson, whereas Bonn, Mainz, and Bates are better suited to explore medium and low Q^2. Recent measurements at Bates [7,12], Bonn [5,4], and Mainz [8] demonstrate that observables sensitive to the R_{SM} can be obtained, but that the extraction of the R_{SM} is dominated by the model error. The importance of background

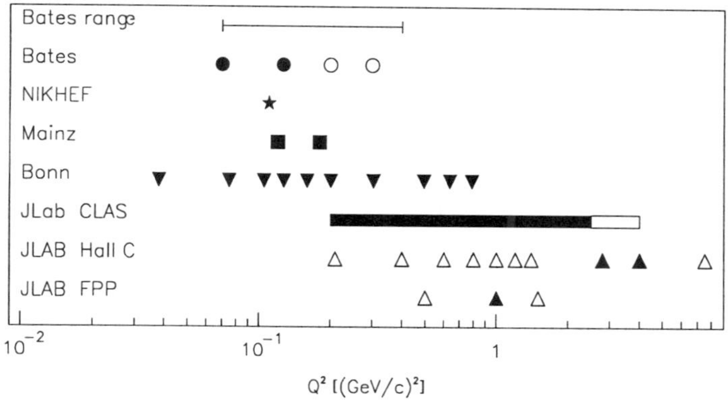

Figure 2. Recently performed (filled symbols) and planned (open symbols) $N \to \Delta$ experiments at different laboratories for different Q^2. CLAS has a continuous coverage for $Q^2 > 0.35$ (GeV/c)^2.

is clearly seen in the W behavior of the responses [7] and the non-vanishing recoil polarization P_n [12,8]. New precise results of R_{EM} are been reported from Bonn [4] and Jefferson [11]. It will be interesting to ascertain whether the R_{EM} at the low and intermediate Q^2 values studied assumes positive values or it stays negative (as at the precisely known photon point [25,26]). We should also point out that the transverse background contributions at finite Q^2 are even less understood than the scalar ones.

2 Theoretical Developments

Nucleon models are continuously being refined providing valuable guidance as to the magnitude of the resonant amplitudes. However, crucial has been the development of phenomenological reaction models which allow the interpretation of experimental data in a meaningful and consistent way. For the photo-pion channel the reaction models of RPI [15], MAID [13] and the Dynamical Model [19] offer a rich and yet flexible phenomenology that allows for the extraction from the data of the amplitudes of interest (albeit model dependent). The reaction model of Sato and Lee (SL) [3,14] goes a step further: It provides the appropriate reaction model consistent with a chiral model of the nucleon. Deficiencies of the models that emerge as a result of the comparison with the data are rectified by re-adjustments of the parameters and/or by modifying the phenomenology. This results in an improved understanding of the underlying physics and gradual reduction of the model error. For the $H(\vec{e}, e'p)\gamma$ channel the new dispersion theory [16] of the Mainz group, taken in conjunction with the previous work of Vanderhaeghen et al. [17] allows to addresses the physics of "deformation" and of nucleon polarizabilities in the region above pion threshold simultaneously.

Of particular importance to the field is the prediction of the Dynamical Model [18,19] and of Sato and Lee [14] that most of the responses and the R_{EM} and R_{SM} exhibit a distinct structure at very low Q^2 values (below 0.10 GeV^2/c^2) which arises as manifestation of the mesonic degrees of freedom. We may thus have for the first time a clear signature of the manifestation of the pion cloud at low Q^2.

Finally, it is worth noting that in the next few years we can expect to establish for the first time contact on this same issue with lattice gauge calculations [20]. Important new initiatives in the US, Europe and Australia hold the promise that for the first time in hadronic physics we will be able to relate directly experimental information to the QCD Lagrangian.

3 The Bates $\gamma^* N \to \Delta$ Program

The Bates $\gamma^* N \to \Delta$ relies on a major instrumentation initiative, the Out-Of-Plane Spectrometer (OOPS) system. The OOPS facility is now fully developed and commissioned. It exceeds all design requirements [23,24].

The first $\gamma^* N \to \Delta$ measurements resulted in the precise determination of the cross section in "parallel kinematics" σ_0, of the LT-asymmetry, A_{LT}, and response R_{LT} and the measurement of the induced proton polarization P_n. P_n is proportional to the R^n_{LT} response and it would be identically zero in the absence of background. It was found [12] to be $-0.397 \pm 0.055 \pm 0.009$ which established the importance of the background contributions.

The coincident cross section in parallel kinematics was measured from $W = 1155$ to $W = 1320$ MeV; however it is the R_{LT} results that amply demonstrate the sensitivity of the data to the "deformation"(Fig. 3). All models fail dramatically to predict the behavior of the data unless nonzero resonant quadrupole amplitudes are allowed. Reasonable agreement is achieved if the parameters of the models are allowed to be adjusted. The superb sensitivity of the data to the quadrupole multipoles, allows for their determination either through a variant of the M1 dominance truncated multipole fit or through model extraction. By adjusting the relevant parameters in the models of MAID [13], of RPI [15] and of SL [14] values for R_{EM} and R_{SM} can be derived [7]. The adjusted MAID2000 model is the only available reaction framework that provides fair agreement with all the data (including P_n) and it can therefore

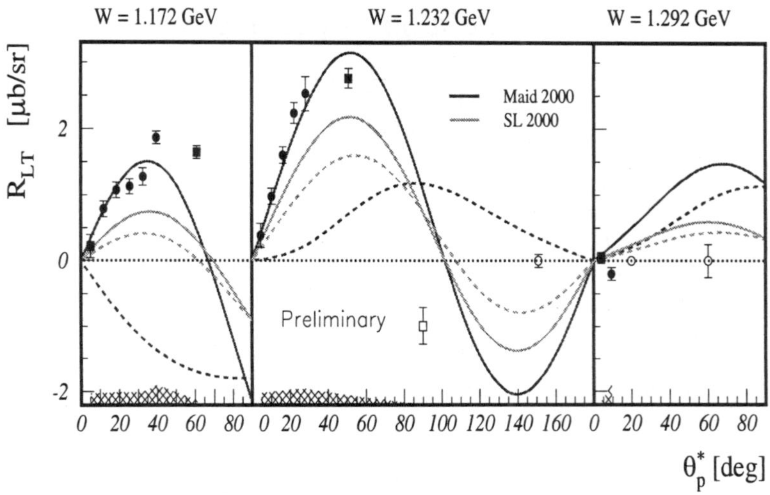

Figure 3. The R_{LT} response at $W = 1172, 1232$ and 1292 GeV. The open square at $W =$ 1232 MeV is a preliminary datum from the 2001 run. For the other 2001 data only the anticipated statistical error is shown as open circles. Bullets and squares are earlier OOPS data. The shaded band represents an estimate of the systematic error.

qualify for extracting R_{EM} and R_{SM} values. Based on this analysis we have derived [7] $R_{SM} = (-6.7 \pm 0.2_{\text{stat+sys}})\%$ and $R_{EM} = (-2.0 \pm 0.2_{\text{stat+sys}})\%$. Unfortunately no theoretical group has yet provided an estimate of the model uncertainty. Based on our own analysis we have attributed very conservatively to R_{SM} and R_{EM} model uncertainties of 2.5% and 2.0% respectively.

$\mathrm{A_{LT}}$ differs noticeably above and below the resonance from $\mathrm{A_{LT}}$ on resonance, they would be equal in the absence of background. The hint of a change of the sign of $\mathrm{A_{LT}}$ from below to above the resonance is especially intriguing. The limited data at W= 1292 MeV motivated new measurements in the spring 2001 at the same W and Q^2 which are being analyzed.

The first out-of-plane $N \to \Delta$ measurements [21] were conducted in 1998. Measurements were performed for proton or pion detection. In addition to measuring the R_{LT} response at a larger angle the data allowed for the extraction of the helicity asymmetry A_h and the R'_{LT} response. A_h has analogous significance to P_n in isolating the background contributions, as they both are an imaginary part of an LT-interference. Preliminary results of the A_h asymmetry in the $n\pi^+$ channel are shown in Fig. 4.

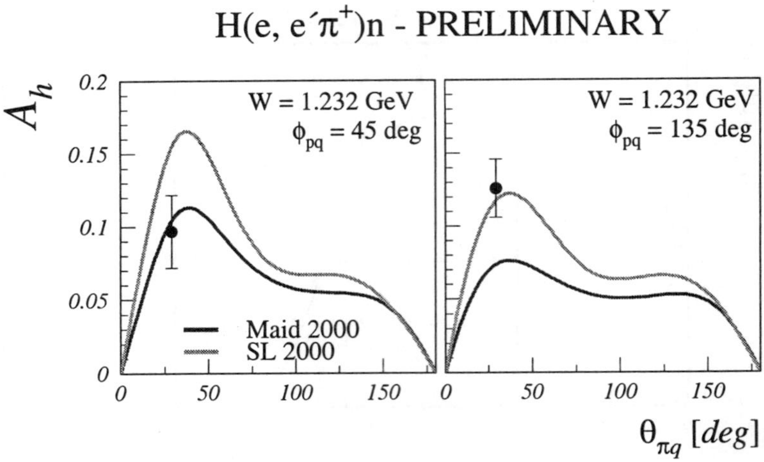

Figure 4. Preliminary results for the A_h beam-helicity asymmetry in the $p\pi^+$ channel. The error is statistical only.

The year 2000/2001 was marked by major technical achievements for the OOPS program which led to productive runs for the VCS and $N \rightarrow \Delta$ experiments. A 950 MeV beam energy, with currents up to 7 μA and a duty-factor in excess of 50% was used in conjunction with the completed and commissioned 4-OOPS cluster. Some preliminary results are presented below.

The TT response which is sensitive to the electric quadrupole amplitude, and of which little is known at non-zero Q^2 was isolated for the first time. The response functions R_T and R_{TT} contain the term $\mathrm{Re}[E^*_{1+}M_{1+}]$ but also the dominant term $|M_{1+}|^2$. The influence of the dominant $|M_{1+}|^2$ term can be diminished by measuring the following combination of the R_T and R_{TT} responses.:

$$R_T(\theta^*_{pq}) + R_{TT}(\theta^*_{pq}) - R_T(0) = 2\mathrm{Re}[E^*_{0+}(3E_{1+} + M_{1+} - M_{1-})](1 - \cos\theta^*_{pq}) \\ -12\mathrm{Re}[E^*_{1+}(M_{1+} - M_{1-})]\sin^2\theta^*_{pq} \quad (1)$$

The term of interest $\mathrm{Re}[E^*_{1+}M_{1+}]$ is enhanced by a factor of twelve (12)! The cross sections needed to derive this quantity were measured; as a result we expect to extract a most precise measurement of R_{EM} at $Q^2 = -0.126\ GeV^2$. The sensitivity to the EMR is maximized at the measured kinematics as shown in Fig. 5.

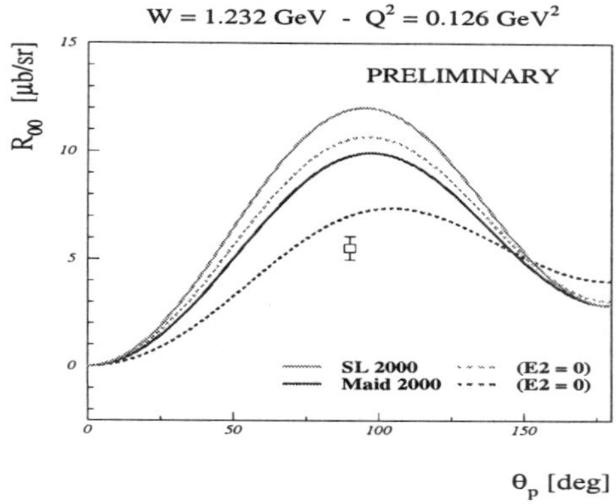

Figure 5. Sensitivity to R_{EM} of the measured σ_{00}. The empty square is the projected datum from 2001 with the expected uncertainty. The model calculations are those of Sato-Lee [14] and MAID.

We were able to access the low momentum branch of the recoiling protons by tuning the OOPS spectrometers to very low momentum detection. We were able to thus access $\theta^*_{pq} = 151°$ and at $\theta^*_{pq} = 180°$ for an R_{LT} measurement.

The high duty-cycle extracted beam also provided a significant advantage for studies of the π^+ channel dramatically reducing the experimental background. The $p(e, e'\pi^+)n$ reaction was measured at W = 1232 MeV, Q^2= 0.127 GeV^2 and $\theta^*_\pi = 44.5°$. All three unpolarized response functions could be determined. In addition we measured the cross section in parallel kinematics.

4 Future Prospects and Acknowledgments

A new level of sophistication and precision is emerging from the experimental programs in pursuit of the issue of hadron deformation through the $N \rightarrow \Delta$ transition. It is apparent that in the next few years the definitive signature will be established and we can hope to achieve firm contact with QCD. Of immediate concern is the role of the pion cloud at low momentum transfers and the quantification of model error in the extracted quantities.

I would like to thank Drs. Stiliaris and Botto and C. Vellides and N. Sparveris for contributing significantly to this paper. I am indebted to the Bates community and to the OOPS collaboration. This work was supported in part by E.U. RTN HPRN-CT-2000-00130 and Athens University.

References

1. S.L. Glashow, *Physica* A **96**, 27 (1979).
2. N. Isgur, G. Karl and R. Koniuk, *Phys. Rev.* D **25**, 2394 (1982); S. Capstick and G. Karl, *Phys. Rev.* D **41**, 2767 (1990).
3. T. Sato and T.-S.H. Lee, *Phys. Rev.* C **54**, 2660 (1996).
4. R.W. Gothe, *these proceedings.*
5. F. Kalleicher *et al., Z. Phys.* A **359**, 201(1997).
6. V. V. Frolov *et al., Phys. Rev. Lett.* **82**, 45 (1999).
7. C. Mertz *et al., Phys. Rev. Lett.* **86**, 2963 (2001).
8. H. Schmieden , *these proceedings*; nucl-ex/0105023.
9. L.C. Smith, *these proceedings.*
10. Bates Proposals 87-09, 89-03, 97-04, 97-05.
11. Approved CEBAF experiments: CLAS: E89-037, E89-042, E94-003; Hall A: E91-011; Hall C: E97-101.
12. G.A. Warren *et al.*, *Phys. Rev.* C **58**, 3722 (1998).
13. D. Drechsel *et al.*, *Nucl. Phys.* A **645**, 145 (1999) and http://www.kph.uni-mainz.de/MAID/.
14. T. Sato and T.-S.H. Lee, *Phys. Rev.* C **63** (2001) 055201.
15. R. M. Davidson *et al., Phys. Rev. Lett.* **56**, 804 (1986); *Phys. Rev.*, D **43**, 71 (1991), *Phys. Lett.* B **353**, 131 (1995).
16. B. Pasquini *et al.*, hep-ph/0102335.
17. M. Vanderhaeghen *Nucl. Phys.* A **595**, 219 (1995).
18. S. S. Kamalov and Shin Nan Yang, *Phys. Rev. Lett.* **83**, 4494 (1999).
19. S.S. Kamalov, *et al.*, nucl-th/0006068.
20. J. Negele and C. Alexandrou (private communication 2001).
21. Christian Kunz, MIT Ph.D. thesis, 1999 unpublished.
22. C. Vellidis, Uni. Athens, Ph.D. thesis, in preparation.
23. S. Dolfini *et al.*, *Nucl. Inst. Meth.* A **344**, 571 (1994).
24. J. Mandeville *et al.*, *Nucl. Inst. Meth.* A **344**, 583 (1994).
25. G. Blanpied *et al.*, *Phys. Rev. Lett.* **79**, 4337 (1997).
26. R. Beck *et al.*, *Phys. Rev. Lett.* **78**, 606 (1997).

PION ELECTROPRODUCTION AT ELSA

R. W. GOTHE

Universität Bonn, Physikalisches Institut, Nussallee 12,
53115 Bonn, Germany
E-mail: gothe@physik.uni-bonn.de

At the Electron Stretcher Accelerator ELSA the four momentum transfer dependence of the N to Δ transition has been investigated by measuring the ϑ_N^* and φ_N^* angular distributions of the double differential pion production cross sections in a series of electron scattering coincidence experiments on hydrogen spanning the $-K^2$-range from $0.04\,GeV^2$ to $0.8\,GeV^2$. In these experiments both final state isospin channels, $p\pi^0$ and $n\pi^+$, have been measured simultaneously. The preliminary results on the separation of the response functions $R_T + \epsilon_L R_L$, R_{LT} and R_{TT} and the subsequently following extraction of the multipoles, both based on the angular dependence of the hadronic cross sections, have been reported in the case of the $p\pi^0$ channel in the previous NSTAR 2000 proceedings[1]. Only the first and preliminary results on the decomposition of the hadronic cross sections in the $n\pi^+$ channel obtained with the large acceptance ELAN time-of-flight spectrometer are described here.

1 Introduction

A more general introduction into the electroproduction multipoles, the extraction method as well as the comparison of experimental and theoretical results for the electric transverse (EMR) and scalar quadrupole to magnetic dipole Ratio (SMR) in the reaction channel $p(e, e'p)\pi^0$ are given in the proceedings of NSTAR 2000 [1]. Only in the $p\pi^0$ final state and only at the K-matrix pole of $1232\,MeV$ these multipole ratios represent the resonant isospin $\frac{3}{2}$ contribution of the production amplitudes. To separate the isospin $\frac{3}{2}$ channel from the proton isospin $\frac{1}{2}$ channel at all other invariant masses in the full range of the $\Delta(1232)$ resonance, the angular dependent decomposition of the $n\pi^+$ final state cross sections is needed.

Due to the nature of the electromagnetic coupling to the nucleon and isospin conservation, the corresponding pion production amplitudes of the four reaction channels

$$\begin{aligned}
\gamma_{(v)} + p &\rightarrow p + \pi^0 \Rightarrow A^{p\pi^0} \\
\gamma_{(v)} + p &\rightarrow n + \pi^+ \Rightarrow A^{n\pi^+} \\
\gamma_{(v)} + n &\rightarrow n + \pi^0 \Rightarrow A^{n\pi^0} \\
\gamma_{(v)} + n &\rightarrow p + \pi^- \Rightarrow A^{p\pi^-}
\end{aligned} \tag{1}$$

can be expressed in terms of three independent isospin amplitudes, namely the isoscalar $A_s^{\frac{1}{2}}$ and the two isovector $A_v^{\frac{1}{2}}$, $A_v^{\frac{3}{2}}$ amplitudes,

$$\begin{aligned}
A^{p\pi^0} &= A_s^{\frac{1}{2}} + \tfrac{1}{3}(A_v^{\frac{1}{2}} + 2A_v^{\frac{3}{2}}) &&= A_p^{\frac{1}{2}} + \tfrac{2}{3}A_v^{\frac{3}{2}} \\
A^{n\pi^+} &= \sqrt{2}[A_s^{\frac{1}{2}} + \tfrac{1}{3}(A_v^{\frac{1}{2}} - A_v^{\frac{3}{2}})] &&= \sqrt{2}[A_p^{\frac{1}{2}} - \tfrac{1}{3}A_v^{\frac{3}{2}}] \\
A^{n\pi^0} &= -A_s^{\frac{1}{2}} + \tfrac{1}{3}(A_v^{\frac{1}{2}} + 2A_v^{\frac{3}{2}}) &&= -A_n^{\frac{1}{2}} + \tfrac{2}{3}A_v^{\frac{3}{2}} \\
A^{p\pi^-} &= \sqrt{2}[A_s^{\frac{1}{2}} - \tfrac{1}{3}(A_v^{\frac{1}{2}} - A_v^{\frac{3}{2}})] &&= \sqrt{2}[A_n^{\frac{1}{2}} + \tfrac{1}{3}A_v^{\frac{3}{2}}] \,.
\end{aligned} \tag{2}$$

Here the often used isospin $\frac{1}{2}$ amplitudes for proton and neutron initial states $A_p^{\frac{1}{2}}$, $A_n^{\frac{1}{2}}$ are given by the rearranged isoscalar and isovector $\frac{1}{2}$ amplitudes,

$$A_p^{\frac{1}{2}} = A_s^{\frac{1}{2}} + \frac{1}{3}A_v^{\frac{1}{2}} \quad \text{and} \quad A_n^{\frac{1}{2}} = A_s^{\frac{1}{2}} - \frac{1}{3}A_v^{\frac{1}{2}} \,. \tag{3}$$

For the separation of all three independent amplitudes at least three reaction channels have to be analyzed,

$$\begin{aligned}
A_v^{\frac{3}{2}} &= A^{p\pi^0} - \tfrac{1}{\sqrt{2}}A^{n\pi^+} &&= A^{n\pi^0} + \tfrac{1}{\sqrt{2}}A^{p\pi^-} \\
A_v^{\frac{1}{2}} &= A^{p\pi^0} + \tfrac{1}{\sqrt{8}}A^{n\pi^+} - \tfrac{3}{\sqrt{8}}A^{p\pi^-} &&= \sqrt{2}A^{n\pi^+} - \tfrac{1}{2}A^{p\pi^0} + \tfrac{3}{2}A^{n\pi^0} \\
A_s^{\frac{1}{2}} &= \tfrac{1}{\sqrt{8}}(A^{n\pi^+} + A^{p\pi^-}) &&= \tfrac{1}{2}(A^{p\pi^0} - A^{n\pi^0}) \,,
\end{aligned} \tag{4}$$

but the desired isovector $\frac{3}{2}$ contribution $A_v^{\frac{3}{2}}$ to the multipole amplitudes is already given by two, namely the $p\pi^0$ and $n\pi^+$, reaction channels.

2 Neutron Identification

The classification of particles as neutrons by the ToF spectrometer (see Fig. 1) of the ELAN experimental set-up[2] requires their detection in wall three or four in anti-coincidence with the first two or three walls (no correlated TDC signals) and that the so identified particles deposit no light at all in the corresponding first walls (no correlated ADC signals). Under the same conditions the light deposit of neutrons in the ToF spectrometer and subsequently the neutron detection efficiency are simulated by implemented GEANT subroutines. The simulated neutron detection efficiency of nearly 4 % for wall three is in this experiment mostly energy independent. Whereas a two radiation lengths thick photon converter is (not shown in Fig. 1 but positioned between the third and fourth wall[2]), enhances the now sensitively energy-dependent neutron detection efficiency of wall four to values between 6 % and 12 %. Nevertheless the efficiency corrected count rates of both walls are in best agreement as shown on the right hand side of Fig. 2. On the left hand side the

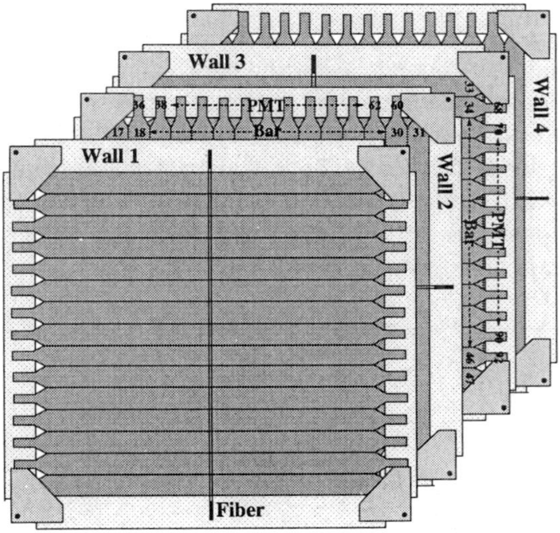

Figure 1. Front view of the time of flight hadron spectrometer that consists of four flight (ToF) walls, each has a surface of $3 \cdot 3\, m^2$ and comprises 15 scintillation bars, each of them $3\, m$ long, $20\, cm$ high and $5\, cm$ thick.

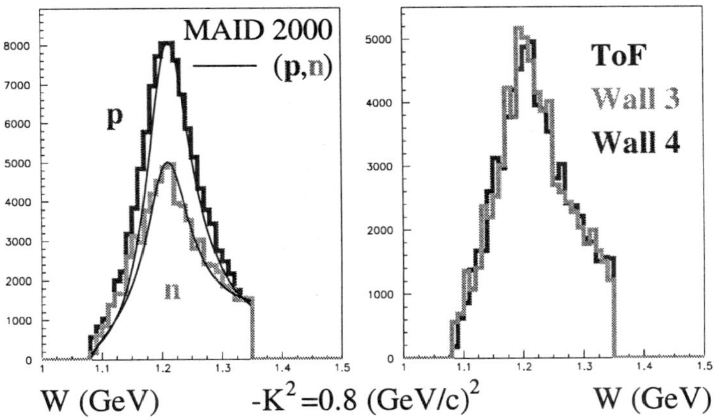

Figure 2. Comparison of the measured proton and neutron count rates with the prediction of the Mainz Isobar Model MAID 2000 and the comparison of the neutron count rate determined by wall three with that of wall four of the Time-of-Flight spectrometer ToF.

proton and neutron count rates are compared to the prediction of the Mainz Isobar Model MAID 2000[3].

3 First Results

The azimuthal angular dependence of the hadronic cross section is only based on the polarization of the virtual photon and separates the response functions $R_T+\epsilon_L R_L$, R_{LT} and R_{TT} which are directly proportional to the fit parameter A^φ, B^φ and C^φ, see Reference[1]. The local fit determines these φ_π^*-dependent parameters for each given ϑ_π^* bin independently, whereas the global fit enforces the ϑ_π^*-dependence of the s- and p-wave truncated cross section. Our Preliminary results are shwon in Fig. 3-5.

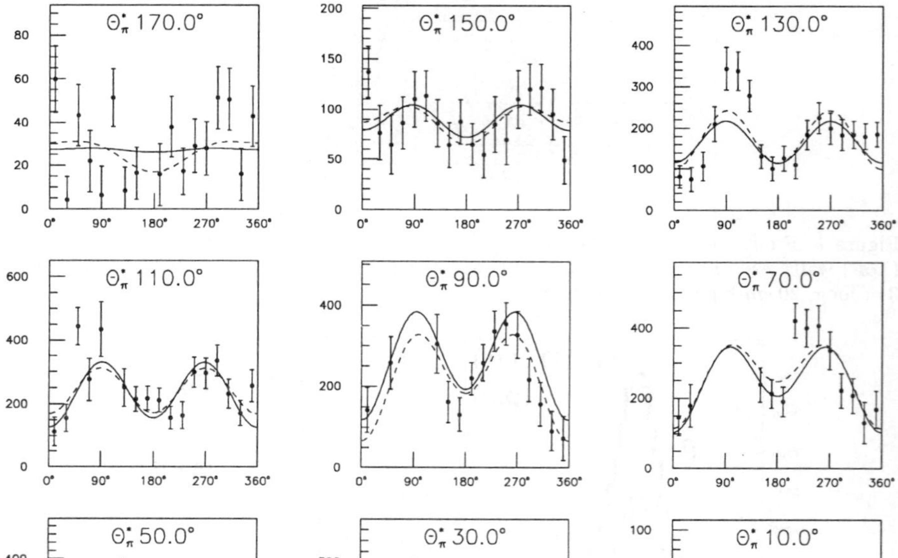

Figure 3. The local (dashed line) and global fit (solid line) to the ϑ_π^*-dependent φ_π^*-modulation of the neutron count rates at the K-matrix pole of 1232 MeV.

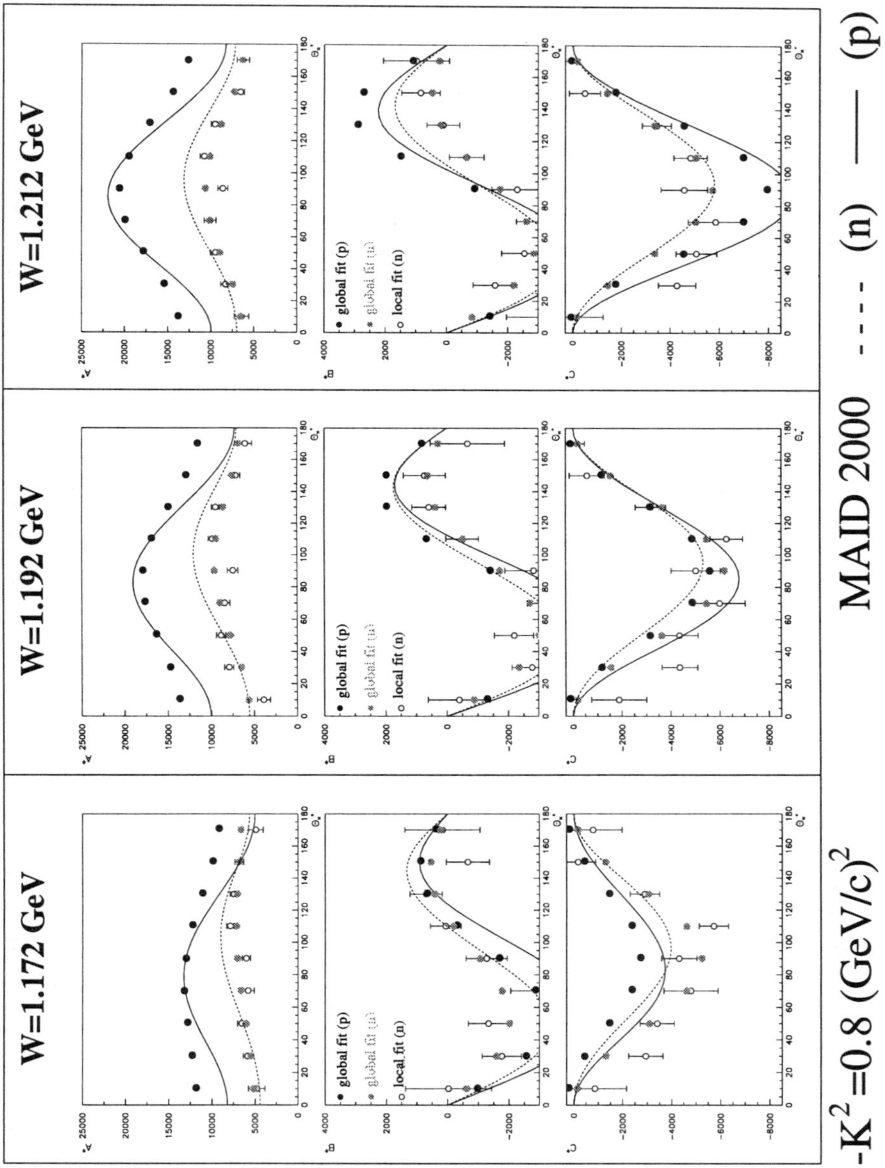

Figure 4. The ϑ_π^*-dependence of the local and global fit parameter A^φ, B^φ and C^φ for the $n\pi^+$, and only the global fit parameters for the $p\pi^0$ reaction channel for the lower invariant masses W compared to MAID 2000.

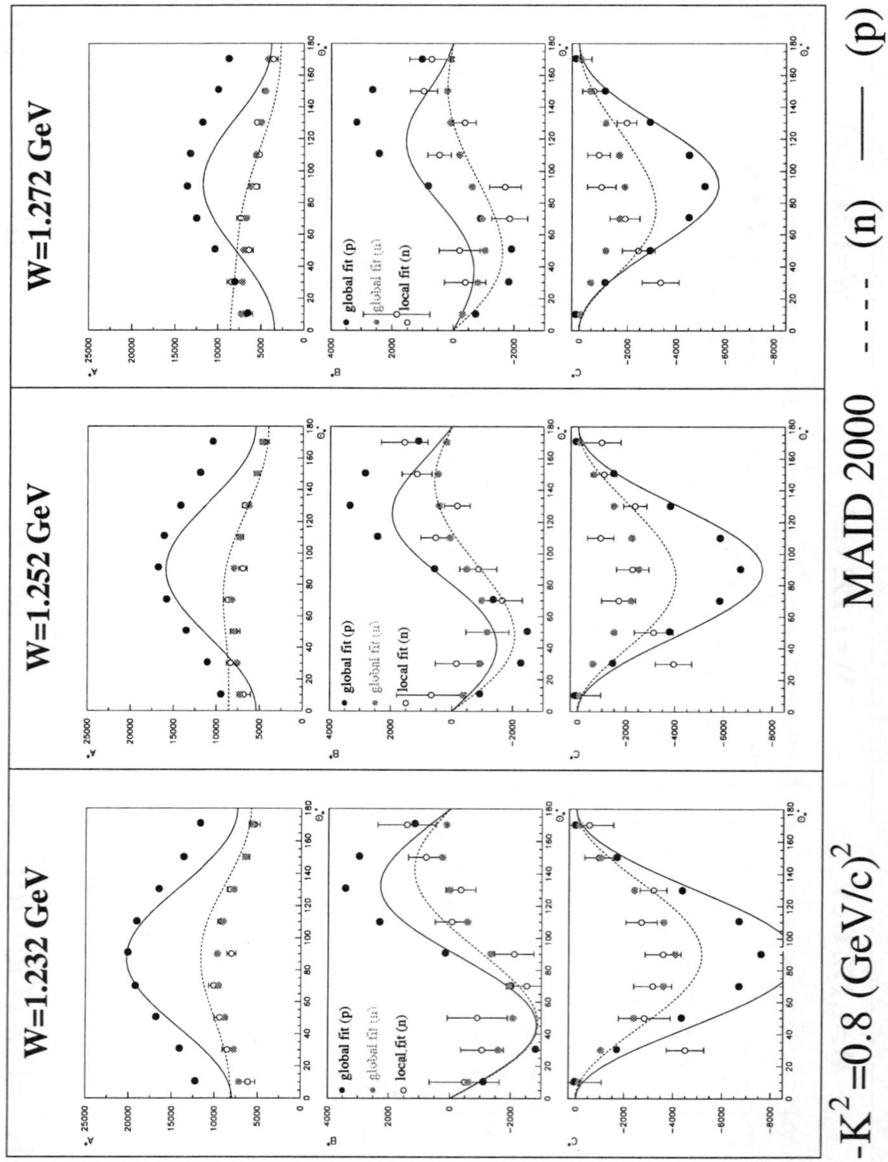

Figure 5. The ϑ_π^*-dependence of the local and global fit parameter A^φ, B^φ and C^φ for the $n\pi^+$, and only the global fit parameters for the $p\pi^0$ reaction channel for the higher invariant masses W compared to MAID 2000.

References

1. R.W. Gothe, Proceedings of NSTAR 2000 - Excited Nucleons and Hadronic Structure, Newport News, USA, World Scientific (2001) 31-41.
2. R.W. Gothe, *Progress in Particle and Nucl. Phys.* **44**, 185-196 (2000).
3. D. Drechsel, O. Hanstein, S.S. Kamalov, and L.Tiator, *Nucl. Phys.* A **645**, 145-174 (1999).

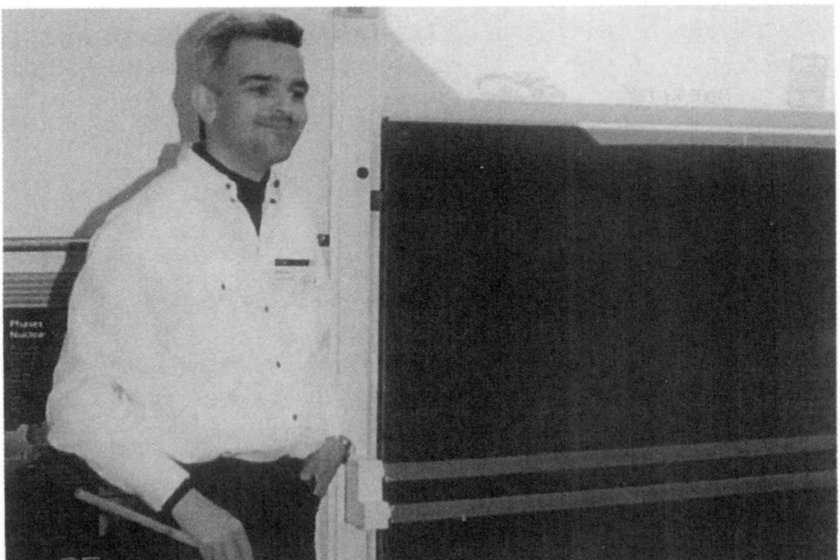

Ralf Gothe

Hartmut Schmieden

Volker Burkert and Lee Smith

π^0 ELECTROPRODUCTION IN THE $\Delta(1232)$ REGION AT MAMI

H. SCHMIEDEN

Institut für Kernphysik, Johannes Gutenberg-Universität, J. J. Becher-Weg 45, D-55099 Mainz, Germany

E-mail: hs@kph.uni-mainz.de

The extraction of the CMR from a $p(\vec{e}, e'\vec{p})\pi^0$ measurement is discussed. Preliminary results from further asymmetry measurements with polarized and unpolarized electron beam indicate that the imaginary background is well under control in the MAID2000 parameterization, but not the real (Born-) amplitudes.

1 Introduction

Similar to the ground state properties of the nucleon, the ratio of quadrupole to dipole strength in the $N \to \Delta(1232)$ transition has been considered as one of the key observables for the understanding of nucleon structure for a long time. Several mechanisms have been proposed for the breaking of spherical symmetry which is required by the non-zero values for that ratio.

While the resonance structure of the nucleon has been revealed almost entirely by π-N scattering, the $\Delta l = 2$ positive parity transitions require electromagnetic excitations. Due to the $\Delta(1232)$ decay branching ratio of 99.4 % into the πN channel, the electric (transverse) and Coulombic (longitudinal) quadrupole to magnetic dipole ratios, EMR and CMR, respectively, are measured in pion photo- and electroproduction. These ratios are defined as EMR $= \Im m\{E_{1+}^{3/2}\}/\Im m\{M_{1+}^{3/2}\}$ and CMR $= \Im m\{S_{1+}^{3/2}\}/\Im m\{M_{1+}^{3/2}\}$, where the pion multipoles, $A^I_{l_\pi \pm}$, are characterized through their magnetic, electric or longitudinal (scalar) nature A, the isospin I, and the pion-nucleon relative angular momentum l_π whose coupling with the nucleon spin is indicated by $\pm$. However, in pion production the $\Delta(1232)$ channel is intimately related to non-resonant mechanisms, which hamper the extraction of the EMR and CMR. The separation of resonant and non-resonant contributions, in principle, remains ambiguous [1]. What can be achieved experimentally is the isospin separation required for the EMR and CMR.

However, a complete experiment with regard to a multipole decomposition seems out of reach yet. It requires the complete angular distributions of at least 7 or 11 independent observables in single pion photo- and electroproduction, respectively. The EMR and CMR thus can only be extracted using a truncated multipole basis or model assumptions. Therefore, it is desirable to

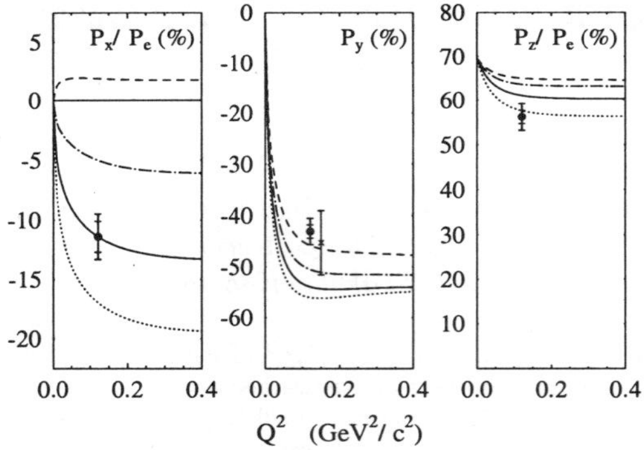

Figure 1. Results for the polarization components measured in the $p(\vec{e},e'\vec{p})\pi^0$ reaction at $W = 1232\,\mathrm{MeV}$, $Q^2 = 0.12\,(\mathrm{GeV/c})^2$ and $\epsilon = 0.71$ in comparison with MAID2000 calculations. The dashed, dot-dashed, full and dotted curves correspond to CMR = 0, −3.2, −6.4, −9.6 %, respectively. The MAMI data (full circles) are shown with statistical and systematical error. For the Bates P_y (cross) only the statistical error is indicated, the value is rescaled in ϵ and, though measured at the same Q^2, slightly shifted for clarity.

measure the small quadrupole amplitudes in different combinations with the unwanted, but unavoidable, non-resonant background. In particular, the measurement of polarization observables is mandatory. A first double-polarization experiment for the extraction of the CMR has been completed at MAMI.

2 Results of the $p(\vec{e},e'\vec{p})\pi^0$ Experiment

High sensitivity to the CMR is provided in the $p(\vec{e},e'\vec{p})\pi^0$ reaction measured in parallel kinematics, both in the proton polarization component P_x and in the polarization ratio P_x/P_z [2]. The axes are defined by $\hat{y} = \vec{k}_i \times \vec{k}_f / |\vec{k}_i \times \vec{k}_f|$, $\hat{z} = \vec{q}/|\vec{q}|$ and $\hat{x} = \hat{y} \times \hat{z}$, where $\vec{k}_i$ and $\vec{k}_f$ are the momenta of incoming and scattered electron, respectively, and $\vec{q} = \vec{k}_i - \vec{k}_f$ denotes the momentum transfer. The experiment was carried out at MAMI with a $15\,\mu\mathrm{A}$ electron beam of $P_e = 75\,\%$ longitudinal polarization impinging on a 5 cm long liquid hydrogen target. The scattered electrons were detected in Spectrometer B of the 3-Spectrometer-Setup [3] in coincidence with the protons in Spectrometer A. This spectrometer was equipped with a focal plane proton polarimeter [4]. Due to the spin precession in the magnetic spectrometer and the helicity flip of the electron beam, it was possible to determine all three proton polarization

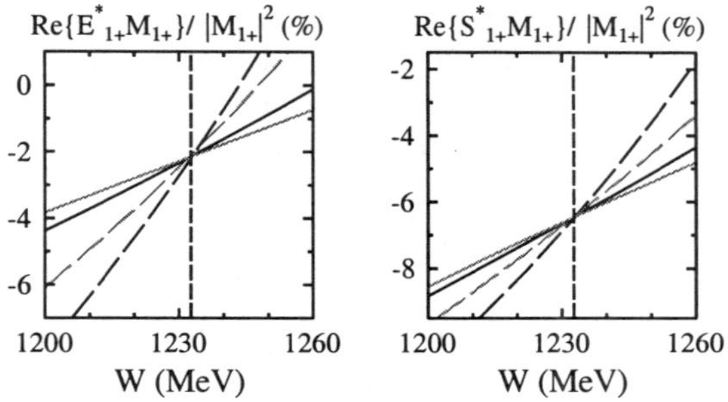

Figure 2. $\Re e\{E_{1+}^{*}M_{1+}/|M_{1+}|^2\}$ (left) and $\Re e\{S_{1+}^{*}M_{1+}/|M_{1+}|^2\}$ (right) as a function of the invariant energy. The full and broken curves represent MAID2000 calculations in the isospin 3/2 and the $p\pi^0$ charge channel, respectively, for $Q^2 = 0.2\,(\mathrm{GeV/c})^2$ (black) and $1.0\,(\mathrm{GeV/c})^2$ (grey). The zero crossing of $\Re e\{M_{1+}^{3/2}\}$ is indicated by the vertical dashed line.

components, P_x, P_y and P_z, simultaneously. The results [5] are depicted in Fig. 1. They were averaged over the experimental acceptance and projected to nominal parallel kinematics ($W = 1232\,\mathrm{MeV}$, $Q^2 = 0.12\,(\mathrm{GeV/c})^2$, $\epsilon = 0.71$, $\Theta_\pi^{\mathrm{cm}} = 180°$) using MAID2000 [6]. A consistency relation among the reduced polarizations [7] seems to be violated [5]. The origin of this fact remains unclear at the present statistical accuracy.

In the MAID analysis CMR is extracted without truncation of the multipole series. Directly at the zero crossing of the real part of the $M_{1+}^{3/2}$ amplitude, i.e. at $W = 1232\,\mathrm{MeV}$, the result obtained in the $p\pi^0$ channel is almost identically with the isospin 3/2 channel. This is demonstrated in Fig. 2, where the relevant interferences are shown as a function of W in both channels. Therefore, the CMR can be reliably extracted from the $p(\vec{e},e'\vec{p})\pi^0$ reaction, without explicit isospin separation. In Fig. 3 our result for $\Re e\{S_{1+}^{*}M_{1+}\}/|M_{1+}|^2$ in the $p\pi^0$ channel is compared to previous results [8,9,10,11,12] and the preliminary results of ongoing experiments [13,14,15,16,17]. The curves show the calculations of Ref. [18,19,20,21].

The double results at high Q^2 are due to different analyses of the same data [12]. The model dependence of the analyses is reduced at smaller Q^2. Here all results agree, except that of Ref. [11] where, in contrast to the recoil polarization measurement, the π^0 was detected in forward direction and all multipoles other than M_{1+} and S_{1+} were neglected. The effect of an additionally non-zero $\Im m S_{0+}/\Im m M_{1+}$ on the extracted $\Im m S_{1+}/\Im m M_{1+}$ is visualized in Fig. 4. The extracted $\Im m S_{1+}/\Im m M_{1+}$ is plotted as a function of $\Im m S_{0+}/\Im m M_{1+}$. Within MAID2000 the imaginary part of S_{0+} has a strong impact on the extraction of the CMR, too. $\Im m S_{0+}$ can be measured via the forward-backward asymmetry of LT-type structure functions [22,23].

3 Forward and Backward LT-Asymmetry in $p(e,e'p)\pi^0$

Due to a different angular weight of $\Re e S_{1+}^* M_{1+}$ and $\Re e S_{0+}^* M_{1+}$, their relative strength can be extracted from the Θ_π^{cm} angular distribution. This requires the LT structure function or the related asymmetry $\rho_{LT} = \frac{d\sigma^L - d\sigma^R}{d\sigma^L + d\sigma^R}$ to be measured over a large angular range [22], where $d\sigma^L$ and $d\sigma^R$ denote the differential cross sections left and right of $\vec{q}$.

At the squared 4-momentum transfer of $Q^2 = 0.2\,(\mathrm{GeV/c})^2$ of the MAMI experiment, the forward ($\Theta_\pi^{cm} = 20°$) and backward ($\Theta_\pi^{cm} = 160°$) kinematics require very different settings of the proton spectrometer both in angle and

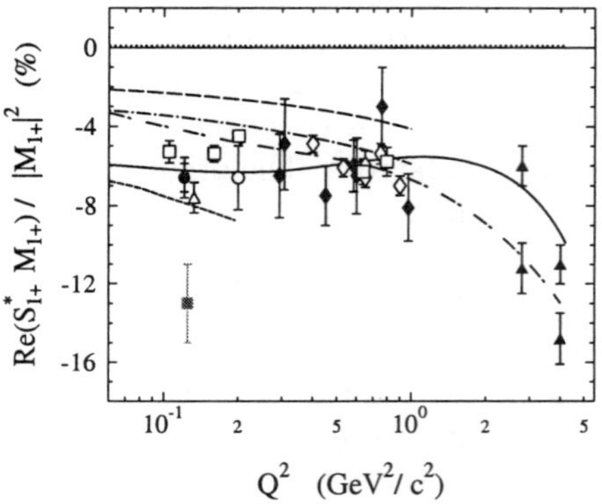

Figure 3. Results for $\Re e\{S_{1+}^* M_{1+}\}/|M_{1+}|^2$ extracted from different experiments. The pre 1980 data are indicated as full diamonds [8,9,10]. The full grey square is from [11]. Full triangles represent the analyses of [12]. The result from the only double polarization measurement is indicated by the full circle with statistical (inner) and quadratic sum of statistical and systematical (outer) error [5]. Preliminary data are also shown from Bates [13] (open triangle), TJNAF/CLAS [14] (open diamonds) ELSA [15] (open squares) and MAMI [16,17] (open circle). The curves are calculations within Heavy Baryon Chiral Perturbation Theory [18] (short dashed), the Chiral Quark Soliton Model [19] SU(2) (dashed-dotted) and SU(3) (long dashed), the dynamical model of Ref. [20] (long-short dashed), and dressed (full) and bare (dotted) solutions of the dynamical model of Ref. [21].

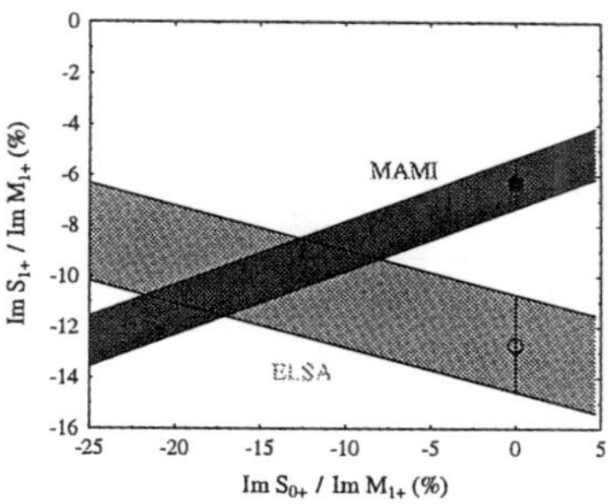

Figure 4. $\Im m S_{1+}/\Im m M_{1+}$ as extracted as a function of $\Im m S_{0+}/\Im m M_{1+}$ from measurements of R_{LT}-like structure functions with the pion in forward direction [11] (light shaded band) and backward direction [5] (dark shaded band). The data points with errors correspond to the respective results at $\Im m S_{0+} = 0$.

momentum. They are summarized in Table 1. A $10\,\mu$A (unpolarized) beam was used on a LH_2 target with a diameter of only 1 cm, in order to minimize the energy loss of the low energy protons of the forward settings (cf. Table 1). The scattered electrons were detected in Spectrometer A and the protons in Spectrometer B.

Table 1. Kinematics of the $p(e,e'p)\pi^0$ experiment. $\Theta_p^{\rm lab}$ and $p_p^{\rm lab}$ are the proton laboratory angles and momenta, respectively, and $\Theta_{pq}^{\rm lab}$ denotes the relative angle between recoiling proton and momentum transfer. For all settings the angle of the momentum transfer is 26.9° relative to the incoming electron beam.

kinematics	$\Theta_\pi^{\rm cm}$	$\Theta_{pq}^{\rm lab}$	$p_p^{\rm lab}$ (MeV/c)	$\Theta_p^{\rm lab}$
b (right)	160°	6.1°	741.7	33.0°
b (left)				20.9°
f (right)	20°	17.2°	265.0	44.2°
f (left)				9.7°

In Fig. 5 the preliminary results [16,17] are shown in comparison with calculations. The full curve corresponds to the full MAID calculation which has the standard ratios $\Im m S_{1+}/\Im m M_{1+} = -6.6\,\%$ and $\Im m S_{0+}/\Im m M_{1+} = +6.0\,\%$. The other curves are due to a truncated multipole basis where only M_{1+}, S_{1+} and S_{0+} partial waves are included. The differences between the full and dashed curves indicate the quality of this approximation. From the result of Ref. [11] the dotted curve would be expected, while the dashed-dotted line

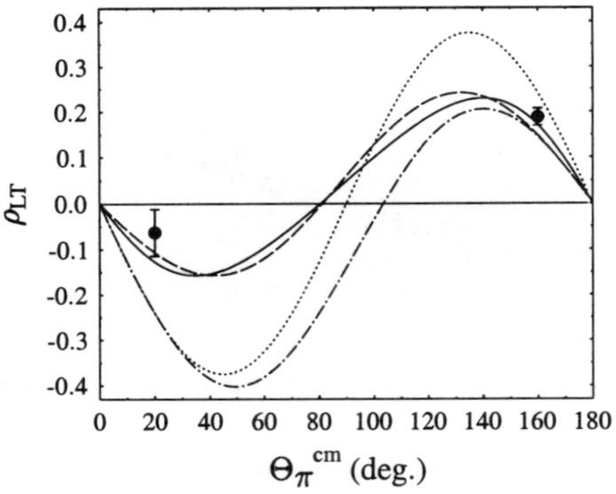

Figure 5. Left-right asymmetries for the $p(e,e'p)\pi^0$ cross section around $W = 1232\,\text{MeV}$, $Q^2 = 0.2\,(\text{GeV/c})^2$ and $\epsilon = 0.6$. Forward and backward proton kinematics, $\Theta_\pi^{\text{cm}} = 160°$ and $\Theta_\pi^{\text{cm}} = 20°$, respectively [16,17], were measured at identical electron kinematics. The full curve is from MAID2000 while the other curves are obtained in a truncated multipole basis (only M_{1+}, S_{1+} and S_{0+} partial waves) with $\Im m S_{1+}/\Im m M_{1+} = -6.6\,\%$, $\Im m S_{0+}/\Im m M_{1+} = +6.0\,\%$ (the MAID standard values — dashed), $-12.5\,\%$, $0\,\%$ (dotted), and $-10.0\,\%$, $-14\,\%$ (dashed-dotted).

corresponds to the overlap region in Fig. 4. The new data do not yet indicate any discrepancy with the standard MAID parameterization concerning the $\Im m S_{1+}/\Im m M_{1+}$ and $\Im m S_{0+}/\Im m M_{1+}$ ratios. Thus, the imaginary parts of the relevant amplitudes seem well under control.

4 Fifth Structure Function in $p(\vec{e},e'p)\pi^0$

In contrast to the imaginary parts, the real background is more problematic. This is indicated by the discrepancy between MAID calculation and data for P_y (cf. Fig. 1, middle). P_y is related to the imaginary part of a longitudinal-transverse interference. In parallel kinematics the leading terms in s and p wave approximation read [2]:

$$P_y \propto \Im m\{(S_{0+} - 4S_{1+} + S_{1-})^* M_{1+}\}. \quad (1)$$

This is very similar to the fifth structure function in $p(\vec{e},e'p)\pi^0$:

$$R_{LT'} \propto P_e \sin\Theta_\pi^{\text{cm}}\, \Im m\{(S_{0+} + 6\cos\Theta_\pi^{\text{cm}} S_{1+})^* M_{1+}\}. \quad (2)$$

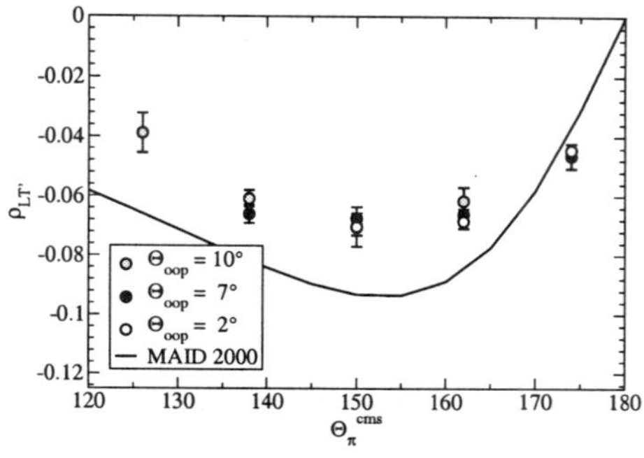

Figure 6. Cross section asymmetry $\rho_{LT'}$ for the $p(\vec{e}, e'p)\pi^0$ reaction around $W = 1232$ MeV and $Q^2 = 0.2\,(\mathrm{GeV/c})^2$, measured at $\Phi = 90°$ (out of plane) [25]. White, grey and black dots indicate 2°, 7° and 10° out of plane settings of the proton spectrometer, respectively. The curve is due to the full MAID2000 calculation.

Therefore, we measured the helicity asymmetry $\rho_{LT'} = \frac{d\sigma^+ - d\sigma^-}{d\sigma^+ + d\sigma^-}$ with regard to the flip of the electron beam helicity between + and −. This required detection of the recoiling proton in Spectrometer B with its 10° out of plane capability. The preliminary result is shown in Fig. 6. Obviously, the MAID parameterization, very similarly to P_y, also overestimates the magnitude of $\rho_{LT'}$. It is therefore concluded that this is not caused by the S_{1-} amplitude, which only contributes to P_y (Eq. (1)), but is rather related to the real (Born) background in the imaginary parts of the interferences of Eqs. (1) and (2). Such a real background does not affect the extraction of the CMR from P_x to first order at the resonance energy.

5 Summary

From a first $p(\vec{e}, e'\vec{p})\pi^0$ double polarization experiment CMR $= (-6.4 \pm 0.7_{\mathrm{stat}} \pm 0.8_{\mathrm{syst}})$ % has been deduced within the MAID2000 parameterization. Imaginary background amplitudes investigated by forward and backward ρ_{LT} measurements are in agreement with MAID. On the contrary, MAID consistently overestimates the magnitude of P_y as well as of $\rho_{LT'}$, which was measured with out of plane detection of the recoiling protons. This seems due to real amplitudes which, at the resonance energy, do not affect the extraction of the CMR to first order. The model uncertainty is thus estimated to be of the order of 1 % absolute.

Acknowledgements

I thank all my colleagues from Basel, Ljubljana, Rutgers and Mainz, in particular the students P. Bartsch, D. Elsner, S. Grözinger, Th. Pospischil, and A. Süle. This work was supported by the Deutsche Forschungsgemeinschaft (SFB 443).

References

1. P. Wilhelm *et al., Phys. Rev.* C **54**, 1423 (1996).
2. H. Schmieden, *Eur. Phys. J.* A **1**, 427 (1998).
3. K.I. Blomqvist *et al., Nucl. Instr. Methods* A **403**, 263 (1998).
4. Th. Pospischil *et al.*, submitted to *Nucl. Instr. Meth.* A and nucl-ex/0010007.
5. Th. Pospischil *et al., Phys. Rev. Lett.* **86**, 2959 (2001).
6. D. Drechsel, O. Hanstein, S.S. Kamalov and L. Tiator, *Nucl. Phys.* A **645**, 145 (1999) and http://www.kph.uni-mainz.de/MAID/maid2000/.
7. H. Schmieden and L. Tiator, *Eur. Phys. J.* A **8**, 15 (2000).
8. R. Siddle *et al., Nucl. Phys.* B **35**, 93 (1971).
9. J.C. Alder *et al., Nucl. Phys.* B **46**, 573 (1972).
10. K. Bätzner *et al., Nucl. Phys.* B **76**, 1 (1974).
11. F. Kalleicher *et al., Z. Phys.* A **359**, 201 (1997).
12. V.V. Frolov *et al., Phys. Rev. Lett.* **82**, 45 (1999).
13. C. Mertz *et al.*, nucl-ex/9902012.
14. V.D. Burkert, proceedings of Few Body Problems in Physics, Taipei (2000), hep-ph/0007297.
15. R. Gothe, proceedings of NSTAR2000, Newport News (2000.)
16. D. Elsner, Diploma thesis, Mainz (2000) (unpublished).
17. A. Süle, Diploma thesis, Mainz (2001) (unpublished).
18. G.C. Gellas *et al., Phys. Rev.* D **60**, 054022 (1999).
19. A. Silva *et al., Nucl. Phys.* A **675**, 637 (2000).
20. T. Sato and T.-S.H. Lee, *Phys. Rev.* C **63**, 055201 (2001).
21. S.S. Kamalov and S.N. Yang, *Phys. Rev. Lett.* **83**, 4494 (1999).
22. H. Schmieden *et al.*, approved MAMI proposal A1-3/98.
23. J.J. Kelly, *Phys. Rev.* C **60**, 054611 (1999).
24. D. Drechsel and L. Tiator, *J. Phys. G: Nucl. Part. Phys.* **18**, 449 (1992).
25. P. Bartsch, Ph. D. thesis, Mainz, in preparation.

PION ELECTROPRODUCTION USING CLAS

L. C. SMITH, FOR THE CLAS COLLABORATION
Department of Physics, University of Virginia
Charlottesville, Virginia, USA
E-mail: cole@nstar.phys.virginia.edu

Measurements of the $p(e,e'p)\pi^o$ reaction in the $\Delta(1232)$ resonance region have been performed with the CLAS detector at Jefferson Lab using beam energies of 1.645 and 2.445 GeV. Cross sections were measured over a range of four-momentum transfer Q^2=0.4-1.8 (GeV/c)^2. Decay angular distributions in the $\pi^o p$ center-of-mass were obtained with full coverage in $cos\,\theta^*_\pi$ and ϕ^*_π. This permits the extraction of the structure functions $\sigma_T+\epsilon_L\sigma_L, \sigma_{TT}$ and σ_{LT} and their partial wave decomposition. Results are shown for the Q^2 and W dependence of the transverse electric and scalar/longitudinal multipole ratios E_{1+}/M_{1+} and S_{1+}/M_{1+}.

1 Introduction

As the most prominent evidence of internal nucleon structure, the $\Delta(1232)$ resonance is well suited for studying the dynamics of quark confinement. A longstanding question is the origin of the non-zero quadrupole strength experimentally observed in the $\gamma^* N \rightarrow \Delta \rightarrow N\pi$ transition. Only the dominant magnetic dipole (M_{1+}) excitation mode is permitted in simple quark models that satisfy $SU(6)$ symmetry, occuring via a single quark spin flip. The $SU(6)$ forbidden electric (E_{1+}) and Coulomb (S_{1+}) quadrupole transitions imply non-spherical forces[1] within the nucleon, which arise through a variety of mechanisms in modern QCD-motivated models: residual tensor interactions between quark pairs (such as gluon[2] and meson exchange [3]) which can introduce d-state admixtures, direct photon coupling to a deformed pion cloud[4] or to $q\overline{q}$ exchange currents (leading to the spin flip of two quarks)[5], and collective rotation of a Skyrmion[6].

A crucial experimental test for these hadron models is the magnitude and Q^2 evolution of the multipole ratios $R_{EM} = E_{1+}/M_{1+}$ and $R_{SM} = S_{1+}/M_{1+}$. Unfortunately, previous measurements from 30 years ago lacked sufficient statistical and systematic precision. A new generation of photo- and electroproduction experiments have begun to publish results. Recent measurements of the $p(\gamma,\pi^o)p$ and $p(\gamma,\pi^+)n$ photo-pion channels at LEGS[7] and MAMI[8] found $R_{EM} \approx -(2.5-3.0)\%$, which is substantially larger than predictions from QCD-inspired constituent quark models (CQM) incorporating one-gluon exchange[2] (color magnetism). A JLAB/Hall C π^0 electroproduction measurement[9] at $Q^2 = 2.8$ and 4.0 (GeV/c)^2 found $R_{EM} < 0$ at the level of a few percent, while R_{SM} was 5-6 times larger and increasingly negative with Q^2. These results appear to rule out the influence of perturbative QCD, which favors helicity conserving amplitudes for which $R_{EM} \to 1$ and $R_{SM} \to constant$ as $Q^2 \to \infty$, and has shifted the theoretical emphasis to models which explicitly incorporate pion degrees of freedom. For example, chiral soliton models[10], in which quark confinement arises through non-linear interactions with the pion cloud, generally predict larger values for R_{EM} at the photon point. Recent progress in chiral effective field theories[11] and dynamical models[12] that incorporate pion rescattering at the $\gamma N \Delta$ vertex suggest meson degree of freedom strongly enhance the quadrupole strength at low Q^2 and affect the shape of the Q^2 dependence overall.

Here we report preliminary CLAS measurements of R_{EM} and R_{SM} over the range $Q^2 = 0.4$-1.8 (GeV/c)^2 obtained from a multipole analysis of the $p(e, e'p)\pi^o$ reaction. Under the one-photon-exchange approximation, the electroproduction cross section factorizes as follows:

$$\frac{d^5\sigma}{dE_{e'}\,d\Omega_{e'}\,d\Omega^*_\pi} = \Gamma_v \frac{d^2\sigma_u}{d\Omega^*_\pi} \tag{1}$$

where Γ_v is the virtual photon flux. For unpolarized beam and target the center-of-mass (c.m.) differential cross section $d^2\sigma_u$ depends on the transverse ϵ and longitudinal ϵ_L polarization of the virtual photon through four structure functions: σ_T, σ_L and their interference terms σ_{LT} and σ_{TT}:

$$\begin{aligned}\frac{d^2\sigma_u}{d\Omega^*_\pi} = \frac{p^*_\pi}{k^*_\gamma}(\sigma_T + \epsilon_L\sigma_L + \epsilon\,\sigma_{TT}\, sin^2\theta^*_\pi\, cos\, 2\phi^*_\pi \\ + \sqrt{2\epsilon_L(\epsilon+1)}\,\sigma_{LT}\, sin\,\theta^*_\pi\, cos\,\phi^*_\pi)\end{aligned} \tag{2}$$

where $(p^*_\pi, \theta^*_\pi, \phi^*_\pi)$ are the π^o c.m. momentum, polar and azimuthal angles, $\epsilon_L = (Q^2/|k^*|^2)\epsilon$ and $|k^*|$ and k^*_γ are the virtual photon c.m. momentum and equivalent energy. Expansion of the structure functions in partial waves up to $l_\pi = 1$ gives:

$$\sigma_T + \epsilon_L\sigma_L = A_o + A_1\, cos\,\theta^*_\pi + A_2\, P_2(cos\,\theta^*_\pi) \tag{3}$$

$$\sigma_{TT} = C_o \tag{4}$$

$$\sigma_{LT} = D_o + D_1\, cos\,\theta^*_\pi \tag{5}$$

The weak quadrupole E_{1+} and S_{1+} transitions are accessible only through their interference with the dominant M_{1+}. To simplify the analysis, a truncated multipole expansion (TME) is used, in which only interference terms which include M_{1+} are retained. Thus, $|M_{1+}|^2$ and its projection onto the other s- and p-wave multipoles $E_{1+}, S_{1+}, M_{1-}, E_{0+}, S_{0+}$ are given in terms of the six partial-wave coefficients by [13]:

$$|M_{1+}|^2 = A_0/2 \tag{6}$$
$$Re(E_{1+}\, M^*_{1+}) = (A_2 - 2\, C_0/3)/8 \tag{7}$$
$$Re(M_{1-}\, M^*_{1+}) = -(A_2 + 2\,(A_0 + C_0))/8 \tag{8}$$
$$Re(E_{0+}\, M^*_{1+}) = A_1/2 \tag{9}$$
$$Re(S_{0+}\, M^*_{1+}) = D_0 \tag{10}$$
$$Re(S_{1+}\, M^*_{1+}) = D_1/6 \tag{11}$$

Because the charge multipole is dominated by the isospin 3/2 channel $M^{(3/2)}_{1+}$, by Watson's theorem $Re(M_{1+})$ vanishes at the resonance position, giving $Re(E_{1+}M^*_{1+}) \approx Im(E^{(3/2)}_{1+})Im(M^{(3/2)}_{1+})$, etc.. Thus we define:

$$R_{EM} = Re(E_{1+}\, M^*_{1+})/|M_{1+}|^2 \tag{12}$$
$$R_{SM} = Re(S_{1+}\, M^*_{1+})/|M_{1+}|^2 \tag{13}$$

Data were obtained from the CEBAF Large Acceptance Spectrometer (CLAS) in Hall B at Jefferson Lab, which provides large solid angle coverage for the study of N^* resonance decays. CLAS utilizes a 2 Tesla toroidal magnetic field. Six wedge-shaped sectors between the field coils are individually instrumented with multiwire drift chambers and time-of-flight (TOF) scintillation counters for charged-particle trajectory reconstruction down to momenta of 0.1 (GeV/c)^2 and over a lab polar angle (θ) range of 8-142°.

For this measurement, two electron beam energies were used to cover the Q^2 interval of 0.4-1.8 (GeV/c)^2. Beams were delivered at 100% duty factor with an average current of 2.5 nA onto a cylindrical liquid hydrogen target of 4.0 cm length and 0.8 cm diameter. The scattered electron trigger in each sector was formed from coincidence of a gas threshold Cereknov counter and a sampling electromagnetic calorimeter. After kinematic corrections the invariant mass resolution was $\sigma_W \approx 8$ MeV. A single magnetic field setting was used for both beam energies. Coincident protons were identified using TOF and momentum. A cut on the reconstructed missing mass (see Fig. 1) of $-0.02 \leq M_x^2(GeV^2) \leq 0.08$ was used to select the $p\pi^o$ final state. Background from elastic Bethe-Heitler radiation was suppressed to below 1% using $e'p$ coplanarity cuts. Target window backgrounds and proton multiple scattering were suppressed with cuts on the reconstructed $e'p$ target vertex.

CLAS acceptance, tracking efficiency and resolution were simulated using a GEANT model of the detector geometry which incorporated the magnetic

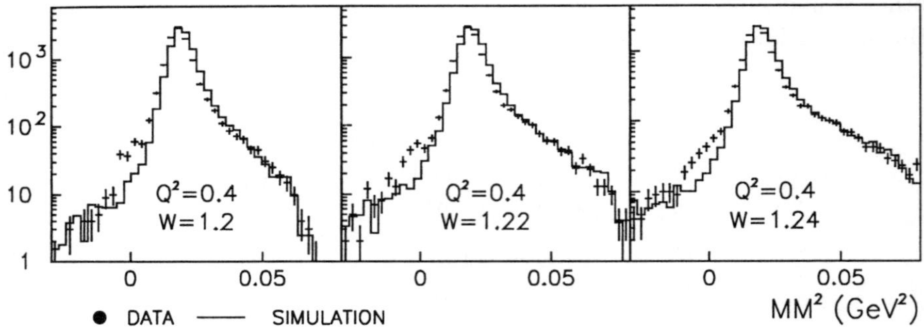

Figure 1. Experimental $p(e, e'p)X$ missing mass distributions for three invariant mass W bins around the peak of the $\Delta(1232)$. Bethe-Heitler photons near $MM^2 = 0$ were suppressed using overdetermined elastic kinematics. Also shown is comparison with GEANT simulation of CLAS response to $p(e, e'p)\pi^o$ reaction (solid line).

field map, surveyed positions of detector elements (including target position relative to coils), dead wires or TOF bars and measured wire chamber drift times. Software fiducial cuts were used to define the solid angle for electrons and hadrons and excluded regions of low Cerenkov efficiency or large multiple scattering from magnet coils. Energy loss in the target and detector was included. Radiative corrections were calculated without the peaking approximation, using a Monte-Carlo integration of the Mo and Tsai formula [15]. Inclusive (e, e') elastic and inelastic cross sections were measured simultaneously with the exclusive data, and agreed to within 5% with previous measurements.

Events were weighted with Γ_v^{-1}, accumulated into bins of $(Q^2, W, cos\,\theta^*_\pi, \phi^*_\pi)$ and corrected for acceptance and radiative effects. Fig. 2 (left) shows an example of the cross section dependence on ϕ^*_π near the $\Delta(1232)$ peak at $Q^2 = 0.9$ $(\text{GeV}/c)^2$, and illustrates the complete out-of-plane coverage using CLAS. The structure functions $\sigma_T + \epsilon_L \sigma_L$, σ_{TT} and σ_{LT} were extracted by fitting a complete ϕ^*_π distribution for each W and $cos\,\theta^*_\pi$ bin with the form in Eq. 2. The solid line in Fig. 2 shows a typical fit.

The $cos\,\theta^*_\pi$ dependence of the extracted structure functions is also shown Fig. 2 (right) for several W bins around the $\Delta(1232)$ peak. The solid line is a fit assuming only s- and p-waves in the $\pi^\circ p$ final state. The data are largely in agreement with this assumption, and fits up to $l_\pi = 2$ showed no consistent d-wave strength around the $\Delta(1232)$ peak. Other curves show the unitary isobar model of Drechsel *et al* (MAID2000)[17] and a pion rescattering model

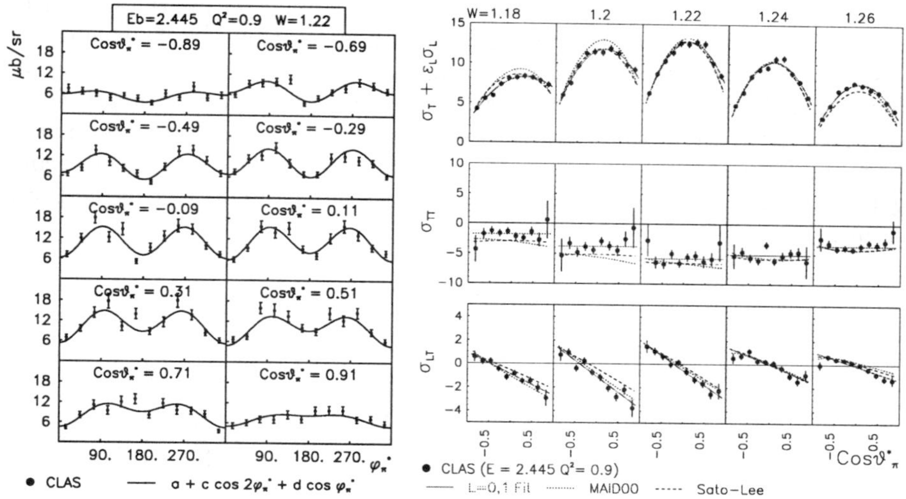

Figure 2. Left: CLAS measurement of $p\pi^0$ C.M. angular distributions. Solid line shows fit used to extract structure functions. Right: Structure functions extracted from $p(e,e'\,p)\pi^o$ data at $Q^2 = 0.9\,(\mathrm{GeV/c})^2$. Each data point represents a fit of Eq. 2 to ϕ^*_π distributions at a fixed W and $cos\,\theta^*_\pi$ bin, such as shown at left. Solid line shows partial wave fit assuming only s and p-waves dominate the $\pi^o p$ final state. Other curves show MAID2000 and Sato-Lee predictions.

of Sato and Lee (SL)[12], both of which reproduce the qualitative features of the data.

These models are compared again to the data in Fig. 3, which shows the W dependence of all s- and p-wave multipoles extracted from the fitted partial wave coefficients using the TME derived solutions of Eqs. 6-11. Here the difference between the models is more evident, and reflects different approaches to unitarization of resonant and background terms. Both measured and predicted non-resonant multipoles M_{1-}, E_{0+} and S_{0+} show appreciable strength below the peak of the Δ due to constructive interference with the real part of the I=3/2 M_{1+} amplitude.

In order to check the model uncertainty and overall error introduced by the TME approximation, we fitted 'pseudo-data' generated from the MAID and SL models using the same binning and statistical error as the CLAS data, and analyzed according to the procedure just described. In this way the extracted observables could be directly compared to those calculated from the model. It was found that interference terms involving the resonant input multipoles E_{1+}, S_{1+} were retrieved in the pseudo-data fits to within the statistical error of the CLAS data, while the extracted non-resonant interference terms displayed systematic discrepancies quite similar to what is shown in Fig. 3, indicating they are largely due to truncation error. The contribution to R_{EM} and R_{SM} from truncation error (including model dependence) was estimated to range from $< 0.5\%$ to 1.0% over the Q^2 range of the experiment, being generally higher at large Q^2 due to the larger relative importance of the Born terms.

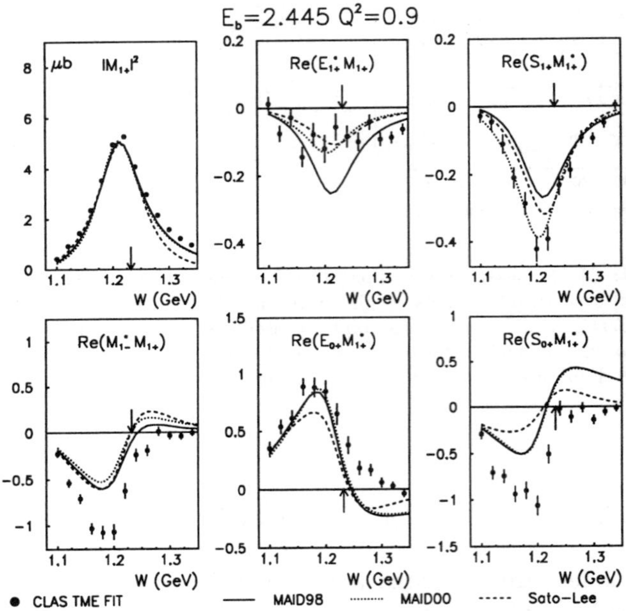

Figure 3. W dependence of $|M_{1+}|^2$ and its interference terms in the $\Delta^+ \to \pi^o p$ charge channel at Q^2= 0.9 (GeV/c)^2, extracted from fits to experimental structure functions using Eqs. 6-11. A separate fit was made for each W bin. Model predictions for $\pi^o p$ multipoles shown for MAID98 (solid line), MAID00 (dashed line) and dynamical pion model of Sato and Lee (dotted line). Arrow indicates Δ resonance position at W=1.232 GeV.

R_{EM} and R_{SM} were determined by a weighted average of the data points at W=1.20,1.22 and 1.24 GeV, as this choice gave the lowest overall truncation error in the pseudo-data fits. The Q^2 dependence of the ratios is shown in Fig. 4 compared to recent model calculations [10,12,14,18,19] and previous

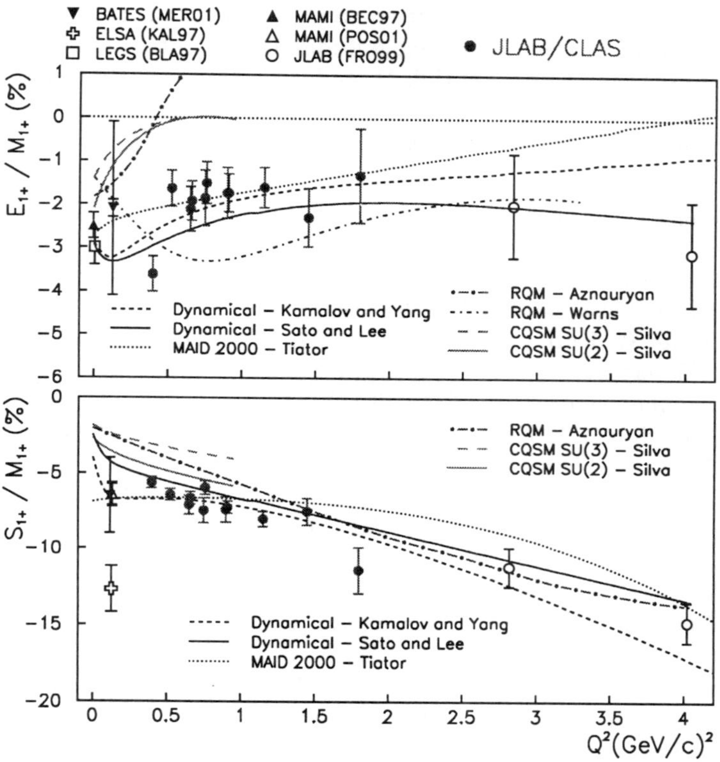

Figure 4. CLAS measurements (•) of Q^2 dependence of multipole ratios R_{EM} and R_{SM} averaged over the W range 1.20-1.24 GeV. Only statistical errors shown. Curves show recent model calculations: chiral quark-soliton[10], dynamical pion cloud[12,14], MAID2000[16], relativistic quark[18,19]. Other R_{EM} points are from BATES[21], LEGS[7], JLAB/Hall C[9] and MAMI[8]. Other R_{SM} points are BATES[21], ELSA[22], JLAB/Hall C[9] and MAMI[20]. Note that the larger error on the BATES points are estimates of model dependence.

measurements [7,8,20,21,22]. The R_{EM} data appear to exclude the relativistic quark[18,19] and chiral quark soliton[10] models, although these models do somewhat better in comparison with R_{SM}. The dynamical and isobar model curves, which are somewhat constrained by fits to the higher Q^2 data of Frolov[9], come closest to describing the Q^2 dependence of our data. The SL model, which is not constrained by the BATES and Mainz R_{SM} measurements around $Q^2 = 0.12$ (GeV/c)^2, clearly misses those data. The Kamalov and Yang DM model, which is fitted to those low Q^2 measurements, comes closest to matching the absolute magnitudes of both our R_{EM} and R_{SM} data.

Thanks go to the JLAB/Hall B staff and CLAS collaboration for their hard work in supporting the experiment. This work was supported in part by DOE contract DEFG02-97ER41018.

References

1. S.L. Glashow, *Physica* A **96**, 27 (1979); V. Vento, G. Baym, and A.D. Jackson, *Phys. Lett.* B **102**, 97 (1981).
2. N. Isgur, G. Karl, and R. Koniuk, *Phys. Rev.* D **25**, 2394 (1982).
3. G. Kälbermann and J. Eisenberg, *Phys. Rev.* D **28**, 71 (1983).
4. S. Kumano, *Phys. Lett.* **214**, 132 (1988).
5. A.J. Buchman, E. Hernandez, U. Meyer, and Amand Faessler, *Phys. Rev.* C **58**,2478 (1998).
6. H. Walliser, G. Holzwarth, *Z. Phys.* A **357**, 317-324 (1997).
7. G. Blanpied *et al.*, *Phys. Rev. Lett.* **79**, 4337 (1997).
8. R. Beck *et al.*, *Phys. Rev. Lett.*, **78**, 606 (1997).
9. V.V. Frolov *et al.*, *Phys. Rev. Lett.* **82**, 45 (1999).
10. A. Silva, D. Urbano, T. Watabe, M. Fiolhais, and K. Goeke, *Nucl. Phys.* A **675**, 637 (2000).
11. Thomas R. Hemmert, nucl-th/0105051 and *these proceedings.*
12. T. Sato and T.-S.H. Lee, *Phys. Rev.* C **63**, 055201 (2001).
13. A.S. Raskin and T.W. Donnelly, *Ann. Phys.* **191**, 78 (1989).
14. S.S. Kamalov and Shin Nan Yang, *Phys. Rev. Lett.* **83**, 4494 (1999); S.S. Kamalov, S.N. Yang, D. Drechsel, O. Hanstein, and L. Tiator, nucl-th/0006068.
15. L.W. Mo and Y.S. Tsai, *Rev. Mod. Phys.* **45**, 205 (1969).
16. O. Hanstein *et al., Nucl. Phys.* A **632**, 561 (1998); L. Tiator *et al*, nucl-th/0012046.
17. D. Drechsel, O. Hanstein, S.S. Kamalov, and L. Tiator, *Nucl. Phys.* A **645**, 145 (1999). Cross sections were calculated from $p\pi^o$ multipoles (up to L=5) downloaded from the MAID web page: http://www.kph.uni-mainz.de/MAID/.
18. M. Warns, H. Schröder, W. Pfeil, and H. Rollnik, *Z. Phys.* C **45**, 627 (1990).
19. I.G. Aznauryan, *Z. Phys.* A **346**, 297 (1993).
20. Th. Pospischil *et al., Phys. Rev. Lett.* **86**, 2959 (2001).
21. C. Mertz *et al., Phys. Rev. Lett.* **86**, 2963 (2001).
22. F. Kalleicher *et al., Z. Phys.* A **359**, 201 (1997).

RECOIL POLARIZATION MEASUREMENTS IN π^0 ELECTROPRODUCTION AT THE PEAK OF THE $\Delta(1232)$

A. J. SARTY
FOR THE JEFFERSON LAB HALL A COLLABORATION
Saint Mary's University
Department of Astronomy & Physics
Halifax, NS, Canada B3H 3C3
E-mail: sarty@ap.stmarys.ca

This talk presents a Status Report, along with some preliminary/on-line results, from the Thomas Jefferson National Accelerator Facility (JLab) experiment E91-011 which was performed in Hall A at JLab during the summer of 2000. The experiment measured angular distributions for the differential cross section and recoil-proton polarizations in the reaction $p(\vec{e}, e'\vec{p})\pi^0$. Kinematics were chosen to be centered at a CMS energy of W = 1232 MeV, and a squared four-momentum transfer of $Q^2 = 1.0\ (GeV/c)^2$. The primary objectives of the experiment are to isolate contributions from the resonant quadrupole $N \rightarrow \Delta$ multipole, S_{1+}, and to clarify the role of other, small non-resonant multipole contributions to the reaction. Details of the experiment itself will be given, along with sample spectra illustrating the quality and coverage of the data obtained.

1 Introduction

The import of using transition form factors for the electroexcitation of nucleon resonances as tools for testing QCD-inspired models of baryon structure is clear simply by inspection of the talk-titles for this Conference. Indeed, attempts to isolate the small quadrupole multipole excitations of the $\Delta(1232)$ resonance (E_{1+} and/or S_{1+}) for just such tests is the topic of the preceding four talks, as well as this present talk. Rather than repeat the details of the physics justifications and motivations for pursuing these small quadrupole multipoles, the reader is referred to these preceding talks for more details (see talks by Papanicolas, Gothe, Schmieden, and Smith).

2 Measurement Philosophy: Exploiting Hall A's Tools

Isolation of specific, small resonant multipoles (such as the S_{1+} for the Δ) is complicated by the presence of nearby resonances, and of nonresonant "background" contributions. A full multipole analysis will require not only extensive cross section data, which will be available from Hall B at JLab (see talk by Smith), but also target and recoil polarization data. Target polarization data will also be acquired in Hall B, but recoil polarization data can be acquired in Hall A. Hall A at JLab has unique tools to carry out a program of recoil polarization measurements for π^0 electroproduction at the $\Delta(1232)$ peak. These tools include: *(i)* two High Resolution magnetic Spectrometers (HRS), each with $\Delta\Omega \approx 6$ msr; *(ii)* a working, calibrated Focal Plane Polarimeter (FPP) to measure recoil proton polarization; *(iii)* a 15 cm liquid-hydrogen cryotarget facility; and *(iv)* a high-current, high-polarization cw electron beam with energies up to 5 GeV. These tools make possible high-resolution recoil polarization measurements at specific, selected kinematics (e.g. at the $\Delta(1232)$ pole energy), which is required for a stringent evaluation of multipole amplitudes. Such polarization measurements will provide access to unique multipole combinations and phase information that is not available with other measurement methods. These types of measurements will complement the Hall B multipole extractions via a more extensive energy-independent analysis. The power of the recoil-polarization technique has already been demonstrated with "single-point" (single angle) measurements at the MIT-Bates Lab [1] and at Mainz-MAMI [2]. The measurements from the experiment reported here will be more far-reaching since a full angular distribution was able to be performed.

3 Specific Measurements for E91-011

The experiment E91-011 ran between May 19 and July 31, 2000, and received an average current of 45 μA for 36.5 days. The longitudinal beam polarization varied between 67 and 79% and was monitored almost continuously by a Compton polarimeter.

The production kinematics are summarized in Tab. 1. The nominal beam energy, $E_0 = 4.535$ GeV, electron scattering angle, $\theta_e = 14.09°$, and scattered electron momentum, $k_f = 3.066$ GeV/c, were constant for the production data. The proton momenta and angles were chosen to give central values of $W = 1.232$ GeV and $Q^2 = 1.0$ (GeV/c)^2 while optimally covering the center-of-mass angular distribution. The table lists nominal center-of-mass angles θ_{cm} where positive (negative) angles refer to nucleon momenta forward (backward) with respect to the virtual photon momentum $\vec{q}$. The total charge collected for each setting is given in Coulombs.

4 Accessible Responses and Expected Sensitivity to S_{1+}

The nuclear response functions accessible in $p(\vec{e}, e'\vec{p})\pi^0$ are defined through a decomposition of the differential cross section in terms of trigonometric dependencies on the azimuthal angle (ϕ) between the electron-scattering plane

Table 1. Kinematics summary for E91-011.

θ_{cm} deg.	θ_{lab} deg.	p_{lab} GeV/c	charge C	θ_{cm} deg.	θ_{lab} deg.	p_{lab} GeV/c	charge C
-155	49.72	0.742	11.5	155	34.71	0.742	11.1
-135	53.64	0.819	20.2	135	30.81	0.819	11.1
-90	54.79	1.066	13.7	90	29.81	1.066	12.0
-50	50.18	1.270	10.5	50	34.29	1.270	13.0
-25	46.33	1.350	4.6	25	38.12	1.350	6.2
0	42.30	1.378	16.7	180	42.28	0.703	3.8

(containing the incoming and scattered electron momentum vectors) and the reaction plane (containing the momentum-transfer vector and the detected proton momentum vector), and in terms of dependence on the beam polarization and on components of the recoil-proton's polarization vector. A standard decomposition for defining the responses is given by: [3]

$$\begin{aligned}\frac{d^5\sigma}{d\Omega_e d\Omega_p^{cm} d\omega} &= \Gamma\{(R_T + P_N R_T^N) + \epsilon(R_L + P_N R_L^N) \\ &+ \sqrt{2\epsilon(1+\epsilon)}\left[(R_{TL} + P_N R_{TL}^N)\cos\phi + (P_L R_{TL}^L + P_S R_{TL}^S)\sin\phi\right] \\ &+ \epsilon\left[(R_{TT} + P_N R_{TT}^N)\cos 2\phi + (P_L R_{TT}^L + P_S R_{TT}^S)\sin 2\phi\right] \\ &+ h\sqrt{2\epsilon(1-\epsilon)}\left[(R'_{TL} + P_N R_{TL}^{\prime N})\sin\phi + (P_L R_{TL}^{\prime L} + P_S R_{TL}^{\prime S})\cos\phi\right] \\ &+ h\sqrt{1-\epsilon^2}(P_L R_{TT}^{\prime L} + P_S R_{TT}^{\prime S})\} \quad .\end{aligned} \tag{1}$$

The labeling of the 18 responses (R's) with subscripts T, L, TT or TL, and superscripts N, L or S, and primed or unprimed, is according to standard convention [3]. Each response is a unique bilinear combination of multipole amplitudes, and may represent either the real or imaginary part of a combination. For a given θ_{cm} angle, the responses can be accessed experimentally via either a "left/right" subtraction (in-plane, either side of $\vec{q}$), "out-of-plane (OOP)" measurements (with $\phi \neq 180°, 0°$), or "Rosenbluth" separation (more than one beam energy). A summary of the response-function types and method of experimental access is given in Tab. 2.

In order to unambiguously determine the multipole amplitudes contributing to the reaction, it is necessary to have information sensitive to the relative phase of the multipoles - meaning that measurements are required of at least some of the "imaginary part" responses. In a recoil polarization experiment such as this one, the 14 non-Rosenbluth responses listed in Tab. 2 are potentially accessible. Since the HRS spectrometers of Hall A are restricted to being in the same plane, angular distributions of the 6 "left/right" responses will

Table 2. Grouping of response functions according to experimental method of access, and to whether response is a real or imaginary part of a multipole combination.

Experimental Method of Accessing	REAL part of Multipole Combination	IMAGINARY part of Multipole Combination
Left/Right	$R_LT, R'^{L,S}_{TL}, R'^{L,S}_{TT}$	R^N_{TL}
Out-of-Plane	$R_TT, R'^{N}_{TL}, R^N_{TT}$	$R'_{TL}, R^{L,S}_{TL}, R^{L,S}_{TT}$
Rosenbluth	R_L, R_T	R^N_L, R^N_T

clearly be obtained. However, due to the kinematic focusing of the reaction cone to a laboratory opening angle of about 13°, considerable out-of-plane acceptance is also provided. This OOP acceptance is demonstrated in Fig. 1, which shows contours of normalized yield for the entire experiment, plotted as a function of θ_{cm} and of ϕ. The nearly complete ϕ coverage for $\theta_{cm} < 50°$ will permit extraction of the additional 8 "OOP" response functions. Moreover, at $\theta_{cm} = 0°$, the symmetry properties of the polarized responses will allow for a determination of the ratio R_L/R_T at that one angle [4,5].

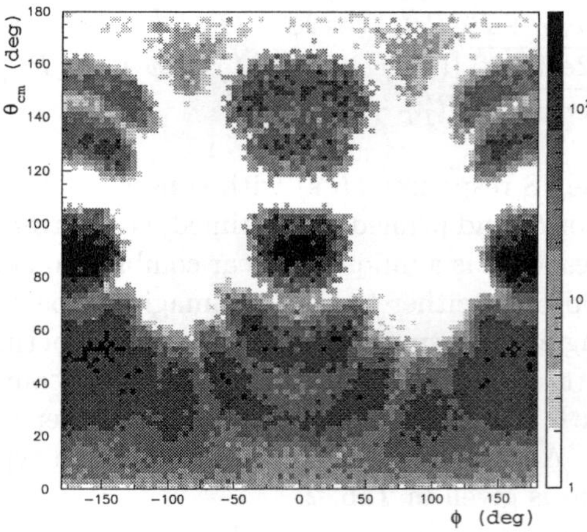

Figure 1. Angular acceptance for E91-011, with cuts of $1.207 \leq W \leq 1.257$ GeV and $0.9 \leq Q^2 \leq 1.1$ (GeV/c)^2. Contours of normalized yield are shown for (θ_{cm}, ϕ), combining the various spectrometer settings.

To demonstrate the expected sensitivity of the various accessible response functions to the size of the interesting S_{1+} multipole, Fig. 2 shows the effect of changing the magnitude of the "SMR" (ratio of S_{1+} to the dominant M_{1+} multipole). The calculations shown in Fig. 2 were performed using a simple isobar-model [6], and allowing the SMR to vary away from the nominal values contained in the SAID [7] database. This figure thus shows, for our chosen kinematics, differing degrees of sensitivity to the small S_{1+} multipole in the various responses. A similar study of the "EMR" sensitivity (ratio of E_{1+} to M_{1+}) shows essentially no change at all in any response as the EMR is varied from +2% to -4%. Given this situation, employing a multipole analysis of the complete angular distributions for all measured quantities should allow a relatively model-independent extraction of S_{1+}.

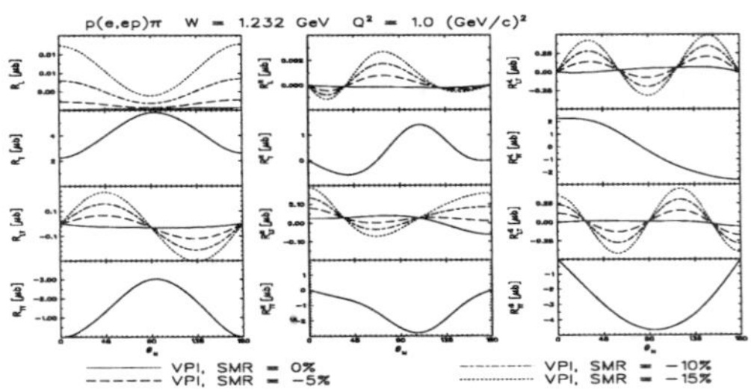

Figure 2. Evaluation of the expected sensitivity of $p(\vec{e}, e'\vec{p})\pi^0$ responses to the SMR under the kinematic conditions of E91-011.

It should also be noted that, if one examines the multipole decomposition of the responses in a simplified M_{1+}-dominance model, the responses show symmetries about $\theta_{cm} = 90°$ that provide redundancies in the identification of small multipoles. For example:

$$ReS^*_{0+}M_{1+} = \frac{R'^S_{TL}(\theta_{cm}) + R'^S_{TL}(\pi - \theta_{cm})}{3\cos 2\theta_{cm} - 1} = \frac{R'^L_{TL}(\theta_{cm}) + R'^L_{TL}(\pi - \theta_{cm})}{3\sin 2\theta_{cm}} \tag{2}$$

This thereby allows for the isolation of the same multipole combination – under the model assumption of M_{1+} dominance – through combination of separate, unique measurements of differing quantities. Ultimately, this feature will reduce the overall model dependence by potentially illuminating the failures of the expected redundancies in the given model.

5 Data Analysis - Sample Spectra

Since the data analysis is still in a preliminary stage, this section is meant to provide a sample of our data quality. The high singles rates on the forward side of $\vec{q}$, and relatively large π^+ contamination, necessitate careful particle identification and background rejection for E91-011. Figure 3 illustrates the effectiveness of several selection criteria. A cut on the correlation between vertex position and coincidence time eliminates most of the accidental coincidences. A cut on mean pulse height in the scintillators versus β eliminates most of the pion contamination. An additional cut on missing energy versus missing momentum also reduces background, particularly on the large-angle side of $\vec{q}$ where the elastic radiative tail is responsible for the strong peak at small missing mass; this peak is cut off on its lower side by acceptance. Very little background remains under the final missing-mass peak. Note that the missing-mass resolution is better on the large-angle than on the small-angle side of $\vec{q}$ where the sensitivity to angular resolution is greater.

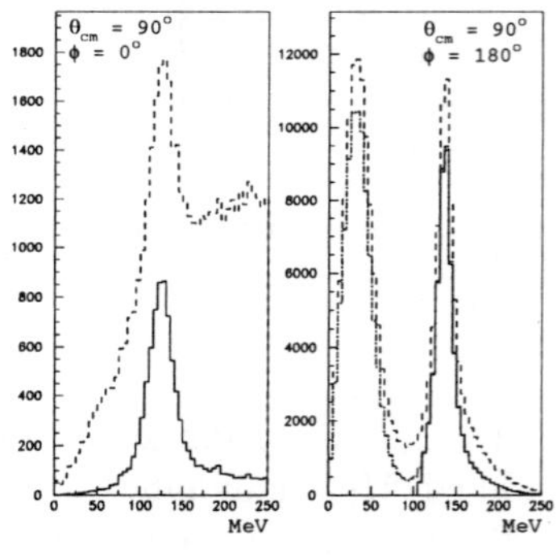

Figure 3. The effectiveness of event selection criteria is illustrated for $\theta_{cm} = \pm 90°$. The dashed curves have only a loose cut on coincidence time; the dash-dotted curves have cuts applied to both coincidence time versus reaction-vertex and β versus pulse-height particle identification; and the solid curves add a cut to select pion production in E_m versus p_m.

For final interpretation of our results, we will rely on having a very good model of the HRS spectrometers to deal with issues such as acceptance and spin-precession. Figure 4 demonstrates that such a model is in hand by showing a comparison of data and simulation for the $\theta_{cm} = 0°$ point. The simulation was generated by the code MCEEP [8], which includes our spectrometer optics model, and uses the Mainz unitary isobar model "MAID2000" [9] as the input generator. To check our polarimetry analysis, several dedicated elastic $p(\vec{e}, e'\vec{p})$ runs were conducted, with proton momenta spanning the range of the production settings; these runs allow us to check/calibrate the "false" (instrumental) asymmetries in the FPP, as well as cross check extracted polarizations with Ref. [10]. Similarly, the radiated tail of the elastic peak which is within the acceptance of the $\theta_{cm} = 90°$ setting allows a further check of the same. All of these FPP studies show consistent and understandable results.

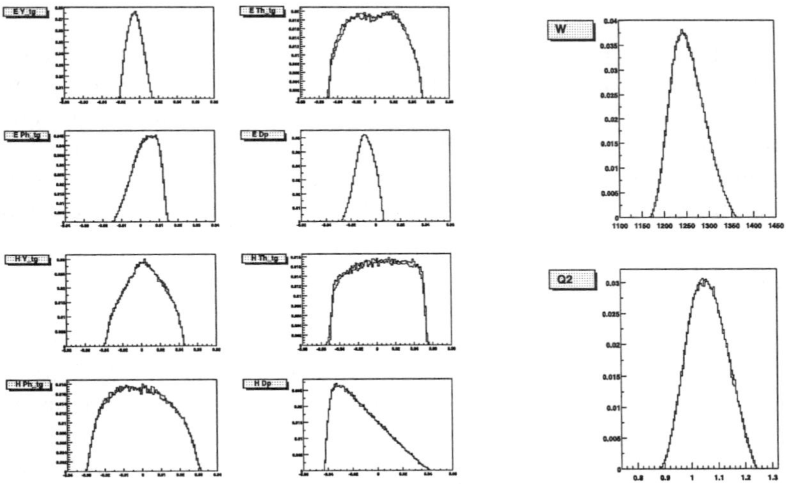

Figure 4. Preliminary comparison of E91-011 data to simulation (MCEEP+MAID) for the $\theta_{cm} = 0°$ setting. The left-most two columns show distributions in spectrometer transport variables y, θ_{target}, ϕ_{target}, δp. The right-hand column shows the comparison for W and Q^2. Data are shown by solid lines, simulation by dotted (essentially indistinguishable). The comparison is for shape comparison only, and the simulation is normalized to the data.

6 Conclusions/Outlook

In the final analysis, we intend to extract the responses through a direct fitting of the measured *focal plane polarizations*, and differential cross sections, according to their ϕ distribution. Note that only the two helicity-dependent transverse components of the polarization, along with the two helicity-independent transverse components, can be measured at the HRS focal plane. However, information regarding the longitudinal polarization at the reaction proper can be obtained due to the mixing generated by precession through the HRS magnetic fields. Thus, the measured focal plane polarizations (two helicity-indep., two helicity-dep.) can be most directly connected to the desired R's of equation 1 by constructing a matrix that accounts for: the combination of R's in each reaction polarization component, the Wigner rotation from cms to lab, rotation of coordinate systems from reaction to beam to spectrometer frame, and (finally) spin-precession through the HRS. This direct-fitting technique has been tested with simulated data (MCEEP+MAID) to show that the R's extracted match with the model input.

Our on-line focal plane polarizations show marked deviations from what is predicted when using either MAID [9] or SAID [7] input multipoles to generate the polarizations. While it is necessary to await completion of the analysis to make firm statements or conclusions about the nature of our observed discrepancies, this does give the first hint at the sensitivity of this recoil-polarization technique for accessing otherwise hidden multipole information.

Acknowledgments

I would like to thank the three hard-working graduate students who are analyzing various aspects of E91-011 for Ph.D. dissertations: Z. Chai (MIT), R. Roché (FSU), and S. Escoffier (CEA-Saclay). Further, the invaluable contributions of our postdoctoral researchers (M. Jones, D. Meekins, and D. Higinbotham), my fellow co-spokespeople (S. Frullani, J.J. Kelly and R. Lourie), and the entire Hall A Collaboration, staff, and technical support crews, are gratefully acknowledged. The Southeastern Universities Research Association (SURA) operates the Thomas Jefferson National Accelerator Facility for the United States Department of Energy under contract DE-AC05-84ER40150.

References

1. G.A. Warren *et al.*, *Phys. Rev.* C **58**, 3722 (1998).
2. Th. Pospischil *et al.*, *Phys. Rev. Lett.* **86**, 2959 (2001).
3. D. Drechsel and L. Tiator, *J. Phys.* G **18**, 449 (1992).
4. H. Schmieden and L. Tiator, *Eur. J. Phys.* A **8**, 15 (2000).
5. J.J. Kelly, *Phys. Rev.* C **60**, 054611 (1999).
6. J.J. Kelly, www.nscp.umd.edu/ kelly/EPIPROD/epiprod.html .
7. R.A. Arndt *et al.*, *Phys. Rev.* C **53**, 430 (1996).
8. P. Ulmer, www.physics.odu.edu/ ulmer/mceep/mceep.html .
9. D. Drechsel *et al.*, NPA **645**, 145 (1999).
10. M. Jones *et al.*, PRL **84**, 1398 (2000).

CHIRAL EFFECTIVE FIELD THEORIES WITH EXPLICIT SPIN 3/2 DEGREES OF FREEDOM—A STATUS REPORT

T. R. HEMMERT
Physik Department T39, TU München,
James-Franck-Straße, D-85747 Garching, Germany
Email: themmert@physik.tu-muenchen.de

Recent developments in calculations of low energy nucleon properties utilizing effective chiral field theories with explicit spin 3/2 matter fields are addressed.

1 Introduction

The chiral symmetry of QCD is spontaneously broken at low energies, giving rise to 3 [8] Goldstone Bosons for the case of 2 [3] light quark flavors. In the following, we will concentrate on a world with only 2 light flavors, all other quark degrees of freedom are taken to be infinitely heavy. We identify the 3 Goldstone Boson degrees of freedom with the physical pions, which therefore owe their small (but finite) mass to the additional explicit breaking of $SU(2)_L \times SU(2)_R$ chiral symmetry due to the "small" masses of the up, down quarks. Chiral Perturbation Theory is a successful effective field theory that parameterizes the interactions among these Goldstone Bosons (in the presence of external fields) in the most general form, based solely on the symmetries of the underlying lagrangian of QCD. If this were all, this theory would obviously be not very interesting for the audience of an NSTAR conference. However, very general principles tell us also how these Goldstone bosons interact with "matter"-fields, even in the presence of additional external sources/fields. For an overview on calculations involving Goldstone Bosons + matter fields I refer to the recent review of ref.[1]. Here I will focus on "matter" that consists of spin 1/2 fermions (nucleons) and their spin 3/2 resonance partners (Delta(1232)), constraining myself again to a world of 2 light flavors. In particular, I will discuss the role of Delta resonances in microscopic calculations of the anomalous magnetic moments of the nucleon, the impact of Deltas on the isovector Pauli form factor of the nucleon and the problems one faces if one wants to calculate the isovector nucleon-delta transition form factors. Further topics of recent interest/activity, like the (reduced ?) screening of Delta(1232) generated paramagnetism in the *isoscalar* nucleon magnetic polarizability $\beta_M^{(s)}$ [2] or the impact of Deltas on the momentum-dependence of the generalized spin-polarizabilities of the nucleon[3] cannot be covered here.

2 Chiral Calculations and Power Counting

For systematic calculations with chiral effective field theories one needs a procedure to construct the most general effective chiral lagrangian that contains all possible terms allowed by chiral symmetry, as well as (subsets of) PCT constraints. In addition[a], one needs to specify a power-counting scheme that tells us, which ones of the plethora of possible diagrams and non-linear vertex structures have to be taken into account if one performs a calculation up to a given order. In the following, I will show results obtained in 2 different chiral effective field theories: SU(2) HBChPT—which constitutes a non-relativistic theory with only pion, nucleon degrees of freedom and power-counting $\mathcal{O}(p^n)$—and SU(2) SSE—which contains pion, nucleon and Delta degrees of freedom and power-counting $\mathcal{O}(\epsilon^n)$. For details I have to refer to the literature[1].

Here, I only want to point out that in chiral effective field theories like SSE[4], which contain the nucleon-delta mass splitting Δ_0 as an additional small parameter, there exists some degree of freedom as to how one organizes the power-counting. The example I want to discuss here concerns the leading $\gamma N\Delta$ vertex, which traditionally is written in the (SU(6) quark-model inspired) form

$$\mathcal{L}^{(2)}_{\gamma N\Delta} = \frac{b_1}{2M_0}\bar{T}^\mu_j \, i \, f^j_{+\mu\nu} \, S^\nu N_v + h.c. \, . \tag{1}$$

I do not want to discuss this operator structure in detail, the important point to observe is that the dimension-less coupling b_1 is assumed to scale with a large baryon mass scale M_0, boosting this structure to $\mathcal{O}(\epsilon^2)$ (i.e. NLO) in the SSE lagrangian. Several reasons—for example the resulting large value for the coupling $b_1 \sim 7.7$, the failure of SU(6) symmetry considerations to predict the full strength of the $\gamma N\Delta$ transition, etc.—have compelled us to propose a slight modification for this operator:

$$\mathcal{L}^{(1)}_{\gamma N\Delta} = \frac{c_V}{\Delta_0}\bar{T}^\mu_j \, i \, f^j_{+\mu\nu} \, S^\nu N_v + h.c. \, . \tag{2}$$

Due to the small mass scale Δ_0 in the denominator, the *leading* $\gamma N\Delta$-transition operator is now a part of the $\mathcal{O}(\epsilon)$ (i.e. LO) SSE lagrangian, leading (in some cases) to a substantial reordering of the chiral expansion. Interesting consequences of this rescaled vertex of Eq. (2) will be discussed in the following sections.

---

[a] Furthermore, one needs to specify, how many light quark flavors one wants to consider, whether one utilizes or non-relativistic or relativistic chiral effective field theory and which regularization procedure (consistent with the symmetries) one wants to employ.

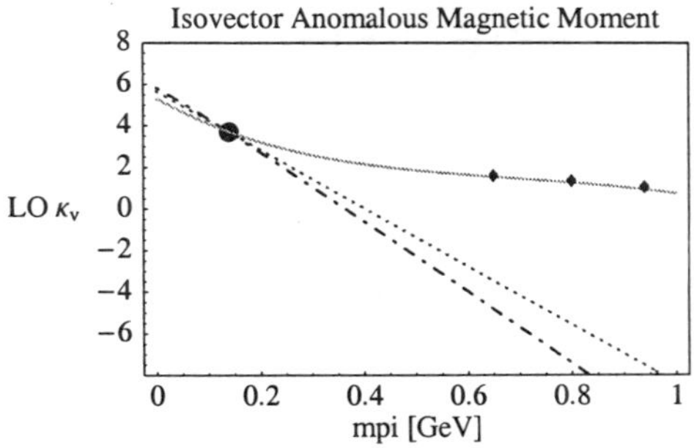

Figure 1. Isovector anomalous magnetic moment of the nucleon (in nucleon magnetons) and its dependence on the effective mass of the pion. The physical value at $m_\pi = 140$ MeV is $\kappa_v = 3.7$ [n.m]. Also shown are 3 (quenched) lattice QCD data points of Leinweber et al. [9]. The curves are explained in the text.

3 Anomalous Magnetic Moments of the Nucleon

Figure 1 shows the leading one-loop order (LO) results obtained in 3 different calculations utilizing chiral effective field theories for the anomalous isovector magnetic moment of the nucleon κ_v (in nucleon magnetons [n.m.]), defined via

$$\kappa_v = \mu_{proton} - \mu_{neutron} - 1\ [n.m.] \tag{3}$$

and plotted as a function of the pion mass m_π. The dotted curve shows the LO HBChPT result of [5]

$$\kappa_v^{(3)} = \kappa_v^{(0)} - \frac{g_A^2 M_N}{4\pi F_\pi^2}\, m_\pi + \mathcal{O}(p^4)\,, \tag{4}$$

with the free parameter κ_v^0 fixed in such a way that $\kappa_v^{(3)}$ reproduces the physical value of $\kappa_v^{(phys.)} = 3.7$ [n.m.] at $m_\pi = 140$ MeV. g_A denotes the axial coupling constant of the nucleon with mass M_N, and F_π is the pion-decay constant. All of these quantities are taken at their physical values, as any implicit quark-mass dependence constitutes an effect of higher order in the chiral expansion. Figure 1 also shows that it is not possible to connect the LO HBChPT result with the results of a quenched lattice QCD calculation by Leinweber et al. [9], employing effective pion masses of 600 MeV and higher. A few years ago [6] the LO influence of explicit Delta(1232) degrees of freedom on the vector (and axial-vector) current of the nucleon was analyzed utilizing the Small Scale Expansion of ref.[4]. As can be clearly seen by the dashed curve

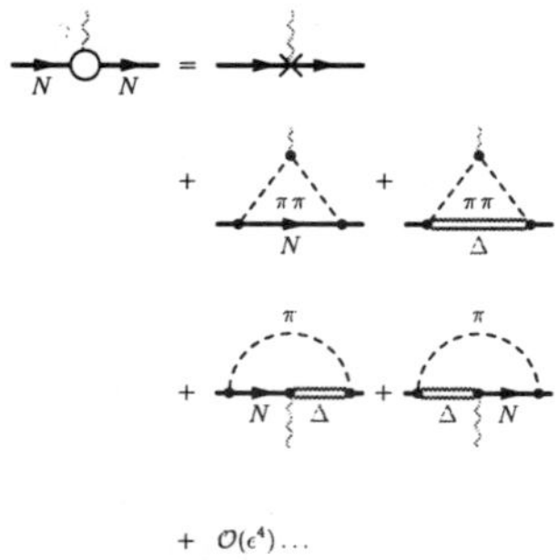

Figure 2. Diagrams contributing to the anomalous magnetic moments of the nucleon to leading one-loop order in the modified Small Scale Expansion.

in Figure 1, there is only a small modification due to the explicit spin 3/2 degrees of freedom for small pion mass and still no satisfying connection to the lattice points can be obtained.

We now discuss the results of a new calculation[7] of the isovector anomalous magnetic moment which utilizes the rescaled leading $\gamma N \Delta$ vertex of Eq. (2) discussed in the previous section. The diagrams taken into account to leading one-loop order (i.e. $\mathcal{O}(\epsilon^3)$) in the now modified Small Scale Expansion are shown in Fig.2. The experts in the audience will notice that there are 2 additional diagrams[b] (in the last row) of Fig. 2 compared to the calculation of [6]. The result of this calculation is shown by the full curve in Fig. 1. It contains 3 free parameters—κ_v^0, the isovector anomalous magnetic moment of the nucleon *in the chiral limit*, c_V, the new leading order $\gamma N \Delta$ coupling constant introduced in Eq. (2), and one additional (quark-mass dependent) higher order γNN coupling constant $E_1(\lambda)$, which also serves as a counterterm to absorb new divergences coming in due to the 2 extra diagrams. All other parameters can be fixed from known low energy quantities. In Fig. 1 we have fit κ_v^0, c_V, $E_1(1\text{GeV})$ to the 3 (quenched) lattice QCD data points of Leinweber et al.[9]. Surprisingly the fit—although performed for pion masses 600...950 MeV !—suggests a quark-mass dependence for the isovector anomalous magnetic moment of the nucleon which extrapolates down to $\kappa_v^{LO} = 3.7$ at the physical pion mass of $m_\pi = 140$ MeV! Such a stable result obtained from only

[b]We note that even the new set of one-loop diagrams given in Fig.2 obtained to LO in the (modified) Small Scale Expansion contains fewer diagrams than considered in the pioneering one-loop calculation of the baryon magnetic moments by Jenkins et al.[8]. The 3 additional diagrams considered in ref.[8] are part of the (estimated) 21 additional one-loop diagrams coming in at NLO (i.e. $\mathcal{O}(\epsilon^4)$) in our approach, as dictated by the SSE power-counting.

leading order input could not be expected, especially if one compares it with the corresponding HBChPT calculation. Details of this calculation, including an error analysis, will be available soon[7]. Here I only want to point out that the analysis suggests so far that the isovector anomalous magnetic moment has the chiral limit value $\kappa_v^0 \sim 5.7$ [n.m.], which is more than 50% larger than the value for finite up, down quark masses. This dramatic reduction of the anomalous magnetic moment is mainly due to pion-loop effects. Furthermore, one can learn from Fig. 1 that the leading linear result of Eq. (4) looses its range of validity already for values smaller than the physical pion mass when curvature effects come in. This finding is consistent with a chiral extrapolation of the same lattice QCD points analyzed with a Pade approximation[10]. Finally, I want to note[7] that the *isoscalar* anomalous magnetic moment of the nucleon shows quite a different/much simpler chiral behavior when calculated to the same order in SSE:

$$\kappa_s^{LO} = \kappa_s^{(0)} - \frac{E_2 M_N}{4\pi^2 F_\pi^2 \Delta}\, m_\pi^2 + \mathcal{O}(\epsilon^4)\,, \tag{5}$$

where E_2 is again an unknown higher order γNN coupling, now to be fit to the isoscalar lattice data.

4 Isovector Form Factors of the Nucleon

A few years ago the role of explicit Delta(1232) degrees of freedom in calculations of the nucleon form factors at low (i.e. $Q^2 < 0.2$ GeV^2) four-momentum transfer was analyzed[6] within the Small Scale Expansion of ref.[4]. Considering the discussion in the previous section, one can now ask the question how the results are changed (to leading one-loop order) if one allows for the rescaling of the leading $\gamma N\Delta$-transition as defined in Eq. (2), which leads to the two additional (i.e. the last two) Feynman diagrams at $\mathcal{O}(\epsilon^3)$ in Fig. 2 and to such a dramatic change in the LO chiral behavior of the isovector anomalous magnetic moment as shown in Fig. 1. However, it turns out that to leading order the momentum-dependence of the SSE curves of ref.[6] is not modified by the additional diagrams for a physical pion mass $m_\pi = 140$ MeV. The results for the isovector Dirac and Pauli form factor of the nucleon are shown in Fig.3. The Dirac form factor turns out to be completely dominated by the radius in this momentum-range. Both the HBChPT ($\mathcal{O}(p^3)$, dot-dashed curve) and the (modified) SSE calculation ($\mathcal{O}(\epsilon^3)$, full curve), as well as an empirical parameterization[11] of the data (dashed lines) cannot be distinguished. Matters are different for the isovector Pauli form factor. Both the HBChPT and the modified SSE curves drop slower than suggested by the empirical param-

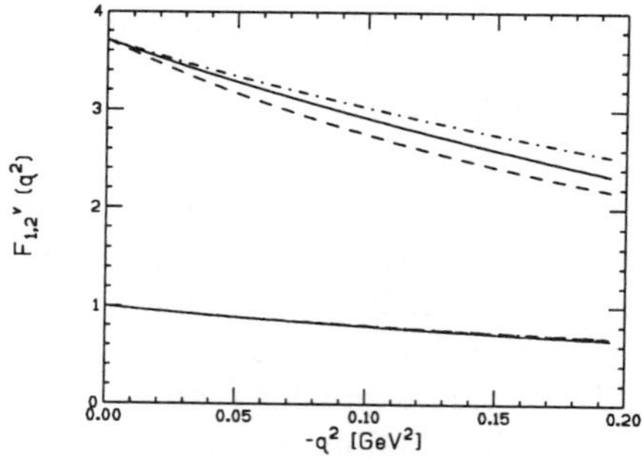

Figure 3. Isovector Dirac and Pauli form factor of the nucleon, calculated to leading one-loop order with (full curve) and without (dot-dashed curve) explicit spin 3/2 degrees of freedom. Also shown for comparison is a parameterization of the data[11] (dashed curve).

eterization. However, the LO SSE calculation arising from the diagrams of Fig. 2 provides an isovector Pauli radius of 0.61 fm^2—arising solely from the pion-cloud around a spin 1/2 or spin 3/2 intermediate baryon—which amounts to more than 75% of the physical isovector Pauli radius and therefore leads to a qualitatively better description of this form factor than provided by the LO HBChPT calculation of ref.[5] (c.f. Fig. 3). It will be interesting to study the comparison of these form factor calculations between HBChPT and SSE also at next-to-leading order, analyzing how far in four-momentum transfer one can trust/fine-tune chiral effective field theories at the one-loop level.

5 Isovector Nucleon-Delta Transition Form Factors

Having discussed the impact of explicit Delta(1232) degrees of freedom in microscopic calculations of the (iso-) vector nucleon current, one can now address the analogous problems in the isovector nucleon-delta transition current. This issue has already been analyzed within SSE in ref.[12]. Again, the rescaling of the leading order $\gamma N\Delta$ transition vertex of Eq. (2) leads to two additional diagrams (displayed in the last row of Fig. 4), compared to the original calculation of ref.[12]. However, one can show that the additional diagrams only lead to a *modified chiral behavior* (i.e. a modified quark mass-dependence of the $\gamma N\Delta$ transition moments[7]), whereas the *leading order* four-momentum dependence for a physical pion-mass of $m_\pi = 140$ MeV of the 3 (M1, E2, C2) isovector nucleon-delta transition form factors is unchanged. As discussed in ref.[12], microscopic calculations of the $N\Delta$-transition using chiral effective field theories suffer (at present) from the fact that there are several unknown

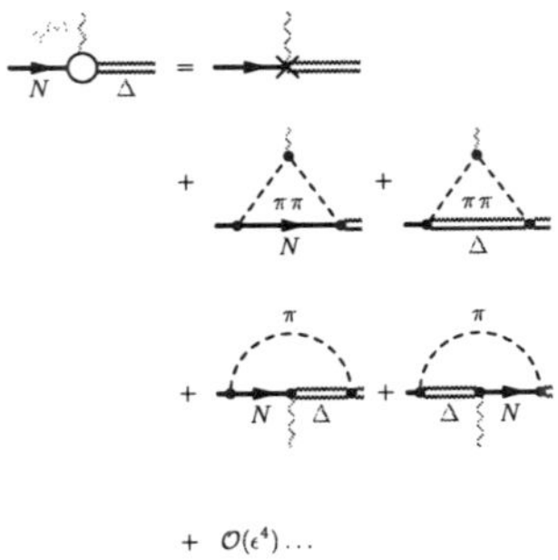

Figure 4. Diagrams contributing to the isovector nucleon-delta transition form factors to leading one-loop order in the modified Small Scale Expansion.

counter-terms, which need to be fixed via external input. As an example, in Fig. 5 we show the so called CMR, i.e. the ratio of the C2 over the M1 nucleon-delta transition strength as a function of the four-momentum transfer q^2. Depending on different ways of fixing the counter-term input as described in [12], one obtains a rather broad band of possible momentum-dependence for the CMR[c]. The most promising way to improve upon these theoretical limitations consists of a full calculation of the pion-electroproduction cross-section *in the resonance region* within a chiral effective field theory like SSE. So far, this has not been attempted[d] due to the high order required for an adequate treatment of the width of the Delta resonance. It is interesting to note that the proposed rescaling of the leading $\gamma N\Delta$ transition vertex Eq. (2) now also provides new hope to get this longstanding problem in chiral effective field theory calculations finally done, as in the modified SSE the order required to include these width effects is lowered. I am looking forward to first results of the exploratory calculations in the new scheme.

6 Summary

I have given a status report regarding some new ideas/open problems in the field of chiral effective field theories with explicit spin 3/2 resonance degrees of freedom. I am convinced that the role and the treatment of baryon resonances continues to be an inspiring and challenging topic in this field.

---

[c]For the status of experimental information on the momentum dependence of CMR I refer to the talks by R.W. Gothe, C.N Papanicolas, and H. Schmieden in these proceedings.

[d]An exception is work of ref.[13], where the authors tried to utilize a low order SSE calculation in the resonance region by averaging over the width of the resonance.

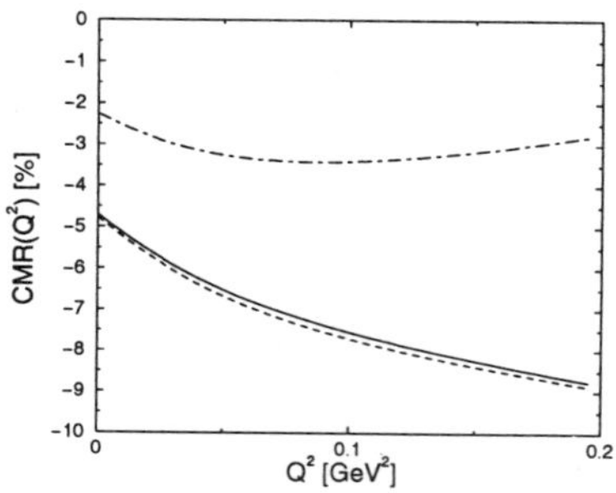

Figure 5. Real part of the four-momentum dependence of the ratio of the C2/M1 $\gamma N\Delta$ transition moments, calculated to leading one-loop order in the Small Scale Expansion. The spread of the curves is discussed in the text.

Acknowledgments

I want to thank the organizers of NSTAR01 for giving me the opportunity to present these results to the community and I gratefully acknowledge their financial support. I also would like to thank A. Thomas and W. Weise for helpful discussions. The research presented here was supported by BMBF.

References

1. see *e.g.* U.-G. Meißner, preprint no. [hep-ph/0007092].
2. H.W. Grießhammer and G. Rupak, preprint no. [nucl-th/0012096].
3. see *e.g.* T.R. Hemmert, preprint no. [nucl-th/0101054].
4. T.R. Hemmert, B.R. Holstein and J. Kambor, *J. Phys.* G **24**, 1831 (1998); *Phys. Lett.* B **395**, 89 (1997).
5. V. Bernard *et al., Nucl. Phys.* B **388**, 315 (1992).
6. V. Bernard *et al., Nucl. Phys.* A **635**, 121 (1998).
7. T.R. Hemmert and W. Weise, forthcoming.
8. E. Jenkins *et al., Phys. Lett.* B **302**, 482 (1993).
9. D.B. Leinweber, R.M. Woloshyn, and T. Draper, *Phys. Rev.* D **43**, 1659 (1991).
10. A.W. Thomas, *these proceedings*; see also D.B. Leinweber, D.H. Lu and A.W. Thomas, *Phys. Rev.* D **60**, 034014 (1999) and references therein.
11. P. Mergell, U.-G. Meißner and D. Drechsel, *Nucl. Phys.* A **596**, 367 (1996).
12. G.C. Gellas *et al., Phys. Rev.* D **60**, 054022 (1999).
13. C.W. Kao and T.D. Cohen, *Phys. Rev.* C **60**, 064619 (1999).

DYNAMICAL MESON-BARYON RESONANCES WITH CHIRAL LAGRANGIANS

A. RAMOS, A. PARREÑO
Departament ECM, Universitat de Barcelona, Diagonal 647, E-08028 Barcelona, SPAIN

J. C. NACHER, E. OSET
Departamento de Física Teórica and IFIC, Institutos de Investigación de Paterna, Aptdo. Correos 22085, E-46071 Valencia, SPAIN

C. BENNHOLD
Center for Nuclear Studies, Department of Physics, The George Washington University, Washington D.C. 20052

The s-wave meson-baryon interaction is studied using the lowest-order chiral Lagrangian in a unitary coupled-channels Bethe-Salpeter equation. In the strangeness $S = -1$ sector the low-energy K^-p dynamics leads to the dynamical generation of the $\Lambda(1405)$ as a $\bar{K}N$ state, along with a good description of the K^-p scattering observables. At higher energies, the $\Lambda(1670)$ is also found to be generated dynamically as a $K\Xi$ quasibound state for the first time. For strangeness $S = 0$, it is the $S_{11}(1535)$ resonance that emerges from the coupled-channels equations, leading to a satisfactory description of meson-baryon scattering observables in the energy region around the $S_{11}(1535)$. We speculate on the possible dynamical generation of Ξ resonances within the chiral $S = -2$ sector as $\bar{K}\Lambda$ or $\bar{K}\Sigma$ quasibound states.

1 Introduction

Gaining insight into the nature and properties of baryon resonances is one of the primary goals of hadron physics. In the last years, intensive theoretical and experimental effort has been devoted to clarify the internal structure of many resonances and establish whether they behave as 3-quark objects, as predicted by the constituent quark model, or they can be described mostly in terms of hadronic degrees of freedom, as meson-baryon quasibound states.

In the last decade, chiral perturbation theory (ChPT) has emerged as a powerful scheme that successfully explains not only low-energy meson-meson dynamics[1,2] but also meson-baryon scattering[3,4], provided the interaction is weak as is the case of πN scattering in the strangeness $S = 0$ sector, or K^+N scattering in the $S = 1$ one. However, the validity of ChPT, which is built as a series expansion in the meson momentum, is limited to *low energies* and, in addition, it is not applicable in the vicinity of *resonances*, which correspond to poles in the T-matrix.

Both difficulties can be overcome within the framework of the chiral Lagrangian, by combining the information contained in it with unitarity using non-perturbative techniques. This has proved to be very successful in both the meson-meson[5,6] and the meson-baryon sector[7,8,9,10]. A review of recent results can be found in Ref. [11].

2 Meson-Baryon Scattering in Coupled Channels

Ref. [7] combined the chiral Lagrangian (to lowest and next-to-leading order) and a non-perturbative resummation by solving a coupled-channels Lippmann-Schwinger equation. In the case of K^-p scattering, the channels included were those opened at threshold and, fitting five parameters, the low-energy scattering observables, as well as the properties of the $\Lambda(1405)$ resonance, were well reproduced. A similar procedure was followed in Ref. [8] in terms of the coupled-channels Bethe-Salpeter equation (BSE) given by

$$T_{ij} = V_{ij} + V_{il} G_l T_{lj} \ . \tag{1}$$

It was shown that the data could be well reproduced taking the amplitudes V_{ij} from the lowest order chiral Lagrangian and only one parameter, the cut-off used to regularize the intermediate meson-baryon propagator G_l. The key difference compared to Ref. [7] was the inclusion of all ten possible meson-baryon states constructed from the octet of pseudoscalar mesons and the octet of $1/2^+$ baryons, thus preserving SU(3) symmetry in the limit of equal baryon and meson masses. Another advantage of the lowest-order Lagrangian is that the amplitudes can be factorized on-shell out of the loop integral, thus reducing the coupled-channels problem to a set of coupled algebraic equations.

The unitarization of the chiral amplitudes through the BSE has been shown to be a particular case of the Inverse Amplitude Method when the choice of the regulator allows one to include the second-order chiral contributions by means of only the s-channel loop[5]. Equivalently, the BSE is a particular case of the N/D method when the effects of the unphysical cut (left-hand singularities) are neglected[6].

The amplitudes V_{ij} in Eq. (1) are easily obtained from the lowest order meson-baryon interaction Lagrangian

$$L_1^{(B)} = \langle \bar{B} i \gamma^\mu \frac{1}{4f^2} [(\Phi \partial_\mu \Phi - \partial_\mu \Phi \Phi) B - B(\Phi \partial_\mu \Phi - \partial_\mu \Phi \Phi)] \rangle \ , \tag{2}$$

where Φ and B are the matrices representing the mesons and baryons, respectively. A key ingredient in the BSE is the loop integral, G_l, which in Ref. [8]

reads

$$G_l = i \int \frac{d^4q}{(2\pi)^4} \frac{M_l}{E_l(\vec{q})} \frac{1}{k^0 + p^0 - q^0 - E_l(\vec{q}) + i\epsilon} \frac{1}{q^2 - m_l^2 + i\epsilon}$$
$$= \int_{|\vec{q}|<q_{\rm max}} \frac{d^3q}{(2\pi)^3} \frac{1}{2\omega_l(\vec{q})} \frac{M_l}{E_l(\vec{q})} \frac{1}{\sqrt{s} - \omega_l(\vec{q}) - E_l(\vec{q}) + i\epsilon} , \quad (3)$$

and has been regularized by means of a cut-off, $q_{\rm max}$. An alternative approach, followed in Ref. [9], is obtained by making use of dimensional regularization

$$G_l = i2M_l \int \frac{d^4q}{(2\pi)^4} \frac{1}{(P-q)^2 - M_l^2 + i\epsilon} \frac{1}{q^2 - m_l^2 + i\epsilon}$$
$$= \frac{2M_l}{16\pi^2} \left\{ a_l(\mu) + \ln \frac{M_l^2}{\mu^2} + \frac{m_l^2 - M_l^2 + s}{2s} \ln \frac{m_l^2}{M_l^2} + \right.$$
$$\left. + \frac{\bar{q}_l}{\sqrt{s}} \ln \frac{M_l^2 + m_l^2 - s - 2\sqrt{s}\bar{q}_l}{M_l^2 + m_l^2 - s + 2\sqrt{s}\bar{q}_l} \right\} , \quad (4)$$

where μ is the regularization scale and $\bar{q}_l$ the on-shell momentum. Obviously, changes in the cut-off value of Eq. (3) can always be accommodated with changes in the regularization scale μ and the corresponding change in the subtraction constant $a_l(\mu)$ in Eq. (4). The dimensional regularization scheme can extend the model to higher energies, while the cut-off method limits the range of applicability to energies whose on-shell momentum $\bar{q}_l$ is smaller than the cut-off value for all channels.

3 Strangeness $S = -1$ Sector

We start this section summarizing the findings of Ref. [8] for $\bar{K}N$ scattering. The cut-off method was used with a value for the decay constant of $f = 1.15 f_\pi$ (chosen between $f_\pi = 93$ MeV and $f_K = 1.22 f_\pi$) and a value $q_{\rm max} = 630$ MeV was obtained by reproducing the threshold branching ratios, $\gamma = \Gamma(K^-p \to \pi^+\Sigma^-)/\Gamma(K^-p \to \pi^-\Sigma^+)$, $R_c = \Gamma(K^-p \to \text{charged})/\Gamma(K^-p \to \text{all})$ and $R_n = \Gamma(K^-p \to \pi^0\Lambda)/\Gamma(K^-p \to \text{neutral})$. The above value for f gave the best agreement for the $\Lambda(1405)$ properties seen in the $\pi\Sigma$ mass spectrum, as shown in Fig. 1. In Fig. 2, we show that the low-energy scattering cross sections, not used in the fit, are well reproduced.

In order to assess the range of validity of our approach we explored the region of higher energies, using the dimensional regularization scheme with $\mu = 630$ MeV. In order to maintain the low-energy results, we adjust the subtraction constant a_l to reproduce the value of the loop function G_l at

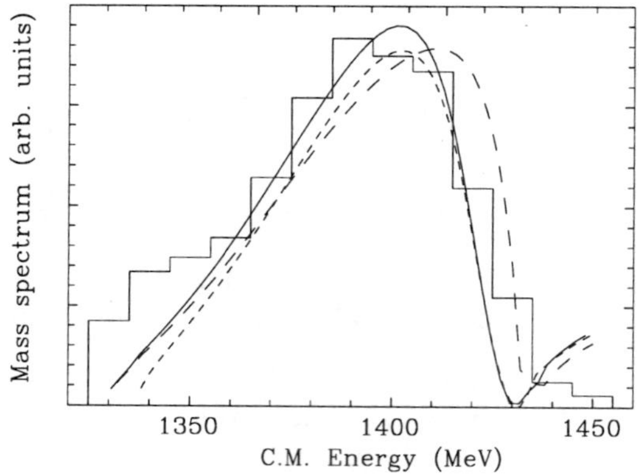

Figure 1. $\Lambda(1405)$ resonance as obtained from the invariant $\pi\Sigma$ mass distribution, with the full basis of physical states (solid line), omitting the η channels (long-dashed line) and with the isospin-basis (short-dashed line). Experimental histogram from Ref. [12]. Figure taken from Ref. [8].

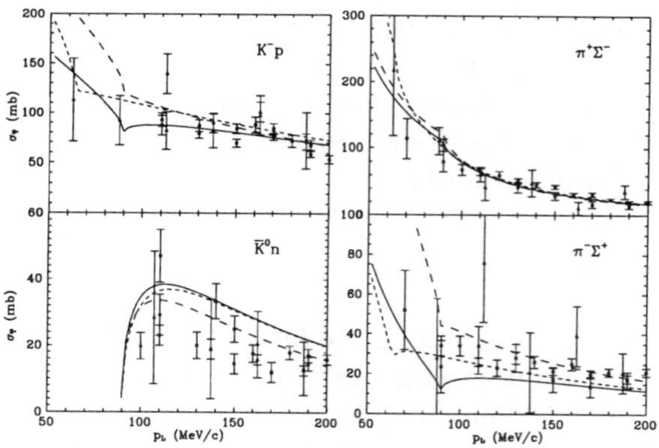

Figure 2. K^-p scattering cross sections as functions of the K^- momentum in the lab frame: with the full basis of physical states (solid line), omitting the η channels (long-dashed line) and with the isospin-basis (short-dashed line). Taken from Ref. [8].

threshold ($\sqrt{s} = m_K + m_N$) calculated with the cut-off, and we find:

$$\begin{array}{lll} a_{\bar{K}N} = -1.84 & a_{\pi\Sigma} = -2.00 & a_{\pi\Lambda} = -1.83 \\ a_{\eta\Lambda} = -2.25 & a_{\eta\Sigma} = -2.38 & a_{K\Xi} = -2.52 \; . \end{array} \tag{5}$$

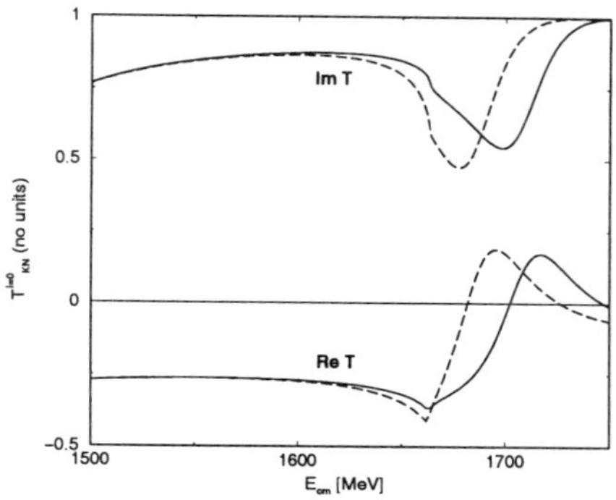

Figure 3. Real and imaginary parts of the $\bar{K}N$ scattering amplitude in the isospin $I = 0$ channel in the region of the $\Lambda(1670)$ resonance.

In Fig. 3, we show the real and imaginary parts of the $I = 0$ scattering amplitude, normalized as in the Partial Wave Analysis of Ref. [13]. Remarkably, the amplitudes shown by the solid lines, which are obtained using the low-energy parameters in Eq. (5), show the resonant structure of the $\Lambda(1670)$ appearing at about the right energy and with a similar size compared to the experimental analysis[13]. The position of the resonance is quite sensitive to the parameter $a_{K\Xi}$ and moderately sensitive to $a_{\eta\Lambda}$. Hence, without spoiling the nice agreement at low energies, which is not sensitive to $a_{K\Xi}$, we exploit the freedom in the parameters by choosing $a_{K\Xi} = -2.70$, moving the resonance closer to its experimental position (dashed lines).

SU(3) symmetry, partly broken here due to the use of physical masses, demands a singlet and an octet of resonances. Within $S = -1$, we have already identified the singlet $\Lambda(1405)$ and the $I = 0$ member of the octet, the $\Lambda(1670)$. Since we found the partial decay widths and couplings[14] of the $\Lambda(1405)$ to $\bar{K}N$ states and the $\Lambda(1670)$ to $K\Xi$ states to be very large, one is naturally led to identify these two resonances as a "quasibound" $\bar{K}N$ and $K\Xi$ state, respectively.

Searching for the $I = 1$ member of the octet, we find that the $I = 1$ amplitudes in our model are smooth and show no trace of resonant behavior, in line with experimental observation. To explore this issue further we conducted a search for the poles of the $\bar{K}N \to \bar{K}N$ amplitudes in the second Riemann sheet and find two poles in the $I = 0$ amplitude $(1426 + i16,\ 1708 + i21)$, corresponding to the $\Lambda(1405)$ and the $\Lambda(1670)$, and one in the $I = 1$ amplitude $(1579 + i296)$, corresponding - most likely - to the resonance $\Sigma(1620)$. The large width found for this resonance may explain why we saw no trace of it in the scattering amplitudes.

4 Strangeness $S = 0$ Sector

The success in the s-wave $S = -1$ sector with the lowest-order chiral Lagrangian and only one-parameter[8] encouraged us to study the $S = 0$ sector in the vicinity of the $N(1535)$ resonance. In this case, however, it was not possible to reproduce the elastic πN scattering observables and a few inelastic cross sections with only one parameter. In ref.[15], we adopted a method similar to the dimensional regularization scheme, by using a fixed physical cut-off of 1 GeV, large enough to study the energy region of the $N(1535)$ resonance, and adding a subtraction constant a_l to the loop function G_l. The values of the decay constants were taken different for πN channels ($f_\pi = 93$ MeV), KY channels ($f_K = 1.22 f_\pi$) and ηN channels ($f_\eta = 1.3 f_\pi$). Our $S = 0$ model has then four parameters, $a_{\pi N}$, $a_{\eta N}$, $a_{K\Lambda}$ and $a_{K\Sigma}$, which are fitted to the phase-shifts and inelasticities for πN scattering in isospin $I = 1/2$ around the energy region of the $N(1535)$ resonance, as shown in Fig. 4, and to the inelastic cross section data, as shown in Fig. 5. The analysis performed in Ref. [15] leads to a dynamical generation of the $N(1535)$ state with a total decay width of $\Gamma \simeq 110$ MeV, divided between $\Gamma_\pi \simeq 43$ MeV and $\Gamma_\eta \simeq 67$ MeV, compatible with present data within errors. This simplified model allows one to tackle other issues, such as the evaluation of the $\pi^0 N^* N^*$ and $\eta N^* N^*$ couplings, which elucidates the question of whether the positive parity N and the negative parity N^* are chiral partners when chiral symmetry is restored[15].

We note that a recent work, which also uses the chiral Lagrangian together with a Bethe-Salpeter unitarization scheme, achieves a satisfactory description of the $S = 0$ s-wave observables in the $I = 1/2$ sector from πN threshold up to 2 GeV (similar in quality to our $S = -1$ description) by fitting 12 parameters[10], generating the both the $N(1535)$ and $N(1650)$ S_{11} resonances in the process.

5 Strangeness $S = -2$ Sector

Within SU(3) the chiral meson-baryon Lagrangian naturally extends to the $S = -2$ sector. Thus, the dynamics of the $\bar{K}\Lambda$ or $\pi\Xi$ interaction can be predicted within the same approach. Due to the experimental difficulties there are very few scattering data on such reactions. Furthermore, little is known about Ξ resonances since they can only be produced as part of a multiparticle final state with small production cross sections. Nevertheless, given the success of the approach presented here in generating most of the $(70, 1_1^-)$ octet of $1/2^-$ excited baryon states in the $S = 0$ and $S = -1$ sector, it is reasonable to assume that its missing member in the $S = -2$ realm,

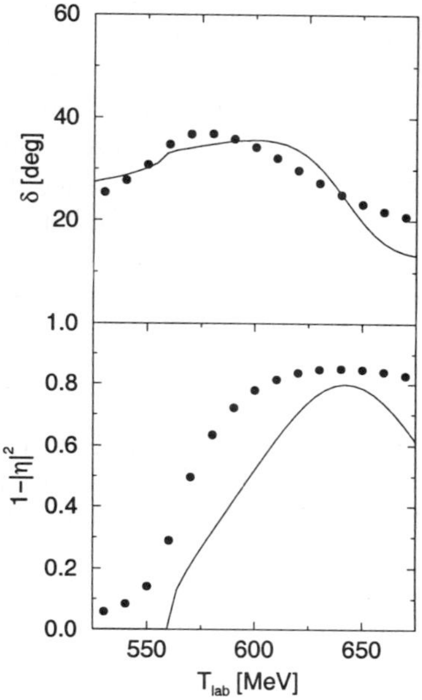

Figure 4. Phase-shifts and inelasticities for πN scattering in the isospin $I = 1/2$ channel. Taken from Ref. [15].

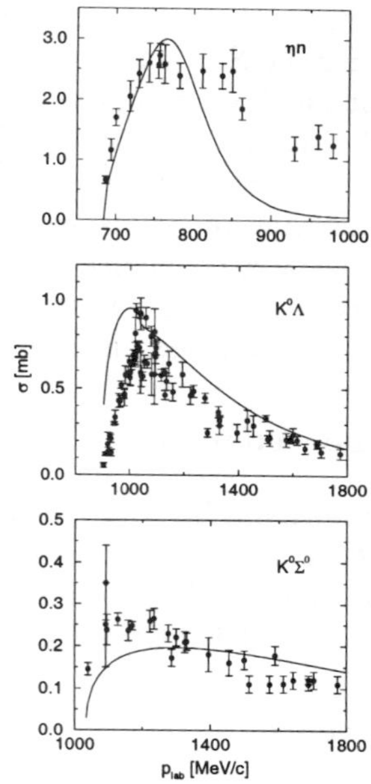

Figure 5. Cross sections for the $\pi^- p \to \eta n$, $K^0\Lambda$ and $K^0\Sigma^0$ reactions as function of the π^- laboratory momentum. Taken from Ref. [15].

a Ξ^* resonance, will be produced as well. Among the observed states, two "suspects" stand out: the $\Xi(1620)$ and the $\Xi(1690)$, with unknown spin and parity. The $\Xi(1690)$ might be the more plausible candidate since it has strong couplings to the $\bar{K}\Lambda(\Sigma)$ channels which may serve as identifying features.

6 Conclusions

We have extended the previously reported $S = -1$ meson-baryon scattering approach[8] to higher energies of up to 2 GeV. This study has revealed the appearance of two other resonances, the $\Lambda(1670)$ and the $\Sigma(1620)$, both belonging to the octet, dynamically generated as $K\Xi$ states.

The $S = 0$ sector has been more difficult to describe with only a few parameters, although the model presented here is able to reproduce the πN data moderately well around the energy of the $N(1535)$ and allows one to tackle other issues related to the $N(1535)$, such as its couplings to π and η

mesons[15]. A complete description from low to high energies, as presented here for the $S = -1$ sector, has not been possible within our simple approach, but it has been achieved in the $I = 1/2$ sector using an extended parameter set[10]. Given the success of this method of producing the members of the $1/2^-$ octet of excited states in the $S = 0$ and $S = -1$ sectors, it is reasonable to assume that an extension to the $S = -2$ realm will generate the missing octet member, a Ξ resonance in the energy region of 1600-1700 MeV. Confirmation of such a state would further demonstrate the power of combining the chiral meson-baryon Lagrangian with non-perturbative unitarization techniques.

Acknowledgments

This work has been supported by DGICYT contract numbers BFM2000-1326, PB98-1247, by the Generalitat de Catalunya project SGR2000-24, by the EU TMR network Eurodaphne, contract no. ERBFMRX-CT98-0169, and by US-DOE grant DE-FG02-95ER-40907.

References

1. J. Gasser and H. Leutwyler, *Nucl. Phys.* B **250**, 465 (1985).
2. A. Pich, *Rep. Prog. Phys.* **58**, 563 (1995).
3. G. Ecker, *Prog. Part. Nucl. Phys.* **35**, 1 (1995).
4. V. Bernard, N.Kaiser, and U.G. Meissner, *Int. J. Mod. Phys.* **E** 4, 193 (1995).
5. J. A. Oller, E. Oset, and J. R. Peláez, *Phys. Rev. Lett.* **80**, 3452 (1998); *Phys. Rev.* D **59**, 074001 (1999); erratum *Phys. Rev.* D **60**, 099906 (1999).
6. J.A. Oller and E. Oset, *Phys. Rev.* D 60, 074023 (1999).
7. N. Kaiser, P. B. Siegel, and W. Weise, *Nucl. Phys.* A **594**, 325 (1995); N. Kaiser, T. Waas, and W. Weise, *Nucl. Phys.* A **612**, 297 (1997).
8. E. Oset and A. Ramos, *Nucl. Phys.* A **635**, 99 (1998).
9. J.A. Oller and U.G. Meissner, *Phys. Lett.* B **500**, 263 (2001).
10. J. Nieves and E. Ruiz Arriola, hep-ph/0104307.
11. J.A. Oller, E. Oset, A. Ramos, *Prog. Part. Nucl. Phys.* **45**, 157 (2000).
12. R.J. Hemingway, *Nucl. Phys.* B **253**, 742 (1985).
13. G.P. Gopal *et al., Nucl. Phys.* B **119**, 362 (1977).
14. E. Oset, A. Ramos and C. Bennhold, in preparation.
15. J.C. Nacher, A. Parreño, E. Oset, A. Ramos, A. Hosaka, and M. Oka, *Nucl. Phys.* A **678**, 187 (2000).

CHIRAL LAGRANGIANS WITH RESONANCES

J. A. OLLER, U.-G. MEIßNER
Forschungszentrum Jülich, Institut für Kernphysik (Theorie)
D-52425 Jülich, Germany
E-mails: j.a.oller@fz-juelich.de, u.meissner@fz-juelich.de

After introducing the general techniques used to construct chiral Lagrangians with matter fields, e.g. nucleons and/or resonances, we concentrate on their usefulness to parameterize experimental data up to rather high energies. It is emphasized how to properly unitarize such Lagrangians in a scheme that respects both the chiral counting and the real parts of the unitarity loops. As practical applications we discuss $N\pi$ and $N\bar{K}$ scattering.

1 Chiral Symmetry

Consider the QCD Lagrangian with N_F left(L)/right(R)-handed massless quarks:

$$\mathcal{L}_{QCD} = i\sum_{A=1}^{N_F} \bar{q}_L^A \gamma^\mu D_\mu q_L^A + i\sum_{A=1}^{N_F} \bar{q}_R^A \gamma^\mu D_\mu q_R^A + ... \tag{1}$$

where the superscript A refers to the quark flavour and the ellipses indicate the pure gluon, heavy quark, gauge fixing and ghost terms. As a result of the previous equations the QCD Lagrangian is invariant under the set of chiral transformation $\gamma \in SU(N_F)_L \otimes SU(N_F)_R$ such that $q_{L,R} \to \gamma_{L,R} q_{L,R}$, respectively.

In the following we will fix $N_F = 3$ taking the u, d and s as light since their masses are much smaller than any typical hadronic mass of around 1 GeV. Nevertheless the whole chiral symmetry $SU(3)_L \otimes SU(3)_R$ is not observed in nature as it is clear from the absence of particles with opposite parity and nearly degenerate in mass. This fact, together with the approximately conserved $SU(3)_{L+R} \equiv SU(3)_V$ symmetry in nature, can be described by stating that the chiral symmetry present in the QCD Lagrangian with massless quarks spontaneously breaks to the diagonal subgroup $SU(3)_V$.

The former symmetry breaking pattern implies, because of the Goldstone theorem, the appearance of eight massless pseudoscalar mesons from the breaking of the eight axial generators. Finally these pseudoscalars acquire 'small' masses because the explicit breaking of the chiral symmetry in the QCD Lagrangian due to the quark mass matrix.

Chiral Perturbation Theory (CHPT)[1] is the low energy effective field theory of QCD implementing in a systematic way the requirements of chiral symmetry and its breakings order by order in a momentum expansion. Since the masses of the pseudoscalars are much smaller than 1 GeV the chiral Lagrangian can be given as a derivative expansion with additional insertions of the quark mass matrix valid for small energies. Nevertheless it would be desirable to be able to use the chiral Lagrangians with its associated chiral constrains for higher energies. One step to accomplish this goal is to include explicit resonance fields and then construct a chiral Lagrangian with the relevant degrees of freedom for higher energies. The standard techniques to do this are known since long[2] and have been extensively used to build up the chiral Lagrangians with nucleons as matter fields.[3] The Lagrangians with meson resonances were worked out in Ref.[4] and for the meson-baryon case they were given in Ref.[5]. A special mention deserves the so-called small scale expansion for the Δ resonance,[6] since it is the only one that allows to implement order by order chiral loops with resonances.

A general transformation matrix $\xi(\phi) \in SU(3)_L \otimes SU(3)_R$ can always be written as $\xi = u_A(\pi)h_V$ with $h_V \equiv h_R \times h_L$ in $SU(3)_{L+R}$ and $u_A(\pi) = u_R(\pi) \times u_L(\pi)^\dagger$. Under a chiral transformation $\gamma(\varphi)$, $\xi \to \gamma\xi = \tilde{u}\tilde{h}_V \equiv \tilde{\xi}$ with $\tilde{u}_R\tilde{h}_R = \gamma_R u_R h_R$ and $\tilde{u}_L^\dagger \tilde{h}_L = \gamma_L u_L^\dagger h_L$. Then we can write:

$$\begin{aligned} \tilde{u}_R &\equiv u_R(\pi') = \gamma_R(\varphi)u_R(\pi)h_R(\varphi,\pi), \\ \tilde{u}_L &\equiv u_L(\pi') = h_L(\varphi,\pi)^\dagger u_L(\pi)\gamma_L(\varphi)^\dagger. \end{aligned} \tag{2}$$

We finally make a mapping of our $SU(3)$ matrices to a common three dimensional vector space where we have $u_L(\pi) = u_R(\pi) = u(\pi)$, this matrix $u(\pi)$ being the basic object to build up chiral Lagrangians with its chiral transformation law given by Eq.(2). As indicated by the notation, the coordinates π can be identified with the octet of lightest pseudoscalars mesons (π, K or η). When working only with the pseudo-Goldstone bosons, it is more appropriate to consider the matrix $U(\pi) = u(\pi)^2$ because from Eq.(2) it obeys the transformation law $U(\pi) \to \gamma_R U(\pi)\gamma_L$ without the presence of the matrix h_V, the so-called compensator.

When matter fields, e.g., N, $N^\star$, Δ, ρ, etc..., are present one can make use of their well defined $SU(3)_V$ transformation laws by taking into account the presence in Eq.(2) of the induced $h_V(\varphi,\pi)$ transformation. Hence, under a transformation γ the matter fields, generally denoted by R, transform as:

$$R \to h(\varphi,\pi)\, R\, h(\varphi,\pi)^\dagger. \tag{3}$$

Note that the π coordinates are finally fields and hence the previous transformation will necessarily require the corresponding covariant derivatives $\nabla_\mu R$, see any of the refs.[3,4,5,6] for explicit expressions.

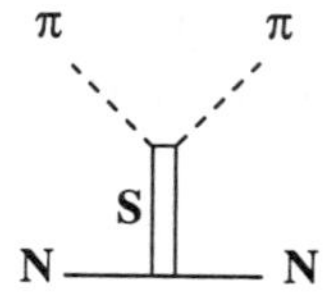

Figure 1. πN scattering diagram mediated by a scalar isoscalar resonance, S.

Since Eq.(3) is the same as a usual $SU(3)_V$ transformation, vertices connecting resonances between themselves will be the same as those derived by just $SU(3)_V$ symmetry. However, differences will arise when comparing the vertices deduced from just the linear representation of the $SU(3)_V$ symmetry to those obtained by exploiting the full non-linear representation of the chiral symmetry $SU(3)_L \otimes SU(3)_R$, Eq.(2). As a simple illustration consider the diagram in Fig. 1. One obtains, see, e.g., Ref.[7]:

$$-2g_S \frac{2c_m M_\pi^2 + c_d(t - 2M_\pi^2)}{F_\pi^2(t - M_S^2)} = -2g_S \frac{2c_m M_\pi^2 + c_d(M_S^2 - 2M_\pi^2)}{F_\pi^2(t - M_S^2)} - 2g_S \frac{c_d}{F_\pi^2} \quad (4)$$

where M_S is the mass of the scalar resonance, $F_\pi = 92.4$ MeV the pion decay constant, c_m and c_d are constants that determine the coupling of the scalar to the pions and g_s is the coupling of the former to the nucleons. Notice that the first term on the right-hand-side is the one obtained in a standard way by $SU(3)_V$ symmetry while the second term is a background required by chiral symmetry and determines that the contribution of Fig. 1 is order p^2. The leading order πN scattering is order p and does not receive any resonance contribution.

2 Elastic πN Scattering and Unitarization

One can ask the question whether the inclusion of matter fields is enough to describe resonance physics or in other words, how much further in energies we can arrive with respect to plain CHPT. The trivial answer is that it depends of course on the process one is looking at. For instance, the usual prescription to dress bare poles of resonances coming from a tree level Lagrangian $1/(M^2 - s) \to 1/(M^2 - s - i\Gamma\sqrt{s})$ (insufficient to obtain unitarized amplitudes when more than one resonance is present and at odds with the chiral counting), or similar ones, leads to acceptable results for the P-wave channels of the resonances $\rho(770)$, $K^\star(890)$ or $\Delta(1232)$. However, it completely fails for some S-waves in which the rescattering effects between the scattered particles requires a more detailed analysis of the unitarity contributions. This occurs in the case of the $f_0(980)$, $a_0(980)$ or of the $\Lambda(1405)$.

We will follow the unitarization scheme first developed in the meson–meson sector in Ref.[8] making use of the N/D method. This method has two main advantages: first, it conserves the chiral counting so that one can match at any desired order with the chiral expansion and second it is exact when the crossed channel singularities in the $T-$matrix are neglected. In the end one obtains amplitudes with a resummed right-hand cut (also known as unitarity cut) with the crossed channel singularities treated perturbatively as given by the chiral inputs made up of CHPT plus resonances (from the Lagrangians just discussed in the previous section).

In Ref.[7] the method was extended to the meson-baryon case and it was applied to πN scattering. The final answer is that one can write the $T-$matrix under consideration in the form:

$$T = \left[\mathcal{T}^{-1} + g(s)\right]^{-1} \tag{5}$$

where $g(s)$ is the meson-baryon loop:

$$\begin{aligned} g(s)_i &= i\int \frac{d^4q}{(2\pi)^4}\frac{1}{(q^2-M_i^2+i\epsilon)((P-q)^2-m_i^2+i\epsilon)} \\ &= \frac{1}{16\pi^2}\left\{a_i(\mu)+\log\frac{m_i^2}{\mu^2}+\frac{M_i^2-m_i^2+s}{2s}\log\frac{M_i^2}{m_i^2}\right. \\ &\left. + \frac{q_i}{\sqrt{s}}\log\frac{m_i^2+M_i^2-s-2\sqrt{s}q_i}{m_i^2+M_i^2-s+2\sqrt{s}q_i}\right\}, \end{aligned} \tag{6}$$

with M_i and m_i, respectively, the meson and baryon masses in the state i and $s = W^2$ the usual Mandelstam variable. On the other hand, $\mathcal{T}$ is fixed by matching with a CHPT input with/without resonances, $\mathcal{T}_\chi$. In Ref.,[7] $\mathcal{T}_\chi$ consists of $\mathcal{O}(p^3)$ Heavy Baryon CHPT (HBCHPT)[9] plus resonances.[7] The matching can be done straightforwardly by expanding Eq.(5) up to one loop calculated at order p^3 as in Ref.[9], then one has:

$$\mathcal{T}_\chi = \mathcal{T} - \mathcal{T}_1 \cdot g(s) \cdot \mathcal{T}_1 \tag{7}$$

where $\mathcal{T}_1$ is the lowest order HBCHPT amplitude. From the above expression one can immediately obtain $\mathcal{T}$ and plug it in Eq.(5). Still the subtraction constant $a(\mu)$ in $g(s)$, see Eq.(6), has to be fixed. This is done by requiring that $g(s)$ vanishes for $W = \sqrt{s} = m + xM_\pi$. Because Im$g(s)$ is fixed by unitarity, x must be below 1 because otherwise $a(\mu)$ would be complex. Furthermore, since HBCHPT converges for the energy region close to πN threshold and the $T-$matrix given by Eq.(5) reduces to the chiral input $\mathcal{T}_\chi$ when $g(s) = 0$,

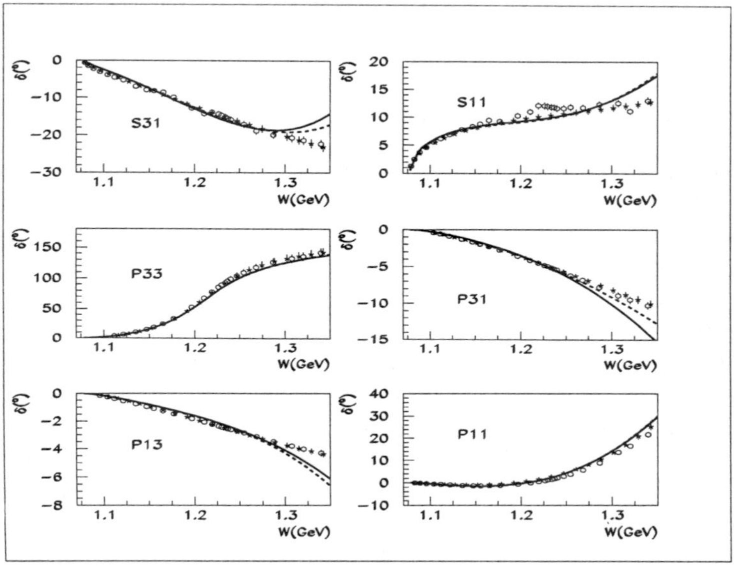

Figure 2. Fit to the phase shifts of several πN partial waves.

one would expect $x \backsim \mathcal{O}(1)$. In fact we obtain in all our fits a value of $x \simeq 0.95$, very close to one. We have 8 more free parameters (corresponding basically to masses and couplings of resonances) although 3 of them are tightly constrained from phenomenology and the reproduction of its expected value is more a check than a real fit. We fit simultaneously all the πN S- and P-waves and the results are summarized in Fig. 2 and in Table 1.

Obs.	result	KA85[10]	SP98[11]	EM98[12]
a_{0+}^{+}	0.30	−0.83	0.0 ± 0.1	0.41 ± 0.09
a_{0+}^{-}	8.28	9.17	8.83 ± 0.07	7.73 ± 0.06
a_{1-}^{+}	−5.07	−5.53	-5.33 ± 0.17	-5.46 ± 0.10
a_{1+}^{+}	13.6	13.27	13.6 ± 0.1	13.13 ± 0.13
a_{1-}^{-}	−1.37	−1.13	-1.00 ± 0.10	-1.19 ± 0.08
a_{1+}^{-}	−8.04	−8.13	-7.47 ± 0.13	-8.22 ± 0.07

Table 1. Values of the S– and P–wave threshold parameters. Units are appropriate inverse powers of the pion mass times 10^{-2}.

It is clear that a good reproduction of the experimental data is achieved both at threshold as in a broad energy region. In particular the pole position of the $\Delta(1232)$ is obtained at $(1210 - i\,53)$ MeV in perfect agreement with dispersive analysis of pion-nucleon scattering [13] or pion photoproduction [14].

3 Coupled Channel $N\bar{K}$ Scattering

In this section we discuss the study done in Ref.[15] of the $N\bar{K}$ S-wave in coupled channels. In this case, Eq.(5) is applied in a matrix notation considering the whole set of $SU(3)$ coupled channels $N\bar{K}$, $\Lambda\pi$, $\Sigma\pi$, $\Sigma\eta$ and ΞK. Here $\mathcal{T}_\chi$ is just the lowest order $SU(3)$ CHPT amplitude. The use of Eq.(5) to lowest order CHPT simply gives $\mathcal{T} = \mathcal{T}_\chi^{(1)}$ where the superscript refers to the chiral order, in this case $\mathcal{O}(p)$. Notice that in contrast with previous studies for the $N\bar{K}$ scattering we keep a complete covariant treatment. It is also worth to note that the work of Ref.[15] is a first step in a complete chiral treatment of this problem since in the same way one could increase the order of the chiral input to be matched. As a result, one has a systematic scheme to study such process since resumming the unitarity loops is the essential ingredient to reproduce the non–perturbative physics behind the appearance of the $\Lambda(1405)$. Thus, the chiral expansion is finally done on $\mathcal{T}$ instead on the whole T−matrix.

In Ref.[15] six cross sections and three ratio parameters measured at threshold were fitted simultaneously. The fit was done in terms of three parameters highly constrained. Because of that one could speak of a set of fixed natural parameters (this corresponds to the dashed line in Fig. (3)) in comparison with the case where all these parameters are fitted (solid line). The parameters involved are the mass of the lightest octet of baryons in the chiral limit m_0, the value of the pion decay constant in the same limit, F, and the subtraction constant a present in the definition of the $g(s)$ function. The natural values are: $m_0 = 1.15$ GeV from an average of the physical masses of the baryons in the corresponding octet, $F = 86.4$ MeV,[1] although in the $SU(2)$ chiral limit. Note the value of the physical pion decay constant is $F_\pi = 92.4$ MeV. Finally, $a = -2$ as estimated from the value of the cut-off used in a previous analysis.[16] Although we use dimensional regularization one can, however, connect both schemes in a expansion over the cut-off and the mass of the baryons. The fitted values are $m_0 = 1.29$ GeV, $F = 74$ MeV and $a = -2.23$ rather close to their estimated values. The pole structure of the S-wave $N\bar{K}$ is also investigated in the unphysical Riemann sheets. Interestingly it is found that together with an $I = 0$ pole close to the $N\bar{K}$ threshold, that could be identified with the $\Lambda(1405)$, one also has another pole with $I = 0$ close to the $\Sigma\pi$ threshold and another one with $I = 1$ is found as well close to the threshold of the $N\bar{K}$ channel. Hence one could think of a nonet of $J = 1/2$ and strangeness=-1 meson-baryon resonances but still one should study the

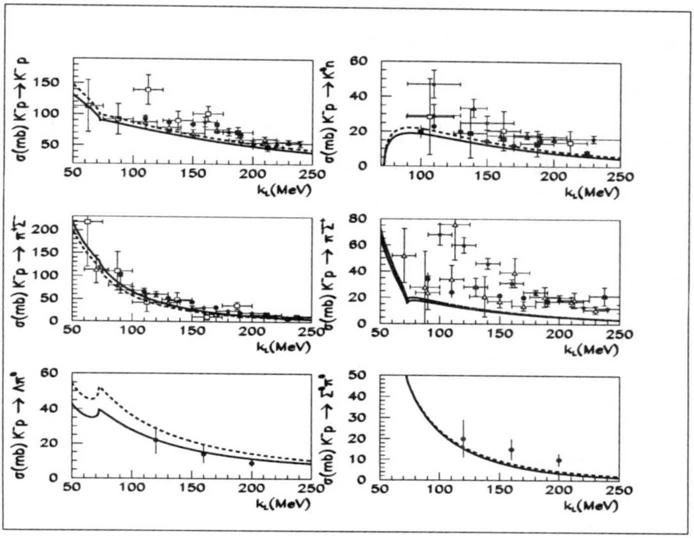

Figure 3. Fit to several cross sections coupling the $N\bar{K}$ channel.

I=1/2 channel with strangeness 0 and -2 in the present energy interval. Furthermore, the method of Ref.[15] was used in Ref.[17] to calculate the $\Lambda\pi$ elastic scattering relevant for the decay $\Xi \to \Lambda\pi \to p\pi\pi$, suited to study direct CP violation.

4 Conclusions

We have presented an approach that incorporates the chiral constraints at low energies by matching with (HB)CHPT and by considering chiral Lagrangians with resonances. These chiral inputs are unitarized by a suitable procedure which is exact in the limit of non-crossed channel dynamics, the latter being incorporated in a perturbative way as given by the chiral input $\mathcal{T}_\chi$. This way of proceeding, the so called Chiral Unitary Approach, provides a very suitable and improved theoretical scheme to analyze experimental data by incorporating the new advances in hadron physics coming from chiral effective field theories. In particular it has been used to analyze πN and $N\bar{K}$ scattering up to rather high energies $\sqrt{s} \lesssim 1.5$ GeV paying special attention to the resonance content of the amplitudes. In the former the parameters of the $\Delta(1232)$ resonance were precisely reproduced and in the latter a set of poles close to the $N\bar{K}$ and $\Sigma\pi$ channels were found. These poles could signal a nonet of meson-baryon resonances in that energy region, including the $\Lambda(1405)$.

Acknowledgments

J.A.O. would like to acknowledge partial financial support from DGICYT under contract no. BFM2000-1326 and from the EU TMR network Eurodaphne, contract no. ERBFMRX-CT98-0169.

References

1. S. Weinberg, *Phys.* A **96**, 327 (1979); J. Gasser, and H. Leutwyler, *Ann. Phys. (NY)* **158**, 142 (1984); J. Gasser, and H. Leutwyler, *Nucl. Phys.*, B **250**, 465, 517, 539 (1985).
2. S. Coleman, J. Wess, and B. Zumino, *Phys. Rev.* **177**, 2239 (1969); C. Callan, S. Coleman, J. Wess, and B. Zumino, *Phys. Rev.* **177**, 2247 (1969).
3. See, e.g., U.-G. Meißner, hep-ph/0007092 and references therein.
4. G. Ecker, J. Gasser, A. Pich, and E. de Rafael, *Nucl. Phys.* B **321**, 311 (1989).
5. B. Borasoy, and U.-G. Meißner, *Int. J. Mod. Phys.* A **11**, 5183 (1996).
6. T. R. Hemmert, B. R. Holstein, and J. Kambor, *Phys. Lett.* B **395**, 89 (1997).
7. U.-G. Meißner and J. A. Oller, *Nucl. Phys.* A **673**, 311 (2000); J. A. Oller and U.-G. Meißner, 'Chiral Unitary Approach to Pion-Nucleon Scattering', The Chiral Dynamics 2000 Proceedings to be published by World Scientific.
8. J. A. Oller and E. Oset, *Phys. Rev.* D **60**, 074023 (1999); J. A. Oller, *Phys. Lett.* B **477**, 187 (2000).
9. N. Fettes, U.-G. Meißner, and S. Steininger, *Nucl. Phys.* A **640**, 199 (1998).
10. R. Koch, *Nucl. Phys.* A **448**, 707 (1986).
11. E. Matsinos, *Phys. Rev.* C **56**, 3014 (1997).
12. SAID-on line program , R. A. Arndt *et al.*, see website http://said.phys.vt.edu/.
13. G. Höhler, in: H. Schopper (Ed.), Landolt-Börnstein, Vol. 9b2, Springer, Berlin, 1983.
14. O. Hanstein, D. Drechsel, and L. Tiator, *Phys. Lett.* B **385**, 45 (1996).
15. J. A. Oller and U.-G. Meißner, *Phys. Lett.* B **500**, 263 (2001).
16. E. Oset and A. Ramos, *Nucl. Phys.* A **635**, 99 (1998).
17. U.-G. Meißner and J. A. Oller, hep-ph/0011293, *Phys. Rev.* D, in press.

PION PHOTO- AND ELECTROPRODUCTION IN DYNAMICAL MODEL

T. SATO

Department of Physics, Graduate School of Science, Osaka University, Toyonaka, Osaka 560-0043 Japan
E-mail: tsato@phys.sci.osaka-u.ac.jp

The dynamical approach for investigating the structure of nucleon resonances in the GeV photon and electron reactions is discussed. The $\gamma N \to \Delta$ transition form factors extracted from the recent data of (γ, π) and $(e, e'\pi)$ reactions are presented.

1 Introduction

An important issue of the GeV electromagnetic reaction study is to understand the quark substructure of nucleons and nucleon resonances. In the Delta resonance region, the $\gamma N \to \Delta$ transition strength and its dependence on the virtual photon momentum provide us a insight into the role of the quark core and the meson cloud on the electromagnetic structure of baryon. Electric and Coulomb quadrupole form factors are particularly interesting in connection with the deformation of the baryons and the helicity conservation law on the exclusive reaction at large Q^2. Recently a detailed study of such $\gamma N \to \Delta$ transition strength became possible due to the precious data including polarization observables for pion photoproduction at LEGS[1] and Mainz[2], and for pion electroproduction at JLAB[3,4], MIT-Bates[5] and NIKHEF[6].

There are essentially three approaches to extract resonance parameters from the data. One of the approaches is based on the dispersion theory, which has been used by Mainz group[7] for the Δ resonance region, and the other is the K-matrix method[8], which is based on the tree approximation of the effective Lagrangian. The latter one is the dynamical approach, which has been investigated by several groups[9–16]. The main objective of a dynamical approach is to separate the reaction mechanism from the excitation of the internal structure of the hadrons. In the model developed by Sato and Lee[15,16] (SL model), this is achieved by applying the reaction theory within the Hamiltonian formalism. The off-shell non-resonant contribution to the $\gamma N \to \Delta$ form factors can be calculated explicitly and then the determined 'bare' form factors can be compared with the prediction of the hadron models.

In sec. 2, we discuss the construction of reaction models in SL approach and their relation to hadron quark models. The consequence of the model applied to the recent pion electroproduction data is discussed in sec. 3.

2 Effective Hamiltonian and Electromagnetic Current

In the dynamical approaches pion photoproduction and electroproduction reactions are described in terms of photon and hadron degrees of freedom. We start from the Hamiltonian and the electromagnetic current given as

$$H = H_0 + \Gamma_{B' \leftrightarrow BM}, \tag{1}$$

$$J^\mu = J^\mu_{B'B} + J^\mu_M + J^\mu_{B' \leftrightarrow BM}. \tag{2}$$

The interaction Hamiltonian $\Gamma_{B' \leftrightarrow BM}$ describes the creation and absorption of a meson (M) by a baryon (B). The electromagnetic currents include baryon current ($J^\mu_{B'B}$), meson current(J^μ_M) and seagull current ($J^\mu_{B' \leftrightarrow BM}$).

It is a nontrivial many body problem to calculate the reaction amplitudes from the above Hamiltonian. To obtain a manageable reaction model, a unitary transformation method[17] is used up to the second order of the strong interaction Hamiltonian. The unitary transformation is chosen to eliminate the 'virtual' processes. Physical meaning of the unitary transformation is that it partially diagonalizes the single baryon state and rewrites the Hamiltonian in terms of the baryon dressed with a meson cloud. Then the eliminated 'virtual' processes appear as effective interactions in the transformed Hamiltonian. Within the πN, Δ and γN Fock space, the effective Hamiltonian has the following form:

$$\begin{aligned} H_{eff} &= U^\dagger H U \\ &= H_0 + v_{\pi N} + \Gamma_{\Delta \leftrightarrow \pi N}, \end{aligned} \tag{3}$$

where $v_{\pi N}$ is the non-resonant πN potential and Delta excitation is described by $\Gamma_{\Delta \leftrightarrow \pi N}$. The effective electromagnetic current is given by using the same unitary transformation as

$$\begin{aligned} J^\mu_{eff} \cdot \epsilon &= U^\dagger J^\mu U \cdot \epsilon \\ &= v_{\gamma\pi} + \Gamma_{\gamma N \to \Delta}, \end{aligned} \tag{4}$$

where $v_{\gamma\pi}$ is the non-resonant pion production interaction and the $\gamma N \to \Delta$ interaction is given by $\Gamma_{\gamma N \to \Delta}$. An important feature of this procedure is that the effective Hamiltonian and current are independent of the total energy of the system and hermitian, therefore the unitarity can be trivially satisfied. One can start from the Hamiltonian determined from the quark model of the hadron[18], or one can fix the off-shell behavior of the strong interaction Hamiltonian by studying the πN reaction. In either case, it is noted that the effective current can be treated in a consistent way as the strong interaction Hamiltonian.

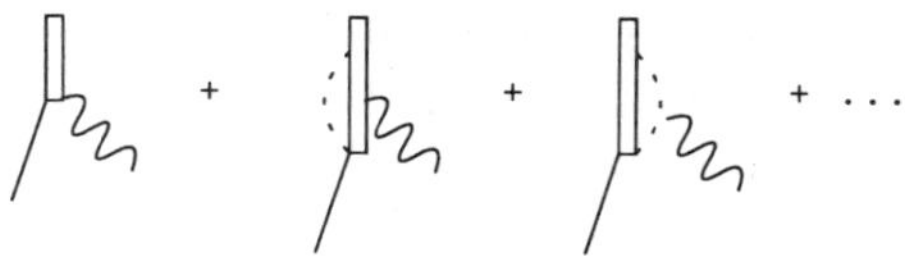

Figure 1. $\gamma N \to \Delta$ Vertex.

The $\Gamma_{\gamma N\to\Delta}$ interaction can be parametrized by three coupling strength G_M, G_E and G_C corresponding to the magnetic dipole (M1), electric quadrupole(E2) and Coulomb quadrupole(C2), respectively. In the work of the SL model[15,16], we have treated those transition strength as a phenomenological constant and determined by studying the pion photoproduction and electroproduction. However, one could start from the Hamiltonian and current in Eqs. (1) and (2) of the constituent quark model or the cloudy bag model. Then the unitary transformation method in one-loop approximation gives the $\Gamma_{\gamma N\to\Delta}$ interaction including off-shell pionic loop correction with $\pi\Delta$ intermediate states as shown schematically in Fig. 1. Our approach provides a well defined interpretation of the the bare coupling constants in terms of the quark model[19].

From the effective Hamiltonian in Eqs. (3) and (4), it is straightforward to derive a set of coupled equations for πN and γN reactions. The resulting pion photoproduction amplitude can be written as

$$T_{\gamma\pi}(E) = t_{\gamma\pi}(E) + \frac{\bar{\Gamma}_{\Delta\to\pi N}\bar{\Gamma}_{\gamma N\to\Delta}}{E - m_\Delta - \Sigma_\Delta(E)}. \tag{5}$$

The non-resonant amplitude $t_{\gamma\pi}$ is calculated from $v_{\gamma\pi}$ by

$$t_{\gamma\pi}(E) = v_{\gamma\pi} + t_{\pi N}(E)G_{\pi N}(E)v_{\gamma\pi}. \tag{6}$$

The dressed vertices in Eq. (5) are defined by

$$\bar{\Gamma}_{\gamma N\to\Delta} = \Gamma_{\gamma N\to\Delta} + \bar{\Gamma}_{\pi N\to\Delta}G_{\pi N}(E)v_{\gamma\pi}, \tag{7}$$

$$\bar{\Gamma}_{\Delta\to\pi N} = [1 + t_{\pi N}(E)G_{\pi N}(E)]\Gamma_{\Delta\to\pi N}. \tag{8}$$

The Δ self-energy in Eq. (5) is then calculated from

$$\Sigma_\Delta(E) = \Gamma_{\pi N\to\Delta}G_{\pi N}(E)\bar{\Gamma}_{\Delta\to\pi N}. \tag{9}$$

As seen in the above equations, an important consequence of the dynamical model is that the influence of the non-resonant mechanisms on the resonance properties can be identified and calculated explicitly. The bare vertex $\Gamma_{\gamma N\to\Delta}$ is modified by the non-resonant interaction $v_{\gamma\pi}$ to give the dressed vertex $\bar{\Gamma}_{\gamma N \to \Delta}$, as defined by Eq. (7). Similar modification is also for the bare $\pi N \to \Delta$ vertex. The resonance position of the amplitude defined by Eq. (5) is shifted from the bare mass m_Δ by the self-energy $\Sigma_\Delta(E)$.

The helicity amplitude of delta is usually defined by the imaginary part of the multipole amplitude of pion photoproduction at resonance energy as

$$A_M = \sqrt{\frac{8\pi E k \Gamma_\Delta}{3 m_N q}} Im[M_{1+}^{3/2}(E)]_{E=E_R}. \tag{10}$$

In the SL model, the same quantity is given as

$$A_M^{SL} = \frac{1}{\sqrt{1 - d\Sigma_\Delta/dE}} \bar{\Gamma}^K_{\gamma N \to \Delta}|_{E=E_R}, \tag{11}$$

where principal value integration of Green function(K-matrix) has to be taken to calculate the dressed $\gamma N \to \Delta$ vertex.

It will be useful to see the close relation between our dynamical model and the quark model. In the cloudy bag model, the baryon dressed with meson cloud $|B>$ is given from the bare three-quark baryon state$|B_0>$ as

$$|B>= \sqrt{Z}[1 + \frac{1}{M_B - H_0 - \Lambda H_I \Lambda} \Lambda H_I]|B_0> . \tag{12}$$

The helicity amplitude is then calculated by the matrix element of the electromagnetic current between the dressed baryon state[20,21]:

$$A_M^{Bag} = < \Delta | J^{em} \cdot \epsilon | N > . \tag{13}$$

If we start from the Hamiltonian and current of the cloudy bag model in Eqs. (1) and (2), we can show that the helicity amplitude of SL model A_M^{SL} coincides with A_M^{Bag} in one-loop approximation[19]. In SL model, the on-shell πN state is explicitly treated in the scattering equation, while off-shell $\pi \Delta$ state is included as a loop correction to the bare $\gamma N \to \Delta$ interaction.

3 Pion Photoproduction and Electroproduction

We apply the formalism outlined in the previous section to the pion photoproduction and electroproduction. The first step is to investigate the πN scattering from threshold up to the Δ excitation region. By fitting the πN phase shift the strong interaction parameters are determined. The obtained bare $\pi N \Delta$ coupling constant is about 20% larger than the SU(6) quark model value and the non-resonant mechanism enhance the strength of the $\pi N \to \Delta$ vertex about 30 %. The pion photoproduction data are then used to determine 'bare' $\gamma N \to \Delta$ coupling constants, G_M and G_E and $g_{\omega NN}$. Because

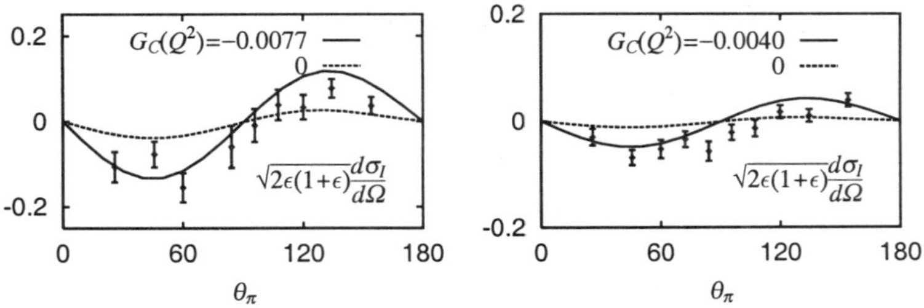

Figure 2. The longitudinal and transverse interference term of the calculated differential cross section of $(e, e'\pi^0)$ reaction compared with the JLAB data at $Q^2 = 2.8(GeV/c)^2$(Left) and $4(GeV/c)^2$(right) .

of our consistent description of the πN and γN reaction, no further parameters are allowed to vary. We consider the differential cross section and the photon asymmetry Σ^2 and find that the data is best reproduced by setting $G_M(0) = 1.85, G_E(0) = 0.025$ and $g_{\omega NN} = 11.5$. The extracted M1 strength of the bare vertex $\Gamma_{\gamma N \to \Delta}$ is very close to the values predicted by the constituent quark models[22,23,24], while the empirical values given by the Particle Data Group(PDG) can only be identified with the dressed vertex $\bar{\Gamma}_{\gamma N \to \Delta}$. The dressed $E2/M1$ ratio is $R_{EM} = -2.7\%$, where the non-resonant meson cloud plays a crucial role and the bare value is $R_{EM}(Bare) = -1.3\%$.

To proceed to pion electroproduction, we need to determine the C2 form factor of $\gamma N \to \Delta$ transition. We use the long wave length limit relation between the E2 and C2 form factors and set $G_C(0) = -0.238$ from $G_E(0) = 0.025$. We find that an overall good description of all the recent data from JLAB and MIT-Bates can be obtained when we choose the following form of the $\gamma N \to \Delta$ form factors

$$G_\alpha(Q^2) = G_\alpha(0)G_D(Q^2)(1 + aQ^2)\exp(-bQ^2), \tag{14}$$

with $a = 0.154$ and $b = 0.166(GeV/c)^2$ for $\alpha = M, E$ and C, and $G_D(Q^2) = [1 + Q^2/0.71(GeV/c)^2]^{-2}$.

The longitudinal C2 $\gamma N \to \Delta$ form factor is best studied by the interference term between transverse and longitudinal currents. Our predicted interference cross sections $d\sigma_I/d\Omega$ at $Q^2 = 2.8$ and $4(GeV/C)^2$ are compared with the data extracted from the JLAB experiment[3] in Fig. 2. The sensitivity to the 'bare' C2 form factor can be seen by comparing the results with full(solid) and $G_C = 0$(dotted). At $Q^2 = 0.126(GeV/c)^2$, where the

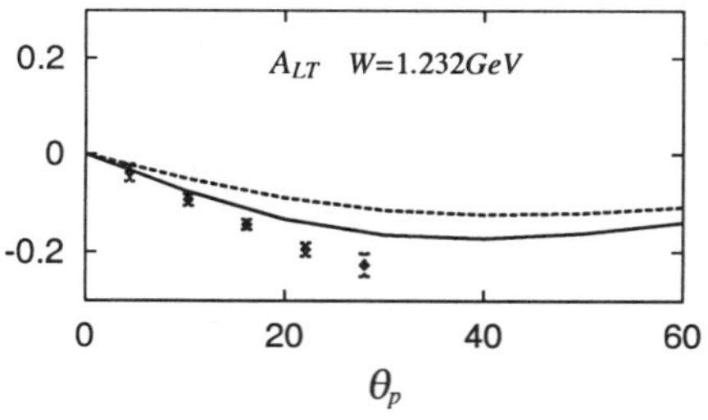

Figure 3. The predicted symmetry A_{LT} of the $(e, e'\pi^0)$ reaction at $Q^2 = 0.126(GeV/c)^2$ are compared with the data from MIT-Bates. The dotted curve is obtained from setting $G_C = G_E = 0$.

non-resonant mechanism plays a more important role, the predicted A_{LT} is compared with MIT-Bates data[5] in Fig. 3. In both cases, the agreement of our results with the data is satisfactory.

Within the present model, we can predict the Q^2 dependence of the $\gamma N \to \Delta$ form factors. In the top panel of Fig. 4, the dressed magnetic dipole form factor normalized by the dipole form factor of proton is plotted, which shows good agreement with the values extracted in previous works. It is clear that both the dressed and bare form factors drop faster than the dipole form factor. The meson cloud effect accounts for about 40% of the dressed form factor at $Q^2 = 0$ but becomes much weaker at higher momentum transfer.

The Q^2 dependence of the $E2/M1$ ratio R_{EM} and the $C2/M1$ ratio R_{SM} of the dressed $\gamma N \to \Delta$ form factors are also shown in Fig. 4 together with the extracted 'data'[25]. R_{SM} drops significantly with Q^2 and reaches almost -14% at $4(GeV/c)^2$, while R_{EM} remains about -3%. It is noticed that the contribution of the pion cloud on the total $S_{1+}^{3/2}$ mutipole amplitude reaches about 70% around $Q^2 \sim 0.1(GeV)^2$ and decreases with increasing momentum transfer. Clearly the $Q^2 < 4(GeV/c)^2$ region is still away from the perturbative QCD region, where R_{EM} is expected to approach unity.

4 Summary

A dynamical model has been developed to study the meson and quark substructure of the baryon resonances. The model has been applied to the pion photoproduction and electroproduction in the Delta region and has described the recent data successfuly. As a result, the non-resonant interaction was found to dress the $\gamma N \to \Delta$ interaction and to enhance its strength at low Q^2

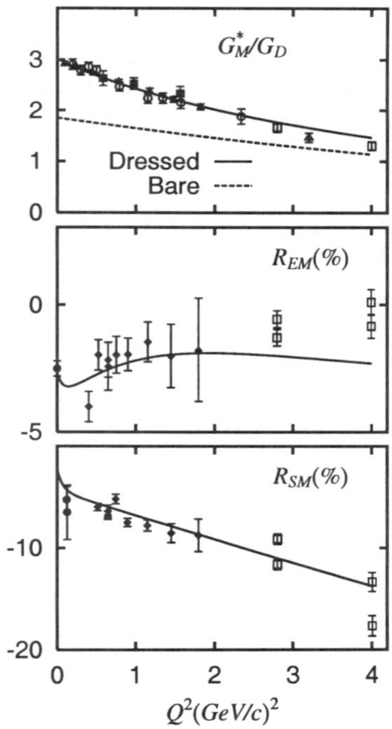

Figure 4. The Q^2 dependences of the ratio between the M1 form factor and dipole form factor of proton, E2/M1 ratio R_{EM} and C2/M1 ratio R_{SM} of the dressed $\gamma N \to \Delta$ transition.

very strongly but much less at high Q^2. Further investigations of the form factors in terms of the models of the hadron, which were briefly discussed, will clarify the structure of the electromagnetic excitation of Delta resonance.

Acknowledgments

This work was partly supported by Japan Society for Promotion of Science, Grant-in-Aid for Scientific Research(C) 12640273.

References

1. G. Blanpied *et al., Phys. Rev. Lett.* **79**, 4337 (1997).
2. R. Beck *et al., Phys. Rev.* C **61**, 035204 (2000).
3. V. V. Frolov *et al., Phys. Rev. Lett.* **82**, 45 (1999).
4. V. D. Burkert, *Nucl. Phys.* A **684**, 16c (2001).
5. C. Mertz *et al., Phys. Rev. Lett.* **86**, 2963 (2001).
6. L. D. van Buuren, *Nucl. Phys.* A **684**, 324c (2001).
7. O. Hanstein, D. Drechsel and L. Tiator, *Nucl. Phys.* A **632**, 561 (1998).
8. R. M. Davidson, N. C. Mukhopadyay, and R. S. Wittman, *Phys. Rev.* D **43**, 71 (1990).
9. H. Tanabe and K. Ohta, *Phys. Rev.* C **31**, 1876 (1985).
10. S. N. Yang, *J. Phys.* G **11**, L205 (1985).
11. S. Nozawa, B. Blankleider, and T.-S. H. Lee, *Nucl. Phys.* A **513**, 459 (1990).
12. S. Nozawa and T.-S. H. Lee, *Nucl. Phys.* A **513**, 511 (1990).
13. S. Nozawa and T.-S. H. Lee, *Nucl. Phys.* A **513**, 543 (1990).
14. S. S. Kamalov and S. N. Yang, *Phys. Rev. Lett.* **83**, 4494 (1999).
15. T. Sato and T.-S. H. Lee, *Phys. Rev.* C **54**, 2660 (1996).
16. T. Sato and T.-S. H. Lee, *Phys. Rev.* C **63**, 055201 (2001).
17. M. Kobayashi, T. Sato, and H. Ohtsubo, *Prog. Theor. Phys.* **98**, 927 (1997).
18. T. Yoshimoto, T. Sato, M. Arima, and T. -S. H. Lee, *Phys. Rev.* C **61**, 065203 (2000).
19. T. Nakamura, T. Sato, and T. -S. H. Lee, in preparation.
20. K. Bermuth, D. Drechsel, and L. Tiator, *Phys. Rev.* D **37**, 89 (1988).
21. D. H. Lu, A. W. Thomas, and A. G. Williams, *Phys. Rev.* D **55**, 3108 (1997).
22. S. Capstick, *Phys. Rev.* D **46**, 1965 (1992); D **46**, 2864 (1992).
23. R. Bijker, F. Iachello, and A. Leviatan, *Ann. Phys. (N.Y.)* **236**, 69 (1994)
24. A. Buchmann, E. Hernandez, and A. Faessler, *Phys. Rev.* C **55**, 448 (1997).
25. *The DNP Town Meeting on Hadronic Physics*, Jefferson Laboratory, 2001.

RECENT DEVELOPMENTS IN THE DYNAMICAL AND UNITARY ISOBAR MODELS FOR PION ELECTROMAGNETIC PRODUCTION

S. N. YANG AND G. Y. CHEN

Department of Physics, National Taiwan University, Taipei, Taiwan

S. S. KAMALOV

Laboratory of Theoretical Physics, JINR, 141980 Dubna, Russia

D. DRECHSEL AND L. TIATOR

Institut für Kernphysik, Universität Mainz, 55099 Mainz, Germany

$\gamma^* N\Delta$ transition form factors and threshold π^0 photo- and electroproduction are studied with the new version of MAID and a dynamical model. By re-analyzing the recent Jlab data on $p(e,e'p)\pi^0$ at $Q^2 = 2.8$ and 4.0 $(\text{GeV/c})^2$, we find that the hadronic helicity conservation is not yet observed in this region of Q^2. The extracted R_{EM}, starting from a small and negative value at the real photon point, actually exhibits a clear tendency to cross zero and change sign as Q^2 increases, while the absolute value of R_{SM} is strongly increasing. Our analysis indicates that $A_{1/2}$ and $S_{1/2}$, but not $A_{3/2}$, starts exhibiting the pQCD scaling behavior at about $Q^2 \geq 2.5$ $(\text{GeV/c})^2$. For the π^0 photo- and electroproduction near threshold, results obtained within the dynamical model with the use of a meson-exchange πN model for the final state interaction are in as good agreement with the data as ChPT.

1 Introduction

Pion photo- and electroproduction reactions are powerful tools to probe the nucleon structure. The most prominent example is the $\Delta(1232)$ which decays almost exclusively into the πN channel. The interest there lies in the possibility of observing a deformation of the Δ. A deformed Δ would indicate that the Δ contains a D-state and the photon can excite a nucleon through electric $E2$ and Coulomb $C2$ quardrupole transitions. The study of $E_{1+}^{(3/2)}$ and $S_{1+}^{(3/2)}$ multipoles is expected to shed light on the structure of the nucleon and its first excited state. Recent experiments[1] give a nonvanishing ratio $R_{EM} = E_{1+}^{(3/2)}/M_{1+}^{(3/2)}$ between magnetic dipole $M_{1+}^{(3/2)}$ and electric quadrupole $E_{1+}^{(3/2)}$ multipoles lying between -2.5% and -3.0% at $Q^2 = 0$. In addition, the reaction $p(e,e'p)\pi^0$ provides us with the possibility of determining the range of photon four-momentum transfer squared Q^2, where perturbative QCD (pQCD) would become applicable. In the limit of $Q^2 \to \infty$, pQCD predicts the dominance of helicity-conserving amplitudes[2] and scaling results[3]. The hadronic helicity conservation would have the consequence that R_{EM} approaches 1. It is an intriguing question how R_{EM} would evolve from a very small negative value at $Q^2 = 0$ to +100% at sufficiently high Q^2.

Another interesting issue is the π^0 photo- and electroproduction in the threshold region. At present time chiral perturbation theory (ChPT) is the basic theory for the description of these reactions in this energy region. It predicts a large one pion loop correction to the low energy theorem (LET) value of the E_{0+} multipole for the neutral pion photoproduction[4]. On the other hand the role of the one-pion loop correction can be related to the effect of final state interaction (FSI) developed within dynamical models[5,6]. It is of interest to compare the predictions of these two different approaches.

In this talk, we present the results we have recently obtained with a new version (hereafter called MAID) of the unitary isobar model developed at Mainz[7], and the dynamical model developed recently in Ref. [8] (hereafter called Dubna-Mainz-Taipei (DMT) model), both of which give an excellent description of most of the existing pion photo- and electroproduction data, on the $\gamma^* N\Delta$ transition and the π^0 threshold production.

2 Formalism

We start with the description of the basic elements of our models. The main equation for the pion photo- and electroproduction t-matrix is

$$t_{\gamma\pi}(E) = v_{\gamma\pi} + v_{\gamma\pi}\, g_0(E)\, t_{\pi N}(E)\,, \tag{1}$$

where $t_{\pi N}$ is the full pion-nucleon scattering matrix with total πN c.m. energy E and $g_0(E)$ is the free πN propagator.

In resonant channels like (3,3), the transition potential $v_{\gamma\pi}$ consists of two terms, $v_{\gamma\pi}(E) = v^B_{\gamma\pi} + v^R_{\gamma\pi}(E)$, where $v^B_{\gamma\pi}$ is the background transition potential and $v^R_{\gamma\pi}(E)$ corresponds to the contribution of the bare resonance. The resulting t-matrix can be expressed as a sum of two terms[8],

$$t_{\gamma\pi} = t^B_{\gamma\pi} + t^R_{\gamma\pi}\,, \tag{2}$$

where

$$t^B_{\gamma\pi} = v^B_{\gamma\pi} + v^B_{\gamma\pi}\, g_0\, t_{\pi N}\,, \qquad t^R_{\gamma\pi} = v^R_{\gamma\pi} + v^B_{\gamma\pi}\, g_0\, t_{\pi N}\,. \tag{3}$$

For physical multipoles in channel $\alpha = \{l, j\}$, Eq. (3) gives

$$t^B_\alpha(q_E, k; E + i\epsilon) = \exp(i\delta_\alpha)\, \cos\delta_\alpha \times \left[v^B_\alpha(q_E, k) + P \int_0 dq' \frac{q'^2 R_\alpha(q_E, q')\, v^B_\alpha(q', k)}{E - E(q')} \right], \tag{4}$$

where δ_α and R_α are the πN scattering phase shift and reaction matrix, in channel α, respectively; q_E is the pion on-shell momentum and $k =| \mathbf{k} |$ is the photon momentum. The procedure which we use for the off-shell extrapolation and for maintaining the gauge invariance is given in Ref. [8].

In the new version of MAID, the S, P, D and F waves of the t_α^B amplitude is defined in accordance with the K-matrix approximation

$$t_\alpha^B(\mathrm{MAID}) = \exp(i\delta_\alpha)\,\cos\delta_\alpha v_\alpha^B(q, W, Q^2)\,, \tag{5}$$

where $W \equiv E$ is the total πN c.m. energy. From Eqs. (4) and (5), one finds that the difference between the background terms of the MAID and DMT models is that pion off-shell rescattering contributions (principal value integral) are not included in MAID's background term.

3 Pion Photo- and Electroproduction in the $\Delta(1232)$ Region

For the resonance contribution t_α^R in Eq. (3), the following Breit-Wigner form is assumed[7] in both models,

$$t_\alpha^R(W, Q^2) \;=\; \bar{\mathcal{A}}_\alpha^R(Q^2)\,\frac{f_{\gamma R}(W)\Gamma_R\, M_R\, f_{\pi R}(W)}{M_R^2 - W^2 - iM_R\Gamma_R}\, e^{i\phi_R}\,, \tag{6}$$

where $f_{\pi R}$ is the usual Breit-Wigner factor describing the decay of resonance R with total width $\Gamma_R(W)$ and physical mass M_R. The phase $\phi_R(W)$ in Eq. (6) is introduced to adjust the phase of t_α^R to have the correct phase δ_α.

The main subject of our study in the resonance region is the strengths of the electromagnetic transitions described by amplitudes $\bar{\mathcal{A}}_\alpha^R(Q^2)$. In general, they are considered as free parameters to be extracted from analysis of the experimental data. In the present talk we will consider only results pertinent to the $\Delta(1232)$ resonance. In this case we impose the following parametrization for the electric ($\alpha = E$), magnetic ($\alpha = M$) and Coulomb ($\alpha = S$) amplitudes

$$\bar{\mathcal{A}}_\alpha^\Delta(Q^2) = X_\alpha^\Delta(Q^2)\,\bar{\mathcal{A}}_\alpha^\Delta(0)\,\frac{k}{k_W}\,F(Q^2), \tag{7}$$

where $k_W = (W^2 - m_N^2)/2W$. The form factor F is taken to be $F(Q^2) = (1 + \beta\, Q^2)\, e^{-\gamma Q^2}\, G_D(Q^2)$, where $G_D(Q^2) = 1/(1 + Q^2/0.71)^2$ is the usual dipole form factor. The values of $\bar{\mathcal{A}}_M^\Delta(0)$ and $\bar{\mathcal{A}}_E^\Delta(0)$ are obtained by setting $X_\alpha(0) = 1$ and fitting the pion photoproduction data. The parameters β and γ can be determined by setting $X_M^\Delta = 1$ and fitting the $\bar{\mathcal{A}}_M^\Delta(Q^2)$ to the existing data for G_M^* form factor.

Note that the physical meaning of the resonant amplitudes in different models is different[8]. In MAID, background contribution does not contain

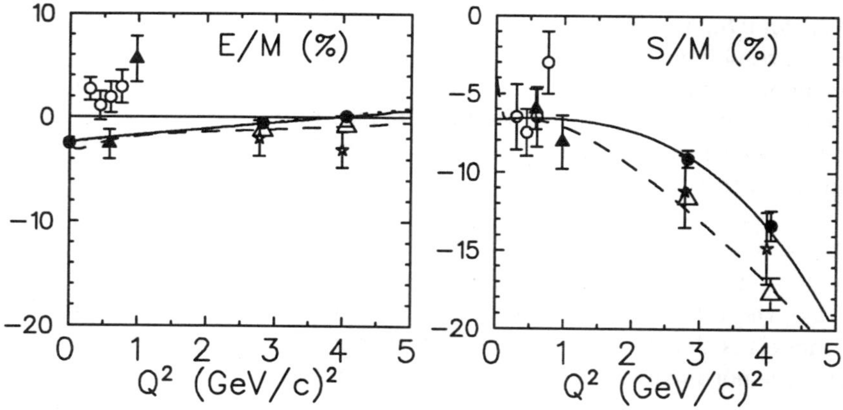

Figure 1. The Q^2 dependence of the ratios $R_{EM}^{(p\pi^0)}$ and $R_{SM}^{(p\pi^0)}$ at $W = 1232$ MeV. The solid and dashed curves are the MAID and DMT models results, respectively, obtained with a violation of the scaling assumption. Data at $Q^2 = 2.8$ and 4.0 $(GeV/c)^2$ are from Ref. [9] (stars). Results of our analysis at $Q^2 = 2.8$ and 4.0 $(GeV/c)^2$ are obtained using MAID (•) and the DMT models (△). Other data are the same as in Ref. [8].

effects from the off-shell pion rescattering. Instead, they are effectively included in the resonance amplitude and this leads to the dressing of $\bar{\mathcal{A}}_\alpha^\Delta(Q^2)$. Thus, using MAID we can extract information about so the called "dressed" $\gamma N\Delta$ vertex. However, in the dynamical model the background excitation is included in t_α^B and the electromagnetic vertex $\bar{\mathcal{A}}_\alpha^\Delta(Q^2)$ corresponds to the "bare" vertex.

We determine the Q^2 dependence both for X_E and X_S, normalized to 1 at $Q^2 = 0$, from the recent $p(e,e'p)\pi^0$ experiment[9] at $Q^2 = 2.8$ and 4.0 (GeV/c)^2. Note that deviations from $X_\alpha^\Delta = 1$ value at finite Q^2 will indicate a violation of the scaling law. The extracted Q^2 dependence of the X_α^Δ parameters is $X_E^\Delta(\text{MAID}) = 1 - Q^2/3.7\,, X_E^\Delta(\text{DMT}) = 1 + Q^4/2.4$, and for the Coulomb $X_S^\Delta(\text{MAID}) = 1 + Q^6/61\,, X_S^\Delta(\text{DMT}) = 1 - Q^2/0.1$. The obtained R_{EM} and $R_{SM} = S_{1+}^{(3/2)}/M_{1+}^{(3/2)}$ ratios are compared with other results in Fig. 1. The main difference between our results and those of Ref. [9] is that our values of R_{EM} show a clear tendency to cross zero and change sign as Q^2 increases. This is in contrast with the results obtained in the original analysis[9] of the data which concluded that R_{EM} would stay negative and tend towards more negative values with increasing Q^2. Furthermore, we find that the absolute value of R_{SM} is strongly increasing.

Finally in Fig. 2, we show our results for $Q^3 A_{1/2}, Q^5 A_{3/2}$, and $Q^3 S_{1/2}$ to check the scaling behavior predicted by pQCD: $A_{1/2} \sim Q^{-3}$, $A_{3/2} \sim Q^{-5}$ and

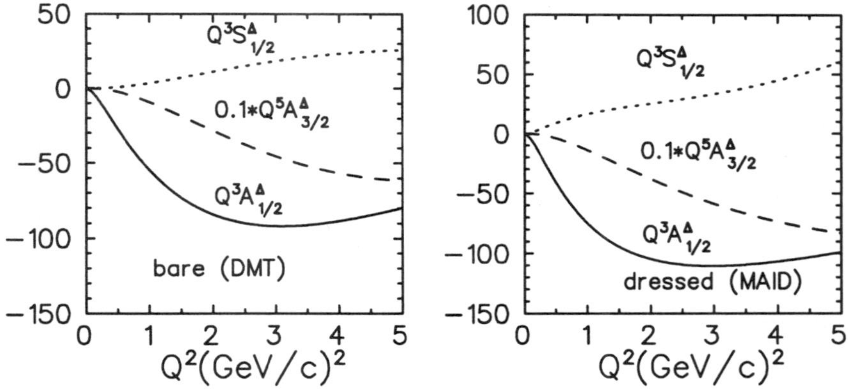

Figure 2. Asymptotic pQCD behaviors for the $A_{1/2}$, $A_{3/2}$ and $S_{1/2}$ helicity amplitudes obtained with DMT (left figure) and MAID (right figure).

$S_{1/2}^{(1/2)} \sim Q^{-3}$. It is interesting to see that both the bare $A_{1/2}$ and $S_{1/2}$ clearly start exhibiting the pQCD scaling behavior at about $Q^2 \geq 2.5\,(GeV/c)^2$, while $A_{3/2}$ does not. From these results, one might be tempted to speculate that scaling will set in earlier than the helicity conservation predicted by pQCD.

4 Threshold π^0 Photoproduction

Let us start with the π^0 photoproduction in the threshold region. Since the reaction can proceed through the $\pi^0 p$ and $\pi^+ n$ intermediate states, we may write the π^0 photoproduction t-matrix as

$$\begin{aligned} t_{\gamma\pi^0}(E) &= v^B_{\gamma\pi^0} + v^B_{\gamma\pi^0}\, g_{\pi^0 p}(E)\, t_{\pi^0 p\to\pi^0 p}(E) \\ &\quad + v^B_{\gamma\pi^+}\, g_{\pi^+ n}(E)\, t_{\pi^+ n\to\pi^0 p}(E)\,, \end{aligned} \tag{8}$$

where $t_{\pi^0 p\to\pi^0 p}$ and $t_{\pi^+ n\to\pi^0 p}$ are the πN scattering t-matrices for the elastic and charge exchange channels, respectively. These are obtained by solving coupled channels equations for the πN scattering using the meson exchange model[10]. The results obtained in such an exact calculation and without the inclusion of FSI are depicted in Fig. 3 by dashed-dotted and dotted curves, respectively. These results clearly indicate that FSI effects are very important in order to achieve agreement with the data. We also find that the main FSI contribution (about 99%) comes from the principal value integral contribution in the charge exchange channel.

On the other hand, if the FSI effect is evaluated with the assumption of isospin symmetry (IS), i.e., with averaged masses in the free πN propaga-

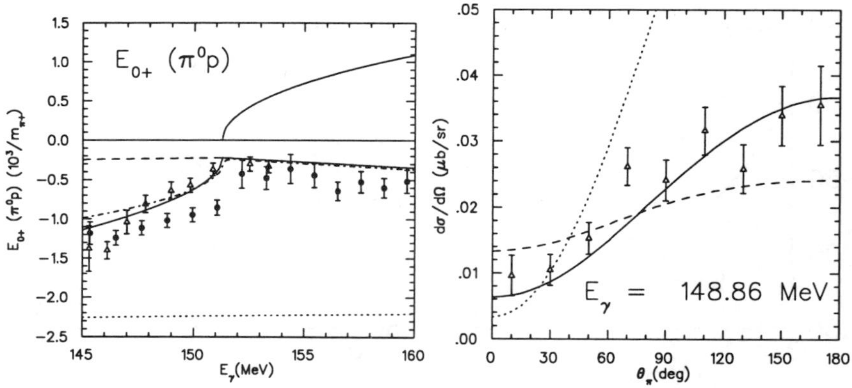

Figure 3. The E_{0+} multipole (left figure) and differential cross section (right figure) for the π^0 photoproduction on the proton. Dotted curves are the results obtained without FSI. Dashed-dotted and solid curves are the results obtained with coupled-channels and K-matrix approaches, respectively. The dashed curve is the result without the cusp effect. Data points are from Ref. [12]($\triangle$) and Ref.[13]($\bullet$).

tor, then the energy dependence in $Re\, E_{0+}$ near threshold would be smooth as given in the dashed curve of Fig. 3. Below π^+ threshold the strong energy dependence (cusp effect) appears only due to the pion mass difference and, as mentioned above, it arises mostly from the coupling with the charge exchange channel. In the literature, effects from the pion-mass difference below π^+ production channel have been taken into account using the K-matrix approach[4,11],

$$Re\, E_{0+}^{\gamma\pi^0} = Re\, E_{0+}^{\gamma\pi^0}(IS) - a_{\pi N}\, \omega_c\, Re E_{0+}^{\gamma\pi^+}(IS)\sqrt{1-\frac{\omega^2}{\omega_c}}\,, \qquad (9)$$

where ω and ω_c are the π^+ c.m. energies corresponding to the $W = E_p + E_\gamma$ and $W_c = m_n + m_{\pi^+}$, respectively, $a_{\pi N} = 0.124/m_{\pi^+}$ is the pion charge exchange amplitude, and $E_{0+}^{\gamma\pi^{0,+}}(IS)$ is the π^0 or π^+ photoproduction amplitude at threshold obtained without the pion mass difference using Eq. (4). The corresponding results are given by the solid curves in Fig. 3. We can see that above π^+ threshold the difference between results obtained in coupled-channels calculation and K-matrix approach is very small. This is consistent with the finding of Ref. [11]. At lower energies the difference is around 10% and becomes visible only very close to the $\pi^0 p$ threshold. In general, we can conclude that Eq. (9) is a good approximation for the pion-mass difference effect. The threshold energy dependence for the imaginary part can be eas-

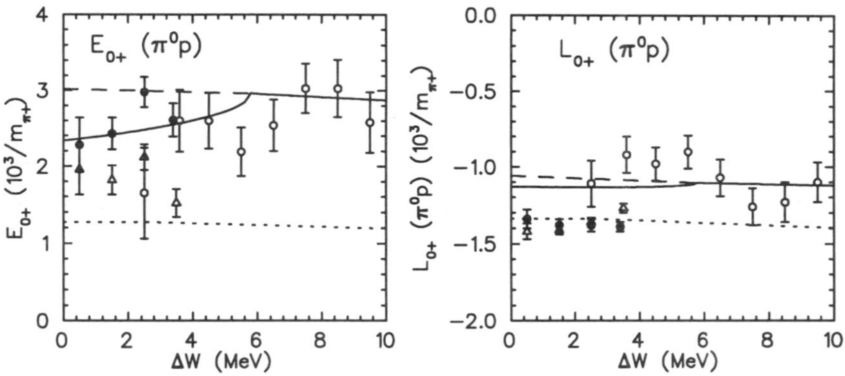

Figure 4. Real parts of the E_{0+} (left figure) and L_{0+} (right figure) multipoles for the π^0 electroproduction on the proton at Q^2=0.1 (GeV/c)^2. Notations for the curves are the same as in Fig. 1. Data points are from Ref.[14]($\circ$) and Ref.[15]($\triangle$). Results of the present work obtained using p-waves from the DMT model are given by ($\bullet$).

ily obtained via the Fermi-Watson theorem if in the threshold region the πN phase shift is taken as a linear function of the π^+ momentum, i.e. approaching 0 at $q_{\pi^+} \to 0$.

In Fig. 3, we also compare our prediction for the differential cross section with the data from Mainz[12]. We see that both off-shell pion rescattering and cusp effects substantially improve the agreement with the data. Moreover this result indicates that our model, without any new free parameter, also gives reliable predictions for the threshold behavior of the p-waves.

5 Threshold π^0 Electroproduction

Pion electroproduction provides us with information on the Q^2 dependence of the transverse E_{0+} and longitudinal L_{0+} multipoles in the threshold region. The "cusp" effects in the L_{0+} multipole is taken into account in a similar way as in the case of E_{0+},

$$Re\, L_{0+}^{\gamma\pi^0} = Re\, L_{0+}^{\gamma\pi^0}(IS) - a_{\pi N}\,\omega_c\, Re\, L_{0+}^{\gamma\pi^+}(IS)\sqrt{1-\frac{\omega^2}{\omega_c}}\,. \qquad (10)$$

At threshold, the Q^2 dependence is given mainly by the Born plus vector meson contributions in $v_{\gamma\pi}^B$, as described in Ref. [7]. In Fig. 4 we show our results for the cusp and FSI effects in E_{0+} and L_{0+} multipoles for the π^0 electroproduction at $Q^2 = 0.1$ (GeV/c)^2, along with the results of the multipole analysis from NIKHEF[14] and Mainz[15]. Note that the results of both groups were obtained using the p-wave predictions given by ChPT. We have made a new analysis of the Mainz data[15] for the differential cross sections, using our DMT p-wave multipoles instead. The s-wave multipoles extracted this way are also shown in Fig. 2 (solid circles). Note that the results of our new analysis for the E_{0+} multipole are closer to the NIKHEF data and in better agreement with DMT model prediction. However, the results for the longitudinal L_{0+} multipole stay practically the same as in the previous Mainz analysis. We see that in this case the DMT model prediction is in better agreement with the NIKHEF data.

In Fig. 5, DMT model predictions (dashed curves) are compared with the Mainz experimental data for the unpolarized cross sections $d\sigma/d\Omega = d\sigma_T/d\Omega + \epsilon\, d\sigma_L/d\Omega$, and longitudinal-transverse cross section $d\sigma_{TL}/d\Omega$. Overall, the agreement is good. If the L_{0+} multipole in the DMT model is replaced with that extracted from the Mainz data, then the agreement with the Mainz data is further improved as given by the solid curves in Fig. 5.

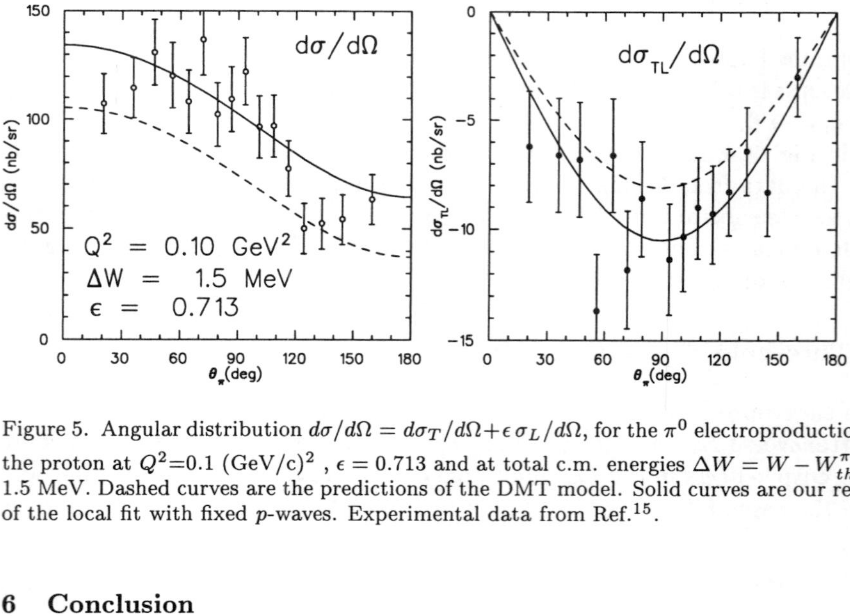

Figure 5. Angular distribution $d\sigma/d\Omega = d\sigma_T/d\Omega + \epsilon\, \sigma_L/d\Omega$, for the π^0 electroproduction on the proton at Q^2=0.1 (GeV/c)^2 , $\epsilon = 0.713$ and at total c.m. energies $\Delta W = W - W_{thr}^{\pi^0 p} =$ 1.5 MeV. Dashed curves are the predictions of the DMT model. Solid curves are our results of the local fit with fixed p-waves. Experimental data from Ref.[15].

6 Conclusion

We have studied the $\gamma^* N\Delta$ transition form factors and threshold π^0 photo- and electroproduction with the new version of MAID and a dynamical model. By re-analyzing recent Jlab data on $p(e,e'p)\pi^0$ at $Q^2 = 2.8$ and $4.0\,(GeV/c)^2$, we find that $A_{3/2}$ is still as large as $A_{1/2}$ at $Q^2 = 4$ (GeV/c)^2, which implies that hadronic helicity conservation is not yet observed in this region of Q^2. Accordingly, our extracted values for R_{EM} are still far from the pQCD predicted value of +100%. However, in contrast to previous results we find that R_{EM}, starting from a small and negative value at the real photon point, actually exhibits a clear tendency to cross zero and change sign as Q^2 increases, while the absolute value of R_{SM} is strongly increasing. With regard to scaling, our analysis indicates that $A_{1/2}$ and $S_{1/2}$, but not $A_{3/2}$, start exhibiting the pQCD scaling behavior at about $Q^2 \geq 2.5(GeV/c)^2$. It appears likely that the onset of scaling might take place at a lower momentum transfer than of hadron helicity conservation. For π^0 photo- and electroproduction near threshold, results obtained within the dynamical model by use of a meson-exchange πN model for the final state interaction, are in as good agreement with the data as ChPT.

References

1. R. Beck *et al., Phys. Rev. Lett.* **78**, 606 (1997); G. Blanpied *et al., Phys. Rev. Lett.* **79**, 4337 (1997).
2. S.J. Brodsky and G.P. Lepage, *Phys. Rev.* D **23**, 1152 (1981); *ibid.* **24**, 2848 (1981).
3. C.E. Carlson, *Phys. Rev.* D **34**, 2704 (1986); C.E. Carlson and N.C. Mukhopadhyay, *Phys. Rev. Lett.* **81**, 2446 (1998).
4. V. Bernard, J. Gasser, N. Kaiser, and U.-G Meissner, *Phys. Lett.* B **268**, 291 (1991); *Z. Phys.* C **70**, 483 (1996).
5. S. N. Yang, *Phys. Rev.* C **40**, 1810 (1989); C. Lee, S. N. Yang, and T.-S. H. Lee, *J. Phys. G: Nucl. Part. Phys.* **17**, L131 (1991).
6. S. Nozawa, T.-S.H. Lee, and B. Blankleider, *Phys. Rev.* C **41**, 213 (1990).
7. D. Drechsel, O. Hanstein, S.S. Kamalov, and L. Tiator, *Nucl. Phys.* A **645**, 145 (1999).
8. S. S. Kamalov and S. N. Yang, *Phys. Rev. Lett.* **83**, 4494 (1999).
9. V.V. Frolov *et al., Phys. Rev. Lett.* **82**, 45 (1999).
10. C. T. Hung, S.N. Yang, and T.-S.H. Lee, *J. Phys.* G **20**, 1531 (1994); to appear in *Phys. Rev.* C, 2001.
11. J. M. Laget, *Phys. Rep.* **69**, 1 (1981).
12. M. Fuchs *et al., Phys. Lett.* B **368**, 20 (1996).
13. J. C. Bergstrom *et al., Phys. Rev.* C **53**, R1052 (1996); *ibid.* 1509 (1997).
14. H.B. van den Brink *et al., Phys. Rev. Lett.* **74**, 3561 (1995); *Nucl. Phys.* A **612**, 391 (1997).
15. M. O. Distler *at al., Phys. Rev. Lett.* **80**, 2294 (1998).

Toru Sato

Shin-Nan Yang

TWO PION PRODUCTION IN THE JÜLICH MODEL

S. KREWALD, S. SCHNEIDER, J. SPETH
Institut für Kernphysik, Forschungszentrum, D-52425 Jülich
E-mail: s.krewald@fz-juelich.de

Two-pion production off the nucleon near threshold is investigated in a meson-exchange model.

1 Introduction

The study of the one- and two-meson decays of excited baryons is one of the challenges in N^*−physics. We are faced with the problem how to separate the resonance decay from the underlying physical background due to final state interactions. One possible approach is to combine both resonances and final state interactions between the hadrons in a theoretical model. This has been done in the Jülich meson-exchange model[1] by combining the pion-nucleon channel with other reactions channels which can decay into two-pion nucleon final states. In the energy range up to 2 GeV, the πN,ηN,σN,ρN, and $\pi\Delta$ reaction channels are considered explicitly. Here, the σ-meson is an abbreviation for a correlated two-pion state. The widths of the σ and the Δ are considered in the intermediate two-body propagators by working out the corresponding self-energies. Details can be found in Ref.[1].

The model describes both the πN phase shifts and inelasticities up to 2 GeV. The inelasticities are quite sensitive to rescattering effects. In the case of the Roper resonance $N^*(1440)$, the final state interactions are so strong that the Jülich model does not need an explicit pole diagram. In order to test this feature of the Jülich meson-exchange model, predictions for the two-meson decay of the Roper, and ultimately, the electro-excitation of the Roper have to be worked out. This is a technically quite involved project within the framework of the Jülich model. Therefore it makes sense to simplify the model.

2 The Model

The πN$\rightarrow$ $\pi\pi$N reactions have been reproduced quite succesfully in the vicinity of the threshold by chiral perturbation theory. It turned out that a lowest order calculation is insufficient[2], but the inclusion of higher order terms provides a good fit to the data, even at tree-level[3,4]. The low-energy constants

c_i introduced in chiral perturbation theory have been found to be largely saturated by hadronic resonances[5,3]. For our purpose, it therefore makes sense to replace the processes involving low energy constants by processes involving explicit mesons and baryons. In a first step, we perform a tree-level calculation similar to a calculation by Oset and collaborators[6]. In a second step, one has to unitarize the tree-level model in order to be able to handle larger energies. In the present communication, we limit ourselves to the first step in order to check whether the approach is viable.

A first class of tree-level diagrams considered in the present approach is shown in Fig. 1. Diagrams (1.1) till (1.3) are identical with the corresponding diagrams employed in the lowest order chiral perturbation theory. Since in the Jülich meson-exchange model the pion-nucleon form factors depend on the momentum of the pion only, these diagrams also coincide with the corresponding tree-level diagrams of Ref.[1].

Diagrams (2.1) till (2.4) show tree-level processes with explicit rho-meson propagators. In diagram (3.2), a rho-meson is exchanged between two pions. In the meson-meson scattering model developped in Jülich, the t-channel exchange of a rho between two pions is strong and generates most of the attraction in the scalar-isoscalar two-pion channel which, when iterated, leads to a quantitative description of the δ_0^0 two-pion phase shift which sometimes is interpreted as the "sigma-meson". In order to include this strong rescattering effect, we replace diagram (3.2) by a diagram like (3.1) with a σ-meson instead of the ρ-meson. For consistency, we also consider the σ-meson in processes like (2.1) till (2.4). In this way, we include the most important meson-meson final state interactions. Non-resonant meson-meson final state interactions and unitarity corrections due to three-particle intermediate states are neglected so far. We also include diagrams with internal nucleons replaced by Δ's or the Roper-resonance. The Roper resonance is parameterized in the present model by a simple Feynman pole diagram; the corresponding masses and coupling constants are chosen such as to reproduce the corresponding T-matrices of our full model, Ref.[1].

3 Results

The total cross sections obtained for the total cross sections of the five reactions $\pi^+p \to \pi^+\pi^+n$, $\pi^+p \to \pi^+\pi^0p$, $\pi^-p \to \pi^0\pi^-p$, $\pi^-p \to \pi^+\pi^-n$, $\pi^-p \to \pi^0\pi^0n$ are shown as functions of the kinetic energy of the incoming pion in Fig. 2.

Elementary isospin considerations suggest that the Δ-resonance should play a dominant role in the reaction $\pi^+p \to \pi^+\pi^+n$ and $\pi^+p \to \pi^+\pi^0p$. This

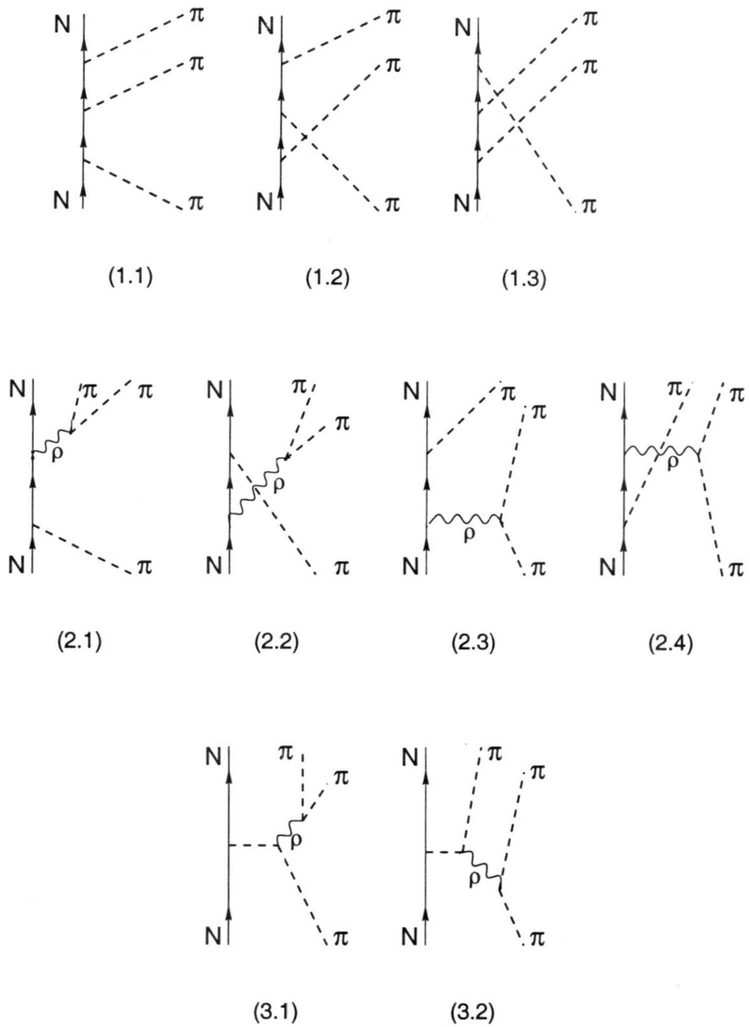

Figure 1. Some tree-level diagrams considered for the $\pi N \to \pi\pi N$ reaction.

is corroborated by our calculation (see the dashed-dotted line). For these reactions, the model works quantitatively close to threshold, but starts to overestimate the cross section for kinetic energies of the incident pion larger than 0.25 GeV. Clearly, one has to unitarize the simple tree-level model.

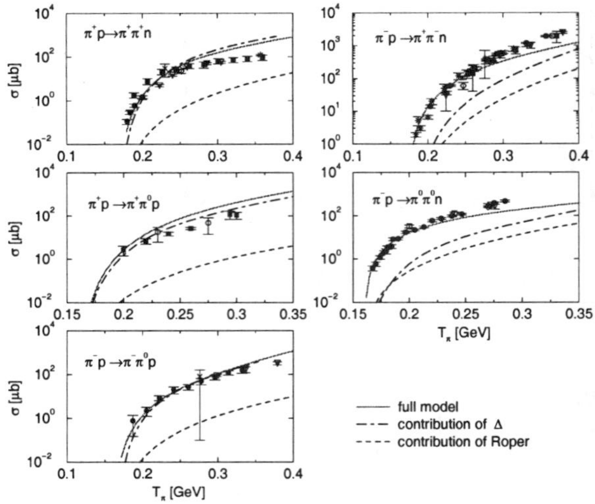

Figure 2. Total cross sections for the reactions $\pi N \to \pi\pi N$ near threshold are shown as functions of the kinetic energy of the incoming pion in the LAB frame. The solid lines refer to the full model. The contributions of diagrams with intermediate Δ's are given by the dashed-dotted line, while diagrams with an intermediate Roper resonance are represented by the dashed lines.

For the reactions $\pi^-p \to \pi^+\pi^-n$ and $\pi^-p \to \pi^0\pi^0n$, intermediate baryon resonances do not play an important role in the vicinity of the threshold (see the dashed-dotted and dashed lines), but meson-meson resonances do. We have checked that diagram (3.1) of Fig. 1 and the corresponding one with the intermediate ρ replaced by the σ play a dominant role in these reactions. In the vicinity of the threshold, the contributions of the Roper resonance are negligible.

In Fig. 3, we compare the differential cross sections for the reactions $\pi^+p \to \pi^+\pi^+n$ and $\pi^-p \to \pi^+\pi^-n$ with the data from Ref. [7].

The t-distributions of both reactions are reproduced satisfactorily, the absolute heights of the cross sections are overestimated with increasing energy, however. We expect that a proper unitarization will reduce the strong rise of the cross section with energy. The t-distribution for the reaction $\pi^-p \to \pi^+\pi^-n$ follows the phase space very closely. This has to be expected because the dominant meson-meson correlation is in the relative S-state. A strong deviation from the phase space distribution is found for the reaction $\pi^+p \to \pi^+\pi^+n$. Our model is able to reproduce the shape of the distribution which

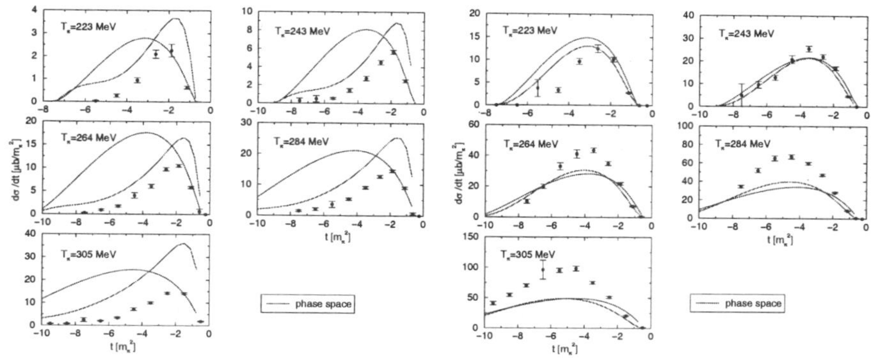

Figure 3. Differential cross section for the reaction $\pi^+p \rightarrow \pi^+\pi^+n$ (left panel) and $\pi^-p \rightarrow \pi^+\pi^-n$ (right panel) as a function of the square t of the momentum transfer between the proton and the outgoing neutron. The solid lines refers to the full model, the dotted line represents the phase space.

suggests that the explicit consideration of resonances (the Δ) represents the most important physical processes.

The angular correlation function W is defined as the following ratio between triple and double differential cross sections:

$$W(\Theta_1, \Phi_1) = 4\pi \frac{\frac{d^3\sigma}{d\Omega_1 d\Omega_2 dT_2}}{\frac{d^2\sigma}{d\Omega_2 dT_2}}. \tag{1}$$

In Ref. [8], experimental angular correlation functions are shown for the reaction $\pi^-p \rightarrow \pi^+\pi^-n$ at a center-of-mass energy $\sqrt{s} = 1.301 GeV$. These data are compared with our model in Fig. 4.

The explicit inclusion of meson resonances in our model allows to understand the shape of the distributions qualitatively: for a fixed angle Θ_{π^-}, the angular correlation is maximal for the angle $\Phi_{\pi^-} = 180^O$. The coordinate system is chosen such that $\Phi_{\pi^+} = 0^O$. This means that the two produced pions tend to separate predominantly back-to-back in the $xy-$plane. We can interprete this finding as follows. Close to threshold, the intermediate meson-resonances ρ and σ are almost at rest in the center-of-mass frame. Therefore the produced pions are indeed expected to be emitted back-to-back. We can test this interpretation by simply including only those diagrams with intermediate meson resonances. The contribution of only these diagrams indeed reproduces most of the experimental cross section. There is some deviation between the dashed line and the full model because the data of Ref. [8] were

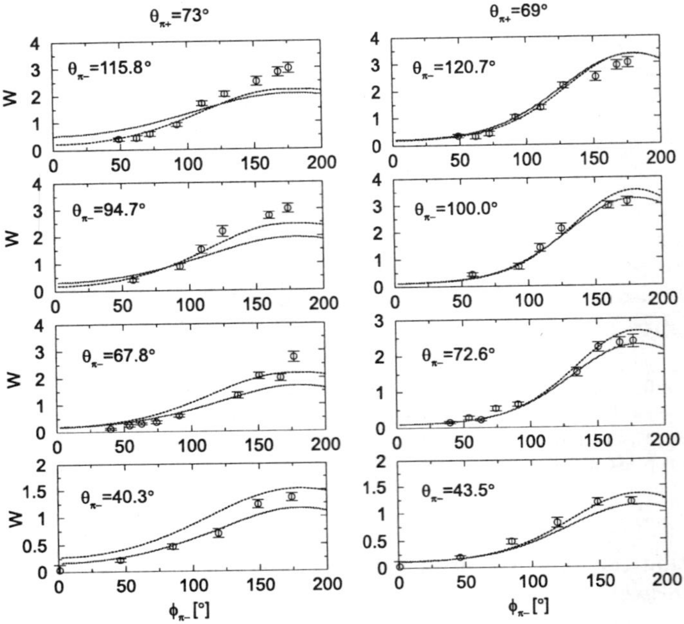

Figure 4. Angular correlation functions for the reaction $\pi^- p \to \pi^+\pi^- n$ are shown as a function of the angle Φ_{π^-} for different angles Θ_{π^-}. The full model is represented by the solid line, while the dashed line includes only those diagrams with intermediate meson resonances.

taken for a kinetic energy of the incident pion of $T_{\pi^-} = 0.284 GeV$ which is large enough to excite the Δ noticeably.

In chiral perturbation theory, the lowest order calculation has to be supplemented by higher order terms in the Langrangian proportional to the so-called low energy constants. In Tab. 1, we compare the values of the low energy constants employed in two recent chiral perturbation calculations with the values determined from the present model. A reasonable agreement is found.

4 Summary

In the present communication, we have presented a simple tree-level model for two-pion production. The model includes intermediate ρ- and σ-mesons as well as intermediate baryon resonanes (Δ, Roper). The experimental data

Table 1. The low energy constants c_i of chiral pertubation theory and the present model

LEC	Ref[3]	Ref[4]	This work
c_1	-0.64	-1.42	0
c_2	1.78	3.13	1.79
c_3	-3.90	-5.85	-4.46
c_4	2.25	3.50	3.16

close to threshold can be reproduced satisfactorily. The model also produces chiral low energy constants which agree with the ones obtained in chiral calculations. It is perhaps premature to talk about the decay properties of the Roper. The present model suggests a branching ratio of the Roper into two pions and a nucleon which is about a factor three larger than the values given by the particle data group. In the immediate future, we have to concentrate on the unitarization of the present model.

Acknowledgments

We would like to thank T. Barnes and H.P. Morsch for many valuable discussions.

References

1. O. Krehl, C. Hanhart, S. Krewald, and J. Speth, *Phys. Rev.* C **62**, 025207 (2000).
2. J. Beringer, πN *Newsletter* **7**, 33 (1992).
3. V. Bernard, N. Kaiser, and U.-G. Meißner, *Nucl. Phys.* A **619**, 261 (1997).
4. N. Fettes, V. Bernard, and U.-G. Meißner, *Nucl. Phys.* A **669**, 269 (2000).
5. G. Ecker *et al., Nucl. Phys.* B **321**, 311 (1989).
6. E. Oset and M.J. Vicente-Vacas, *Nucl. Phys.* A **446**, 584 (1985).
7. M. Kermani *et al., Phys. Rev.* C **58**, 3419 (1998).
8. R. Mueller *et al., Phys. Rev.* C **48**, 981 (1993).

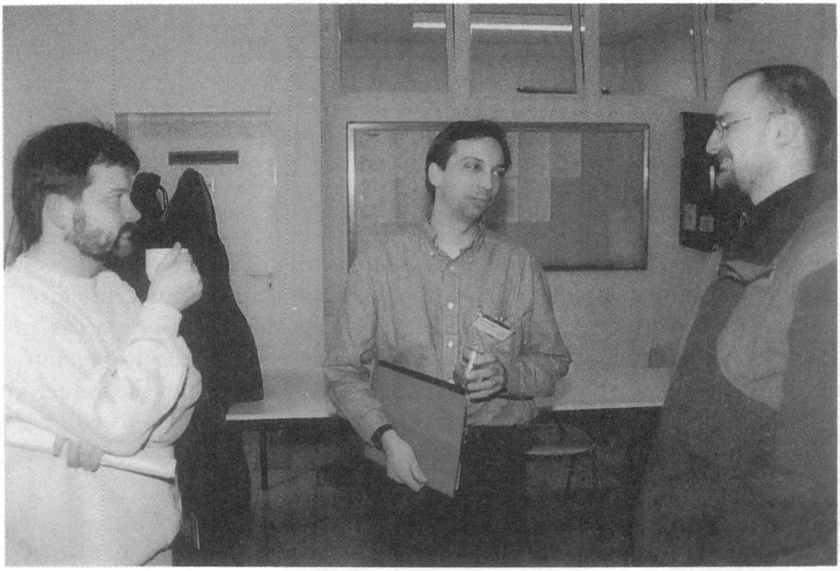

Reinhard Beck, Adam Sarty, David Hornidge

Siegfried Krewald

UPDATE ON PARTIAL-WAVE ANALYSIS

R. WORKMAN
Dept. of Physics, The George Washington University, Washington D.C. 20052
E-mail: rworkman@gwu.edu

Partial-wave analysis is one step in a process connecting experimental measurements to the N* states we are studying. Progress has been made in the area of 'model-independent' analysis. However, more model-dependent approaches are needed to cover broader energy ranges. An example of the problems faced by these more ambitious analyses is given.

1 Overview

The majority of recent partial-wave analyses (PWA), applied to the study of N* physics, have focused on photo- or electroproduction processes[1]. In the following, I will discuss only this topic and further restrict my comments to pseudoscalar-meson photoproduction. I will first mention fits that have focused on the near-threshold region, where a single resonance is dominant, and one can argue for a truncated multipole expansion. Here the goal is a 'model-independent' extraction of amplitudes. Other more ambitious fits have extended over wider energy ranges where model-dependence is inevitable. As a result, in such cases, it becomes difficult to distinguish between models, with fitted parameters, and data-fits (with theoretical constraints). The addition of Born contributions will be discussed as an example illustrating the problems one encounters in fits to the full resonance region.

Before moving on to examples, it is worth repeating that the requirements for an 'amplitude' analysis and a 'multipole' analysis are not equivalent. This has been pointed out in the past[2], but the distinction is rarely noted in the literature. Amplitude analyses have been performed for pion-nucleon and nucleon-nucleon scattering. However, in most[3] reactions of interest, arguments concerning the requirements for a 'complete experiment' are academic. We will be dealing with incomplete experimental information and should concentrate on finding methods which are as model-independent as possible.

In a workshop preceding this meeting, the Baryon Resonance Analysis Group (BRAG) compared the results of seven different fits to a benchmark dataset, with an agreed upon method for handling errors[4]. These fits will be refined and repeated over an expanded dataset (of pion photoproduction data), giving an objective measure of the model-dependence in multipole amplitudes extracted from this reaction.

2 Near-Threshold Studies

Much of the recent work on pseudoscalar-meson photoproduction has focused on the 'first bump' region. In pion photoproduction, this structure is due to the $\Delta(1232)$ resonance. Analyses of π^0 production (both photo and electro-production) have used truncated (S- and P-wave) multipole expansions with simplifications following from the dominance of the $M_{1+}^{3/2}$ contribution. In η photoproduction, one again has a dominant ($E_{0+}^{1/2}$) multipole near threshold and higher waves are playing a minor role[5]. A similar fitting strategy has been applied to η' photoproduction[6], with less conclusive results. While the total cross section displays a near-threshold peak, there is little agreement on the underlying resonance components[7,8].

In π and η photoproduction, interesting results follow from fits including measurements sensitive to the interference of dominant and sub-dominant amplitudes. In both cases, the photon asymmetry (Σ) has played an important role, providing information on the E2/M1 ratio (for π) and the $D_{13}(1520)$ contribution[5] (for η). This observable has also had a significant impact on pion photoproduction at higher energies[9] and the 'second bump' in K^+ photoproduction[10].

These fits have suggested a few new states (in K^+ and η' photoproduction), and have given new determinations of the N(1520) and N(1535) photo-decay amplitudes (in η photoproduction)[5]. Unfortunately, the new determinations disagree with older ones obtained from fits to π photoproduction data. As a result, the status of photo-couplings is probably *less* certain[11] than we would have claimed prior to this renewed experimental effort.

More ambitious analyses, covering broader energy ranges, will be required to convincingly establish the existence (or absence) of the many 'missing resonances' predicted by quark models. Clearly, we must also have *mutually consistent* photo-couplings from fits to the different reactions. This important issue is being studied by a number of groups[12].

3 Fitting the Full Resonance Region

Most photo-couplings have been determined from fits to π photoproduction data. This reaction has been extensively studied, with measurements covering the full resonance region, and at least an order of magnitude more data than exists for the photoproduction of other pseudoscalar mesons. In π photoproduction one also has the advantage of knowing where resonances should occur, given the equally extensive study of elastic $\pi^{\pm}p$ scattering.

Moving to other channels, we must rely either on quark model predictions or much less reliable hadronic information (for example, $\pi N \to \eta N$ and $\pi N \to KY$) for a guide to possible resonance content. The total cross section is of little help, generally having a rapid increase near threshold, and a relatively smooth fall-off at higher energies. As a result, photo-decay amplitudes will be much more difficult to determine (unambiguously) in these reactions.

One method gaining popularity is the use of a multi-channel formalism which incorporates information from many channels simultaneously. This approach requires a single formalism capable of describing all included channels. An ingredient common to each of these processes is the Born term. In the following, I will briefly describe how this contribution behaves and explain why its inclusion is problematic.

3.1 The Born Terms: General Features

The most obvious influence of Born terms in photoproduction is the forward peak seen in the $\pi^+ n$ differential cross section[1]. For example, at a photon energy of 1 GeV, both the data and Born term contribution have a forward peak and a dip near $t = -\mu^2$. However, away from this forward peak, the Born contribution has entirely the wrong shape. In addition, the total cross section from this piece alone exceeds the experimental value just a few hundred MeV above threshold. This divergence of the Born contribution from the experimental total cross sections is also seen in K^+ and η' photoproduction[13,14].

At higher energies (10 GeV), a simple Born approximation continues to give a qualitative description of the forward peak in $\pi^+ n$ photoproduction. This rather unexpected behavior has been explained in terms of finite energy sum rules[15]. Thus we have a Born contribution which is 'correct' for the forward peak, but clearly is problematic for the total cross section. These features, together with the constraints of gauge invariance and proper analytic structure, can be used to assess the model input adopted in fitting data.

3.2 Taming the Born Contribution

Born term plus Breit-Wigner fits are adequate to describe cross section data. However, extrapolation beyond the fitted energy range tends to be very unstable. This problem is linked to the behavior of the Born component described above. One simple way to compensate for the Born contribution, is to multiply by an overall form factor. This approach clearly violates crossing symmetry and reduces the full angular distribution uniformly, including the forward peak which requires no suppression. It can even result in a 'resonancelike' Born contribution to the total cross section[13], as demonstrated in a fit to η' photoproduction.

Some of these problems can be avoided if each strong vertex, in the Born terms, is modified by a form factor depending on the off-shell 4-momentum. In this case a contact term is required to restore gauge invariance. This can be done in a way which preserves crossing symmetry. The effect of these form factor recipes, at the multipole level, is very different. An overall form factor, $F(s)$, having no angular dependence, reduces each multipole by the same amount as s increases. If the recipe of Ref.[16] is used instead, the effect is most important for low partial waves, and becomes negligible as the angular momentum increases. However, using simple and common (real) form factors, we have found that the recipe of Ref.[16] also tends to kill the forward peak seen in $\pi^+ n$. As a result, it is difficult to see how the use of real form factors can be justified, based both on formal arguments and in comparison with general features seen in the data.

3.3 Results from Dispersion Relations

All of the above approaches to the Born contribution implicitly separate the 'resonance' and 'background' contributions. If one examines this problem within the context of dispersion relations, an entirely different picture emerges. In a fixed-t dispersion relation, the real part is given by the Born term plus a weighted integral over imaginary parts. In π photoproduction, it has been shown that the Born term is damped primarily through the inclusion of $M_{1+}^{3/2}$ in the integral[17]. Adding this contribution cancels the non-forward Born contribution and enhances the forward peak[18]. Other resonance contributions appear to play a much smaller role, and this holds far above the energy of the $\Delta(1232)$ resonance.

It is possible to write a Born term with form factors in terms of an unmodified Born term plus a contact term. This contact term must grow with energy to cancel the unmodified Born contribution, and therefore represents a significant contribution to the real part of the invariant amplitude. However, in fits such as the one reported by Crawford at this meeting, the dispersion integral appears to be well approximated by resonance contributions alone (over the resonance region).

3.4 SAID versus MAID

How then are the Born terms handled in the SAID[1] and MAID[19] fits over the resonance region? In both analyses, the Born terms are added (without form factors). In MAID, the full Born contribution is added to resonance contributions. In SAID, the low partial waves are parametrized and the high partial waves are given by a (real) Born contribution.

In Figure 13 of Ref.[19], a cancellation of the unitarized Born and P_{33} multipoles was noted in π^+n photoproduction. We have checked this behavior in SAID and find a very similar result. In fact, this simple model is remarkably successful in reproducing the forward π^+n peak up to the limit of the MAID analysis (1 GeV). We have also found that this model results in a qualitative description of very forward (4°) $\pi^\circ p$ differential cross sections over the same energy region. We did not anticipate this result and haven't found any specific mention of it in the literature.

In SAID and MAID the Born contribution is damped both by the (K-matrix) unitarization and a cancellation with the $P_{33}(1232)$ multipoles. Both contributions are required and important in the observed cancellation.

4 Conclusions and Further Questions

Many recent studies have focused on reactions and kinematic regions with simplifying features. These have generally been near-threshold fits where a large signal is present, the final meson is neutral, and only a few partial-waves are important. If our goal is to find the many missing states predicted by quark models, these exploratory fits must be followed by more complete treatments of the full resonance region.

We have seen that the SAID and MAID fits have features similar to those mentioned long ago in the context of fixed-t dispersion relations. Aspects of the forward peaking and suppression of Born contributions at non-forward angles are largely due to the $\Delta(1232)$, with smaller contributions from other states. These results weigh against approaches which attempt to separate Born (or generic background) and resonance contributions, suppressing the Born contribution through the use of phenomenological form factors.

Whether the photoproduction of other pseudoscalar mesons can be treated in an analogous manner is less clear. High energy K^+ photoproduction has a forward dip, rather than the peak seen in π^+n. [Aside: This forward region was, however, used by Dombey to estimate the $K\Lambda N$ coupling constant[15].] Some cancellation of the nucleon pole term and resonance contributions has been mentioned in the work of Zhao[8] applied to η' photoproduction.

It would be interesting to see how well the various models, used in fitting data (or partial-wave amplitudes), reproduce the forward peak seen in π^+n photoproduction. A second test would involve determining the consistency of each approach with the constraints imposed by fixed-t dispersion relations. Checks of this kind may help to select the most promising approach to pursue.

Acknowledgments

Thanks are due to R.A. Arndt for modifying SAID specifically for the NSTAR and BRAG meetings. Discussions with R.M. Davidson are also gratefully acknowledged. This work was supported in part by the U.S. Department of Energy Grant DE-FG02-99ER41110 and a contract from Jefferson Lab, which is operated by the Southeastern Universities Research Association under the U.S. Department of Energy Contract DE-AC05-84ER40150.

References

1. Many databases and fits associated with the reactions considered in this review are available at the website: http://gwdac.phys.gwu.edu.
2. R. Workman, *Few-Body Systems Suppl.* **11**, 94 (1999); Eq. 5 contains a sign error, but the argument remains intact. See also V.F. Grushin, A.A. Shikanyan, E.M. Leikin, and A. Ya. Rotvain, *Sov. J. Nucl. Phys.* **38**, 881 (1984).
3. Kaon photoproduction experiments proposed at JLab may provide a 'nearly complete' set of observables.
4. The benchmark datasets and fits are available on the BRAG website: http://http://cnr2.kent.edu/ manley/BRAG.html.
5. L. Tiator, D. Drechsel, G. Knöchlein, and C. Bennhold, *Phys. Rev.* C **60**, 035210 (1999).
6. R. Plötzke *et al.*, *Phys. Lett.* B **444**, 555 (1998).
7. J.-F. Zhang, N.C. Mukhopadhyay, and M. Benmerrouche, *Phys. Rev.* C **52**, 1134 (1995); N.C. Mukhopadhyay, J.-F. Zhang, R.M. Davidson, and M. Benmerrouche, *Phys. Lett.* B **410**, 73 (1997); Z.P. Li, *J. Phys* G **23**, 1127 (1997).
8. Q. Zhao, *Phys. Rev.* C **63**, 035205 (2001).
9. F.V. Adamian *et al.*, *Phys. Rev.* C **63** (in press).
10. T. Mart and C. Bennhold, *Phys. Rev.* C **61**, 012201(R) (1999).
11. R. Workman, R.A. Arndt, and I.I. Strakovsky, *Phys. Rev.* C **62**, 048201 (2000).
12. T.P. Vrana, S.A. Dytman, and T.-S.H. Lee, *Phys. Rept.* **328**, 181 (2000); C. Bennhold and S.A. Dytman, *these proceedings*; T. Feuster and U. Mosel, *Phys. Rev.* C **59**, 460 (1999); D.M. Manley, private communications.
13. R.M. Davidson and R. Workman, *Phys. Rev.* C **63**, 025210 (2001).
14. H. Tanabe, M. Kohno, and C. Bennhold, *Phys. Rev.* C **39**, 741 (1989). This paper considers the effect of final-state interactions in reducing the

effective Born contribution.
15. See contributions of N. Dombey and H. Harari to *Hadronic interactions of electrons and photons*, eds. J. Cummings and H. Osborn (Academic Press, London, 1971). Given the current interest in local duality, it is worth noting that Harari suggested this as a possible mechanism to explain why N^* states contribute very little to the FESR relevant to the forward peak in π photoproduction.
16. R.M. Davidson and R. Workman, *Phys. Rev.* C, in press.
17. J. Engels, G. Schwiderski, and W. Schmidt, *Phys. Rev.* **166**, 1343 (1968).
18. I. Barbour, W. Malone, and R.G. Moorhouse, *Phys. Rev.* D **4**, 1521 (1971). This reference claims that the $\Delta(1232)$ contribution plays a special role, at forward angles, within a simple quark model picture.
19. D. Drechsel, O. Hanstein, S.S. Kamalov, and L. Tiator, *Nucl. Phys.* A **645**, 145 (1999).

Ron Workman

Thomas Hemmert

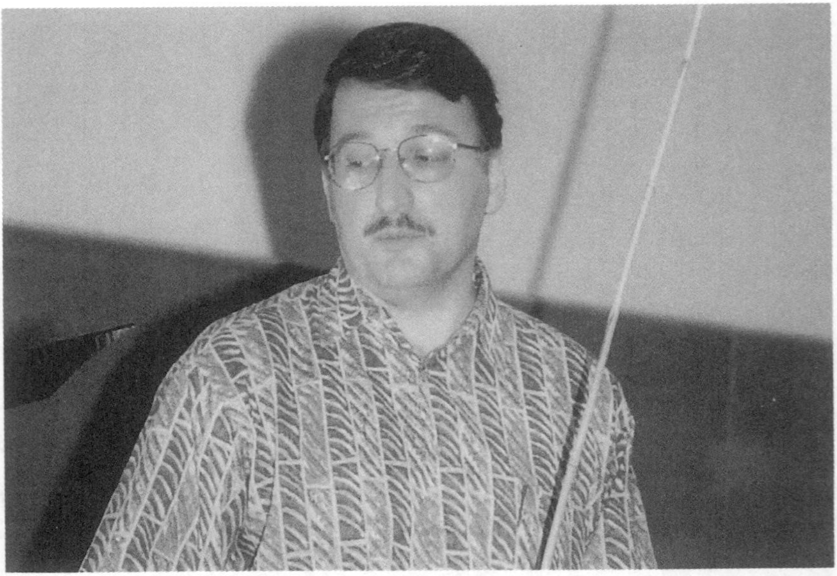

Cornelius Bennhold

NUCLEON RESONANCE PROPERTIES IN MULTICHANNEL APPROACHES

C. BENNHOLD, A. WALUYO, H. HABERZETTL, AND F.X. LEE
Center for Nuclear Studies, Department of Physics, The George Washington University, Washington, D.C. 20052

S.A. DYTMAN, J. GREENWALD
Department of Physics and Astronomy, University of Pittsburgh, Pittsburgh, PA 15260

D.M. MANLEY
Department of Physics, Kent State University, Kent, OH 44242

T. MART
Jurusan Fisika, FMIPA, Universitas Indonesia, Depok 16424, Indonesia

G. PENNER AND U. MOSEL
Institut für Theoretische Physik, Universität Giessen, D-35392 Giessen, Germany

Nucleon resonance properties are compared in different multichannel analyses. The approaches incorporate unitarity and as a minimum include the πN, $\pi\pi N$, ηN and γN asymptotic states. They differ in their descriptions of the background, resonance parameterizations and degree of incorporating theoretical constraints and symmetries, such as analyticity, gauge invariance and chiral symmetry. We find that in most cases the properties of the first resonance in each partial wave are in fair agreement with each other. For each second resonance in a given partial wave the extracted masses may be similar but its properties vary widely. Any evidence for a third resonance in a particular partial wave is tenuous.

1 Introduction

In contrast to atomic physics the excited states of the nucleon cannot simply be inferred from cleanly separated spectral lines. Rather, a "spectral analysis" in nucleon resonance (N^*) physics is complicated by the fact that the resonances are broadly overlapping states which decay into a multitude of final states involving mesons and baryons. For this purpose, a number of laboratories like MAMI, ELSA, BATES, GRAAL and TJNAF have developed extensive experimental programs to address the issue of N^* physics, and have begun to deliver new experimental data with unprecedented accuracy. In order to provide a link between the new and improved data on one side and the results from lattice QCD, QCD sum rules and QCD-inspired quark models on the other side, dynamical descriptions using hadronic degrees of freedom are required that can analyze the data in the various asymptotic reaction channels, like $\gamma N, \pi N, \pi\Delta, \rho N, \eta N, K\Lambda$, and others.

2 Coupled-channels Analyses

In order to provide a consistent and complete picture of an individual nucleon resonance, the various possible production and decay channels must be treated in a multichannel framework that permits separating resonance from background contributions. In this paper, we compare the results of three such analyses from the GW-Giessen collaboration[1,2], the Pitt-ANL collaboration[3] and the Kent State University group[4]. To our knowledge, these are the only presently ongoing multichannel studies that include as a minimum the πN, $\pi\pi N$, ηN and γN asymptotic states. All three studies incorporate unitarity and allow the separation of background from resonance contributions. Although model dependent and subject to conceptual uncertainties, these resonance couplings can be compared to results from quark models, QCD sum rule calculations and QCD lattice results. We emphasize that the comparison is not rigorous at this time since the fits were not performed with the identical database. The conclusions drawn in this paper should therefore be regarded as preliminary.

2.1 The GW-Giessen model

The GW-Giessen approach solves a Bethe-Salpeter equation in the K-matrix Born approximation. The driving terms consist of s-, u- and t-channel Feynman diagrams defined through effective Lagrangians. The K-matrix approximation allows the resonance widths to be generated dynamically, while the principal-value part of the meson-baryon propagator is neglected, thus analyticity is not maintained. Because the full Lorentz and Dirac spin structure of the resonance vertices and propagators is included, the model is presently limited to resonances with spin 1/2 and 3/2; the extension to spin 5/2 states is in progress. The field-theoretical basis of the model allows straightforward implementation of gauge invariance and chiral symmetry constraints. Its asymptotic states are $\gamma N, \pi N, \pi\pi N, \eta N, K\Lambda, K\Sigma$ and $\eta' N$; the $\pi\pi N$ channel is approximated by a fictitious isovector, scalar meson with the mass of two pions. Because the background is given by a small number of physical diagrams all partial waves must be fitted simultaneously. This very feature limits the range of applicability to $W = 2.0$ GeV at the present time.

2.2 The Pitt-ANL model

The Pitt-ANL fitting results come from an updated and extended version of the CMB model[5]. The basic model provides a unitary, analytic representation of s-channel resonances that couple to an arbitrary number of asymptotic states; t-channel diagrams must be added by hand. In the Pitt-ANL implementation, photoproduction of pions and etas is included with both resonant and Born diagrams, though gauge invariance is not assured. The main strength of the model is the good description of the features of the analytic plane, particularly the inelastic threshold openings. Each resonance is a bare pole which becomes a pole in the complex plane through inelastic processes (coupling to open channels). Its asymptotic states are $\gamma N, \pi N, \eta N, \rho N, \pi\Delta, \sigma N$ and $\pi N^*(1440)$. The main approximations are the incomplete representation of the left-hand cuts and the quasi-two-body treatment of 3-body final states. Since the background is parametrized by distant unphysical poles specific to each partial wave, separate fits are performed for each partial wave.

2.3 The KSU model

The Kent State approach is a generalization of the multichannel Breit-Wigner representation to include an arbitrary number of resonances plus nonresonant background. The method treats background empirically by using a product S-matrix so that the total S matrix is given by $S = B^T RB$, where B is a unitary background matrix (not generally symmetric), R is the [symmetric] S-matrix in the absence of all background, and B^T is the transpose of B. Background couplings having the proper threshold behavior (analogous to partial widths) may be introduced for each channel, providing a great deal of flexibility. Resonance poles can be extracted but dispersion relations are not incorporated explicitly; gauge invariance is not assured in this approach. The Kent State method has been successfully applied to the simultaneous description of πN elastic scattering, pion photoproduction, and various quasi-two-body $\pi\pi N$ channels, such as $\rho N, \pi\Delta, \sigma N$ and $\pi N^*(1440)$. Additional channels such as ηN and $K\Lambda$ can be included when necessary to satisfy unitarity.

3 Results and Discussion

In general, the fits performed by the Pitt-ANL and KSU groups achieve an excellent description of the experimental phase shifts and multipoles, reflecting the flexibility of their phenomenological background with separate parameter sets for each partial wave. The GW-Giessen fits are very good for most partial waves but do encounter significant deviations from experiment for a few partial waves in selected energy regimes. This can be traced to the nonresonant terms that define the background through a small set a Feynman diagrams with few parameters, thus providing less flexibility. The Pitt-ANL analysis extracts properties of 27 N^* states, the KSU includes 24 and the GW-Giessen effort analyzes 16 N^* resonances. Here we point out some general patterns identified and discuss a few specific partial waves. The N^* states naturally sort themselves into separate categories when discussed in terms of the first, second and third N^* *in each partial wave.* Most properties of most first-tier resonances, such as their mass, total width, partial widths and helicity amplitudes were found to be in agreement with each other at the 20-30% level. In the I=1/2 sector this includes, for example, the $P_{11}(1440)$, $D_{13}(1520)$, $D_{15}(1675)$, and $F_{15}(1680)$ states; in the I=3/2 sector one finds, e.g., the $P_{33}(1232)$, $D_{33}(1700)$, $P_{31}(1910)$ and $F_{37}(1950)$ states (note that for spin 5/2 and 7/2 states only the Pitt-ANL and KSU results can be compared with each other). In most cases, these first-tier (virtually all of them 4-star according to the PDG rating system) states are fairly isolated, cleanly separated resonances with typical Breit-Wigner behavior and small background; exceptions are the $S_{31}(1620)$ and $P_{13}(1720)$ states. For second-tier states, rough agreement is found only for the resonance mass, while the extracted total and partial widths vary widely in most cases. Examples are the $P_{33}(1600)$, $P_{11}(1710)$, $D_{13}(1700)$ and $P_{13}(1900)$. Finally, the very existence of third-tier states in each partial wave is tenuous at best. Since for most second- and third-tier resonances the coupling to the elastic πN channel is weak, they require significant input from inelastic channels. For second-tier states, available inelastic data are of poor quality and for third-tier states, the data are almost nonexistent.

3.1 Resonances in the S_{11}, D_{13} and P_{13} partial waves

The S_{11} partial wave have long been recognized as an unusual partial wave, with two strongly interfering resonances in close proximity to each other and to the ηN threshold. Since any multi-resonance, coupled-channels framework naturally includes these features, the three analyses are in fair agreement with each other for most of the $S_{11}(1535)$ properties, listed in Table 1. Exceptions are the total width where the GW-Giessen result around 250 MeV is considerably larger (closer to single-channel extractions[6]), and the KSU $A^p_{1/2}$ helicity amplitude which with 0.042 $\text{GeV}^{-1/2}$ is much smaller compared to the results of the other two analyses. The latter is due to the fact that KSU has not yet included the eta photoproduction data in their fit. It is known that analyses based solely on pion photoproduction consistently underestimate this quantity

Table 1. Selected nucleon resonance parameters for the S_{11} partial wave. The first (second, third) line shows the results from GW-Giessen (KSU, Pitt-ANL). For the first set, estimated ranges are given representing systematic model uncertainties, while the latter two sets show standard statistical uncertainties. GW-Giessen did not find a third S_{11} state. (^{a}) *obtained without ηN production data in the fit.*

	$S_{11}(1535)$	$S_{11}(1650)$	$S_{11}(2090)$
M (MeV)	1542-1556 1528(1) 1542(3)	1679-1701 1649(1) 1689(12)	- 2000(1) 1822(43)
Γ_{tot} (MeV)	213-252 111(9) 112(19)	194-219 108(14) 202(40)	- 132(16) 248(185)
Γ_π (%)	29-31 41(3) 35(4)	63-77 32(5) 74(2)	- 10(3) 17(7)
Γ_η (%)	60-66 $50(4)^a$ 51(5)	10-12 $5(3)^a$ 6(1)	- $3(2)^a$ 41(4)
$A^p_{1/2}$ (10^{-3} GeV$^{-1/2}$)	101-111 $42(6)^a$ 87(3)	35-46 $17(7)^a$ 48(9)	- $39(17)^a$ -

(around 0.060 GeV$^{-1/2}$ or lower) while extractions which include the (γ, η) data place the value closer to 0.10 GeV$^{-1/2}$. The results are less stable for the $S_{11}(1650)$; here the KSU results for Γ_{tot}, Γ_π, and $A^p_{1/2}$ are considerably lower than the ranges found by GW-Giessen and Pitt-ANL. The new (γ, η) GRAAL data at higher energy[7,8], included in the GW-Giessen analysis, suggest a Γ_η of 10-12%, higher than previously quoted. Finally, there is clearly no agreement with regard to the existence and location of a third S_{11} state. Figure 1 shows that all three analyses produce surprisingly similar backgrounds for the E_{0+}, despite the rather different model inputs. For the imaginary part the KSU background is larger in the region of the second S_{11}, which explains the lower KSU values for the $S_{11}(1650)$.

The $D_{13}(1520)$ state is a first-tier resonance with a clear signal in both πN phase and (γ, π) multipoles, as shown in Fig. 1. The three analyses agree fairly well on its hadronic properties (see Table 2), whereas for both helicity couplings the GW-Giessen values lie significantly below the results from the two other groups. We point out that recent single-channel (γ, η) fits[9] find a much larger (negative) value of $A^p_{1/2} = -0.078$ GeV$^{-1/2}$ when the new GRAAL data are included. The second D_{13} decouples almost totally from the πN channel, as found in all three analyses. For this state, the $\pi\Delta$ channel represents the dominant decay mode. All three groups find some evidence for a third D_{13} state around 1900-2000 MeV with a very large total width. This state has recently been implicated[10] to play a major role in the reaction $\gamma p \to K^+\Lambda$, where new cross-section data revealed a structure around 1900 MeV. Displayed in Fig. 1, the M_{2-} background terms are very similar below 1500 MeV but significant differences appear in the real part at higher energies between GW-Giessen and KSU results compared to the Pitt-ANL results.

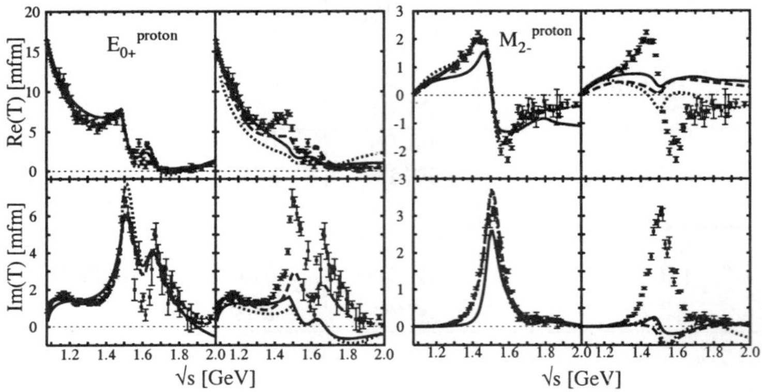

Figure 1. The pion photoproduction E_{0+} and M_{2-} proton multipoles. The solid (dashed, dotted) curves show the results of the GW-Giessen (KSU, Pitt-ANL) analysis. For each multipole the left-hand panels show the full calculation while the right-hand panels show the background contributions only.

As an exception we discuss the $P_{13}(1720)$, designated as a 3-star resonance by the PDG. Figure 2 displays this partial wave with an unusual background and a weak resonance signal. Large differences appear between the three analyses, but no clear pattern emerges. For the real part of the M^{proton}_{1+}(I=1/2) the GW-Giessen and KSU backgrounds closely follow the data, quite different from the Pitt-ANL background. On the other hand, for the real part of the E^{proton}_{1+}(I=1/2) the KSU and Pitt-ANL results tend to agree with each other while the GW-Giessen result follows the data. The imaginary parts of both multipoles display large differences between all three models, especially at higher energies. These variations are reflected in the wide range of values extracted for the $P_{13}(1720)$ properties, listed in Table 2. This level of instability for the fitting results resembles the behavior of second-tier, rather

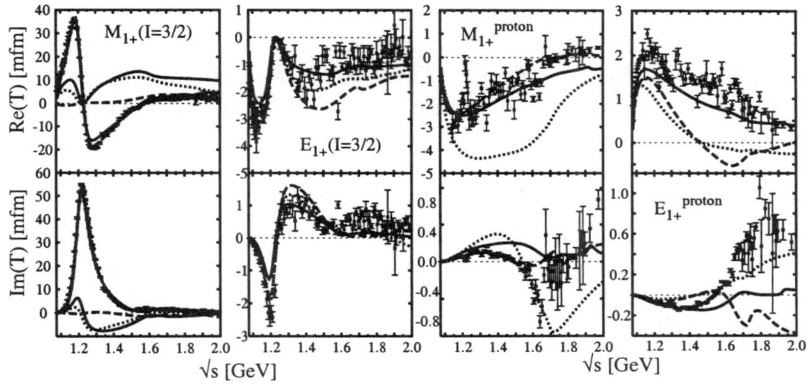

Figure 2. The pion photoproduction M_{1+} and E_{1+} multipoles. Notation of the curves as in Fig. 1. Background contributions are shown for I=3/2 (left two plots) and I=1/2 (right two plots). The full calculations are also shown for the M_{1+} (I=3/2) multipole.

Table 2. Selected nucleon resonance parameters for the D_{13} and P_{13} partial wave. The first (second, third) line shows the results from GW-Giessen (KSU, Pitt-ANL). The helicity amplitudes are given in units of 10^{-3} GeV$^{-1/2}$. Pitt-ANL did not find a second P_{13} state. (^{b}) *preliminary results.*

	$D_{13}(1520)$	$D_{13}(1700)$	$D_{13}(2080)$	$P_{13}(1720)$	$P_{13}(1900)$
M (MeV)	1506-1512 1521(1) 1518(3)	1690 1662(1) 1736(33)	1912 2003(1) 2003(18)	1771-1809 1720(18) 1716(112)	1987 1919(26) -
Γ_{tot} (MeV)	84-110 120(4) 124(4)	202 509(30)^b 175(133)	598 698(35) 1070(858)	344-637 246(52) 121(39)	531 569(127) -
Γ_π (%)	53-59 56(2) 63(2)	<1 <1^b 4(2)	8 26(2) 13(3)	20-22 17(1) 5(5)	10 16(3) -
$A^p_{1/2}$	(−9)-(+2) −18(5) −25(6)	3 −121(45)^b −47(9)	−15 −20(12) −6(33)	23-36 70(11) 50(16)	10 122(21) -
$A^p_{3/2}$	134-152 207(13) 171(7)	42 −458(89)^b 74(7)	7 −99(22) 174(36)	23-75 −45(10) 17(26)	43 −75(19) -

than first-tier, resonances. This impression is supported when one examines the $P_{13}(1900)$, which behaves like a third-tier state: not even its existence is confirmed in all three analyses. These exceptions to the "first-tier rule" clearly demand a deeper understanding of the differences between the various background models. To emphasize this point, Fig. 2 also displays the differences between backgrounds for the extensively studied partial waves M_{1+} and E_{1+} with I=3/2. Especially for the M_{1+} around W=1232 MeV KSU finds a very small brackground in contrast to GW-Giessen and Pitt-ANL. Since the $P_{33}(1232)$ has the "luxury" of Watson's theorem as a way to enforce unitarity, all backgrounds vanish at the resonance position; this is clearly not the case for the higher-lying P_{33} states.

4 Conclusions and Outlook

We conclude by posing a number of questions that future refinements of these models and their comparison with each other will have to address:

- *How many N^* states do we need to identify?* The answer clearly depends on the partial wave and on the theoretical description being used. On average, we would suggest that three N^* per partial wave should be sufficient, with lower partial waves like the S_{11} and P_{11} possibly requiring 4-5 states in order to settle the question of the "missing resonances", while only two resonances may need to be studied for higher partial waves.

- *What form of resonance vertices and propagators are needed for high-spin resonances?* For N^* states with spin greater than 1/2, do we need the full Lorentz and Dirac spin structure as demanded by field theory or do we "get away" with simple Breit-Wigner parametrizations? Indications from first-tier states with spin 3/2 like the $P_{33}(1232)$ and $D_{13}(1520)$ are that one extracts similar properties either way. However, it is not clear yet if differences for some of the higher-lying states are due to this feature. Should the full Dirac structure be required, this would complicate the extraction of the properties of states with spin 5/2 and higher.

- *What is the proper background description?* Since the extraction of resonance properties is inexorably linked to the separation of background from resonance terms this question has an immediate impact. The concern here is that in the case of a very weak signal from a broad resonance superposed on a large, complicated background, too flexible a parametrization might lead to extracting wrong N^* properties or even erase the resonance signal altogether. If this problem were valid one would have expected to find large differences between the field-theoretical background from GW-Giessen on the one side, and the empirical backgrounds of the KSU and Pitt-ANL groups on the other side. What was found in this study, however, are rather similar background contributions in many cases, with some notable exceptions that did not follow any clear pattern.

Future studies will focus in more detail on the question of data dependency by using a uniform data base. The model dependencies to resonance and background parametrizations within and across different analysis methods will also be studied more closely; this issue has already been looked at by Pitt-ANL[3] for the S_{11} channel. We expect that such efforts will lead to a solid understanding of first-tier resonances, using exceptions like the $P_{13}(1720)$ to provide a clearer picture of the background. These findings, together with an improved data base for the inelastic channels, should help pave the way for a deeper understanding of the second- and third-tier N^* states as well.

Acknowledgments

This work has been supported by US-DOE grant DE-FG02-95ER-40907 (CB, AW, HH, FXL), NSF grant PHY-9901529 (SAD), NSF grant PHY-0072240 (DMM), and the BMBF, DFG and GSI Darmstadt in Germany (GP, UM).

References

1. T. Feuster and U. Mosel, Phys. Rev. C **58**, 457 (1998); **59**, 460 (1999).
2. C. Bennhold *et al.*, nucl-th/9901066, nucl-th/0008024, nucl-th/0008023.
3. T. P. Vrana, S. A. Dytman, and T.-S. H. Lee, Phys. Rep. **328**, 181 (2000); and to be published.
4. D.M. Manley, Few-Body Systems Suppl. **11**, 104 (1999); and to be published.
5. R. E. Cutkosky *et al.*, Phys. Rev. D **20**, 2839 (1979).
6. B. Krusche *et al.*, Phys. Lett. **B397**, 171 (1997).
7. F. Renard *et al.*, hep-ex/0011098 and A. D'Angelo, these Proceedings.
8. A. Waluyo *et al.*, these Proceedings.
9. L. Tiator *et al.*, Phys. Rev. C **60**, 035210 (1999).
10. T. Mart and C. Bennhold, Phys. Rev. C **61**, 012201(R) (1999).

Angels Ramos

Anthony Thomas

VARIATION OF HADRON STRUCTURE WITH QUARK MASS

A. W. THOMAS

CSSM, Adelaide University, Adelaide SA 5005, AUSTRALIA
E-mail: athomas@physics.adelaide.edu.au

We outline some of the insights into hadron structure that have resulted from recent studies of chiral corrections to lattice data.

1 Introduction

In the non-perturbative regime appropriate to the study of hadron structure QCD is a highly non-linear problem with no possibility of an exact solution. Lattice QCD involves the brute force numerical computation of hadron properties on a discrete grid of space-time points. In spite of tremendous advances in computer power (up to one tera-flop at present) and clever improvements in algorithms, including various improved actions, one still cannot obtain solutions for realistic quark masses.

The essential problem in performing calculations at realistic quark masses (of order 5 MeV) is the approximate chiral symmetry of QCD. Goldstones theorem tells us that chiral symmetry is dynamically broken and that the non-perturbative vacuum is highly non-trivial, with massless Goldstone bosons in the limit $m_q \to 0$. For finite quark mass these bosons are the three charge states of the pion with a mass $m_\pi^2 \propto m_q$. Although this result strictly holds only for m_π^2 near zero (the Gell Mann-Oakes-Renner relation), lattice simulations show it is a good approximation for m_π^2 up to 0.8 GeV^2 and we shall use m_π^2 here as a measure of the deviation from the chiral limit.

From the point of view of lattice simulations with dynamical quarks (i.e. unquenched) the essential point is that the time taken goes as $m_q^{-2.5}$. The state-of-the-art for hadron masses is m_q above 60 MeV, although there is a preliminary result from CP-PACS at about 40 MeV! In general the quark masses for which simulations currently exist are at masses a factor of 8-20 too high. This means an increase of computing power to roughly 500 tera-flop is needed to calculate realistic hadron properties. Even with the current remarkable rate of increase this will take a long time.

Faced with such a serious difficulty physicists (like all other people facing a tough challenge) fall into two classes:

(A) Those who believe that "the cup is half empty":

In this case the emphasis is on the uncertainties associated with having to make big extrapolations to the chiral limit.

(B) Those who believe that "the cup is half full":

In this case we realize that the lattice data obtained so far represents a wealth of information on the properties of hadrons within QCD itself over a range of quark masses. Just as the study of QCD as a function of N_c has taught us a great deal, so the behaviour as a function of m_q can give us great insight into hadronic physics and guide our model building. Furthermore, as a bonus, approach (B) leads us to resolve most of the difficulty identified under (A)!

One of the major aims in establishing the CSSM was to bring together, into the same offices, seminars and workshops, lattice practitioners, builders of models and phenomenologists actively engaged with experimental groups around the world. The expectation was that this unique atmosphere would lead to significant advances in our understanding of hadron structure. One of the more important outcomes of this philosophy so far is the progress made on this problem of connecting lattice data with the real world.

In light of the brief space available to outline a great deal of evidence we first summarise the conclusions which emerge from the work of the past three years. We then choose just a couple of illustrations of the reasoning behind these conclusions.

Summary:

- In the region of quark masses $\bar{m} > 60$ MeV or so (m_π greater than typically 400-500 MeV) hadron properties are smooth, slowly varying functions of something like a constituent quark mass, $M \sim M_0 + c\bar{m}$ (with $c \sim 1$).

- Indeed, $M_N \sim 3M, M_{\rho,\omega} \sim 2M$ and magnetic moments behave like $1/M$.

- As $\bar{m}$ decreases below 60 MeV or so, chiral symmetry leads to rapid, non-analytic variation, with $\delta M_N \sim \bar{m}^{3/2}, \delta\mu_H \sim \bar{m}^{1/2}$ and $\delta < r^2 >_{\rm ch} \sim \ln \bar{m}$.

- Chiral quark models like the cloudy bag provide a natural explanation of this transition. The scale is basically set by the inverse size of the pion source – the inverse of the bag radius in the bag model.

- When the pion Compton wavelength is smaller than the size of the composite source chiral loops are strongly suppressed. On the other hand, as soon as the pion Compton wavelength is larger than the source one begins to see rapid, non-analytic chiral corrections.

These are remarkable results that will have profound consequences for our further exploration of hadron structure within QCD as well as the analysis of the vast amount of data now being taken concerning unstable resonances. In terms of immediate results for the structure of the nucleon, we note that the careful incorporation of the correct chiral behaviour of QCD into the extrapolation of its properties calculated on the lattice has produced:

- The best values of the proton and neutron magnetic moments from QCD.
- The best value of the sigma commutator.
- Improved values for the charge radii of the baryon octet.
- Improved values for the magnetic moments of the hyperons.
- Good agreement between the extrapolated moments of the non-singlet distribution $u - d$ and the experimentally measured moments.
- An understanding of the failures of the assumption of universality of quark electromagnetic properties and an improved lattice estimate of the strangeness magnetic moment of the proton G_M^s [1].

2 Electromagnetic Form Factors

It is a general consequence of quantum mechanics that the long-range charge structure of the proton comes from its π^+ cloud ($p \to n\pi^+$), while for the neutron it comes from its π^- cloud ($n \to p\pi^-$). However, it is not often realized that the LNA contribution to the nucleon charge radius goes like $\ln m_\pi$ and diverges as $\bar{m} \to 0$ [3]. This cannot be reproduced by a constituent quark model. Figure 1 shows the latest data from Mainz and NIKHEF for the neutron electric form factor, in comparison with CBM calculations for a confinement radius between 0.9 and 1.0 fm. The long-range π^- tail of the neutron plays a crucial role.

While there are only limited (and indeed quite old) lattice data for hadron charge radii, recent experimental progress in the determination of hyperon charge radii has led us to examine the extrapolation procedure for obtaining charge data from the lattice simulations [4]. Figure 2 shows the extrapolation of the lattice data [5] for the charge radius of the proton. Clearly the agreement with experiment is much better once the log required by chiral symmetry is correctly included, than if, for example, one simply made a linear extrapolation in the quark mass (or m_π^2). Full details of the results for all the octet baryons may be found in Ref. [4].

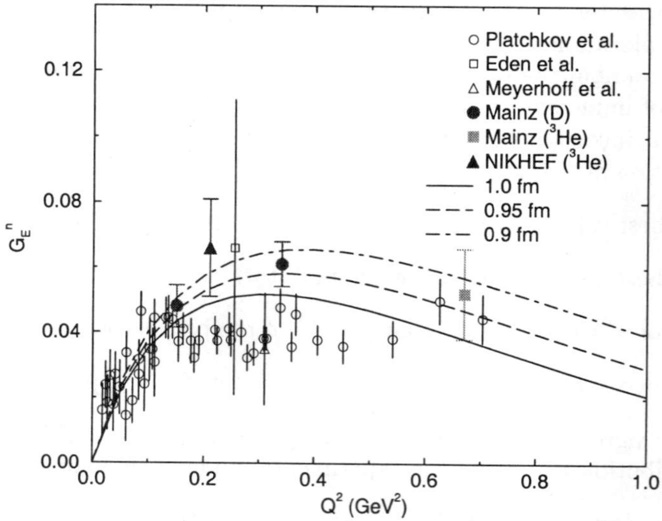

Figure 1. Recent data for the neutron electric form factor in comparison with CBM calculations for a confining radius around 0.95fm – from Ref. [2].

The situation for baryon magnetic moments is also very interesting. The LNA contribution in this case arises from the diagram where the photon couples to the pion loop. As this involves two pion propagators the expansion of the proton and neutron moments is $\mu^{p(n)} = \mu_0^{p(n)} \mp \alpha m_\pi + \mathcal{O}(m_\pi^2)$. Here $\mu_0^{p(n)}$ is the value in the chiral limit and the linear term in m_π is proportional to $\bar{m}^{\frac{1}{2}}$, a branch point at $\bar{m} = 0$. The coefficient of the LNA term is $\alpha = 4.4\mu_N \mathrm{GeV}^{-1}$. At the physical pion mass this LNA contribution is $0.6\mu_N$, which is almost a third of the neutron magnetic moment. *No constituent quark model can or should get better agreement with data than this.*

Just as for M_N, the chiral behaviour of $\mu^{p(n)}$ is vital to a correct extrapolation of lattice data. One can obtain a very satisfactory fit to some rather old data, which happens to be the best available, using the simple Padé [6]:

$$\mu^{p(n)} = \frac{\mu_0^{p(n)}}{1 \pm \frac{\alpha}{\mu_0^{p(n)}} m_\pi + \beta m_\pi^2} \tag{1}$$

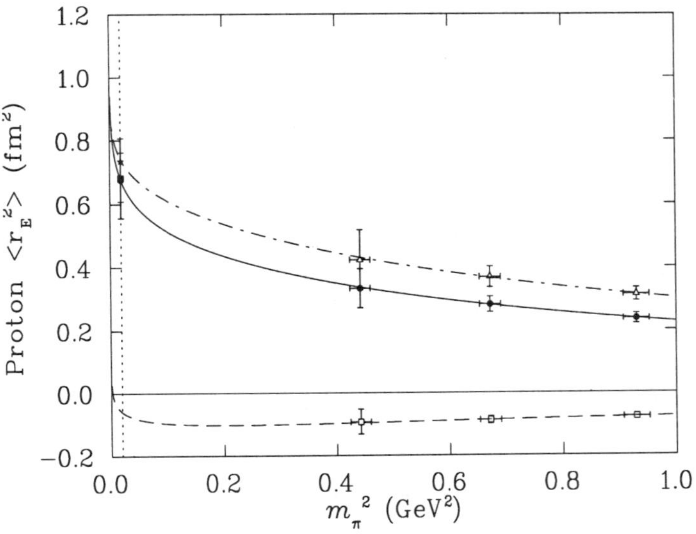

Figure 2. Fits to lattice results for the squared electric charge radius of the proton – from Ref. [4]. Fits to the contributions from individual quark flavours are also shown: the u-quark sector results are indicated by open triangles and the d-quark sector results by open squares. Physical values predicted by the fits are indicated at the physical pion mass, where the full circle denotes the result predicted from the first extrapolation procedure and the full square denotes the baryon radius reconstructed from the individual quark flavor extrapolations. (N.B. The latter values are actually so close as to be indistinguishable on the graph.) The experimental value is denoted by an asterisk.

The data can only determine two parameters and Eq.(1) has just two free parameters while guaranteeing the correct LNA behaviour as $m_\pi \to 0$ **and** the correct behaviour of heavy quark effective theory (HQET) at large m_π^2. The extrapolated values of μ^p and μ^n at the physical pion mass, $2.85 \pm 0.22\mu_N$ and $-1.90 \pm 0.15\mu_N$, respectively, are currently the best estimates from non-perturbative QCD [6]. For more details of this fit we refer to Ref. [6], while for the application of similar ideas to other members of the nucleon octet we refer to Ref. [7], and for the strangeness magnetic moment of the nucleon we refer to Ref. [1].

Incidentally, from the point of view of the naive quark model it is interesting to examine the ratio of the absolute values of the proton and neutron magnetic moments as a function of m_π^2. The agreement of the constituent quark result, namely 3/2, with the experimental value to within a few percent is usually taken as a major success. However, it has been recently pointed out that this close agreement is little more than an accident of the quark mass [8].

3 Structure Functions

The parton distribution functions (PDFs) of the nucleon are light-cone correlation functions which, in the infinite momentum frame, are interpreted as probability distributions for finding specific partons (quarks, antiquarks, gluons) in the nucleon. They have been measured in a variety of high energy processes, ranging from deep-inelastic lepton scattering to Drell-Yan and massive vector boson production in hadron–hadron collisions. A wealth of experimental information now exists on spin-averaged PDFs, and an increasing amount of data is being accumulated on spin-dependent PDFs [9].

At high momentum transfer (Q^2) the dominant component of the PDFs are determined by non-perturbative matrix elements of certain "leading twist" operators. In principle these matrix elements, which correspond to moments of the measured structure functions, contain vital information about the non-perturbative structure of the target. An extensive phenomenology has been developed over the years within model QCD studies, and in some cases remarkable predictions have been made from the insight gained into the non-perturbative structure of the nucleon. An example is the $\bar{d} - \bar{u}$ asymmetry, predicted [10] on the basis of the nucleon's pion cloud [11], which has been spectacularly confirmed in recent experiments at CERN and Fermilab [12]. Other predictions, such as asymmetries between strange and antistrange [13] and spin-dependent sea quark distributions, $\Delta\bar{u} - \Delta\bar{d}$, still await experimental confirmation. Note that none of these could be anticipated without insight into the non-perturbative structure of QCD.

Early calculations of structure function moments within lattice QCD were performed by Martinelli and Sachrajda [14]. However, the most comprehensive analysis has been performed by the QCDSF Collaboration[14–17] – albeit within quenched QCD. Recently the MIT group has performed the first full (unquenched) QCD calculations of non-singlet moments [18]. The moments from the full QCD simulations are very similar to those from the quenched calculations. This is consistent with the suggestions of chiral quark models, like the CBM, and the general conclusions outlined in the Introduction. That is, in the quark mass region currently accessible quark loops are suppressed.

As for the other nucleon properties discussed above, we propose to extraplate the lattice data to the physical pion mass using a formula which is compatible with the LNA structure of the PDFs. This behaviour was derived recently, with the result that the LNA behaviour involved a term in $m_\pi^2 \ln m_\pi$ [19]. For an initial investigation we concentrate on the non-singlet combination

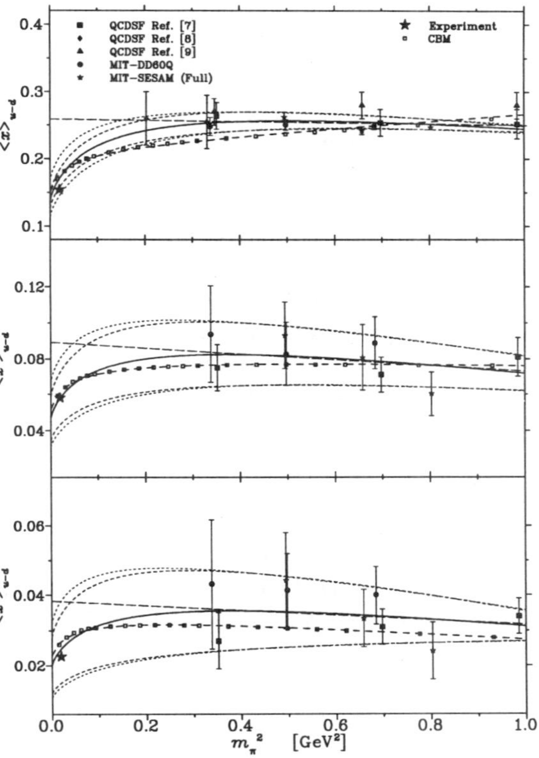

Figure 3. Moments of the $u-d$ quark distribution – from Ref.[20]. The straight (long-dashed) lines are linear fits to the data, while the curves have the correct LNA behavior in the chiral limit. For each moment, the best fit to the lattice data using Eq.(2) is shown by the solid curve (with $\mu = 550$ MeV), while the inner envelope about this represents the statistical errors in the data. The uncertainty in the parameter μ is illustrated by the outer lower (upper) short-dashed curves, which correspond to $\mu = 450$ (650) MeV. The small squares are the CBM results, and the dashed curve through them best fits using Eq.(2). The star represents the phenomenological values taken from NLO fits.

of PDFs, $u - d$, in which "disconnected" quark loops cancel. Calculations based on the CBM (which incorporate the LNA chiral structure just discussed) actually produce quite a reasonable description of the behaviour of the moments of the PDFs as a function of quark mass, as shown in Fig. 3 (open squares). More important from the phenomenological point of view, the CBM calculations (for the n'th moment of the PDFs) can be fit with the simple expansion in m_π:

$$\langle x_u^n - x_d^n \rangle = a_n + b_n m_\pi^2 + a_n c_{\rm LNA} m_\pi^2 \ln\left(\frac{m_\pi^2}{m_\pi^2 + \mu^2}\right) , \qquad (2)$$

where $c_{\rm LNA}$ is model independent.

The scale μ in Eq.(2) is effectively the scale at which the rapid, chiral variation at low m_π turns off. The best fit to the lattice data is obtained with a value $\mu \sim 0.4 - 0.5$ GeV – a very similar scale to that found, for example, for the magnetic moments. Clearly Eq.(2) gives a very good description of the lattice data for the first moment of the non-singlet distribution $d - u$. Taking into account the rapid chiral variation as $m_\pi^2 \to 0$ there is also quite good agreement between the extrapolated value of the first moment and the experimentally determined moment. A similar result holds for the second and third moments too [20].

Acknowledgments

It is a pleasure to acknowledge the work of my many collaborators in the work described here, especially D. Leinweber, W. Detmold, W. Melnitchouk and Stewart Wright. This work was supported by the Australian Research Council and Adelaide University.

References

1. D. B. Leinweber and A. W. Thomas, *Phys. Rev.* D **62**, 074505 (2000) [hep-lat/9912052].
2. D. H. Lu *et al.*, in Proc. Int. Conf. Few Body Problems (Taipei, 2000), to appear in *Nucl. Phys.* A; D. H. Lu *et al.*, *Phys. Rev.* C **60**, 068201 (1999) [nucl-th/9807074].
3. D. B. Leinweber and T. D. Cohen, *Phys. Rev.* D **47**, 2147 (1993) [hep-lat/9211058].
4. E. J. Hackett-Jones, D. B. Leinweber, and A. W. Thomas, *Phys. Lett.* B **494**, 89 (2000) [hep-lat/0008018].
5. D.B. Leinweber, R.M. Woloshyn, and T. Draper, *Phys. Rev.* D **43** (1991) 1659.
6. D. B. Leinweber *et al.*, *Phys. Rev.* D **60**, 034014 (1999) [hep-lat/9810005].
7. E. J. Hackett-Jones, D. B. Leinweber, and A. W. Thomas, *Phys. Lett.* B **489**, 143 (2000) [hep-lat/0004006].
8. D. B. Leinweber, A. W. Thomas, and R. D. Young, hep-ph/0101211.

9. M. Erdmann, Talk given at 8th International Workshop on Deep Inelastic Scattering and QCD (DIS 2000), Liverpool, England, 25-30 Apr. 2000, hep-ex/0009009; E.W. Hughes and R. Voss, *Ann. Rev. Nucl. Part. Sci.* **49**, 303 (1999).
10. A. W. Thomas, *Phys. Lett.* B **126**, 97 (1983).
11. E.M. Henley and G.A. Miller, *Phys. Lett.* B **251**, 453 (1990); A.I. Signal, A.W. Schreiber and A.W. Thomas, *Mod. Phys. Lett.* A **6**, 271 (1991); W. Melnitchouk, A.W. Thomas, and A.I. Signal, *Z. Phys.* A **340**, 85 (1991); S. Kumano, *Phys. Rev.* D **43**, 3067 (1991); S. Kumano and J.T. Londergan, *Phys. Rev.* D **44**, 717 (1991); W.-Y.P. Hwang, J. Speth, and G.E. Brown, *Z. Phys.* A **339**, 383 (1991).
12. P. Amaudraz *et al.*, *Phys. Rev. Lett.* **66**, 2712 (1991); A. Baldit *et al.*, *Phys. Lett.* B **332**, 244 (1994).
13. A.I. Signal and A.W. Thomas, *Phys. Lett.* B **191**, 206 (1987); X. Ji and J. Tang, *Phys. Lett.* B **362**, 182 (1995); S.J. Brodsky and B.-Q. Ma, *Phys. Lett.* B **381**, 317 (1996); W. Melnitchouk and M. Malheiro, *Phys. Rev.* C **55**, 431 (1997).
14. G. Martinelli and C.T. Sachrajda, *Phys. Lett.* B **196**, 184 (1987); *Nucl. Phys.* B **306**, 865 (1988).
15. M. Göckeler, R. Horsley, E. M. Ilgenfritz, H. Perlt, P. Rakow, G. Schierholz, and A. Schiller, *Phys. Rev.* D **53**, 2317 (1996).
16. M. Göckeler, R. Horsley, E. M. Ilgenfritz, H. Perlt, P. Rakow, G. Schierholz, and A. Schiller, *Nucl. Phys. Proc. Suppl.* **53**, 81 (1997).
17. C. Best *et al.*, hep-ph/9706502.
18. D. Dolgov *et al.*, hep-lat/0011010.
19. A.W. Thomas, W. Melnitchouk, and F.M. Steffens, *Phys. Rev. Lett.* **85**, 2892 (2000).
20. W. Detmold, W. Melnitchouk, J. W. Negele, D. B. Renner, and A. W. Thomas, hep-lat/0103006, to appear in *Phys. Rev. Lett.* (2001).

José Oller

Dan-Olof Riska

THE ROLE OF THE PION IN NUCLEON RESONANCE STRUCTURE

D.-O. RISKA
Helsinki Institute of Physics
POB 64, 00014 University of Helsinki

An overview of what is known about the coupling of pions to light constituent quarks is given. The pion decay rates of excited charm mesons indicate that the standard chiral coupling of pions to quarks applies, if augmented by the two-quark axial current contribution. This supports the satisfactory description of the empirical baryon spectrum that obtains if the hyperfine interaction between constituent quarks obtain is described in terms of pion and two pion exchange.

1 Pion Coupling to Constituent Quarks

The role of the spin as a Goldstone boson associated with the spontaneously broken approximate chiral symmetry of QCD suggests that its coupling to hadrons be of the form

$$\mathcal{L} = -\frac{1}{f_\pi}\vec{A}_\mu \cdot \partial_\mu \vec{\pi}, \tag{1}$$

where f_π is the pion decay constant and $\vec{A}_\mu$ is the axial current of the hadron. The axial current of a light flavor constituent quark is

$$\vec{A}_\mu = i g_A^q \bar{\psi}_q \gamma_\mu \gamma_5 \vec{\tau}/2 \psi_q, \tag{2}$$

where ψ_q are the quark fields. The axial coupling constant of the quark g_A^q would be 1, if the quarks were pure Dirac particles, but is probably somewhat less due to screening by sea-quarks [1].

The pseudovector pion-quark coupling constant $f_{\pi qq}$ is then obtained as

$$f_{\pi qq} = \frac{g_A^q}{2}\frac{m_\pi}{f_\pi} \simeq 0.65, \tag{3}$$

if Weinberg's value for g_A^q (0.87) [2] is used. This value is close to the static quark model value

$$f_{\pi qq} = \frac{3}{5} f_{\pi NN} \simeq 0.6. \tag{4}$$

The next question is to what extent this single quark-pion coupling applies quantitatively for the constituent quarks that form the baryons. The single quark-pion coupling would imply that the ratios of the pion decay widths of the decuplet baryons be proportional to the square of the number of light flavor constituent quarks in the decaying state. Hence the ratios of the widths for $\Delta \to N\pi$, $\Sigma^* \to \Sigma\pi$ and $\Xi^* \to \Xi\pi$ should be 9:4:1. The empirical ratios are 12:4:1, and hence fairly close to these theoretical ratios, especially if one takes into account the larger phase space available to the pion in $\Delta \to N\pi$ decay. This however does not prove that the pion coupling to the decuplet and octet baryons has to be a single quark mechanism, because two-quark contributions that involve a point (seagull) coupling of the pion and the two-quark interaction gives rise to the same ratios.

The recent analysis based on the large N_c limit of QCD in Ref. [3] of the π decay rates of the radially excited 56-plet does suggest that two-quark mechanisms may be significant. The empirical values for the widths of the measured transitions of the form $56' \to 56 + \pi$ indicate that the corresponding transition matrix elements depend on the momentum k of the emitted pion as $f(k) = 2.8/k$, with a χ^2 per degree of freedom of 1.1. A single quark mechanism would require that the function $f(k)$ vanish for $k = 0$ by the orthogonality of the qqq states in the 56 and 56'. The empirical function $f(k)$ does not, however, reveal such behavior.

Finally it has long been known that the static quark model underpredicts the width of $\Delta \to N\pi$ with the single quark coupling model. This width may be expressed as

$$\Gamma(\Delta \to N\pi) = \frac{1}{3}\frac{f_{\pi N\Delta}^2}{4\pi}\frac{E_N + m_N}{m_\Delta}\frac{k^3}{m_\pi^2} \tag{5}$$

where $f_{\pi N\Delta}$ is the $\pi N\Delta$ transition coupling. From the empirical value for the width, 120 MeV, this equation yields $f_{\pi N\Delta} = 2.2 \pm 0.04$. The static quark model value is

$$f_{\pi N\Delta} = 2\sqrt{2}\frac{g_A^q}{2f_\pi}m_\pi \tag{6}$$

With $g_A^q = 0.87$ this expression gives $f_{\pi N\Delta} = 1.6$, which is too small by $\sim 30\%$. The empirical value for $f_{\pi N\Delta}$ would obtain with $g_A^q = 1$.

2 Model Calculations of Resonance Decay Widths

If pions couple to constituent quarks by the chiral coupling (1), they will inevitably also mediate a pion exchange interaction between two constituent quarks, which is described by the Yukawa potential

$$V(r) = -\frac{1}{3}(\frac{g_A^q}{2f_\pi})^2\vec{\sigma}^1 \cdot \vec{\sigma}^2\vec{\tau}^1 \cdot \vec{\tau}^2\{4\pi\delta^{(3)}(\vec{r}) - m_\pi^3\frac{e^{-m_\pi r}}{m_\pi r} + \text{tensor pot.}\}. \tag{7}$$

An interaction of this form - i.e. a flavor-spin dependent interaction, which is attractive at short range - has been shown to provide an overall satisfactory description of the baryon spectrum, with correct ordering of positive and negative parity states [4]. In this aspect it differs from other phenomenological dynamical models for the interaction between constituent quarks.

Reproduction of the nucleon and Δ-spectrum is not, however, sufficient to ensure a satisfactory description of the hadronic decay widths of the nucleon resonances. A recent systematic comparative study of the pionic decay widths of the nucleon and Δ resonances [5] shows that the single quark (actually the $^3P_0$) model for nucleon resonance decays does not describe the systematics of the empirical decay widths. In that study 4 different Hamiltonians for the 3 quark systems were considered: semirelativistic and nonrelativistic Hamiltonians, which model the interquark interaction as a linear confining interaction and a hyperfine interaction based either on meson exchange or gluon exchange. The former model describes the spectrum, while the latter does not. (A notable feature of the gluon exchange, as well as the alternate instanton induced interaction model for the quark-quark interaction, is that they place the $\Delta(1600)$ resonance above 1900 MeV). None of these models were found to describe the pionic decay widths with the single quark mechanism.

3 Pion Decays of D-Mesons

The ideal systems for studying how pions couple to light constituent quarks are the heavy-light D and B mesons. These are formed of a light flavor constituent u or d quark and a heavy $\bar{c}$ or $\bar{b}$ antiquark. Only the former couples to pions. Hence the pion decay widths of the excited D and B mesons should give fairly direct information on the form of the coupling between pions and u and d quarks.

So far only upper limits are known experimentally for the ground state decays of the form $D^* \to D\pi$, with exception of the D^{*+} [6], while decay widths of the order $\sim$ 20 MeV with wide uncertainty limits have been found for the orbitally excited $D_1(2420)$ and $D_2^*(2460)$ mesons with $L = 1$. Even so this information is close to be constraining for theoretical models. A complicating factor is, however, that as both the $D_1(2420)$ and $D_2^*(2460)$ mesons lie above threshold for $\pi\pi$ decay, which therefore contributes to their total widths.

To calculate the pion decays of the excited D and B mesons one needs a Hamiltonian model that reduces to the well tested heavy quark potential for heavy quarkonia in the equal mass limit. That potential may in turn be matched by a dynamical model, with a scalar confining and a hyper-fine interaction. The conventional model for the latter is gluon exchange.

	$g_A^q = 0.87$	$g_A^q = 1.0$	Expt. [11]
$D^{*\pm} \to D^{\pm}\pi^0$	0.029	0.038	<0.04
$D^{*\pm} \to D^0\pi^{\pm}$	0.064	0.084	<0.09
$D^{*0} \to D^0\pi^0$	0.041	0.054	<1.3

Table 1. Calculated and measured values for the $D^* \to D\pi$ decay widths in MeV (From Ref. [7]). The measured total width of the D^{*+} is $0.096 \pm 0.04 \pm 0.022$ MeV [6].

For heavy quarkonia there is a delicate balance between the spin-orbit components of the scalar confining interaction and the gluon exchange interaction. In the simplest static version of these interaction models this balance is lost for heavy light systems, with empirically contraindicated spin-orbit inversion of the $L = 1$ states as the consequence. Relativistic treatment of the gluon exchange interaction allows avoidance of this problem [7].

Because of the small mass of the light flavor constituent quarks in the charm mesons a relativistic description of the bound quark-antiquark system is called for. This is most readily achieved by means of a covariant three-dimensional reduction of the Bethe-Salpeter equation, which admits employment of instantaneous interactions. One such reduction, which is natural for heavy-light systems is the Gross equation, which has been used for the D-mesons in Ref. [8], and which has recently been used to calculate their pionic decay widths with the chiral quark coupling (1) [9]. Another quasi-potential reduction is the Blankenbecler-Sugar equation, which has the intuitively appealing feature of formal similarity with the Schrödinger equation. The latter has also recently been used to calculate both the single and double pion decay widths of the excited D-mesons [7,10].

In Table 1 the calculated pion decay widths obtained in Ref.[7] with a relativistic treatment of the chiral π-quark coupling are shown for $g_A^q = 0.87$ and 1. The calculated values include an enhancement of $\sim 10\%$ due to the axial exchange current which involves coupling to negative energy intermediate states and and the scalar confining interaction. For $g_A^q = 1$ the calculated values are close to the present empirical upper limits for the charged D-mesons.

The positive parity D-mesons with $L = 1$ lie so high above the D and the D^* mesons, that they can decay with the emission of both one and two pions. For the single pion decay both the spatial current and the charge part of the axial current in (1) contributes. In the non-relativistic limit the single quark operator is then $\sim \vec{\sigma} \cdot k_\pi + \vec{\sigma} \cdot (\vec{p}' + \vec{p})/2m_q$, where $\vec{p}$ and $\vec{p}'$ are the initial final quark momenta.

The axial charge component of the pion emission operator by itself would lead to a large overprediction of the decay width of the D_1 meson. The

	$D\pi + D^*\pi$	$D\pi\pi + D^*\pi\pi$	Tot	Expt [11]
D_1	9.6	1.3	10.9	$18.9^{+4.6}_{-3.5}$
D_2^*	14.9	3.0	17.9	25^{+8}_{-7}

Table 2. Calculated and measured values for the pion decay widths in MeV of the positive parity D-mesons for $g_A^q = 1$.

required suppression of this part of the operator is brought about by the two-body contribution, which involves coupling to negative energy intermediate states of the pion and the scalar confining interaction [7]. The calculated single pion decay widths of the D_1 and D_2^* mesons, with inclusion of this suppression of the axial charge, are given in Table 2 [7].

The decay widths for two-pion decay may be calculated by similar methods: the two-pion amplitude for a single quark then consists of the Born terms given by (1) supplemented by the Weinberg-Tomozawa term for quarks. In this case also two-quark operators involving intermediate negative energy terms appear, which strongly reduce the amplitude [10]. The thus calculated two-pion decay widths for the D_1 and D_2^* mesons obtained with $g_A^q = 1$ are also given in Table 2. Combination of the calculated single pion and two-pion decay widths yield total widths, which reach the lower end of the empirical uncertainty range with $g_A^q = 1$. Given the large uncertainty is the present empirical widths the conclusion is that the chiral quark model does provide an acceptable description of the presently known widths of the D-mesons. The theoretical uncertainty in the calculation is also substantial: the calculation with static interactions in the framework of the Gross equation, using much smaller constituent quark mass values obtained the best agreement with the empirical widths with $g_A^q = 0.75$ [9].

4 Conclusion

The results reported above for the pion decay widths of the excited D-mesons as well as the comparable results obtained in Ref.[9] show that the chiral coupling (1) for pions to light flavor constituent quarks provides a realistic starting point. It then follows that light constituent quarks interact by means of the pion exchange Yukawa interaction (7). Whether this interaction is valid at short range as well, and hence significant for the explanation of the baryon spectrum still remains a contentious issue in the literature, however.

What is known is that a phenomenological interaction of the form

$$V_\chi = -C_\chi \sum_{i<j} \vec{\tau}_i \cdot \vec{\tau}_j \vec{\sigma}_i \cdot \vec{\sigma}_j, \tag{8}$$

with $C_\chi \sim 30$ MeV, describes the hyperfine interaction of the empirical baryon spectrum well [4]. A combination of the one-pion exchange and the two-pion exchange interactions between quarks does indeed give rise to a phenomenological interaction of this type [12]: the spin-spin components of the one- and two-pion interactions add, while the corresponding tensor interactions tend to cancel out.

The calculation of the π decays of the excited D-mesons is interesting in that the relativistic corrections that are associated with the small components of the Dirac spinors are so important. This is so irrespective of whether these effects are treated in the coupled channel approach implied by the Gross equation as in Ref.[9] or by means of two-quark exchange currents as in Ref.[7,10]. In the latter approach the most important such effect is the suppression of the axial charge operator, which is required to avoid large overpredictions of the decay width of the $D_1(2420)$ meson. This result is similar and closely related to the corresponding suppression of M1 transition strength seen in the electromagnetic decay $J/\psi \to \eta_c\gamma$[13].

Acknowledgments

Research supported in part by the Academy of Finland through grant No. 43982.

References

1. D.A. Dicus *et al., Phys. Lett.* B **284**, 384 (1992).
2. S. Weinberg, *Phys. Rev. Lett.* **67**, 3473 (1991).
3. C.E. Carlson and C.D. Carone, *Phys. Lett.* B **484**, 260 (2000).
4. L. Glozman and D.O. Riska, *Phys. Rep.* **268**, 263 (1996).
5. L. Theussl *et al.*, nucl-th/0010099.
6. T.E.Coan *et al.*, CLEO collaboration, hep-ex/0102007.
7. K.O.E. Henriksson *et al.*, Nucl. Phys. A **686**, 355 (2001).
8. J. Zheng, J.W. van Orden, and W. Roberts, *Phys. Rev.* D **52**, 5229 (1995).
9. J.L. Goity and W. Roberts, *Phys. Rev.* D **60**, 034001 (1999).
10. T.A. Lähde and D.O. Riska, hep-ph/0102039.
11. D.E. Groom *et al., Eur. Phys. J.* C **15**, 1 (2000).
12. D.O. Riska and G.E. Brown, *Nucl. Phys.* A **679**, 577 (2001).
13. T.A. Lähde, C. Nyfält, and D.O. Riska, *Nucl. Phys.* A **645**, 587 (1999).

RELATIVISTIC QUARK MODELS

S. SIMULA
Istituto Nazionale di Fisica Nucleare, Sezione di Roma III,
Via della Vasca Navale 84, I-00146 Roma (Italy)

The application of relativistic constituent quark models to the evaluation of the electromagnetic properties of the nucleon and its resonances is addressed. The role of the pair creation process in the Feynmann triangle diagram is discussed and the importance both of choosing the light-front formalism and of using a Breit frame where the *plus* component of the four-momentum transfer is vanishing, is stressed. The nucleon elastic form factors are calculated free of spurious effects related to the loss of rotational covariance. The effect of a finite constituent size is considered and the parameters describing the constituent form factors are fixed using only the nucleon data up to $Q^2 \simeq 1$ $(GeV/c)^2$; a constituent charge radius of $\simeq 0.45$ fm is obtained in this way. Above $Q^2 \simeq 1$ $(GeV/c)^2$ our parameter-free predictions are in good agreement with the experimental data, showing that soft physics may be dominant up to $Q^2 \sim 10$ $(GeV/c)^2$. Our new results for the longitudinal and transverse helicity amplitudes of the $N - P_{11}(1440)$ transition are presented.

1 Introduction

The electromagnetic (e.m.) properties of the nucleon and its resonances contain important pieces of information on the internal structure of the nucleon, and therefore an extensive experimental program aimed at their determination is currently undergoing and planned at several facilities around the world. As it is well known, a successful phenomenological model for the description of the hadron structure is based on the concept of constituent quarks (CQ's) as effective degrees of freedom in hadrons. The CQ model assumes that the degrees of freedom associated to gluons and $q\bar{q}$ pairs are frozen, while their effects are hidden in the CQ mass and (possible) size as well as in the effective interaction among CQ's. Originally the CQ model was formulated in a non-relativistic way, and indeed its first success was the reproduction of the mass spectra of mesons with two heavy quarks [1]. However, the non-relativistic CQ model is unexpectedly successful also in the light-quark sector. As a matter of fact it suffices to remind that the simple assumption of an $SU(6)$ symmetry of the nucleon wave function leads immediately to several predictions, which are in qualitative agreement with experimental data; among them we mention: i) the neutron electric form factor $G_E^n(Q^2)$ is vanishing; ii) the neutron to proton ratio of magnetic form factors $G_M^n(Q^2)/G_M^p(Q^2)$ is independent of Q^2 and equal to $-2/3$; iii) all the radiative transitions among members of the $[56, 0^+]$ and $[70, 1^-]$ $SU(6)$-multiplets can be written in terms of only three independent parameters [2].

The non-relativistic CQ model has several flaws, which were clearly pointed out in [3]. Here we limit ourselves to mention the most relevant inconsistency, namely the fact that the average momentum of a light CQ in a hadron is definitely larger than its mass (see, e.g., the CQ momentum distribution in the nucleon and in the most prominent electroproduced resonances reported in [4]). Therefore, relativized versions of the CQ effective Hamiltonian have been proposed in the literature, trying mainly to take into account the relativistic form of the kinetic energy and the relativistic smearing of the CQ positions, as it is the case of the one-gluon- exchange (OGE) model [3] and of the chiral model based on Goldstone-boson-exchange (GBE) mechanism [5]. Nevertheless, for both relativized models the proton size turns out to be much smaller ($\simeq 0.3\ fm$) than its experimental charge radius ($\simeq 0.8\ fm$), showing the need for a proper treatment of relativistic effects in the evaluation of the nucleon e.m. properties even close to the photon point (see [6]).

2 Relativistic Constituent Quark Models

There are three main ways to include the interaction in the generators of the Poincaré group and to maximize at the same time the number of kinematical (interaction-free) generators. The three relativistic forms are: i) the instant form, where the interaction is present in the time component P^0 of the four-momentum and in the Lorentz boost operators; ii) the point form (PF), where all the components of the four-momentum operator depend on the interaction, and iii) the light-front (LF) form, in which the interaction appears in the "minus" component of the four-momentum ($P^- \equiv P^0 - P^z$) and in the transverse rotations around the x and y axes (with the null-plane being defined by $x^+ \equiv t + z = 0$). Since in the instant form it is not possible to define kinematical boost operators, in what follows we will limit ourselves only to the PF and to the LF formalism.

In recent years relativistic CQ models employing the two above-mentioned forms of the dynamics have been used for the investigation of hadron e.m. properties. In such applications the one-body approximation for the e.m. current has been widely considered, because it allows to address the important issue to know to what extent the e.m. properties of a composite system can be understood in terms of the e.m. properties of its constituents. The one-body approximation can be generally written as

$$J^\mu \approx J_1^\mu = \sum_{j=1}^{N} \left[f_1^{(j)}(Q^2)\gamma^\mu + f_2^{(j)}(Q^2)\frac{i\sigma^{\mu\nu}q_\nu}{2m_j} \right] \qquad (1)$$

where $Q^2 \equiv -q \cdot q$ is the squared four-momentum transfer and $f_{1(2)}^{(j)}(Q^2)$ are Dirac (Pauli) form factors of the j-th constituent quark with mass m_j. While the exact current J^μ is Poincaré covariant in all the three forms, the approximate current J_1^μ is covariant only in the PF.

However, if a connection between relativistic quantum models and covariant quantum field theories is searched, a minimal requirement is the matching with the Feynmann one-loop triangle diagram. As it is well known [7], the latter is given by the sum of two terms, namely the spectator-pole and the pair creation diagram (the Z-graph). The spectator term corresponds to a one-body process, while the Z-graph is a many-body process. In the PF the two terms are separately covariant, while in the LF formalism only their sum is covariant. This means that only in the LF formalism one may find a reference frame where the Z-graph is suppressed. As a matter of fact, it is well known [7] that such a reference frame is the Breit frame where $q^+ \equiv q^0 + q^z = 0$.

In the PF the contribution of the one-body current (1) does not correspond to the spectator term of the triangle diagram, so that in order to match the latter one needs to include explicitly the Z-graph. A manifestation of the mismatch of the spectator term with the one-body PF approximation is the fact that the squared four-momentum transfer seen by the constituent $\bar{Q}^2$ differs from Q^2 (see [8]). On the contrary in the LF formalism the use of a Breit frame where $q^+ = 0$ makes the one-body contribution equivalent to the spectator term [7], while at the same time the Z-graph is suppressed. Note that $\bar{Q}^2 = Q^2$ only when $q^+ = 0$. However, performing the limit $q^+ \to 0$ of the Z-graph requires some care. As properly pointed out in [9], the longitudinal zero-momentum mode may contribute to few matrix elements of the current components even when $q^+ = 0$ and such contributions turn out to be essential in order to recover the full covariance of the calculations. Therefore, the strategy to evaluate the form factors encoded in the triangle diagram is to identify the subset of amplitudes characterized by the matching of the triangle diagram with the one-body LF approximation at $q^+ = 0$. To carry out this strategy a useful formalism is the covariant LF approach developed in [10]. There, it has been argued that the loss of covariance for an approximate current implies the dependence of its matrix elements upon the choice of the null four-vector ω defining the spin-quantization axis. The particular direction defined by ω is irrelevant only if the current satisfies covariance. Therefore, the covariant decomposition of an approximate current can still be done provided the four-vector ω is explicitly considered in the construction of the relevant covariant structures. If we indicate by $\Gamma^{\mu;\alpha\ldots\beta\ldots}$ the amplitude tensor corresponding to the triangle diagram, where μ is the index for the current and α... (β...) are the Lorentz indices describing the spinorial structure of initial (final) bound-state vertex, one has (at $q^+ \equiv \omega \cdot q = 0$)

$$\Gamma_{LF}^{\mu;\alpha...\beta...}(\omega) = \Gamma^{\mu;\alpha...\beta...} + B^{\mu;\alpha...\beta...}(\omega) \tag{2}$$

where $\Gamma_{LF}^{\mu;\alpha...\beta...}$ corresponds to a given approximate LF current, like the one-body approximation (1), and $B^{\mu;\alpha...\beta...}(\omega)$ is the tensor depending on ω and describing the mismatch between the triangle diagram and the given LF approximation. Note that ω is proportional to the difference between the Feynmann and LF momenta of the struck constituent, viz. $p - p_{LF} \propto \omega$, where p is the constituent (off-mass-shell) momentum entering the triangle diagram (i.e., $p^2 \neq m^2$), while p_{LF} is by definition constructed on-mass-shell (i.e., $p_{LF}^2 = m^2$). In Eq. (2) the tensor $B^{\mu;\alpha...\beta...}(\omega)$ is responsible for the so-called violations of the angular condition, which have been explicitly estimated in [11] and [12] in case of the ρ-meson and the $N - \Delta(1232)$ transitions, respectively, and also in [13] and [6(b)] in case of the nucleon elastic form factors.

3 Nucleon Elastic Form Factors and the $N - P_{11}(1440)$ Transition

Let us now consider explicitly the case of the e.m. transition to a nucleon resonance with same quantum number $J^P = 1/2^+$ of the nucleon and indicate by $\bar{u}(p^*\nu')\Gamma^\mu u(p\nu)$ the matrix elements of the e.m. current corresponding to the triangle diagram for an initial (final) helicity equal to ν (ν'). One has

$$\Gamma^\mu = F_1^{N^*}(Q^2)\left[\gamma^\mu + q^\mu \frac{M^* - M}{Q^2}\right] + F_2^{N^*}(Q^2)\frac{i\sigma^{\mu\rho}q_\rho}{M^* + M} \tag{3}$$

where p and p^* are the four-momentum of the nucleon and its resonance with mass M and M^*, respectively, and $F_j^{N^*}(Q^2)$ (with $j = 1, 2$) are the physical form factors. Following Eq. (2) the LF approximation can be written as: $\Gamma_{LF}^\mu = \Gamma^\mu + B^\mu(\omega)/(\omega \cdot P)$ with

$$\begin{aligned} B^\mu(\omega) = {} & B_1^{N^*}(Q^2)\left[\not{\omega} - \frac{2\omega \cdot P}{(M^* + M)(1 + \eta)}\right]\left[P^\mu + q^\mu \frac{M^{*2} - M^2}{2Q^2}\right] \\ & + B_2^{N^*}(Q^2)\omega^\mu + B_3^{N^*}(Q^2)\omega^\mu \frac{\not{\omega}}{\omega \cdot P} + (M^* - M)B_4^{N^*}(Q^2)q^\mu \frac{\not{\omega}}{\omega \cdot P} \\ & + (M^* - M)B_5^{N^*}(Q^2)i\sigma^{\mu\rho}\omega_\rho \ , \end{aligned} \tag{4}$$

where $\eta \equiv Q^2/(M^* + M)^2$, $P = (p + p^*)/2$ and $B_j^{N^*}(Q^2)$ (with $j = 1, ..., 5$) are (unphysical) spurious form factors. Note that, while $\Gamma \cdot q = 0$, one has $\Gamma_{LF} \cdot q \neq 0$ because $B \cdot q \neq 0$. However, in the elastic case ($M^* = M$) the structures proportional to $B_4^{N^*}(Q^2)$ and $B_5^{N^*}(Q^2)$ are absent thanks to time reversal symmetry, so that one gets $\Gamma_{LF} \cdot q = 0$. Generalizing the results obtained in [6(b)] and adopting our standard Breit frame where $q = (q_x, q_y, q^+, q^-) = (Q, 0, 0, 0)$, one gets

$$G_E^{N^*}(Q^2) \equiv F_1^{N^*}(Q^2) - \eta F_2^{N^*}(Q^2) = \mathrm{Tr}\left\{\frac{\Gamma_{LF}^+}{4P^+}\right\} - \frac{iQ}{M^* + M}\mathrm{Tr}\left\{\frac{\Gamma_{LF}^+}{4P^+}\sigma_y\right\}$$
$$G_M^{N^*}(Q^2) \equiv F_1^{N^*}(Q^2) + F_2^{N^*}(Q^2) = -\frac{i}{2Q}\mathrm{Tr}\left\{\Gamma_{LF}^y \sigma_z\right\} \tag{5}$$

with σ_y and σ_z being Pauli matrices. Therefore, the charge form factor $G_E^{N^*}(Q^2)$ can be extracted from the matrix elements of the *plus* component of the current, while the magnetic form factor $G_M^{N^*}(Q^2)$ has to be determined from the y-component, as already obtained in the elastic case in Ref. [6(b)].

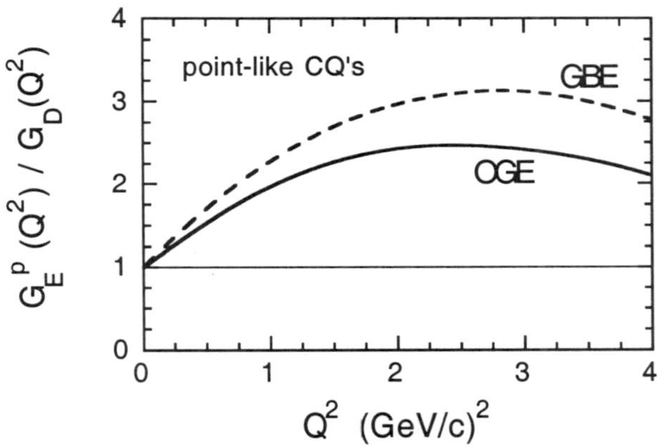

Figure 1. The proton charge form factor $G_E^p(Q^2)$, divided by the standard dipole form factor $G_D(Q^2) = 1/(1 + Q^2/0.71)^2$, versus Q^2. The solid and dashed lines correspond to our LF result (5) obtained assuming point-like CQ's in Eq. (1) and adopting the nucleon wave functions corresponding to the OGE [3(b)] and GBE [5] models, respectively.

In Fig. 1 we have reported the results obtained for the proton charge form factor $G_E^p(Q^2)$ assuming point-like CQ's in Eq. (1) and adopting the nucleon wave functions corresponding to the OGE and GBE models which provides a proton size of $\simeq 0.3$ fm. It can clearly be seen that the introduction of boost effects increase remarkably the calculated proton charge radius (up to $\simeq 0.6$ fm), but this is not sufficient to get agreement with its experimental value. Therefore, we now consider the effects of a finite CQ size by introducing CQ form factors in Eq. (1). The functional forms of $f_{1(2)}^{(j)}(Q^2)$ (with $j = u, d$)

are the same as those adopted in [14]; however, in this work the parameters are fixed by the requirement of reproducing the nucleon and pion data of [14] only up to $Q^2 \simeq 1\ (GeV/c)^2$, i.e. up to the scale of chiral symmetry breaking. We stress that the basic difference with [14] is that in this work by means of Eq. (5) the nucleon form factors are evaluated free of the spurious effects arising from the loss of rotational covariance. The CQ form factors, obtained from our least-χ^2 procedure, provide a CQ charge radius equal to $\simeq 0.45\ fm$, representing a clear improvement with respect to the findings of [14].

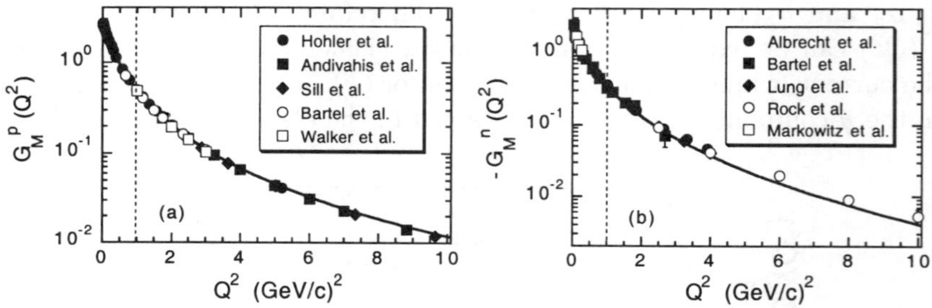

Figure 2. Proton (a) and neutron (b) magnetic form factors versus Q^2. The solid lines are our LF results (5) obtained adopting the OGE model and including CQ form factors in the one-body approximation (1). The dotted vertical lines correspond to $Q^2 = 1\ (GeV/c)^2$, which limits the range of values of Q^2 used for fixing the parameters of the CQ form factors (see text). The data set is the same as in [14].

Our results for all the nucleon form factors are reported in Figs. 2 and 3 in an extended range of values of Q^2. Note that above $Q^2 \simeq 1\ (GeV/c)^2$ our results should be considered as parameter-free predictions. It can be seen that a qualitative agreement with the data holds in a wide range of values of Q^2, and this is a clear indication that soft physics may play an important (even dominant) role also above the scale of chiral symmetry breaking, at least up to $Q^2 \simeq 10\ (GeV/c)^2$. Note that the Q^2-behavior obtained for $G_E^n(Q^2)$ [see Fig. 3(b)] is in overall agreement with available experimental data, but the neutron charge radius is remarkably underestimated (by $\simeq 50\%$). Such a failure might be ascribed to the lack of explicit many-body currents in our present calculations.

In Fig. 4 our results are presented for the electric to magnetic proton ratio $\mu_p G_E^p(Q^2)/G_M^p(Q^2)$, where $\mu_p \equiv G_M^p(Q^2 = 0)$, and compared with experimental data, including also the recent results from $JLAB$ [16]. It can be seen that above $Q^2 \simeq 1\ (GeV/c)^2$ our approach predicts a sharp suppression from the dipole law in nice agreement with the $JLAB$ data. Moreover, for

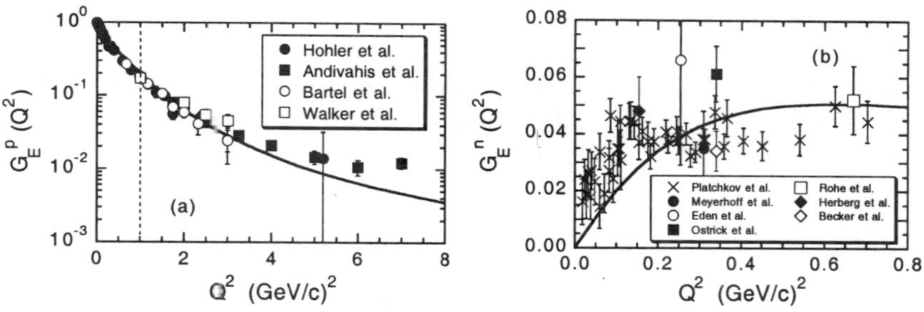

Figure 3. The same as in Fig. 2, but for the nucleon electric form factors. The data are from [14] (a) and [15] (b).

$Q^2 \gtrsim 3$ $(GeV/c)^2$ there is an interesting sensitivity to the inclusion of CQ form factors, which could be further checked by future experiments.

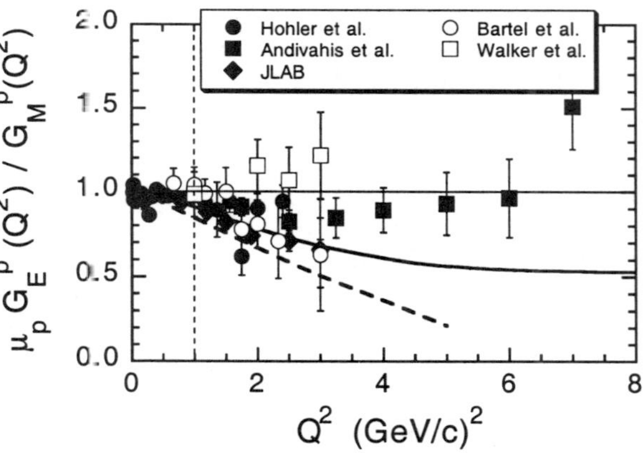

Figure 4. Electric to magnetic ratio $\mu_p G_E^p(Q^2)/G_M^p(Q^2)$ versus Q^2. The solid line is the LF result including CQ form factors, while the dashed line is taken from Ref. [6(b)] and represents our LF calculation assuming point-like CQ's. The dotted vertical line has the same meaning as in Fig. 2. The data set is from [14] and [16].

We have applied Eq. (5) to the calculation of the helicity amplitudes of the $N - P_{11}(1440)$ transition, which is of great interest because of a possible presence of a hybrid component in the Roper $[P_{11}(1440)]$ resonance (see [17]). Our predictions are reported in Fig. 5 and compared with both the scarce experimental data and few theoretical predictions of other approaches. It can be seen that the boost effects reduce significantly both the transverse $A_{1/2} \propto G_M(Q^2)$ and the longitudinal $S_{1/2} \propto G_E(Q^2)/Q^2$ amplitudes (see [18] for their precise definitions). However, the fastest fall-off is exhibited by the hybrid q^3G model. Finally, note that our LF result strongly underestimates the PDG result for $A_{1/2}$ at the photon point.

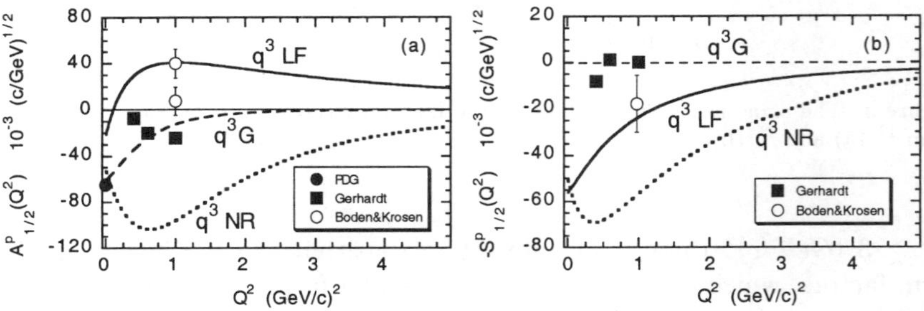

Figure 5. Transverse (a) and longitudinal (b) helicity amplitudes for the $p - P_{11}(1440)$ transition. The solid lines correspond to our LF prediction, while the dotted and dashed lines are the result of the non-relativistic q^3 and of the hybrid q^3G models obtained in [17]. The meaning of the markers is the same as in [18].

4 Conclusions

The application of relativistic CQ models to the evaluation of the e.m. properties of the nucleon and its resonances has been addressed. The role of the pair creation process in the Feynmann triangle diagram has been discussed and the importance both of choosing the LF formalism and of using a Breit frame where $q^+ = 0$, has been stressed. The nucleon elastic form factors have been calculated free of the spurious effects related to the loss of rotational covariance. The effect of a finite CQ size has been considered and the parameters describing the CQ form factors have been fixed using only the nucleon data up to $Q^2 \simeq 1\ (GeV/c)^2$; a CQ charge radius of $\simeq 0.45\ fm$ has been obtained in this way. Above $Q^2 \simeq 1\ (GeV/c)^2$ our parameter-free predictions are found to be in remarkable agreement with the experimental data, showing that soft physics may be dominant up to $Q^2 \sim 10\ (GeV/c)^2$. Our results for the longitudinal and transverse helicity amplitudes of the $N - P_{11}(1440)$ transition have been presented. Finally, let us stress that the results presented in this work supersede those of Refs. [14] and [18].

References

1. For a recent review see G.S. Bali, *Phys. Rept.* **343**, 1 (2001).
2. V.D. Burkert, *Czech. J. Phys.* **46**, 627 (1996).
3. S. Godfrey and N. Isgur, *Phys. Rev.* D **32**, 185 (1985); S. Capstick and N. Isgur, *Phys. Rev.* D **34**, 2809 (1986).
4. F. Cardarelli *et al.*, *Few Body Syst. Suppl.* **11**, 66 (1999).
5. L. Glozmann *et al.*, *Phys. Rev.* C **57**, 3406 (1998); *Phys. Rev.* D **58**, 094030 (1998).
6. F. Cardarelli and S. Simula, *Phys. Lett.* B **467**, 1 (1999); *Phys. Rev.* C **62**, 065201 (2000).
7. L.L. Frankfurt and M.I. Strikman, *Nucl. Phys.* B **148**, 107 (1979); G.P. Lepage and S.J. Brodsky, *Phys. Rev.* D **22**, 2157 (1980); T. Frederico and G.A. Miller, *Phys. Rev.* D **45**, 4207 (1992); N.B. Demchuk *et al.*, *Phys. of Atom. Nuclei* **59**, 2152 (1996).
8. W.H. Klink, *Phys. Rev.* C **58**, 3587 (1998).
9. J.P.B.C. de Melo *et al.*, *Phys. Rev.* C **59**, 2278 (1999); J.P.B.C. de Melo and T. Frederico, *Phys. Rev.* C **55**, 2043 (1997); J.P.B.C. de Melo *et al.*, *Nucl. Phys.* A **660**, 219 (1999).
10. V.A. Karmanov and A.V. Smirnov, *Nucl. Phys.* A **546**, 691 (1992); *ib.* A **575**, 520 (1994). For a recent review see J. Carbonell *et al.*, *Phys. Rept.* **300**, 215 (1998).
11. F. Cardarelli *et al.*, *Phys. Lett.* B **349**, 393 (1995); V.A. Karmanov, *Nucl. Phys.* A **608**, 316 (1996).
12. F. Cardarelli *et al.*, *Phys. Lett.* B **371**, 7 (1996); *Nucl. Phys.* A **623**, 361 (1997); E. Pace *et al.*, *Nucl. Phys.* A **666&667**, 33c (2000).
13. V.A. Karmanov and J.-F. Mathiot, *Nucl. Phys.* A **602**, 338 (1996).
14. F. Cardarelli *et al.*, *Phys. Lett.* B **357**, 267 (1995).
15. S. Platchkov *et al.*, *Nucl. Phys.* A **510**, 740 (1990); M. Meyerhoff *et al.*, *Phys. Lett.* B **327**, 201 (1994); T. Eden *et al.*, *Phys. Rev.* C **50**, R1749 (1994); M. Ostrick *et al.*, *Phys. Rev. Lett.* **83**, 276 (1999); D. Rohe *et al.*, *Phys. Rev. Lett.* **83**, 4257 (1999); C. Herberg *et al.*, *Eur. Phys. J.* A **5**, 131 (1999); J. Becker *et al.*, *Eur. Phys. J.* A **6**, 329 (1999).
16. M.K. Jones *et al.*, *Phys. Rev. Lett.* **84**, 1398 (2000).
17. Z. Li and V.D. Burkert, *Phys. Rev.* D **46**, 70 (1992).
18. F. Cardarelli *et al.*, *Phys. Lett.* B **397**, 13 (1997).

Stefan Scherer

Valery Lyubovitskij

GENERALIZED POLARIZABILITIES IN A CONSTITUENT QUARK MODEL

B. PASQUINI,[1] S. SCHERER,[2] AND D. DRECHSEL[2]
[1] *ECT*, Strada delle Tabarelle 286, I-38050 Villazzano (Trento), and INFN, Trento, Italy*

[2] *Institut für Kernphysik, Johannes Gutenberg-Universität, J. J. Becher-Weg 45, D-55099 Mainz, Germany*

We discuss low-energy virtual Compton scattering off the proton within the framework of a nonrelativistic constituent quark model. A simple interpretation of the spin-averaged generalized polarizabilities is given in terms of the induced electric polarization (and magnetization). Explicit predictions for the generalized polarizabilities obtained from a multipole expansion are presented for the Isgur-Karl model and are compared with results of various models.

1 Introduction

In recent years, low-energy virtual Compton scattering (VCS) off the proton as tested in, e.g., the reaction $e^-p \to e^-p\gamma$, has attracted considerable interest from both the experimental[1] and theoretical[2] side. In fact, real Compton scattering (RCS) has a long history of providing important theoretical and experimental tests for models of hadron structure. For example, the famous low-energy theorem (LET) of Low and Gell-Mann and Goldberger[3] predicts the scattering amplitude for a spin-$\frac{1}{2}$ system in terms of the charge, mass, and magnetic moment to the first two orders in the photon energy. In principle, any model respecting the symmetries entering the derivation of the LET should reproduce the constraints of the LET. It is only terms of second order which contain new information on the structure of the nucleon specific to Compton scattering. For a general target, these effects can be parametrized in terms of two constants, the electric and magnetic polarizabilities $\bar{\alpha}$ and $\bar{\beta}$, respectively[4]. A generalized low-energy theorem (GLET) including virtual photons was given in Refs. [5,6]. The use of virtual photons considerably increases the possibilities to investigate the structure of the target, because on the one hand the energy and three-momentum of the virtual photon can be varied independently, and on the other hand also longitudinal components of current operators become accessible. In Ref. [5], the model-dependent response beyond the GLET was analyzed by means of a multipole expansion. Only terms contributing to first order in the energy of the outgoing real photon were kept, and the result was expressed in terms of ten generalized polarizabilities (GPs) which are functions of the three-momentum of the initial-state virtual photon. If charge conjugation and particle crossing are symmetries of the underlying model, only six of the ten GPs are independent[7]. A somewhat different generalization in terms of generalized dipole polarizabilities was obtained in Ref. [8], where a covariant starting point was chosen.

2 Compton Tensor and Low-Energy Theorem

In a covariant framework, the nucleon Compton tensor is defined as

$$(2\pi)^4\delta^4(p_f + q' - p_i - q)M_{fi}^{\mu\nu}(p_f, q'; p_i, q) = \int d^4x\, d^4y\, e^{-iq\cdot x}e^{iq'\cdot y}\langle N(p_f)|T[J^\mu(y)J^\nu(x)]|N(p_i)\rangle, \tag{1}$$

where T refers to the covariant time-ordered product. As a special case, the RCS amplitude ($q^2 = q'^2 = 0$) is obtained by contracting Eq. (1) with the polarization vectors ϵ_ν and ϵ'^*_μ of the initial and final photons, respectively,[a]

$$\mathcal{M}_{\rm RCS} = -ie^2\epsilon'^*_\mu\epsilon_\nu M_{fi}^{\mu\nu}(p_f, q'; p_i, q). \tag{2}$$

The invariant amplitude for $\gamma^* N \to \gamma N$ reads

$$\mathcal{M}_{\rm VCS} = ie^2\left(\vec{\epsilon}_T\cdot\vec{M}_T + \frac{q^2}{\omega^2}\epsilon_z M_z\right), \tag{3}$$

where $\epsilon_\nu = e\bar{u}\gamma_\nu u/q^2$ is the polarization vector of the virtual photon, and where we made use of current conservation. In the center-of-mass (c.m.) frame, using the Coulomb gauge for the final (real) photon, the transverse and longitudinal parts of $\mathcal{M}_{\rm VCS}$ can be expressed in terms of eight and four independent structures, respectively,

$$\vec{\epsilon}_T\cdot\vec{M}_T = \vec{\epsilon}\,'^*\cdot\vec{\epsilon}_T A_1 + \cdots, \quad M_z = \vec{\epsilon}\,'^*\cdot\hat{q}A_9 + \cdots, \tag{4}$$

where the functions A_i depend on the three kinematical variables $|\vec{q}|$, $|\vec{q}\,'|$, and $z = \hat{q}\cdot\hat{q}\,'$.

Model-independent predictions for the functions A_i, based on Lorentz invariance, gauge invariance, crossing symmetry, and the discrete symmetries were obtained in Ref. [6]. For example, the result for A_1 up to second order in the momenta $|\vec{q}|$ and $|\vec{q}\,'|$ reads

$$A_1 = -\frac{1}{M} + \frac{z}{M^2}|\vec{q}| - \left(\frac{1}{8M^3} + \frac{r_E^2}{6M} - \frac{\kappa}{4M^3} - \frac{4\pi\bar{\alpha}}{e^2}\right)|\vec{q}\,'|^2$$

---

[a] We use natural units $\hbar = c = 1$, $e > 0$, $e^2/(4\pi) \approx 1/137$.

$$+\left(\frac{1}{8M^3}+\frac{r_E^2}{6M}-\frac{z^2}{M^3}+\frac{(1+\kappa)\kappa}{4M^3}\right)|\vec{q}|^2. \tag{5}$$

To this order, all functions A_i are completely specified in terms of quantities which can be obtained from elastic electron-proton scattering and RCS, namely M, κ, G_E, G_M, r_E^2, $\bar{\alpha}$, and $\bar{\beta}$.

3 Generalized Polarizabilities and their Interpretation

A systematic analysis of the structure-dependent terms specific to VCS off the nucleon was performed by Guichon *et al.*[5]. The nonpole terms were expressed in terms of three spin-independent and seven spin-dependent GPs which were introduced in the framework of a multipole analysis, restricted to first order in $|\vec{q}\,'|$, but for arbitrary $|\vec{q}|$.

For a spin-0 system such as the pion or for the spin-averaged nucleon amplitude the GPs can, to a certain extent, be interpreted in terms of simple classical concepts[8]. To that end, let us consider a spherical charge distribution which is exposed to a uniform static electric field. The connection between the electric field and the induced electric polarization is, to first order, given by the polarizability tensor $\alpha_{ij}(\vec{r})$ (see Fig. 1),

$$P_i(\vec{r}) = 4\pi\alpha_{ij}(\vec{r})E_j. \tag{6}$$

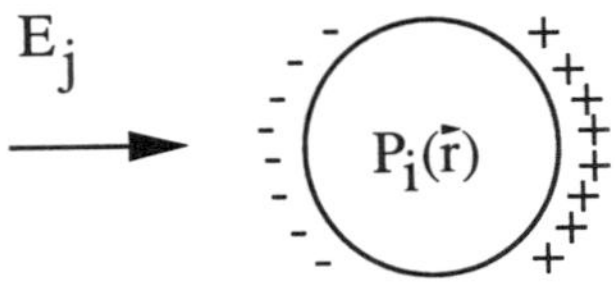

Figure 1. The electric field induces an electric polarization.

Defining

$$\alpha_{ij}(\vec{r}) = \int \frac{d^3q}{(2\pi)^3}\exp(-i\vec{q}\cdot\vec{r})\alpha_{ij}(\vec{q}), \tag{7}$$

spherical symmetry imposes the general form

$$\alpha_{ij}(\vec{q}) = \alpha_L(q)\underbrace{\hat{q}_i\hat{q}_j}_{\text{longitudinal}} + \alpha_T(q)\underbrace{(\delta_{ij}-\hat{q}_i\hat{q}_j)}_{\text{transverse}}, \tag{8}$$

where $q = |\vec{q}|$. We now consider the form factor

$$F(\vec{q}) = \int d^3r\exp(i\vec{q}\cdot\vec{r})\rho_{\text{eff}}(\vec{r}) \tag{9}$$

associated with the effective charge density

$$\rho_{\text{eff}}(\vec{r}) = \rho_0(r) - \vec{\nabla} \cdot \vec{P}(\vec{r}). \tag{10}$$

Inserting Eq. (10) into Eq. (9), performing a partial integration, and finally applying the definition of Eq. (7) with the specific form of Eq. (8), one obtains

$$\begin{aligned} F(\vec{q}) &= F_0(q) - \int d^3r \exp(i\vec{q}\cdot\vec{r})\vec{\nabla}\cdot\vec{P}(\vec{r}) \\ &= F_0(q) + i\vec{q}\cdot \int d^3r \exp(i\vec{q}\cdot\vec{r})\vec{P}(\vec{r}) \\ &= F_0(q) + 4\pi i q_i \int d^3r \exp(i\vec{q}\cdot\vec{r})\alpha_{ij}(\vec{r})E_j \\ &= F_0(q) + 4\pi i\vec{q}\cdot\vec{E}\alpha_L(q). \end{aligned} \tag{11}$$

Equation (11) nicely displays the connection between the longitudinal generalized polarizability $\alpha_L(q)$ and the modification in the charge distribution. The general connection between the polarizability tensor $\alpha_{ij}(\vec{r})$ and the generalized electric polarizabilities $\alpha_L(q)$ and $\alpha_T(q)$ of Eq. (8) reads[8]

$$\begin{aligned} \alpha_{ij}(\vec{r}) &= \alpha_L(r)\hat{r}_i\hat{r}_j + \alpha_T(r)(\delta_{ij} - \hat{r}_i\hat{r}_j) \\ &\quad + \frac{3\hat{r}_i\hat{r}_j - \delta_{ij}}{r^3}\int_r^\infty \left[\alpha_L(r') - \alpha_T(r')\right] r'^2\, dr'. \end{aligned} \tag{12}$$

In particular, both generalized electric polarizabilities are required to fully describe the induced electric polarization. The transverse generalized electric polarizability $\alpha_T(q)$ is associated with rotational displacements of charges which do not contribute to $\delta\rho(\vec{r}) = -\vec{\nabla}\cdot\vec{P}(\vec{r})$. It is *not* contained in the approximation of Guichon *et al.*[5].

The above classical considerations can easily be extended to nonrelativistic quantum mechanics. For example, for a simple harmonic oscillator model of the type

$$(M \to \infty, e) \bullet - - - - - - - \bullet (m, -e): \quad V = \frac{1}{2} m\omega_0^2 r^2, \tag{13}$$

the form factor without external field reads

$$F_0(q) = e\left[1 - \exp\left(-\frac{q^2}{4\omega_0 m}\right)\right]. \tag{14}$$

It is straightforward to obtain an analytical expression for the form factor in the presence of a static uniform electric field $\vec{E}$,

$$F(\vec{q}) = e\left[1 - \exp\left(-\frac{q^2}{4\omega_0 m}\right)\exp\left(-ie\frac{\vec{q}\cdot\vec{E}}{m\omega_0^2}\right)\right]$$

$$= F_0(q) + 4\pi i\vec{q}\cdot\vec{E}\underbrace{\frac{e^2}{4\pi m\omega_0^2}\exp\left(-\frac{q^2}{4\omega_0 m}\right)}_{\alpha_L(q)} + \cdots. \tag{15}$$

As $q \to 0$, $\alpha_L(q)$ reduces to the ordinary electric polarizability $\alpha = e^2/(4\pi m\omega_0^2)$ which is commonly interpreted as a measure of the stiffness or rigidity of a system[9].

4 Compton Tensor in a Nonrelativistic Framework

Before presenting the results for the GPs in the framework of Guichon *et al.*[5], we provide a short outline of the ingredients entering the calculation of the Compton tensor[10]. The starting point is the nonrelativistic interaction Hamiltonian in the Schrödinger picture,

$$H_I = H_{I,1} + H_{I,2}, \tag{16}$$

$$H_{I,1} = \int d^3x\, J^\mu(\vec{x})A_\mu(\vec{x}), \tag{17}$$

$$H_{I,2} = \frac{1}{2}\int d^3x \int d^3x' B^{\mu\nu}(\vec{x},\vec{x}\,')A_\mu(\vec{x})A_\nu(\vec{x}\,'), \tag{18}$$

where $A^\mu(\vec{x})$ is the second-quantized photon field and

$$\vec{J}(\vec{x}) = \sum_{\alpha=1}^{N}\frac{e_\alpha}{2m_\alpha}\delta^3(\vec{x}-\vec{r}_\alpha)\left(\frac{\overrightarrow{\nabla}_\alpha}{i} - \vec{\sigma}_\alpha\times\overrightarrow{\nabla}_\alpha\right) + h.c., \tag{19}$$

$$\rho(\vec{x}) = \sum_{\alpha=1}^{N} e_\alpha\delta^3(\vec{x}-\vec{r}_\alpha), \tag{20}$$

$$B^{\mu 0} = B^{0\nu} = 0, \quad B^{ij}(\vec{x},\vec{x}\,') = \delta_{ij}\sum_{\alpha=1}^{N}\frac{e_\alpha^2}{m_\alpha}\delta^3(\vec{x}-\vec{r}_\alpha)\delta^3(\vec{x}\,'-\vec{r}_\alpha). \tag{21}$$

The Compton tensor is obtained by calculating the contributions of $H_{I,2}$ and $H_{I,1}$ in first-order and second-order perturbation theory, respectively,

$$M_{fi}^{\mu\nu}(q',q,\vec{p}) = S_{fi}^{\mu\nu}(q',q) + T_{fi}^{\mu\nu}(q',q,\vec{p}), \tag{22}$$

where $\vec{p} = (\vec{p}_i + \vec{p}_f)/2$. The contribution from the "seagull" terms is given by

$$S_{fi}^{\mu 0} = S_{fi}^{0\nu} = 0, \quad S_{fi}^{ij} = \delta_{ij}\langle 0'|\sum_\alpha \frac{e_\alpha^2}{m_\alpha}e^{i(\vec{q}-\vec{q}\,')\cdot\vec{r}\,'_\alpha}|0\rangle, \tag{23}$$

where $\vec{r}'_\alpha$ refers to the intrinsic coordinates of particle α relative to the center of mass. Note that $S^{\mu\nu}_{fi}$ is symmetric under photon crossing, $q^\mu \leftrightarrow -q'^\mu$ and $\nu \leftrightarrow \mu$. The explicit form of the direct and crossed channels reads

$$T^{\mu\nu}_{fi} = \sum_X \frac{\langle 0'|J^\mu(-\vec{q}\,', 2\vec{p}_f + \vec{q}\,')|X\rangle\langle X|J^\nu(\vec{q}, 2\vec{p}_i + \vec{q})|0\rangle}{E_f(\vec{p}_f) + \omega' - E_X(\vec{p}_f + \vec{q}\,')}$$
$$+\text{crossed-channel sum}, \tag{24}$$

where every single intermediate state X gives rise to a contribution which is explicitly symmetric under photon crossing. This will be of some importance in checking the predictions for the GPs. For further evaluation, the current operator is divided into intrinsic and c.m. contributions[11,12]

$$\vec{J}(\vec{q},\vec{p}) = \vec{\jmath}^{\,in}(\vec{q}) + \frac{\vec{p}}{M}\rho(\vec{q}), \quad J_0(\vec{q},\vec{p}) = \rho(\vec{q}). \tag{25}$$

5 Generalized Polarizabilities in a Constituent Quark Model

Neither $S^{\mu\nu}_{fi}$ nor $T^{\mu\nu}_{fi}$ of Eqs. (23) and (24), respectively, is separately gauge invariant. Thus, before calculating the GPs we divide $M^{\mu\nu}_{fi}$ into two separately gauge-invariant contributions. The first consists of the ground-state propagation in the direct and crossed channels together with an appropriately chosen term to satisfy gauge invariance. The residual part contains the relevant structure information from which one obtains the GPs.

For the calculation of the GPs we make use of two different expansion schemes. The first consists of a conventional nonrelativistic $1/M$ expansion. The advantage of this approach is that, at leading order in $1/M$, the result does not depend on the average nucleon momentum $\vec{p} = (\vec{p}_i + \vec{p}_f)/2$. In the c.m. frame $\vec{p} = -(\vec{q} + \vec{q}\,')/2$ such that naive photon crossing—implying the replacement of *any* $\vec{q} \leftrightarrow -\vec{q}\,'$—would turn $\vec{p}$ into $-\vec{p}$. However, at leading order in $1/M$, naive photon crossing is the same as true photon crossing which, at that order, leads to constraints for the GPs as $|\vec{q}| \to 0$ (see Ref. [10] for details).

Here we present the results of a second scheme used by Liu *et al.* in Ref. [13]. There, the crossed-channel contribution is divided into a leading plus a recoil part which is defined via matrix elements depending on both $\vec{q}$ and $\vec{q}\,'$. Furthermore, in order to incorporate, to some extent, effects due to relativity, relativistic expressions for the energies of the intermediate states were used. Since particle crossing is not a symmetry of the nonrelativistic constituent quark model (NRCQM), the results do *not* satisfy the relations among the ten GPs which were found in Refs. [7].

The calculation is performed in the framework of the model of Isgur and Karl [14], where the $\Delta(1232)$ resonance and low-lying negative-parity baryons

$D_{13}(1520)$, $S_{11}(1535)$, $S_{31}(1620)$, $S_{11}(1650)$, $S_{13}(1700)$, $D_{33}(1700)$ have been included. In this model gauge invariance is violated first because of a mismatch between the resonance masses of the intermediate states and the corresponding wave functions and second because of a truncation of the model space. On the other hand, photon crossing symmetry is respected, because for each diagram the crossed diagram is taken into account.

6 Results

The numerical results for six of the GPs are shown in Fig. 2 together with the predictions of the linear sigma model[15] (dashed lines), heavy-baryon chiral perturbation theory[16] (dashed-dotted lines), an effective Lagrangian model[17] (dotted lines), and dispersion relations[18] (full thin lines). For $|\vec{q}| \to 0$, the GPs α and β reduce to the ordinary RCS polarizabilities $\bar{\alpha}$ and $\bar{\beta}$. In the Isgur-Karl model, these polarizabilities receive their main contributions from the resonances $D_{13}(1520)$ and $\Delta(1232)$, respectively. In case of the spin GPs $P^{(01,01)1}$ and $P^{(11,02)1}$, the NRCQM generates much smaller values than the linear σ model[15] and chiral perturbation theory[16]. This is interpreted as a consequence of missing pionic degrees of freedom.

For the generalized electric polarizability

$$\alpha(|\vec{q}|) = -\frac{e^2}{4\pi}\sqrt{\frac{3}{2}}P^{(01,01)0}(|\vec{q}|),$$

our result is in agreement with Ref. [13]. Discrepancies in $P^{(01,01)1}$, $P^{(11,11)1}$, and $P^{(01,12)1}$ were found to originate from the contributions of the crossed channel relative to the direct one (for details, see Ref. [10]).

The results for those structure functions that can be accessed in an unpolarized experiment, are shown in Table 1. The values of $|\vec{q}|$ correspond to the RCS limit, MIT-Bates[19] and MAMI[1] kinematics, respectively. Table 2

Table 1. Structure functions P_{LL}, P_{TT}, and P_{LT} in GeV^{-2}.

	P_{LL}	P_{TT}	P_{LT}
$\lvert\vec{q}\rvert = 0$ MeV	37.0	−0.1	−11.2
$\lvert\vec{q}\rvert = 240$ MeV	28.7	−2.8	−8.8
$\lvert\vec{q}\rvert = 600$ MeV	9.9	−5.8	−3.2

contains a comparison of our results with experiment[1].

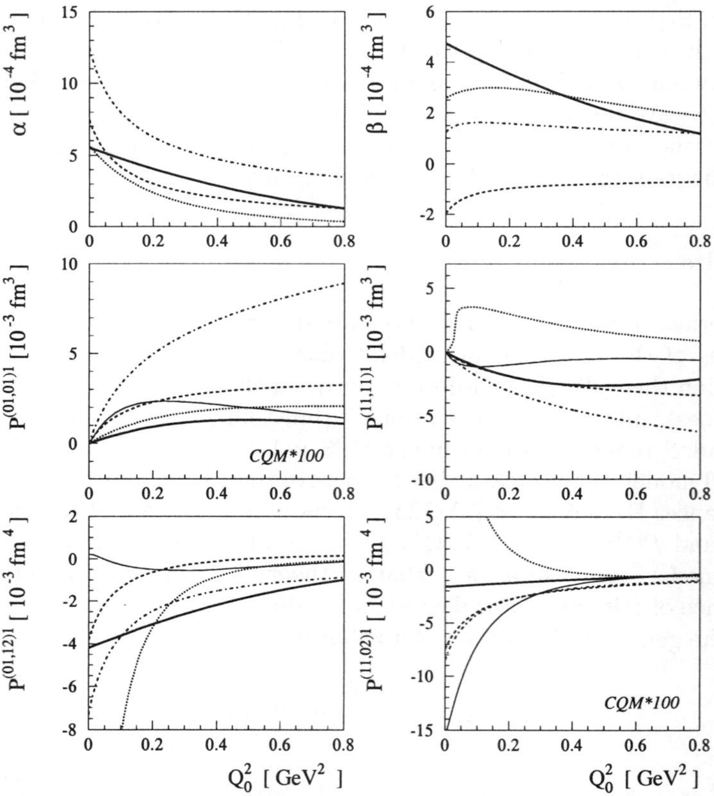

Figure 2. Results for GPs in different model calculations as function of squared momentum transfer at $\omega' = 0$, $Q_0^2 = Q^2|_{\omega'=0}$. Full thick lines: our results in the NRCQM with the scheme of Ref. [13]; dashed lines: linear sigma model[15]; dashed-dotted lines: heavy-baryon chiral perturbation theory[16]; dotted lines: effective Lagrangian model[17]; full thin lines: dispersion-relation calculation[18]. Note the scaling of the CQM value for two spin polarizabilities by a factor 100.

7 Summary

Low-energy virtual Compton scattering opens the door to a new and rich realm of issues regarding the structure and dynamics of composite systems—namely, what is the *local* deformation of a system exposed to a (soft) external stimulus.

Table 2. Structure functions $P_{LL} - P_{TT}/\epsilon$ and P_{LT} for $|\vec{q}| = 600$ MeV in GeV^{-2} ($\epsilon = 0.62$).

	$P_{LL} - P_{TT}/\epsilon$	P_{LT}
Our calculation	19.2	-3.2
Experiment[1]	23.7 $\pm 2.2 \pm 0.6 \pm 4.3$	-5.0 $\pm 0.8 \pm 1.1 \pm 1.4$

In an expansion in terms of the energy of the final-state real photon and the three-momentum of the initial-state virtual photon, the GLET specifies the accuracy needed to be sensitive to higher-order terms in order to distinguish between different models. This structure information beyond the GLET can be parametrized in terms of generalized polarizabilities. To some extent, these quantities have their analogies in classical physics and can be interpreted in terms of the induced electric polarization and magnetization. We discussed the virtual Compton scattering tensor in the framework of nonrelativistic QM and made use of the model of Isgur and Karl to obtain predictions for the GPs of the proton. The model does not provide the relations among the GPs due to nucleon crossing in combination with charge conjugation. Clearly, the calculation has its limitations regarding relativity, gauge invariance and chiral symmetry, leaving room for improvement in any of the above-mentioned directions. Despite these shortcomings, the predictions provide an order-of-magnitude estimate for the nucleon resonance contributions and as such are complementary to the results of the linear sigma model and chiral perturbation theory that emphasize pionic degrees of freedom and chiral symmetry.

Acknowledgments

S. Scherer would like to thank A. I. L'vov for numerous interesting and fruitful discussions. This work was supported by the Deutsche Forschungsgemeinschaft (SFB 443).

References

1. J. Roche *et al.*, *Phys. Rev. Lett.* **85**, 708 (2000).
2. See, e.g., S. Scherer, *Czech J. Phys.* **49**, 1307 (1999), and references therein.
3. F. E. Low, *Phys. Rev.* **96**, 1428 (1954); M. Gell-Mann and M. L. Goldberger, *ibid.*, 1433 (1954).
4. A. Klein, *Phys. Rev.* **99**, 998 (1955).

5. P. A. M. Guichon, G. Q. Liu, and A. W. Thomas, *Nucl. Phys.* A **591**, 606 (1995).
6. S. Scherer, A. Yu. Korchin, and J. H. Koch, *Phys. Rev.* C **54**, 904 (1996).
7. D. Drechsel, G. Knöchlein, A. Metz, and S. Scherer, *Phys. Rev.* C **55**, 424 (1997); D. Drechsel, G. Knöchlein, A. Yu. Korchin, A. Metz, and S. Scherer, *ibid.* **57**, 941 (1998); *ibid.* **58**, 1751 (1998).
8. A. I. L'vov, S. Scherer, B. Pasquini, C. Unkmeir, and D. Drechsel, hep-ph/0103172, to appear in *Phys. Rev.* C **64**, (2001).
9. B. R. Holstein, *Comments Nucl. Part. Phys.* **19**, 221 (1990).
10. B. Pasquini, S. Scherer, and D. Drechsel, *Phys. Rev.* C **63**, 025205 (2001).
11. J. L. Friar, *Ann. Phys. (N.Y.)* **95**, 170 (1975).
12. H. Arenhövel, NATO Advanced Study Institute on New Vistas in Electro-Nuclear Physics, 1985, Banff, Alberta (Plenum Press, New York, 1986).
13. G. Q. Liu , A. W. Thomas, and P. A. M. Guichon, *Aust. J. Phys.* **49**, 905 (1996).
14. N. Isgur and G. Karl, *Phys. Rev.* D **18**, 4187 (1978).
15. A. Metz and D. Drechsel, *Z. Phys.* A **356**, 351 (1996); *ibid.* **359**, 165 (1997).
16. T. R. Hemmert, B. R. Holstein, G. Knöchlein, and S. Scherer, *Phys. Rev.* D **55**, 2630 (1997); *Phys. Rev. Lett.* **79**, 22 (1997); T. R. Hemmert, B. R. Holstein, G. Knöchlein, and D. Drechsel, *Phys. Rev.* D **62**, 014013 (2000).
17. A. Yu. Korchin and O. Scholten, *Phys. Rev.* C **58**, 1098 (1998).
18. B. Pasquini, D. Drechsel, M. Gorchtein, A. Metz, and M. Vanderhaeghen, *Phys. Rev.* C **62**, 052201 (2000); hep-ph/0102335, submitted to *Eur. Phys. J.*
19. J. Shaw *et al.*, MIT–Bates proposal No. 97-03, 1997.

NUCLEON PROPERTIES IN THE PERTURBATIVE CHIRAL QUARK MODEL

V. E. LYUBOVITSKIJ[†,*], TH. GUTSCHE[†] AND A. FAESSLER[†]

[†]*Institut für Theoretische Physik, Universität Tübingen, Auf der Morgenstelle 14, D-72076 Tübingen, Germany,*

[*]*Department of Physics, Tomsk State University, 634050 Tomsk, Russia,*

E-mail: Valeri.Lyubovitskij@uni-tuebingen.de, Thomas.Gutsche@uni-tuebingen.de, Amand.Faessler@uni-tuebingen.de

We apply the perturbative chiral quark model (PCQM) to analyse low-energy nucleon properties: electromagnetic form factors, meson-nucleon sigma-terms and pion-nucleon scattering. Baryons are described as bound states of valence quarks surrounded by a cloud of Goldstone bosons (π, K, η) as required by chiral symmetry. The model is based on the following guide lines: chiral symmetry constraints, fulfilment of low-energy theorems and proper treatment of sea-quarks, that is meson cloud contributions. Analytic expressions for nucleon observables are obtained in terms of fundamental parameters of low-energy pion-nucleon physics (weak pion decay constant, axial nucleon coupling constant, strong pion-nucleon form factor) and of only one model parameter (radius of the nucleonic three-quark core). Our results are in good agreement with experimental data and results of other theoretical approaches.

1 Introduction

Hadron models set up to understand the structure of the nucleon should respect the constraints imposed by chiral symmetry. Spontaneous and explicit chiral symmetry breaking require the existence of the pion whose mass vanishes in the limit of zero current quark mass. In turn, nucleon observables must receive contributions from pion loops.

We recently suggested [1] a baryon model, the perturbative chiral quark model (PCQM), which includes relativistic quark wave functions and confinement and respects chiral symmetry. The PCQM was successfully applied to σ-term physics [1] and extended to the study of electromagnetic properties of the nucleon [2] and πN scattering [3]. Although the Lagrangian for this model fulfills the $SU(2)\times SU(2)$ current algebra, it is in general nontrivial to identify the explicit dynamics which leads to well-known predictions for various observables. For instance, S-wave πN scattering at threshold is such a process which should be correctly described by a nucleon model involving the pion cloud.

2 Perturbative Chiral Quark Model

2.1 Effective Lagrangian and zeroth order properties

Following considerations are based on the perturbative chiral quark model (PCQM), a relativistic quark model suggested in [4] and extended in [1] for the study of low-energy properties of baryons. Similar models have also been studied in Refs. [5]. The PCQM is based on the effective, chirally invariant Lagrangian $\mathcal{L}_{inv}$ [1]

$$\mathcal{L}_{inv} = \bar{\psi}\Big\{ i \not\partial - \gamma^0 V(r) - S(r) \Big[\frac{U + U^\dagger}{2} + \gamma^5 \frac{U - U^\dagger}{2} \Big] \Big\} \psi + \frac{F^2}{4} \mathrm{Tr}[\, \partial_\mu U \, \partial^\mu U^\dagger \,] \tag{1}$$

where ψ is the quark field, U is the chiral field and $F = 88$ MeV is the pion decay constant in the chiral limit [1,6]. The quarks move in a self-consistent field, represented by scalar $S(r)$ and vector $V(r)$ components of an effective static potential providing confinement. The interaction of quarks with Goldstone bosons is introduced on the basis of the nonlinear σ-model [7]. We define the chiral field as $U = \exp[i\hat{\Phi}/F]$ using the *exponentional parametrization* of the nonlinear σ-model, where $\hat{\Phi}$ is the matrix of pseudoscalar mesons. When treating mesons fields as small fluctuations we do perturbation theory in the expansion parameter $1/F$. Usually, the Lagrangian (1) is linearized with respect to the field $\hat{\Phi}$. Such an approximation is valid if we consider processes without free pions (e.g. nucleon mass shift due to the pion cloud) or with a single external pion field (e.g. pion-nucleon form factor). The resulting approximate chiral invariance of the linearized Lagrangian [1,8] guarantees both a conserved axial current (or PCAC in the presence of the meson mass term) and the Goldberger-Treiman relation. However, when we study πN scattering the quadratic term in the expansion of the chiral field U should be kept. With the additional nonlinear term we reproduce [3] the model-independent result for the πN S-wave scattering lengths as derived by derived by Weinberg and Tomozawa [9].

Here we concentrate on the application of the PCQM to the meson-nucleon σ-terms [1] and the electromagnetic properties of the nucleon [2]. We use the effective Lagrangian $\mathcal{L}_{eff} = \mathcal{L}_{inv}^{lin} + \mathcal{L}_{\chi SB}$ including the linearized chiral-invariant term $\mathcal{L}_{inv}^{lin}$ and a mass term $\mathcal{L}_{\chi SB}$ which explicitly breaks chiral symmetry

$$\mathcal{L}_{inv}^{lin} = \bar{\psi}[i \not\partial - S(r) - \gamma^0 V(r)]\psi + \frac{1}{2}(\partial_\mu \hat{\Phi})^2 - \bar{\psi} S(r) i\gamma^5 \frac{\hat{\Phi}}{F} \psi,$$

$$\mathcal{L}_{\chi SB} = -\bar{\psi}\mathcal{M}\psi - \frac{B}{8}\mathrm{Tr}\{\hat{\Phi}, \{\hat{\Phi}, \mathcal{M}\}\}, \tag{2}$$

where $\mathcal{M} = \mathrm{diag}\{\hat{m}, \hat{m}, m_s\}$ is the mass matrix of current quarks (here and in the following we restrict to the isospin symmetry limit with $m_u = m_d = \hat{m}$); B is the low-energy constant which measures the vacuum expectation value of the scalar quark densities in the chiral limit [10]. We rely on the standard picture of chiral symmetry breaking [10] and for the masses of pseudoscalar mesons we use the leading term in their chiral expansion (i.e. linear in the current quark mass): $M_\pi^2 = 2\hat{m}B$, $M_K^2 = (\hat{m} + m_s)B$, $M_\eta^2 = \frac{2}{3}(\hat{m} + 2m_s)B$. Meson masses satisfy the Gell-Mann-Oakes-Renner and the Gell-Mann-Okubo relation $3M_\eta^2 + M_\pi^2 = 4M_K^2$. In the evaluation we use the following set of QCD parameters: $\hat{m}$= 7 MeV, $m_s/\hat{m}$=25 and $B = M_{\pi^+}^2/(2\hat{m})$=1.4 GeV. Introduction of the electromagnetic field A_μ into the PCQM is accomplished by standard minimal substitution into Eq. (2).

To describe the properties of baryons which are modelled as bound states of valence quarks surrounded by a meson cloud we formulate perturbation theory. In our approach the mass (energy) m_N^{core} of the three-quark core of the nucleon is related to the single quark energy $\mathcal{E}_0$ by $m_N^{core} = 3 \cdot \mathcal{E}_0$. For the unperturbed three-quark state we introduce the notation $|\phi_0>$ with the appropriate normalization $<\phi_0|\phi_0>= 1$. The single quark ground state energy $\mathcal{E}_0$ and wave function (w.f.) $u_0(\vec{x})$ are obtained from the Dirac equation

$$[-i\vec{\alpha}\vec{\nabla} + \beta S(r) + V(r) - \mathcal{E}_0]u_0(\vec{x}) = 0. \tag{3}$$

The quark w.f. $u_0(\vec{x})$ belongs to the basis of potential eigenstates (including excited quark and antiquark solutions) used for expansion of the quark field operator $\psi(x)$. Here we restrict the expansion to the ground state contribution with $\psi(x) = b_0 u_0(\vec{x}) \exp(-i\mathcal{E}_0 t)$, where b_0 is the corresponding single quark annihilation operator. In Eq. (3) the current quark mass is not included to simplify our calculational technique. Instead we consider the quark mass term as a small perturbation.

For a given form of the potentials $S(r)$ and $V(r)$ the Dirac equation (3) can be solved numerically. Here, for the sake of simplicity, we use a variational *Gaussian ansatz* [1] for the quark wave function given by the analytical form:

$$u_0(\vec{x}) = N \exp\left[-\frac{\vec{x}^2}{2R^2}\right] \begin{pmatrix} 1 \\ i\rho\,\dfrac{\vec{\sigma}\vec{x}}{R} \end{pmatrix}, \tag{4}$$

where $N = [\pi^{3/2}R^3(1 + 3\rho^2/2)]^{-1/2}$ is a constant fixed by the normalization condition $\int d^3x\, u_0^\dagger(x)\, u_0(x) \equiv 1$; χ_s, χ_f, χ_c are the spin, flavor and color quark wave functions, respectively. Our Gaussian ansatz contains two model parameters: the dimensional parameter R and the dimensionless parameter ρ. The parameter ρ can be related to the axial coupling constant g_A calculated in zeroth-order (or 3q-core) approximation:

$$g_A = \frac{5}{3}\left(1 - \frac{2\rho^2}{1+\frac{3}{2}\rho^2}\right) = \frac{5}{3}\frac{1+2\gamma}{3}, \tag{5}$$

where $\gamma = 9g_A/10 - 1/2$. The parameter R can be physically understood as the mean radius of the three-quark core and is related to the charge radius $< r_E^2 >^P_{LO}$ of the proton in the leading-order (LO) approximation as [1]

$$< r_E^2 >^P_{LO} = \frac{3R^2}{2}\frac{1+\frac{5}{2}\rho^2}{1+\frac{3}{2}\rho^2} = R^2\left(2 - \frac{\gamma}{2}\right). \tag{6}$$

In our calculations we use the value g_A=1.25 as obtained in ChPT [6]. We therefore have only one free parameter, R. In the numerical evaluation R is varied in the region from 0.55 fm to 0.65 fm.

2.2 Perturbation theory and nucleon mass

Following the Gell-Mann and Low theorem we define the mass shift of the nucleonic three-quark ground state Δm_N due to the interaction with Goldstone mesons as

$$\Delta m_N \doteq {}^N\!\!<\phi_0|\sum_{n=1}^{\infty}\frac{i^n}{n!}\int i\delta(t_1)\, d^4x_1 \ldots d^4x_n\, T[\mathcal{L}_I(x_1)\ldots\mathcal{L}_I(x_n)]|\phi_0>^N_c \tag{7}$$

where $\mathcal{L}_I = -S(r)\bar{\psi}\, i\gamma^5(\hat{\Phi}/F)\,\psi$ is the interaction Lagrangian between mesons and quarks treated as a perturbation and subscript "c" refers to contributions from connected graphs only. We evaluate Eq. (7) at one loop with $o(1/F^2)$ using Wick's theorem and the appropriate propagators. For the quark field we use a Feynman propagator for a fermion in a binding potential. By restricting the summation over intermediate quark states to the ground state we get

$$\begin{aligned} iG_\psi(x,y) &= <\phi_0|T\{\psi(x)\bar{\psi}(y)\}|\phi_0> \\ &\to u_0(\vec{x})\bar{u}_0(\vec{y})\exp[-i\mathcal{E}_0(x_0-y_0)]\theta(x_0-y_0). \end{aligned} \tag{8}$$

For meson fields we use the free Feynman propagator for a boson field. Superscript "N" in Eq. (7) indicates that the matrix elements are projected on the respective nucleon states.

The total nucleon mass including one-loop corrections is given by

$$m_N = 3(\mathcal{E}_0 + \gamma\hat{m}) + \sum_\Phi d_N^\Phi \Pi(M_\Phi^2);\; d_N^\pi = \frac{171}{400}, d_N^K = \frac{6}{19}d_N^\pi, d_N^\eta = \frac{1}{57}d_N^\pi, \tag{9}$$

where d_N^Φ are the recoupling coefficients defining the partial contribution of the π, K and η-meson cloud to the mass shift of the nucleon. The contribution of a finite current quark mass to the nucleon mass shift is taken into account perturbatively (for details see [1]). The self-energy operator $\Pi(M_\Phi^2)$ is given by

$$\Pi(M_\Phi^2) = -\left(\frac{g_A}{\pi F}\right)^2 \int\limits_0^\infty \frac{dp\,p^4}{w_\Phi^2(p^2)} F_{\pi NN}^2(p^2) \qquad (10)$$

where $w_\Phi(p^2) = \sqrt{M_\Phi^2 + p^2}$ is the meson energy with $p = |\vec{p}|$ and $F_{\pi NN}(p^2)$ is the πNN form factor normalized to unity at zero recoil ($p^2 = 0$):

$$F_{\pi NN}(p^2) = \exp\left(-\frac{p^2R^2}{4}\right)\left\{1 + \frac{p^2R^2}{8}\left(1 - \frac{5}{3g_A}\right)\right\}. \qquad (11)$$

2.3 Meson nucleon sigma-terms and Feynman-Hellmann theorem

The scalar density operators S_i^{PCQM} $(i = u, d, s)$, relevant for the calculation of the meson-baryon sigma-terms in the PCQM, are defined as the partial derivatives of the model χSB Hamiltonian $\mathcal{H}_{\chi SB} = -\mathcal{L}_{\chi SB}$ with respect to the current quark mass of i-th flavor m_i:

$$S_i^{PCQM} \doteq \frac{\partial \mathcal{H}_{\chi SB}}{\partial m_i} = S_i^{val} + S_i^{sea}. \qquad (12)$$

S_i^{val} is the set of valence-quark operators coinciding with the ones obtained from the QCD Hamiltonian. The set of sea-quark operators S_i^{sea} arises from the pseudoscalar meson mass term [1]. To calculate meson-baryon sigma-terms we perform the perturbative expansion for the matrix element of the scalar density operator S_i^{PCQM} between unperturbed 3q-core states and then project it onto the respective baryon wave functions.

As an example, the expression for $\sigma_{\pi N}$ is given by

$$\sigma_{\pi N} = \hat{m} < p|S_u^{PCQM} + S_d^{PCQM}|p >= 3\gamma\hat{m} + \sum_\Phi d_N^\Phi \Gamma(M_\Phi^2) \qquad (13)$$

where the first term of the right-hand side of Eq. (13) corresponds to the valence quark, the second to the sea quark contribution. The vertex function $\Gamma(M_\Phi^2)$ is related to the partial derivative of the self-energy operator $\Pi(M_\Phi^2)$ with respect to $\hat{m}$:

$$\Gamma(M_\Phi^2) = \hat{m}\frac{\partial}{\partial \hat{m}} \Pi(M_\Phi^2). \qquad (14)$$

Using Eqs. (9), (13) and (14) we directly prove the Feynman-Hellmann theorem $\sigma_{\pi N} = \hat{m} \cdot \partial m_N / \partial \hat{m}$.

3 Results

3.1 Meson-nucleon sigma-terms

We start our numerical analysis with the πN sigma-term. First, we restrict to the $SU(2)$ flavor picture. Numerically, the contribution of the valence quarks to the πN sigma-term is 13.1 MeV, the contribution of sea quarks at order $o(M_\pi^2)$ is 66.9 ± 5.7 MeV. Here and in the following the error bars are due to variation of the range parameter R of the quark wave function (4) from 0.55 to 0.65 fm. Taking into account higher-order contributions of the sea quarks we have the following result for the πN sigma-term $\sigma_{\pi N}^{\pi}$ (superscript π refers to the SU(2) flavor picture) of

$$\sigma_{\pi N}^{\pi} = 43.3 \pm 4.4\,\mathrm{MeV}. \tag{15}$$

The contributions of kaon and η-meson loops, $\sigma_{\pi N}^{K}$ and $\sigma_{\pi N}^{\eta}$ (superscripts K and η refer to the respective meson cloud contribution) are significantly suppressed relative to the pion cloud and to the valence quark contributions due to the energy denominators in the structure integrals. The same conclusion regarding the suppression of K and η-meson loops was obtained in the cloudy bag model [11]. Numerically, kaon and η-meson cloud contributions are $\sigma_{\pi N}^{K} = 1.7 \pm 0.4$ MeV and $\sigma_{\pi N}^{\eta} = 0.023 \pm 0.006$ MeV. For the πN sigma-term we have the following final value:

$$\sigma_{\pi N} = \sum_{\Phi} \sigma_{\pi N}^{\Phi} = 45 \pm 5\,\mathrm{MeV}. \tag{16}$$

Our result for the πN sigma-term is in perfect agreement with the value of $\sigma_{\pi N} \simeq 45$ MeV deduced by Gasser, Leutwyler and Sainio [12] using dispersion-relation techniques and exploiting the chiral symmetry constraints.

Next we discuss our prediction for the strangeness content of the nucleon y_N which is defined in the PCQM as

$$y_N = \frac{2 < p|S_s^{PCQM}|p >}{< p|S_u^{PCQM} + S_d^{PCQM}|p >}. \tag{17}$$

The direct calculation of the strange-quark scalar density $< p|S_s^{PCQM}|p >$ is completely consistent with the indirect one applying the Feynman-Hellmann theorem $< p|S_s^{PCQM}|p >= \partial m_N / \partial m_s$. The small value of $y_N = 0.076 \pm 0.012$ in our model is due to the suppressed contributions of kaon and η-meson loops. Our prediction for y_N is smaller than the value $y_N \simeq 0.2$ obtained in [12] from an analysis of experimental data on πN phase shifts. On the other hand, our prediction is quite close to the result obtained in the cloudy bag model $y_N \approx 0.05$ [11].

For the KN sigma-terms we obtain

$$\sigma^u_{KN} = 340 \pm 37\,\mathrm{MeV} \;\; \text{and} \;\; \sigma^d_{KN} = 284 \pm 37\,\mathrm{MeV}, \tag{18}$$

which within uncertainties is consistent with values deduced in HBChPT and lattice QCD (see discussion in Ref. [1]). Hopefully, future DAΦNE experiments at Frascati will allow for a determination of the KN sigma-terms and hence for a better knowledge of the strangeness content of the nucleon.

3.2 Electromagnetic properties of the nucleon

We extended our model analysis to the description of basic electromagnetic properties of the nucleon. Formal details can be found in Ref. [2]. We start with results for the magnetic moments of nucleons, μ_p and μ_n. For our set of parameters we obtain:

$$\mu_p = 2.62 \pm 0.02, \quad \mu_n = -2.02 \pm 0.02, \quad \text{and} \quad \frac{\mu_n}{\mu_p} = -0.76 \pm 0.01. \tag{19}$$

The leading order (LO, three-quark core) μ_p^{LO}, μ_n^{LO} and next-to-leading order (NLO, meson cloud and finite current quark mass) contributions μ_p^{NLO}, μ_n^{NLO} to the magnetic moments are given by

$$\mu_p^{LO} = 1.8 \pm 0.15, \quad \mu_n^{LO} \equiv -\frac{2}{3}\mu_p^{LO}, \tag{20}$$

$$\mu_p^{NLO} = \mu_p - \mu_p^{LO} = 0.82 \pm 0.13, \quad \mu_n^{NLO} = \mu_n - \mu_n^{LO} = -0.82 \pm 0.08.$$

For the electromagnetic nucleon radii we obtain

$$r_E^p = 0.84 \pm 0.05\,\mathrm{fm}, \qquad < r^2 >^n_E = -0.036 \pm 0.003\,\mathrm{fm}^2, \tag{21}$$

$$r_M^p = 0.82 \pm 0.02\,\mathrm{fm}, \qquad r_M^n = 0.85 \pm 0.01\,\mathrm{fm}.$$

The LO contributions to the charge radius of the proton (see Eq. (6)) and to the magnetic radii of proton and neutron are dominant

$$r_E^{p;\,LO} = 0.77 \pm 0.06\,\mathrm{fm}, \quad r_M^{p;\,LO} \equiv r_M^{n;\,LO} = 0.73 \pm 0.06\,\mathrm{fm}. \tag{22}$$

For the neutron charge radius squared we get the observed (negative) sign, but its magnitude is smaller than the experimental value. As in the naive $SU(6)$ quark model, the LO contribution to the neutron charge radius is zero and only one-loop diagrams give nontrivial contributions to this quantity:

$$< r^2 >^n_E = -0.036 \pm 0.003\ \mathrm{fm}^2. \tag{23}$$

In Table 1 we summarize our results for the static electromagnetic properties of the nucleon in comparison to experimental data. Results on the Q^2-dependence of the electromagnetic form factors can be found in Ref. [2].

Table 1. Static nucleon properties.

Quantity	Our Approach	Experiment
μ_p	2.62 ± 0.02	2.793
μ_n	-2.02 ± 0.02	-1.913
μ_n/μ_p	-0.76 ± 0.01	-0.68
r_E^p (fm)	0.84 ± 0.05	0.86 ± 0.01
$< r^2 >_E^n$ (fm^2)	-0.036 ± 0.003	-0.119 ± 0.004
r_M^p (fm)	0.82 ± 0.02	0.86 ± 0.06
r_M^n (fm)	0.85 ± 0.01	0.88 ± 0.07

Acknowledgments

This work was supported by the Deutsche Forschungsgemeinschaft (DFG, grant FA67/25-1). V.E.L. thanks the organizers for the invitation.

References

1. V.E. Lyubovitskij, T. Gutsche, A. Faessler, and E.G. Drukarev, *Phys. Rev.* D **63**, 54026 (2001).
2. V.E. Lyubovitskij, T. Gutsche, and A. Faessler, hep-ph/0105043.
3. V.E. Lyubovitskij, T. Gutsche, A. Faessler, A. Rusetsky, and R. Vinh Mau (in preparation).
4. T. Gutsche and D. Robson, *Phys. Lett.* B **229**, 333 (1989); T. Gutsche, Ph. D. thesis, Florida State University, 1987 (unpublished).
5. A.W. Thomas, *Adv. Nucl. Phys.* **13**, 1 (1984); E. Oset, R. Tegen, and W. Weise, *Nucl. Phys.* A **426**, 456 (1984); S.A. Chin, *Nucl. Phys.* A **382**, 355 (1982).
6. J. Gasser, M.E. Sainio, and A.B. Švarc, *Nucl. Phys.* B **307**, 779 (1988).
7. M. Gell-Mann and M. Lévy, *Nuovo Cim.* **16**, 1729 (1960).
8. A. W. Thomas, *J. Phys.* G **7**, L283 (1981).
9. S. Weinberg, *Phys. Rev. Lett.* **17**, 616 (1966); Y. Tomozawa, *Nuovo Cim.* A **46**, 707 (1966).
10. S. Weinberg, *Physica* A **96**, 327 (1979); J. Gasser and H. Leutwyler, *Ann. Phys. (N.Y.)* **158**, 142 (1984); *Nucl. Phys.* B **250**, 465 (1985).
11. R. E. Stuckey and M. C. Birse, *Nucl. Phys.* B **250**, 465 (1985); *J. Phys.* G **23**, 29 (1997).
12. J. Gasser, H. Leutwyler, and M. E. Sainio, *Phys. Lett.* B **253**, 252,260 (1991).

THE GLASGOW PION PHOTOPRODUCTION PARTIAL WAVE ANALYSIS, 2001

R. L. CRAWFORD
Department of Physics and Astronomy,
University of Glasgow,
Glasgow G12 8QQ,
Scotland, UK
E-mail: r.crawford@physics.gla.ac.uk

Preliminary results from a partial wave analysis of single pion photoproduction using fixed-t dispersion relations are presented. All one to four star resonances below 2.5 GeV were included in the analysis. Couplings, masses and widths were evaluated.

1 Introduction

Fixed-t dispersion relations have been used successfully in the past for the partial wave analysis of pion photoproduction[1,2,3]. Since it is necessary to parametrise only the imaginary parts of the amplitudes and because these are very resonance dominated, the number of free parameters is significantly decreased. Additionally, analyticity with respect to the energy variable is intrinsic to the calculation. As a result, the possibility of ambiguous solutions is reduced. There are also disadvantages. The dispersion integrals require the fitting of data at all possible energies. This analysis uses data from threshold to photon laboratory energies of 16 GeV. Also, because of divergence problems in the partial wave expansions, it is not possible to use experimental data at some values of t. In spite of this, analyses of this type have measured many resonance couplings and, in some cases, resonance masses and widths.

2 The Analysis

The work is based on the old Glasgow analysis[3] and differs mainly due to the inclusion of new data and additional resonances. Work is still in progress and the results are preliminary. At present, only the resonance couplings for proton targets have been fully evaluated although neutron target data has been fitted to get a good estimate of the u channel contributions. A more careful and more refined analysis will be carried out.

The analysis is divided into four energy regions.

The first is from threshold to $W = 1325$ Mev. Here, Watson's theorem fixes the complex phases of many partial waves. For the $\Delta(1232)$, a modified Omnes-Muskhelishvilli method[4,5] is used to fix the complex phase across the resonance without imposing constraints at higher energies. The couplings, mass and width are fitted. Non-resonant contributions from Watson's theorem, mainly in the S-waves, are also included by a simple iterative method.

The second region is from $W = 1.325$ to 2 GeV. In it, the contributions from the resonances to the dispersion integrals are parametrised as simple Breit-Wigner amplitudes with energy independent widths. A small amount of non-resonant background is required for the low angular momentum states. Because of the divergence problem, data for $t \leq -1.5$ $(\mathrm{GeV/c})^2$ is not used.

It is best to describe the fourth region before the third. This region is for $W \geq 2.5$ GeV. and a Regge parametrisation is used for the dispersion integrands [6]. Data is fitted for photon laboratory energies up to 16 GeV and for $t \leq -1.0$ $(\mathrm{GeV/c})^2$.

In the third region, between regions two and four, the resonance parametrisation is switched linearly into the Regge parametrisation. As a result, although all known resonances in this region are included and can give important corrections to the fits, the values of the parameters obtained becomes less meaningful as the energy increases.

3 Results

The results from the analysis are given in the table below. They are still preliminary. Because much of the amplitudes in the upper part of the third energy region are built from the Regge parametrisation, no results are quoted for resonances above 2.2 GeV, although all known resonances up to 2.5 GeV were included since they give important corrections. In the first energy region, the results for the $\Delta(1232)$ couplings are typical of those from other partial wave analysis. The mass is taken to be where complex phase is $\pi/2$ and is 1231.8 ± 0.5 MeV. This is for the Δ^+. The width is 118 ± 3 MeV. The E/M ratio is -0.016 ± 0.006.

Many couplings from the second energy region and the lower part of the next region are similar to those from earlier analyses. Most of the masses and widths have been fitted without constraints. Those values which were not varied or had strong constraints imposed upon them are shown in italics. Typically, the values agree more closely with the pole positions and it seems that what is being seen in this type of analysis are the resonance poles. This is perhaps not surprising since analyticity is an intrinsic feature of the parametrisation. New resonances are the $F_{15}(2000)$, $S_{11}(2090)$, $P_{11}(2100)$, $P_{31}(1750)$, $S_{31}(1900)$, $P_{33}(1920)$, $D_{33}(1940)$, $F_{35}(2000)$ and $S_{31}(2150)$.

Table 1. Resonance γp couplings (in 10^3 GeV$^{-1/2}$) , masses and widths (in Mev) from this analysis. Masses and widths in italics have not been varied or have been severely constrained. The rating is from this analysis.

Resonance	Mass	Width	$A_{1/2}$	$A_{3/2}$	Status
$P_{11}(1440)$	1363	413	-88		***
$D_{13}(1520)$	1502	118	-15	162	****
$S_{11}(1535)$	1532	133	55		****
$S_{11}(1650)$	1659	194	71		***
$D_{15}(1675)$	1658	145	13	38	****
$F_{15}(1680)$	1672	103	-14	135	****
$D_{13}(1700)$	1667	208	-19	14	**
$P_{11}(1710)$	*1670*	*400*	27		***
$P_{13}(1720)$	1729	338	43	-20	***
$P_{13}(1900)$	2170	519	54	86	*
$F_{17}(1990)$	2163	308	32	35	**
$D_{13}(1950)$	1954	365	24	-87	**
$F_{15}(2000)$	1877	461	-48	90	*
$S_{11}(2090)$	1955	422	31		*
$P_{11}(2100)$	*2120*	*300*	-9		*
$G_{17}(2190)$	*2100*	362	-24	93	**
$P_{33}(1232)$	1231.8	118.0	-149	-259	****
$P_{33}(1600)$	*1650*	*350*	23	38	**
$S_{31}(1620)$	1616	90	17		***
$D_{33}(1700)$	1621	200	79	90	****
$P_{31}(1750)$	1715	176	96		**
$S_{31}(1900)$	1858	218	27		**
$F_{35}(1905)$	1848	120	17	-18	**
$P_{31}(1910)$	1708	576	-36		See text
$P_{33}(1920)$	1814	586	-37	-139	*
$D_{35}(1930)$	1912	131	8	8	**
$D_{33}(1940)$	1799	229	16	-64	*
$F_{37}(1950)$	1897	220	-59	-62	****
$F_{35}(2000)$	2099	505	-61	21	*
$S_{31}(2150)$	2167	237	45		*

There is a tendency for widths of the resonances at the top of the energy range to be high. This may be due to the presence of additional, as yet unknown, resonances which are not included in the analysis. There are two notable results.

The first is a very stable mass and width for the $P_{31}(1750)$, currently rated by the PDG[7] as a one star resonance. This indicates that the resonance pole is being seen clearly. In contrast, the parameters for the $P_{31}(1910)$ have ended up with a very low mass and very large width. The interpretation of this is that there is non-resonant background in the imaginary part of this partial wave and that the Breit-Wigner inserted for the $P_{31}(1910)$ has fitted it. The coupling to the $P_{31}(1910)$ for proton targets has always been small and this analysis does not seem to be seeing it at all.

The second result is a stable pole position for a D_{13} resonance at about 1950 MeV. This started as an attempt to measure the $D_{13}(2080)$. Although the signal is not strong, the values have stayed acceptably constant for many iterations of the fitting process. It may be the $D_{13}(2080)$ but the mass and width seem to be in closer agreement with those for a possible D_{13} seen in $K^+\Lambda$ photoproduction[8]. The overall situation is still not clear.

4 Conclusions

This analysis shows that there is still new information to be obtained from pion photoproduction and that it is possible to use the reaction to evaluate resonance masses and widths (or pole positions). New γN decay couplings have been evaluated. The evidence for one resonance has been strengthened and a possible new resonance has been identified. Work is still in progress and the analysis is being refined. There is a need for new experimental data for $W \geq 1.8$ Gev.

References

1. R.L. Crawford, Proceedings of Baryon 80, p. 107, Toronto, 1980.
2. I. Arai and H.Fujii, *Nucl. Phys.* B **194**, 251 (1982).
3. R.L. Crawford and W. Morton, *Nucl. Phys.* B **221**, 1 (1983).
4. J. Engels and W. Schmidt, *Phys. Rev.* **169**, 1296 (1968).
5. R.L. Crawford, *Nucl. Phys.* B **28**, 573 (1971).
6. I.M. Barbour, R.L. Crawford and N. Parsons, *Nucl. Phys.* B **111**, 358 (1978).
7. Particle Data Group, *EPJ* C **15**, 697 (2000).
8. C. Bennhold, A. Waluyo, H. Haberzettl, T. Mart, G. Penner and U. Mosel, nucl-th/0008024 (2000).

RESIDUES OF THE MULTIPOLE AMPLITUDES $M_{1+}^{3/2}, E_{1+}^{3/2}$ AT THE T-MATRIX POLE FOUND IN THE FRAMEWORK OF DISPERSION RELATIONS

I. G. AZNAURYAN

Yerevan Physics Institute, Alikhanian Brothers St.2, Yerevan, 375036 Armenia
(e-mail addresses: aznaur@jerewan1.yerphi.am, aznaury@jlab.org)

Using solutions of integral equations, which follow from dispersion relations for $M_{1+}^{3/2}, E_{1+}^{3/2}$, continuation of these amplitudes to the T-matrix pole is made, and the residues of the amplitudes are obtained.

1 Introduction

There are two characteristics of the multipole amplitudes $M_{1+}^{3/2}, E_{1+}^{3/2}$ which are usually connected with the magnitude of the $P_{33}(1232)$ resonance contribution in the reaction of pion photoproduction on nucleons: these are the values of the imaginary parts of these amplitudes at the K-matrix pole (namely, at the point where $\delta_{1+}^{3/2} = \frac{\pi}{2}$) and their residues at the T-matrix pole. The residues at the T-matrix pole are not directly related to the magnitudes of the amplitudes on the real energy axis, and for their evaluation it is necessary to have some approach for the continuation of the multipole amplitudes into the complex energy plane on the lower half plane of the second Riemann sheet. Such continuation usually done on the basis of the speed plot analysis. This method was applied to pion photoproduction in Refs. [1,2], although for this reaction there are uncertainties connected with the normalization of the amplitudes (Ref. [2]). For the continuation of $M_{1+}^{3/2}, E_{1+}^{3/2}$ to the T-matrix pole the phenomenological parametrization of these amplitudes was also used (Refs. [3,4]).

Analyzing the results obtained in the framework of dispersion relations we found that under some reasonable assumptions dispersion relations used within the approach developed in Refs. [5,6] can be used for the continuation of the amplitudes $M_{1+}^{3/2}, E_{1+}^{3/2}$ to the T-matrix pole and for the evaluation of the residues of the amplitudes at this pole. In the present work we will show this. We will find the residues of the amplitudes $M_{1+}^{3/2}, E_{1+}^{3/2}$ at the T-matrix pole. The results will be compared with the residues obtained in Refs. [1,2,4].

2 Reducing of Dispersion Relations for $M_{1+}^{3/2}$, $E_{1+}^{3/2}$ to Integral Equations and the Solution of these Equations

Let us write dispersion relations for the amplitudes $M_{1+}^{3/2}/kq$, $E_{1+}^{3/2}/kq$ in the form

$$M(W) = M^B(W) + M^\omega(W) + \frac{1}{\pi}\int\limits_{W_{thr}}^{W_{max}} \frac{ImM(W')}{W' - W - i\varepsilon}dW'$$
$$+\frac{1}{\pi}\int\limits_{W_{thr}}^{W_{max}} K(W,W')ImM(W')dW', \qquad (1)$$

where $M(W)$ denotes any of these amplitudes, $M^B(W)$ is the contribution of the Born term to the amplitudes, $K(W,W')$ is a nonsingular kernel arising from the u-channel contribution to the dispersion integral and the nonsingular part of the s-channel contribution. In the last term of the relation (1), we did not take into account the couplings of $M(W)$ to other multipole amplitudes because by our estimation they are small and do not affect our final results. High energy contributions into the dispersion integrals are approximated by direct inserting of the ω-exchange as a simple t-channel pole: $M^\omega(W)$.

From the phase shift analyses of πN scattering in the $P_{33}(1232)$ resonance region [7,8], it is known that the πN amplitude $h_{1+}^{3/2}(W)$ is elastic up to $W = 1.45\ GeV$ and due to the Watson theorem in this region the amplitudes $M(W)$ can be written in the form $ImM(W) = h^*(W)M(W)$. Using the VPI partial-wave analysis of pion photoproduction on the nucleons [3] one can estimate the dispersion integrals (1) over the region $1.45\ GeV \div 2\ GeV$. The estimations show that in the $P_{33}(1232)$ resonance region these integrals are very small: at $W < 1.35\ GeV$ they are smaller than $0.01M^B$ for $M_{1+}^{3/2}(W)$ and $0.02M^B$ for $E_{1+}^{3/2}(W)$. Therefore, the amplitude $M(W)$ obtained from (1) at $W < 1.35\ GeV$ will not be changed, if in the integrands of (1) we will reduce $ImM(W')$ to 0 by continuing the relation $ImM(W) = h^*(W)M(W)$ to the energies $1.45\ GeV < W < 2\ GeV$ and supposing that $\delta_{1+}^{3/2}(W) \rightarrow \pi$, when $W \rightarrow W_{max} = 2\ GeV$. Let us recall that $\delta_{1+}^{3/2} = 157°$ at $W = 1.45\ GeV$. Under these assumptions the dispersion relations for $M_{1+}^{3/2}$, $E_{1+}^{3/2}$ turn into singular integral equations (1), which at $K(W,W') = 0$ have a solution in an analytical form [5,6]:

$$M_{K=0}(W) = M^B_{part,K=0}(W) + M^\omega_{part,K=0}(W) + c_M M^{hom}_{K=0}(W), \qquad (2)$$

where $M^{B}_{part,K=0}(W)$, $M^{\omega}_{part,K=0}(W)$ are the particular solutions of Eq. (1) generated by M^B and M^{ω}:

$$M^{B,\omega}_{part,K=0}(W) = M^{B,\omega}(W) + \frac{1}{\pi}\frac{1}{D(W)}\int\limits_{W_{thr}}^{W_{max}} \frac{D(W')h(W')M^{B,\omega}(W')}{W' - W - i\varepsilon}dW', \tag{3}$$

and

$$M^{hom}_{K=0}(W) = \frac{1}{D(W)} = \frac{W_{max}}{|W_{max} - W|}\exp\left[\frac{W}{\pi}\int\limits_{W_{thr}}^{W_{max}} \frac{\delta(W')}{W'(W' - W - i\varepsilon)}dW'\right] \tag{4}$$

is the solution of Eq. (1) with $M^B = M^{\omega} = 0$. This enters the solution of Eq. (2) with an arbitrary weight, i.e. multiplied by an arbitrary constant c_M. These constants were found using the amplitudes $M^{3/2}_{1+}, E^{3/2}_{1+}$ from Ref. [3].

At $K(W, W') \neq 0$ one can transform the singular integral equation (1) into a nonsingular integral equation, and it can be shown that with high accuracy the solution of this equation is equal to

$$M(W) = M_1(W) + \frac{c_M + M_2(W)}{D(W)}, \tag{5}$$

where

$$M_1(W) = M^B + M^{\omega} + \frac{1}{\pi}\int\limits_{W_{thr}}^{W_{max}} K(W, W')ImM_{K=0}(W')dW',$$

$$M_2(W) = \frac{1}{\pi}\int\limits_{W_{thr}}^{W_{max}} \frac{D(W')h(W')M_1(W')}{W' - W - i\varepsilon}dW'. \tag{6}$$

3 Continuation to the T-Matrix Pole and the Results

In a simple example it can be shown that $\frac{1}{D(W)}$ has a pole at $\Gamma \to 0$, and, therefore, the pole behaviour of Eq. (5) is determined by $\frac{1}{D(W)}$.

Calculations made with $\delta^{3/2}_{1+}$, defined according to Sec. 2, show that $\frac{1}{D(W)}$ can be approximated by the expression:

$$\frac{1}{D(W)} = \frac{1.39\ GeV}{M_{pole} - W - i\frac{\Gamma}{2}}e^{-i21.7^{\circ}}, \tag{7}$$

	Present analysis	SP[1]	SP[2] (I)	SP[2] (II)	[4] (I)	[4] (II)
r_E $(10^{-3}/m_{\pi^+})$	$2.41 \div 2.61$	2.46	2.47	2.24	2.76	2.02
ϕ_E (deg)	$-150 \div -152$	-154.7	-156	-162	-158	-131
r_M $(10^{-3}/m_{\pi^+})$	$41.1 \div 41.9$	42.32	39.9	41.5	41.8	41.4
ϕ_M (deg)	$-19.3 \div -20.3$	-27.5	-26.0	-36.5	-31	-31
$\phi_E - \phi_M$	$-(131 \div 132)$	-127.5	-130	-125.5	-127	-100

Table 1. Residues $r\exp(i\phi)$ of the amplitudes $M_{1+}^{3/2}, E_{1+}^{3/2}$ at the T-matrix pole from various analyses.

where $\Gamma = 0.098\ GeV$, $M_{pole} = 1.209\ GeV$. The residues of $M(W)$ of Eq. (5) are determined by $c_M + M_2(W = M_{pole} - i\frac{\Gamma}{2})$. We have shown that they can be found by linear continuation from the point $W = M_{pole}$ and are equal to $c_M + (1+a)M_2(M_{pole})$, $a = 0.25 \div 0.3$.

The obtained results are presented in Table 1. The given ranges of our results reflect the uncertainties connected with a and with the ω contribution. The results obtained via different procedures of the speed plot analysis (Ref. [1] and versions (I,II) of Ref. [2]) and using different phenomenological parametrizations of the amplitudes [4] (versions (I) and (II)) are also presented. It is seen that for the absolute value of the residue for $M_{1+}^{3/2}$ there is a good agreement of our results with the results of other analyses except the results (II) of Ref. [2]. For the absolute value of the residue for $E_{1+}^{3/2}$, the difference between the results of different analyses is quite large; our results agree well with the results obtained via speed plot analyses (Ref. [1] and version (I) of Ref. [2]). As far as the phases are concerned the difference between the results of all analyses is very large. We found that the results for $\phi_E - \phi_M$ are more stable; here all results are in not bad agreement with each other, except the results (II) of Ref. [4].

References

1. O.Hanstein, D.Drechsel, and L.Tiator, *Nucl. Phys.* A **632**, 561 (1998).
2. Th.Wilbois, P.Wilhelm, and H.Arenhövel, *Phys. Rev.* C **57**, 295 (1998).
3. R.A.Arndt, I.I.Strakovsky, R.L.Workman, *Phys. Rev.* C **56**, 577 (1997).
4. R.L.Workman, and R.A.Arndt, Phys. Rev. C **59**, 1810 (1999).
5. D.Schwela, H.Rolnik, R.Weizel, and W.Korth, *Z. Phys.* **202**, 452 (1967).
6. D.Schwela, and R.Weizel, *Z. Phys.* **221**, 71 (1969).
7. R.A.Arndt *et al.*, *Phys. Rev.* C **52**, 2120 (1995).
8. R.A.Arndt *et al.*, nucl-th/9807087.

AN ISOBAR MODEL FOR η PHOTO- AND ELECTROPRODUCTION ON THE NUCLEON

W.-T. CHIANG AND S.-N. YANG
Department of Physics, National Taiwan University, Taipei 10617, Taiwan

L. TIATOR AND D. DRECHSEL
Institut für Kernphysik, Universität Mainz, 55099 Mainz, Germany

An isobar model containing Born terms, vector meson exchange and nucleon resonances is used to analyze recent η photoproduction data for cross sections and beam asymmetries, as well as JLab electroproduction data. Good overall description is achieved up to $Q^2 = 4.0$ (GeV/c)^2. Besides the dominant $S_{11}(1535)$ resonance, we show that the second S_{11} resonance, $S_{11}(1650)$, is also necessary to be included in order to extract the $S_{11}(1535)$ resonance parameters properly. In addition, the beam asymmetry data allow us to extract very small ($< 0.1\%$) $N^* \to \eta N$ decay branching ratios of the $D_{13}(1520)$ and $F_{15}(1680)$ resonances because of the overwhelming s-wave dominance. The model is implemented as a part of the MAID program.

1 Introduction

Electromagnetic eta production on the nucleon, $\gamma N \to \eta N$, provides an alternative tool to study N^* besides πN scattering and pion photoproduction. The ηN state couples to nucleon resonances with isospin $I = 1/2$ only. Therefore, this process is cleaner and more suitable to distinguish certain resonances than other processes, e.g., pion photoproduction. It provides opportunities to access less studied resonances and the possibly "missing resonances".

Eta photoproduction at low energy is dominated by the $S_{11}(1535)$ resonance, which is the only nucleon resonance with a substantial decay to the ηN channel. Therefore, η photoproduction is an ideal process to study $S_{11}(1535)$ properties. In contrast, πN scattering and pion photoproduction are always interweaved with the ηN channel threshold opening, and often produce inconsistent and/or controversial results.

Recently, precise experimental data of this process have been measured. These data include total and differential cross sections for $\gamma N \to \eta N$ from TAPS (MAMI/Mainz) [1] and GRAAL [2], as well as beam asymmetries from GRAAL [3] and target asymmetries from ELSA (Bonn) [4]. In addition, there are two recent η electroproduction data sets from Jefferson Lab [5,6].

2 Isobar Model

The isobar model used in this work is closely related to the unitary isobar model (UIM) developed by Drechsel *et al.* [7]. The major difference is that in the UIM, which deals with pion photo- and electroproduction, the phases of the multipole amplitudes are adjusted to the corresponding pion-nucleon elastic scattering phases, while in η production the unitarization procedure is not feasible since the eta-nucleon scattering information is not experimentally available.

The nonresonant background contains the usual *Born terms* and *vector meson exchange* contributions, and can be obtained by evaluating the Feynman diagrams derived from an effective Lagrangian. In addition to the dominant $S_{11}(1535)$ nucleon resonance, we also consider resonance contributions from $D_{13}(1520)$, $S_{11}(1650)$, $D_{15}(1675)$, $F_{15}(1680)$, $D_{13}(1700)$, $P_{11}(1710)$, and $P_{13}(1720)$. For the relevant multipoles $A_{\ell\pm}$ ($= E_{\ell\pm}$, $M_{\ell\pm}$, $S_{\ell\pm}$) of resonance contributions, we assume a Breit-Wigner energy dependence of the form

$$A_{\ell\pm}(Q^2, W) = \tilde{A}_{\ell\pm}(Q^2)\, \frac{\Gamma_{tot}\, W_R}{W_R^2 - W^2 - iW_R\Gamma_{tot}}\, f_{\eta N}(W)\, C_{\eta N}\,, \quad (1)$$

where $f_{\eta N}(W)$ is the usual Breit-Wigner factor describing the decay of the N^* resonance [7], and the isospin factor $C_{\eta N}$ is -1. The total width Γ_{tot} is taken as the sum of $\Gamma_{\eta N} + \Gamma_{\pi N} + \Gamma_{\pi\pi N}$.

3 Results and Discussion

3.1 Photoproduction Results

We have fitted recent η photoproduction data including total and differential cross sections from TAPS [1] and GRAAL [2], as well as the polarized beam asymmetry from GRAAL [3] with the isobar model described above. In Fig. 1, we compare our results of differential cross sections with the data from TAPS and GRAAL, and we find very good agreement. In the low energy region the differential cross section is flat, indicating the s-wave dominance. As the energy increases higher, other partial waves start to contribute. Note that our result for $E_\gamma^{\rm lab} > 1$ GeV shows a decrease at forward angles, which is not seen in the GRAAL data.

Our result for the total cross section is shown in Fig. 1, and compared with the TAPS and GRAAL data. Again, there is good agreement except for the bump observed in the GRAAL data in the region $E_\gamma^{\rm lab} = 1050$ - 1100 MeV can not be reproduced by our model. However, note that the total cross section in the GRAAL data is obtained from integration of the differential cross sections,

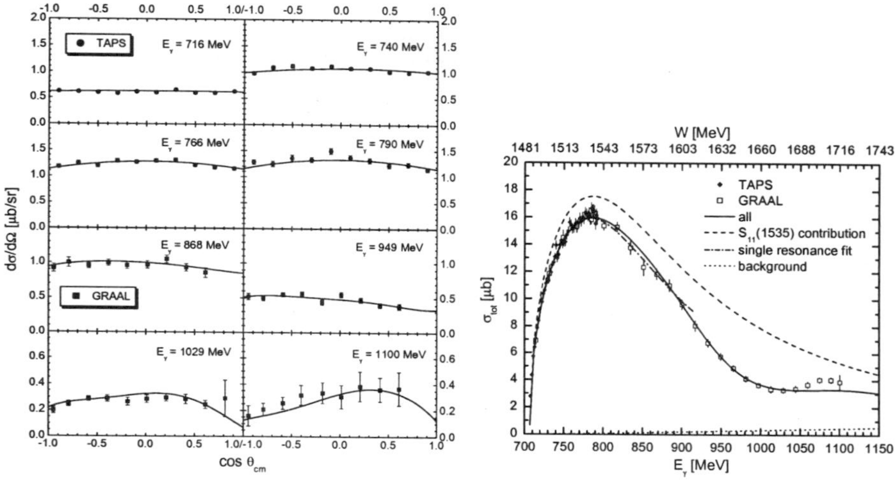

Figure 1. Differential and total cross section for $\gamma p \to \eta p$. Data are from TAPS and GRAAL.

using a polynomial fit in $\cos\theta$ for extrapolation to the uncovered region. We find that the discrepancy is due to the extrapolation of the GRAAL data to forward angles and is not really supported by the data themselves.

Fig. 1 shows that the background contribution is very small, and that the total cross section is dominated by the $S_{11}(1535)$ at low energy. However, the contribution from the second resonance, $S_{11}(1650)$, can not be neglected. Even though a single S_{11} resonance can fit the low energy data nicely (the dashed-dotted curve in Fig. 1), it can by no means describe the higher energy region. Moreover, the single resonance fit yields incorrect resonance parameters. In fact, the decay width (159 vs. 191 MeV) and photon coupling (103 vs. 118×10^{-3} GeV$^{-1/2}$) obtained in the single S_{11} resonance fit are significantly smaller than the full results when both S_{11} resonances are properly included.

One special feature in polarization measurements of η photoproduction is that through the interference of the dominant E_{0+} multipole with smaller multipoles, one can access small contributions from particular resonances. The available beam asymmetry data were measured at GRAAL [3] from threshold to $E_\gamma^{\rm lab} = 1.1$ GeV. Higher energy data up to $E_\gamma^{\rm lab} = 1.5$ GeV are being analyzed and will be available soon [8]. In Fig. 2, we compare our results with the GRAAL data. Good overall agreement has been achieved. At low energies, we observe that the beam asymmetry has a clear $\sin^2\theta$ dependence as a result of interference between s- and d-waves. From these low energy

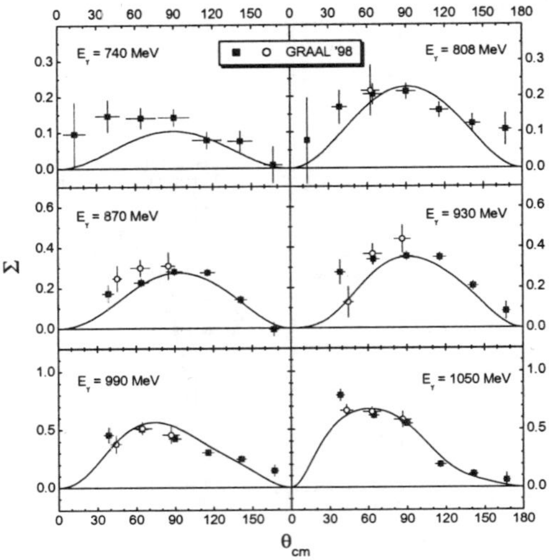

Figure 2. Beam asymmetry for $\gamma p \to \eta p$. The data are from GRAAL.

data, a branching ratio of $\beta_{\eta N} = 0.06\%$ can be determined for the $D_{13}(1520)$. When energies get higher than $E_\gamma^{\text{lab}} = 930$ MeV, the data develop a forward-backward asymmetry behavior, which becomes especially evident at $E_\gamma^{\text{lab}} = 1050$ MeV. The $F_{15}(1680)$ is sensitive to this forward-backward asymmetry in Σ as discussed by Tiator *et al.* [9]. This is the reason why such a small branching ratio (0.06%) can be extracted for this resonance.

3.2 Electroproduction Results

When fitting recent electroproduction data from JLab [5,6], we fix all the parameters determined from the photoproduction data except the Q^2 dependence of the helicity amplitudes $A^p_{1/2,\,3/2}(Q^2)$. The result for the $S_{11}(1535)$ is shown in Fig. 3. In order to avoid large model uncertainties arising from different values of partial and total widths of the $S_{11}(1535)$ employed in other analyses, we choose not to compare the helicity amplitudes $A^p_{1/2}(Q^2)$ extracted from different analyses. Instead, we compare the model-independent quantity introduced by Benmerrouche *et al.* [10], $\xi = \sqrt{\chi\beta_{\eta N}/\Gamma_{tot}}\,A_{1/2}$, where $\chi = kM/(qM_R)$ is a kinematic factor. The ξ quantity covers the uncertainty from $\beta_{\eta N}$ and Γ_{tot} between different analyses and is almost independent of the extraction process. In Fig. 3 we compare our ξ values with the ones extracted from the recent JLab data [5,6] and older data [11]. It is seen that overall good agreement is achieved up to $Q^2 = 4.0$ (GeV/c)^2.

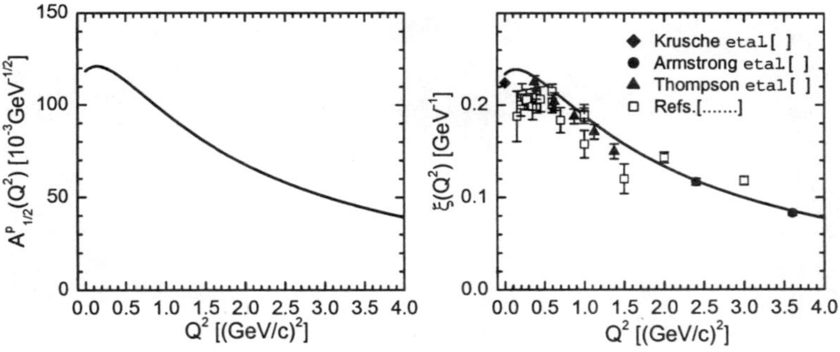

Figure 3. The helicity amplitude $A^p_{1/2}(Q^2)$ for $S_{11}(1535) \rightarrow \gamma p$ is shown in the left figure. On the right, we plot the quantity ξ ($\equiv \sqrt{\chi\beta_{\eta N}/\Gamma_{tot}}\, A^p_{1/2}$), and compare with the values extracted from the data.

Acknowledgments

W.-T. C. would like to thank Universität Mainz for the hospitality extended to him during his visits. This work was supported in parts by the National Science Council of ROC under Grant No. NSC89-2112-M002-078, by Deutsche Forschungsgemeinschaft (SFB 443), and by a joint project NSC/DFG TAI-113/10/0.

References

1. B. Krusche *et al.*, *Phys. Rev. Lett.* **74**, 3736 (1995).
2. F. Renard *et al.*, hep-ex/0011098.
3. J. Ajaka *et al.*, *Phys. Rev. Lett.* **81**, 1797 (1998).
4. A. Bock *et al.*, *Phys. Rev. Lett.* **81**, 534 (1998).
5. C.S. Armstrong *et al.*, *Phys. Rev.* D **60**, 052004 (1999).
6. R. Thompson *et al.*, *Phys. Rev. Lett.* **86**, 1702 (2001).
7. D. Drechsel, O. Hanstein, S.S. Kamalov, and L. Tiator, *Nucl. Phys.* A **645**, 145 (1999).
8. A. D'Angelo, *these proceedings.*
9. L. Tiator, D. Drechsel, G. Knöchlein, and C. Bennhold, *Phys. Rev.* C **60**, 035210 (1999).
10. M. Benmerrouche, N.C. Mukhopadhyay, and J.F. Zhang, *Phys. Rev.* D **51**, 3237 (1995).
11. F.W. Brasse *et al.*, *Nucl. Phys.* B **139**, 37 (1978); F.W. Brasse *et al.*, *Z. Phys.* C **22**, 33 (1984); U. Beck *et al.*, *Phys. Lett.* B **51**, 103 (1974); H. Breuker *et al.*, *Phys. Lett.* B **74**, 409 (1978).

Sergo Gerasimov, Barbara Pasquini

Latifa Elouadrhiri, Angels Ramos

DETAILED ANALYSIS OF η PRODUCTION IN PROTON-PROTON COLLISIONS

A. ŠVARC, S. CECI
Rudjer Boškovic Institute, Bijenička c. 54, 10000 Zagreb, Croatia,
E-mail: svarc@rudjer.irb.hr, ceci@rudjer.irb.hr

Coupled-channel, multiresonance partial wave analysis (PWA), developed and used by the Carnegie-Mellon-Berkeley analysis group (CMU-LBL 79) [1] has been recently very frequently used, and agreed as a potentially safe tool for extracting N* resonance parameters [2,3,4,5]. The newly formed subsection for resonance parameter analyses of Baryon Resonance Analysis Group (BRAG) [6] has decided to re-evaluate relatively old analyses [1,7], as has been suggested even by their own authors recently [8]. BRAG has chosen three independent analyses which repeat the formalism suggested earlier, but with the use of new and improved data [9]. They have come to a certain level of agreement regarding the number of poles and their values. However, some of the "Cutkosky" like analyses have been dropped [10,11], for technical reasons presumably.

However, one possible direction of analysis has been dropped altogether. The question arises whether the obtained PWA can be "relatively" safely taken as input for processes involving more then two particles in the final state. Temporarily forgetting the expected additional complications (initial and final state interactions, off-mass shell behavior of two body amplitudes, etc.) this work tends to estimate whether the present two body T-matrices can account for the observables of a three body process $pp \to pp\eta$, very carefully measured in Uppsala near the threshold [12]. Special attention has been given to understand the apparently inverted shape of the proton-eta differential cross section in the final state [12,13]. The tendency of this work is not to improve the two body fit (which needs a lot of additional observables even to be semi-reliable) but to see if the present T-matrices can explain the $2 \to 3$ body processes without drastic assumptions of the complications by three body physics.

The first, and natural test of the reliability of the two body amplitudes appeared in the carefully measured total and differential cross section for the process $pp \to pp\eta$ [12]. We have developed a simple model based on the exchange of the lowest mesons depicted in Fig. 1.

Unfortunately, there is a number of models which claim to reproduce the results, but differ among themselves drastically [14,16,15,17,18], so it has been left to us to show that our model [19], reduced to the assumptions of the

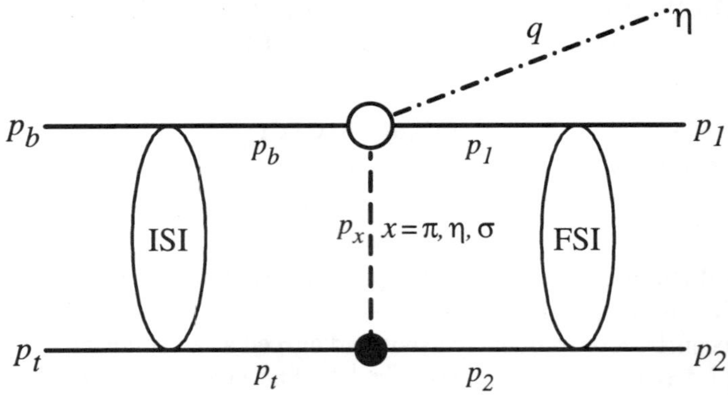

Figure 1. The meson-exchange mechanism of $pp \to pp\eta$ reaction. The initial and final state interactions are denoted as ISI and FSI respectively.

mentioned models, gives a very similar result. The comparison has been made successfully and will be shown elsewhere. The main idea of this presentation is to draw attention to the fact that the differential $p\eta$ cross sections in a three body process tends to show a different curvature when compared to the two body $\pi^- p \to \eta n$ process which should dominate the process [12]. In spite of the additional uncertainties of processes like ISI, FSI and off-mass shell extrapolation of two-body amplitudes, the effects should be extremely high, and acting in the same direction in order to turn the slope of the differential cross section. The disturbing data are shown in Fig. 2.

It is to be expected that the two $\to$ three-body process is dominated by the two body proton-meson $\to$ proton η amplitude, in the vicinity of the threshold in particular. However, it turns out that even the shape of the differential cross sections of the impulse approximation in the two body process [20] and the measured 2$\to$3 body processes are different. Let us just mention that only higher partial waves (like D_{13}) can account for the opposite curvature. Therefore, we are left with only two possibilities: either the ISI, FSI and off-mass shell effects of the higher order processes are responsible for the discrepancy, or the D_{13} partial wave is not confidently extracted in the 2$\to$2 body processes.

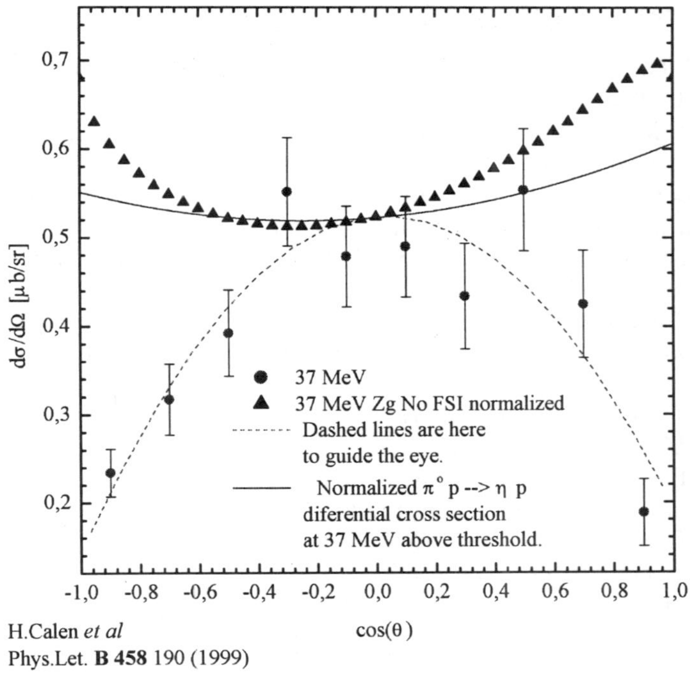

Figure 2. The comparison of the experimental values (full dots with error bars) and the polynomial fit to them (dotted line) with $\pi^- p \to \eta n$ at the comparable energy (full triangles) with the Zagreb calculation (full thin line).

As it is obvious, the lowest partial wave in two body processes which can cause such a curvature are the D_{13} partial waves, and they have been under close scrutiny at the Mainz workshop. Our feeling is that ISI, FSI and off-mass shell effects of higher order processes should be surprisingly strong to account for such a drastic change in the shape of the differential cross section. Therefore, there is an open possibility that something remains hidden in the D_{13} two body partial wave which we have not been able to detect in two body processes. The future goal is to investigate all suggested possibilities, and see whether the ρ exchanged meson domination in the hadron η production channel, which is quite an open problem, can account for the apparent disagreement. The results of further research will be reported.

References

1. R.E. Cutkosky, C.P. Forsyth, R.E. Hendrick, and R.L. Kelly, *Phys. Rev.* D **20**, 2839 (1979).
2. D.M. Manley and E.M. Saleski, *Phys. Rev.* D **45**, 4002 (1992).
3. S. Capstick *et al.*, *Phys. Rev.* C **59**, R3002 (1999).
4. C. Bennhold, nucl-th/0008023, to be published in the Proceedings of the 18th Indonesian National Physics Symposium.
5. T.P. Vrana, S.A. Dytman, and T.-S.H. Lee, *Phys. Rept.* **328** 181 (2000).
6. BRAG on web, http://cnr2.kent.edu/~manley/BRAG.html
7. G. Höhler, in *Elastic and Charge Exchange Scattering of Elementary Particles*, edited by H. Schopper, Landolt-Börnstein, New Series, Group X, Vol.9, Part 2b (Springer-Verlag, Berlin 1983).
8. G. Höhler, *these proceedings.*
9. BRAG, *these proceedings.*
10. M. Batinić, I. Šlaus and A. Švarc, *Phys. Rev.* C **52**, 2188 (1995); M. Batinić *et al.*, *Physica Scripta* **58**, 15 (1998).
11. B. Krusche *et al.*, *Phys. Rev. Lett.* **74**, 3736 (1995); B. Krusche *et al.*, *Phys. Lett.* B **397**, 171 (1997).
12. H.Calén *et al.*, *Phys. Lett.* B **458**, 190 (1999).
13. B.M.K. Nefkens, *these proceedings.*
14. J.M. Laget and F. Wellers, *Phys. Lett.* B **257**, 254 (1991).
15. G. Fäldt and C. Wilkin, TSL/ISV-96-0162 Uppsala University preprint.
16. T. Vetter *et al*, *Phys. Lett.* B **263**, 153 (1991).
17. E. Gedalin, A. Moalem and L. Razdolskaja, *Nucl. Phys.* A **634**, 368 (1998).
18. A.B. Santra and B.K. Jain, *Nucl. Phys.* A **634**, 309 (1998).
19. M. Batinić, A. Švarc, and T.-S.H. Lee, *Physica Scripta* **56** 321 (1997).
20. S. Ceci and A. Švarc, work in progress.

PHENOMENOLOGICAL ANALYSIS OF N* EXCITATION IN CHARGED DOUBLE PION PRODUCTION.

V. MOKEEV[1], M. RIPANI[2], M. ANGHINOLFI[2], M. BATTAGLIERI[2], G. FEDOTOV[4], E. GOLOVACH[1,2], B. ISHKHANOV[3,4], M. OSIPENKO[4], G. RICCO[2,3], V. SAPUNENKO[1,2], M. TAIUTI[2]
AND CLAS COLLABORATION

[1] *NPI, Moscow State University, Russia;* [2] *INFN, Genova, Italy;* [3] *Dipartimento di Fisica, Università di Genova, Italy;* [4] *Physics Department of Moscow State University, Russia*

1 Introduction

Investigations of double pion production by real and virtual photons on the proton, represent an important direction in experimental studies of baryon structure and strong interaction dynamics. This exclusive channel, discussed also in [1] of this Workshop, provides information on high lying N*s, mainly coupled to multipion final states and can give indication about undiscovered or "missing" new states.

We report here on a phenomenological approach to describe this exclusive channel [2–6], relating quantities of physical interest (N* electromagnetic form factors, contributions of different quasi-two-body channels $\pi^-\Delta^{++}$, $\pi^+\Delta^0$, ρp) with measured observables: differential cross-sections, asymmetries.

2 Model Description

The charged double pion production mechanisms were parameterized as superposition of quasi two-body processes: $\gamma p \to \pi^-\Delta^{++}$, $\gamma p \to \pi^+\Delta^0$, $\gamma p \to \rho p$ [4–6]. The amplitudes were evaluated as product of two-body production and decay amplitudes and intermediate particle propagator. Additional mechanisms were treated as 3-body phase space [4–6]

The decay amplitudes were evaluated from effective Lagrangians with form factors depending on the invariant mass of the decay products [5,6]. In $\pi - \Delta$ and $\rho - p$ production, amplitudes for N* excitation as well as non-resonant mechanisms were implemented. The N* contributions were described by a Breit-Wigner ansatz [2]. 12 N* states below 2 GeV with sizeable contribution to this channel were included. The N* strong decay amplitudes were taken from Ref. [7]. As initial step, $A_{1/2}$, $A_{3/2}$ electromagnetic form factors were taken from experimental data analysis, as discussed below.

Non-resonant mechanisms for $\pi - \Delta$ production were described by a set of Reggeized gauge invariant Born terms [2–8]. We developed a special procedure for an effective description of interactions in the initial and final state with open inelastic channels starting from the world data on πN amplitudes [2]. The diffractive approximation [9] was used for non-resonant ρ production in the N* excitation region. Such approximation provides reasonable data description at $W < 2.0$ GeV.

3 Real Photon

We performed calculations of the full charged double pion production and of $\gamma p \to \pi^- \Delta^{++}$ and $\gamma p \to \rho p$ quasi-two-body channels at the photon point in the N* excitation region (W < 2.0 GeV). PDG values for $A_{1/2}$ and $A_{3/2}$ N* amplitudes were used. Free model parameters, pion Regge trajectory coupling, magnitude of ρ diffractive production amplitude and 3-body phase space, were determined from a simultaneous fit of $\pi^+\pi^-$ and $\pi^+ p$ invariant mass distributions [5,10]. Model results in comparison with ABBHHM data [10] are shown in Fig. 1. Our approach provides a good reproduction of data for both total and partial cross-sections.

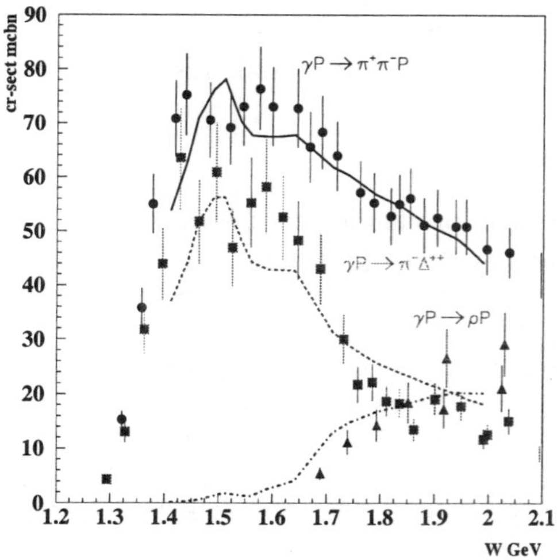

Figure 1. Calculated $\gamma p \to \pi^+\pi^- p$ cross-sections at the photon point in comparison with ABBHHM data [10].

We also evaluated the difference $\sigma_{1/2} - \sigma_{3/2}$ for charged double pion production [11] and its decomposition in resonant and non-resonant mechanisms. Results are presented in Fig. 2. The predicted W-dependence of the asymmetry shows a strong jump around $W = 1.5$ GeV as reported in preliminary results from MAMI [12]. According to our evaluations this jump is produced by the $D_{13}(1520)$ resonance with drastically different values of the $A_{1/2}$, $A_{3/2}$ amplitudes. This is a very important feature in all resonance searches.

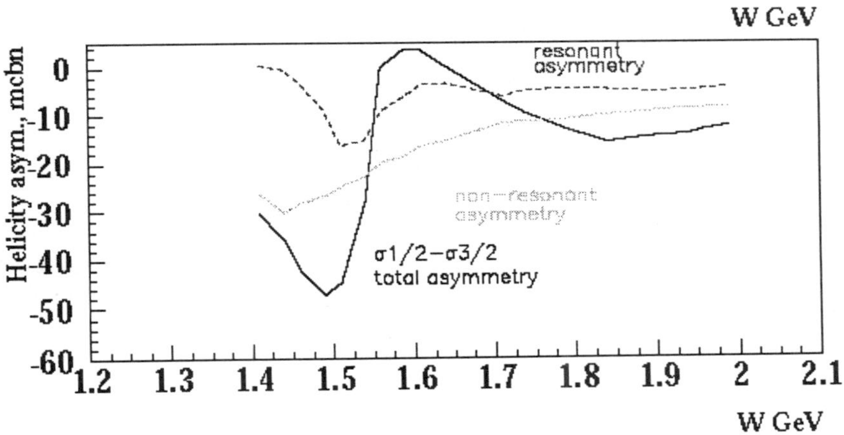

Figure 2. Prediction for asymmetry in the photon point.

4 Virtual Photons

We analysed preliminary CLAS Collaboration data on charged double pion production by virtual photons obtained in E93-006 experiment [13] in two Q^2 bins. As a starting point for the N^* electromagnetic form factors, we used the results of a SU(6)-based fit of world data [14]. As for real photons, the same free model parameters were extracted from the data fit. The calculated cross sections in comparison with CLAS data are shown in Fig. 3. The calculation reproduces the data well, however at W around 1.7 GeV we have a resonance-like excess of measured data respect to calculations for both Q^2 bins. The possible reasons for this could be a missing baryon state contribution as well as deviations of N^* $A_{1/2}$, $A_{3/2}$ couplings from the SU(6) fit prediction. We have strong experimental evidence for a sizeable departure from SU(6) values for electromagnetic couplings of $P_{13}(1720)$ state as discussed in [13]. All these features will be subject to further investigation.

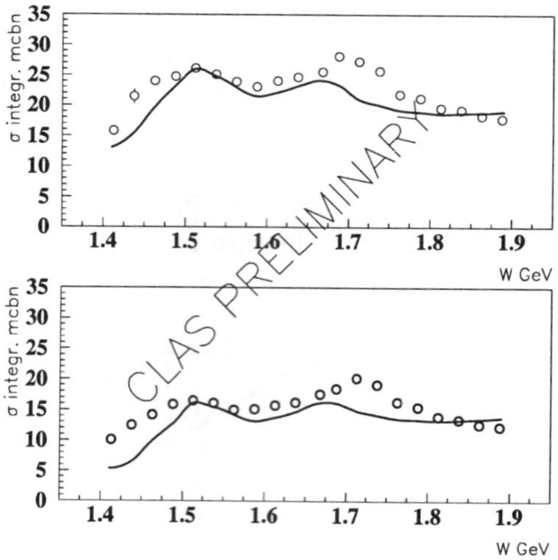

Figure 3. Preliminary results on $\gamma^* p \to \rho p$ cross-section as fuction of W for $Q^2 = 0.65\text{GeV}^2$ (upper figure) and $Q^2 = 0.95\text{GeV}^2$ (lower figure) obtained by the E93-006 CLAS Collaboration experiment [12] with our model evaluation.

References

1. J.C.Nacher, E.Oset, *Nucl. Phys.* A **674**, 205 (2000).
2. M.Ripani, V.Mokeev, *et al., Nucl. Phys.* A**672**, 220 (2000).
3. V.Mokeev, M.Ripani, *et al., Few Body Syst. Suppl. 19* **293** (1999).
4. V.Mokeev, M.Ripani, *et al.* in Proc. of NSTAR 2000, ed by V.D.Burkert, L.Elouadrini, J.J.Kelly, and R.C.Minehart, p. 234.
5. M.Ripani, V.Mokeev, *et al., Physics of Atom. Nucl.* **63**, 1943 (2000).
6. V.Mokeev, M.Ripani, *et al., Physics of Atom. Nucl.* **64**, 1 (2001).
7. D.M.Manley, E.M.Salesky, *Phys. Rev.* D **45**, 4002 (1992).
8. M.Guidal, J-M.Laget, M.Vanderhaegen, *Phys. Lett.* B**400**, 6 (1997).
9. D.G.Cassel *et al., Phys. Rev.* D **24**, 787 (1981).
10. ABBHHM Collaboration, *Phys. Rev.* **175**, 1669 (1968).
11. L. Tiator in Proc. of GDH 2000, eds. by D.Drechsel and L.Tiator, World Scientific (Singapore), p. 57.
12. M.Lang *ibid* p.125.
13. M.Ripani, *these proceedings.*
14. V.D.Burkert, *Czech. Journ. Phys.* **46**, 627 (1996).

RESULTS ON $\Delta(1232)$ RESONANCE PARAMETERS: A NEW πN PARTIAL WAVE ANALYSIS

G. HÖHLER
Institut für Theoretische Teilchenphysik, Universität Karlsruhe
D-76128 Karlsruhe, Germany
E-mail: gerhard.hoehler@physik.uni-karlsruhe.de

The residue of the $\Delta(1232)$ pole derived from a speed plot for the VPI-GWU solution SP00 differs considerably from the value given by the Particle Data Group. An updated version of KH80 is in preparation.

1 Determination of Resonance Parameters from Speed Plots

The $\Delta(1232)$ can be treated as a resonance in the elastic region. Then the *speed* is defined by (W=total energy in the c.m. frame, $s = W^2$)

$$SP(W) = |dT(W)/dW|, \qquad T(W) = \frac{1}{2i}[exp(2i\delta(W) - 1]. \tag{1}$$

$T(W)$ is the dimensionless P33 partial wave amplitude and $\delta(W)$ its phase.

It is 'noncontroversial among theorists' (see Chew[1] and the references in my 'pole-emics', p.697 in Ref.[2] that in S-matrix theory the effects of resonances follow from first order poles in the 2nd sheet. Following many other authors, we consider the pole in the W-plane nearest to the physical real axis.

The resonant parts of $T(W)$ and of $\delta(W)$ are

$$T_R(W) = \frac{\Gamma/2}{M - W - i\,\Gamma/2}, \qquad \tan\delta_R(W) = \frac{\Gamma/2}{M - W}. \tag{2}$$

It follows for the speed of the resonant part of the P33 amplitude

$$SP(M) \equiv H = 2/\Gamma, \qquad SP(M \pm \Gamma/2) = H/2. \tag{3}$$

Fig.1 shows $SP(W)$ from the VPI-GWU solution SP00. The height H and the mass M are well defined, whereas the speed at half height shows a small asymmetry due to the background,

$$M = 1210.8\,MeV, \quad H = 20.2\,GeV^{-1}, \quad \Gamma = 99\,MeV. \tag{4}$$

A comparison with the table of the PDG (p.725 in Ref.[2]) shows an agreement with all earlier determinations of M and Γ. But we obtain $\Gamma/2 = 49.5\,MeV$ for the residue, whereas the value in Ref.[2] is $38\,MeV$. This is not due to a difference of the P33 phases but to the new method used in the determination from SM95. Our value for the residue is in reasonable agreement with the values derived from SM90, KH80 and CMB80.

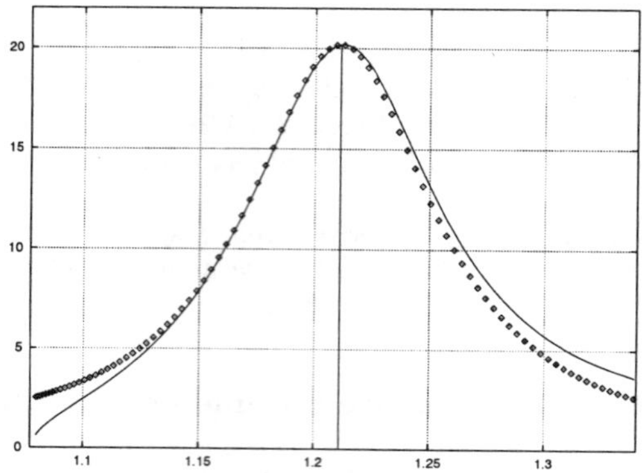

Figure 1. Speedplot for the Resonance $\Delta(1232)$. Solid line: from SP00, dots: from Eq.(2) using Eq.(4).

Fig.1 shows that $SP(W)$ calculated from the P33 phase (SP00) at $W = M \pm \Gamma/2$ almost agrees with the speed calculated from the resonant part of P33 alone. This confirms our result in Eq.(4). Next the background will be discussed[a].

2 The Background in the P33 Partial Wave

2.1 Contribution of the background to the phase

In the elastic region the background can be described by its phase

$$\delta_B(W) = \delta(W) - \delta_R(W). \tag{5}$$

At $W = M$ we have $\delta_R = 90°$ whereas the total P33 phase δ is much smaller: $\delta = 66°$, so $\delta_B = -24°$. Fig. 2 (left panel) shows that the W-dependence of δ_B is almost negligible in the range $W = M \pm \Gamma/2$.

If the background is taken into account, the T-matrix element for elastic scattering can be written

$$T(W) = T_B(W) + T_R(W)\exp(i\phi(W)), \quad T_B(W) = \sin(\delta_B)\exp(i\delta_B). \tag{6}$$

Elastic unitarity demands that $\phi(W) = 2\,\delta_B(W)$. The residue of the pole term is now complex-valued. Our calculation gives $\phi(M) = -48°$. A determination

[a] see Ref.[6] for details and further references

of ϕ from an *Argand plot of the speed vector* dT/dW gives the same result[4]. Again we find a large discrepancy with the value of the PDG[2] from SM95 $\phi = -22°$ and also with the value from SM90, but agreement with KH80 and CMB80.

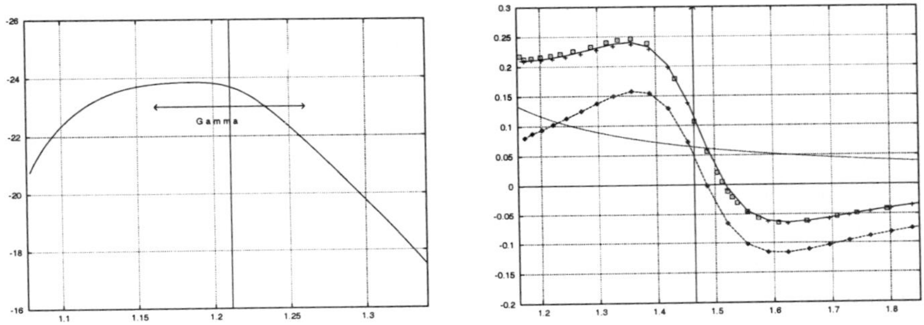

Figure 2. Left: Background phase in degrees vs. W in GeV. Right: Contributions to $ReT/q^3 * m_\pi^{-3}$ vs. s in (GeV^2).

Upper solid line: from SP00 (nearby squares from KA85), nearby circles from the r.h. side of Eq.(7). Background $L(s)$: decreasing line, solid line with circles: dispersion integral, evaluated with $ImT(SP00)$.

2.2 *The dispersion relation for the P33 partial wave*

The dispersion relation was studied in great detail by J. Hamilton et al. who showed that an approximation led to a *relativistic Chew-Low plot*[3]. An improved version was evaluated by R. Koch et al.[7,8], using KH80 and t-channel partial waves of our group[5] as input.

The relation is written for the *reduced amplitude* $F(s) = T(s)/q^3$ in order to suppress the contributions of distant sigularities in the s-plane which are neglected in our simplified calculation

$$ReF(s) = L(s) + \frac{1}{\pi}\int_{sth}^{\infty} \frac{Im\,F(s')}{s'-s} ds' \ . \tag{7}$$

According to table 2 in Ref.[8], the dominant contributions to $L(s)$ come from the u-channel nucleon Born term (Chew-Low) and the t-channel S-wave. The sum can be approximated by an effective pole (m=nucleon mass)

$$L(s) = \frac{0.037}{s - m^2} \quad \text{units: } m_\pi^{-3},\ s \text{ in } GeV^2 \tag{8}$$

An accurate evaluation has recently been made by J. Stahov.

3 An Updated Version of the KH80 Partial Wave Analysis

Since fixed-t analyticity can be proven within the framework of QCD[9], *it is necessary to include this constraint in πN partial wave analysis.* Using the main part of a version of E. Pietarinen's program rewritten for a PC in 1992, H.M. Staudenmaier, C. Hansch and G. H. have produced a program which is running on our workstation alpha, including the graphics and a new data base. Since KH80 has a problem with new spin-rotation data[11], our earlier study of the zero trajectories has been taken up again, taking into account the important papers by I.S. Stefanescu. They include consequences of two-variable analyticity and questions of uniqueness and stability (see[10] for a review). - We hope that the test runs can be finished in 2001.

Acknowledgments

Thanks are due to the organizers for hospitality, to E. Pietarinen for sending the new program, to R.A. Arndt and J. Stahov for a correspondence and to R.L. Workman for a discussion.

References

1. G.F. Chew, Resonances, Particles, and Poles from the Experimenter's Point of View. Berkeley UCRL-16983 (1966).
2. D.E. Groom *et al.*, Particle Data Group, *Eur. Phys. Jour.* C **15**, 1 (2000).
3. J. Hamilton, Pion-Nucleon Interactions in High Energy Physics, **Vol. I**, 193, ed. E. Burhop, (1967).
4. G. Höhler, πN *Newsletter* **9**, 1 (1993).
5. G. Höhler, Pion-Nucleon Scattering in Landolt-Börnstein, **Vol.I/9b2**, ed. H. Schopper, Springer, (1983).
6. G. Höhler, and H.M. Staudenmaier, πN *Newsletter* **12**, 26 (1993).
7. R. Koch, and M. Hutt, *Z. Phys.* C **19**, 119 (1983).
8. R. Koch, *Z. Phys.* C **29**, 597 (1985).
9. R. Oehme, πN *Newsletter* **7**, 1 (1992); Int'l Congress Math. Phys., La Sorbonne, Paris (1994).
10. I.S. Stefanescu, *Prog. Phys.* **35**, 573 (1987).
11. I.G. Alekseev *et al., Phys. Lett.* B **485**, 32 (2000); *Phys. Rev.* C **55**, 2049 (1997); *Phys. Lett.* B **351**, 585 (1995).

SOLVING THE PUZZLE OF THE $\gamma p \to \pi^+\pi^0 n$ REACTION

J. C. NACHER, E. OSET, M. J. VICENTE VACAS AND L. ROCA

Departamento de Física Teórica and IFIC,
Institutos de Investigación de Paterna, Apdo. correos 22085,
46071, Valencia, Spain

Recent information on invariant mass distributions of the $\gamma p \to \pi^+\pi^0 n$ reaction, where previous theoretical models had shown deficiencies, have made more evident the need for new mechanisms, so far neglected or inaccurately included. We have updated a previous model to include new necessary mechanisms. We find that the production of the ρ meson, and the $\Delta(1700)$ excitation, through interference with the dominant terms, are important mechanisms that solve the puzzle of the $\gamma p \to \pi^+\pi^0 n$ reaction without spoiling the early agreement with the $\gamma p \to \pi^+\pi^- p$ and $\gamma p \to \pi^0\pi^0 p$ reactions.

1 Introduction

Recently, new improvements in the experimental techniques have made possible the study of total cross sections with accuracy for the two-pion photoproduction reactions as: $\gamma p \to \pi^+\pi^- p$, $\gamma p \to \pi^0\pi^0 p$, $\gamma p \to \pi^+\pi^0 n$ and $\gamma n \to \pi^-\pi^0 p$ using the large acceptance detector DAPHNE and high intensity tagged photons at Mainz. Some polarization observables are also being measured, like the spin asymmetry $\sigma_{3/2} - \sigma_{1/2}$ and the helicity cross sections $\sigma_{1/2}$ and $\sigma_{3/2}$ with the DAPHNE acceptance [1]. The invariant masses of $\pi^0\pi^0$ [2], $\pi^-\pi^0$ [3] and $\pi^+\pi^0$ [4] have also been measured for different bins of incident photon energies.

Our aim is to improve the model of [5] guided by the new additional experimental results, trying to find the missing mechanisms in the previous description of the $\gamma p \to \pi^+\pi^0 n$ reaction which bring agreement with the new data and do not spoil the agreement reached in other pion charge channels. The work about this topic is found in [6].

2 Formalism

The model [5] describes double pion photoproduction based on a set of tree level diagrams. These Feynman diagrams involve pions, nucleons and nucleonic resonances. Several baryon resonances are included in the model. They are: $\Delta(1232)$ or P_{33} ($J^\pi = 3/2^+$, I=3/2), $N^*(1440)$ or P_{11} ($J^\pi = 1/2^+$, I=1/2) and $N^*(1520)$ or D_{13} ($J^\pi = 3/2^-$, I=1/2). The $N^*(1520)$ has a large coupling to the photons and is an important ingredient due to its interference with the dominant term of the process, the $\gamma N \to \Delta\pi$ transition called the Δ Kroll-Ruderman contact term. No other resonances were considered at that time assuming their contribution to the process would be small in the Mainz range of energies below $E_\gamma = 800 MeV$.

A model for $\Delta\pi$ electroproduction on the proton was presented in [7]. The aim of this work was to extend the model of [5] for the $\gamma N \to \pi\pi N$ reaction to virtual photons selecting the diagrams which have a Δ in the final state. The agreement found with $\gamma_v p \to \Delta^{++}\pi^-$ was good. This reaction selecting the Δ final state was an interesting test for the forthcoming full model of the $\gamma_v N \to \pi\pi N$ reactions.

We must note that the formalism followed in [7] for the vertices of the $N^*(1520)$ is different from that of [5]. We include two additional mechanism in the model of [5], which were introduced in the approach of Ref. [8]. The Feynman diagrams corresponding to these terms involve the $N^*(1520) \to N\rho$ decay mode appearing in both the channels $\gamma p \to \pi^+\pi^- p$ and $\gamma p \to \pi^+\pi^0 n$. This new diagram is zero for $\gamma p \to \pi^0\pi^0 p$ because the intermediate ρ^0 is not allowed to decay to $\pi^0\pi^0$. The other diagram contains a $\gamma N\rho N$ contact interaction or ρ Kroll-Ruderman term. This term will contribute only in the $\pi^+\pi^0$ channel.

3 Discussion and Results

In Fig. 1 we show the contribution of the different terms to the $\gamma p \to \pi^+\pi^0 n$ reaction. We can see that the ρ Kroll-Ruderman term gives by itself a small contribution, something already noted in Ref. [8]. The $N^*(1520)$ excitation followed by ρN decay shows the N^* resonant shape and has a strength at the peak of about 10 μb. The Δ Kroll-Ruderman term provides a background increasing smoothly with the energy. We also show the contribution from the coherent sum of the Δ Kroll-Ruderman and $N^*\rho N$ terms in the case of g_ρ [a] positive (lower solid line) and with g_ρ negative (lower dark dashed line). We can see that g_ρ positive leads to constructive interference between these terms, while g_ρ negative leads to destructive interference. In [6] we show additional analysis for the contribution coming from $\Delta(1700)$ in these reactions. We have included it in our model and the new diagrams which involve the excitation of this resonance are $\Delta(1700) \to \rho N$ and $\Delta(1700) \to \Delta\pi$.

In Fig. 2 we show a set of figures for different bins of photon energies for the invariant mass of $(\pi^+\pi^0)$. The bins are 540-610 MeV, 610-650 MeV, 650-700 MeV, 700-740 MeV, 740-780 MeV, and 780-820 MeV.

---

[a] g_ρ is the coupling of the $N^*(1520)\rho N$ vertex.

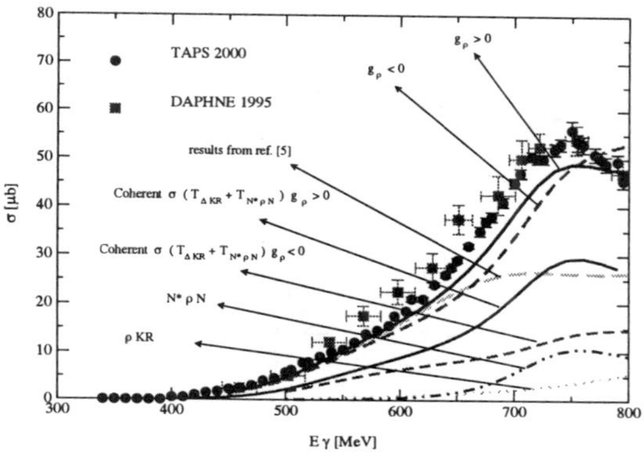

Figure 1. Total cross section for $\gamma p \to \pi^+\pi^0 n$: The labels indicate different partial contributions. Experimental data from Refs. [4] (circles) and [9] (squares).

We show with a dashed line the results of the model without the new resonances and with a continuous line our final results. We find an important contribution to the invariant masses due to the tail of the ρ meson, moving the strength to higher energies of the spectrum. These results are consistent with our predictions for the total cross section and they reassert the influence of the ρ production mechanisms.

4 Summary

In [6] we have made a new analysis of the $\gamma p \to \pi\pi N$ reaction channels. The cross section for $\gamma p \to \pi^+\pi^- p$, $\gamma p \to \pi^+\pi^0 n$ and $\gamma p \to \pi^0\pi^0 p$ were calculated with the new additional contributions of ρ meson production and $\Delta(1700)$ excitation. The calculated cross section and invariant masses show a much better agreement with the experimental results than found in [5] for $\gamma p \to \pi^+\pi^0 n$. The improvements in the $\gamma p \to \pi^+\pi^0 n$ channel have been done without spoiling the agreement found previously in the other channels. The results reported here bring new light to the old problem of the $\gamma p \to \pi^+\pi^0 n$ reaction. The new elements introduced have been stimulated by new experimental measurements that gave clear indications that the ρ production mechanism was important in that reaction.

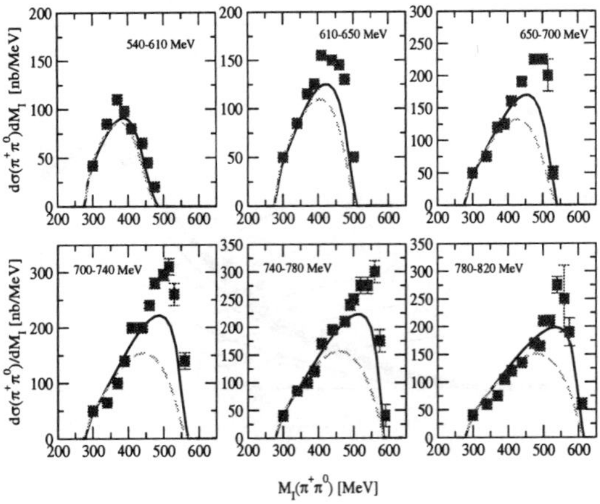

Figure 2. Differential cross section with respect to the invariant mass of the $(\pi^+\pi^0)$ system for different values of E_γ from 540 MeV to 820 MeV for the $\gamma p \to \pi^+\pi^0 n$ reaction. With continuous line we show the final results with ρ meson and $\Delta(1700)$ terms, and with dashed line we show the results of the model without those contributions. Experimental data from ref. [4].

References

1. M. Lang, *private communication, 2000.*
2. M. Wolf *et al., Eur. Phys. J.* A **9**, 5 (2000).
3. A. Zabrodin *et al., Phys. Rev.* C **60**, 055201 (1999).
4. S. Schadmand *et al., Acta Phys. Pol.* B **31**, 2431 (2000); W. Langgaertner *et al.*, submitted to *Phys. Rev. Lett.*
5. J.A. Gómez-Tejedor and E. Oset, *Nucl. Phys.* A **571**, 667 (1994); J.A. Gómez-Tejedor and E. Oset, *Nucl. Phys.* A **600**, 413 (1996).
6. J.C. Nacher, E.Oset, and M.J. Vicente, nucth-th/0012065.
7. J.C. Nacher and E. Oset, *Nucl. Phys.* A **674**, 205 (2000).
8. K. Ochi, M. Hirata, and T. Takaki, *Phys. Rev.* C **56**, 1472 (1997); M. Hirata, K. Ochi, and T. Takaki, nucl-th/9711017.
9. A. Braghieri *et al., Phys. Lett.* B **363**, 46 (1995).

THE GIESSEN MODEL - VECTOR MESON PRODUCTION ON THE NUCLEON IN A COUPLED CHANNEL APPROACH

G. PENNER* AND U. MOSEL
Institut für Theoretische Physik, Universität Giessen, D-35392 Giessen
**Email: gregor.penner@theo.physik.uni-giessen.de*

In Ref.[1] we developed a unitary gauge invariant effective Lagrangian model including the final states γN, πN, $2\pi N$, ηN, $K\Lambda$, and $K\Sigma$ (Ref.[2]) for a simultaneous analysis of all avaible experimental data for photon- and pion-induced reactions on the nucleon. In Ref.[3] this analysis was extended to K^- induced reactions. In this paper we discuss an extension of this method to vector meson nucleon final states, outline a generalization of the standard partial wave formalism that is applicable to any meson-/photon-baryon reaction, and show first results for ωN production.

1 Introduction

The determination of nucleon resonance properties from experiments where the nucleon is excited via either hadronic or electromagnetic probes is one of the major issues of hadron physics. The goal is to be finally able to compare the extracted masses and partial decay widths with predictions from lattice QCD (e.g. Ref.[4]) and/or quark models (e.g. Ref.[5]).

As has been shown in Ref.[1] it is inevitable for a reliable extraction of these properties, to analyze photon and pion induced experimental data simultaneously for as many channels as possible. Therefore our coupled channel model developed in Ref.[1] incorporates the final states γN, πN, $2\pi N$, ηN, and $K\Lambda$, where the $2\pi N$ channel was modeled for simplicity by an isovector 0^+ meson. But as soon as we try to extend the model to CMS energies up to $\sqrt{s} = 2$ GeV for an investigation of higher and so-called missing nucleon resonances, the inclusion of the ωN final state becomes unavoidable due to unitarity. This can be seen from the left panel of Fig. 1. Furthermore, ω production on the nucleon represents a possibility to project out $I = \frac{1}{2}$ resonances in the reaction mechanism. Due to its intrinsic spin the inclusion of the ωN final state in our coupled channel model requires an extension of the standard partial wave decomposition (PWD) method developed for $\pi N/\gamma N \to \pi N$ and $\gamma N \to \gamma N$ (see e.g. Ref.[1]). Such an extension is provided in section 4. In addition, this formalism enables us to achieve a realistic treatment of the most dominant inelastic channel in πN scattering, i.e. the $2\pi N$ state via ρN, $\pi\Delta$, and σN.

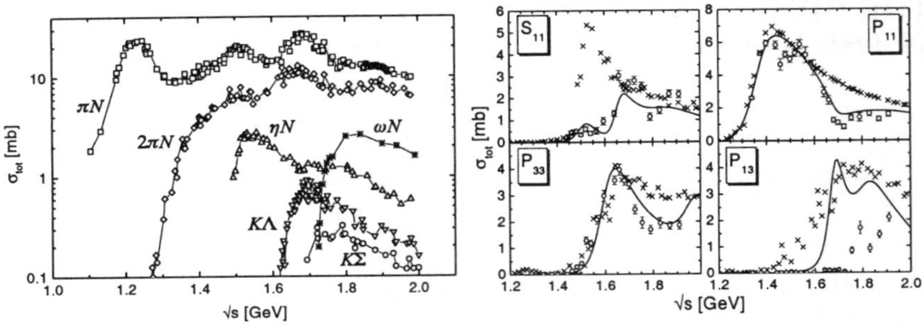

Figure 1. Left: Total cross sections for the reactions $\pi^- p \to X$ with X as given in the figure. Data are from Ref. [6]. Right: Total partial wave cross sections for $\pi N \to 2\pi N$.

2 The Model

Our method to solve the Bethe-Salpeter (BS) equation is the so called K matrix Born-approximation, which is equivalent to setting the intermediate particles in the BS propagator on-shell; for more details see Ref.[1]. Then the reaction matrix T, defined by $S = 1+2\mathrm{i}T$, can be calculated from the potential V after a PWD in total spin J, parity P, and isospin I via matrix inversion:

$$T(p', p; \sqrt{s}) = \frac{V(p', p; \sqrt{s})}{1 - \mathrm{i}V(p', p; \sqrt{s})}. \tag{1}$$

and unitarity is fulfilled as long as V is hermitian.

The potential V_{fi} is built up by s-, u-, and t-channel Feynman diagrams by means of effective Lagrangians which can be found in refs.[1,3]. The advantage of this method is that the background contributions are created dynamically and the number of parameters is greatly reduced, i.e. as compared to Breit-Wigner driven models or those including only pointlike interactions.

3 Results on (Pseudo)scalar Meson Production

As an example for the quality of the calculations we show in the right panel of Fig. 1 the total partial wave cross sections, as extracted by the standard PWD, for $\pi N \to 2\pi N$ in comparison with the inelasticities from $\pi N \to \pi N$ ($\times$) and a $\pi N \to 2\pi N$ analysis[7] ($\circ$). The necessity of the inclusion of a large set of final states in a coupled channel calculation can be seen in various partial waves. In the S_{11} partial wave the difference between the inelasticity and the 2π analysis around $\sqrt{s} = 1.5$ GeV is easily explained by the opening of the ηN final state, the same holds true for P_{11} above $\sqrt{s} = 1.6$ GeV and $K\Lambda$ and for P_{33} above $\sqrt{s} = 1.7$ GeV and $K\Sigma$. However, there still is a discrepancy left in the P_{13} partial wave arising around 1.7 GeV. Since this particular calculation did not include the ωN final state yet, this problem might be solved upon a reanalysis including the ωN final state.

4 Vector Meson Production

Since the orbital angular momentum ℓ is not conserved in, e.g., $\pi N \to \omega N$ the standard PWD becomes really clumsy for many of the channels that have to be included. A more elegant and in particular uniform PWD for all channels would be desirable. Hence we use here a generalisation of the standard PWD method which represents a tool to analyze any meson- and photon-baryon reaction on an equal footing.

We start with the decomposition of a two-particle momentum states into states characterized by the total spin J and its z-component M (see Ref.[8]):

$$|p; JM, \lambda_1\lambda_2\rangle = \sqrt{\frac{2J+1}{4\pi}} \int e^{i(M-\lambda)\varphi} d^J_{M\lambda}(\vartheta)|p\vartheta\varphi; \lambda_1\lambda_2\rangle \mathrm{d}\Omega, \quad \lambda = \lambda_1 - \lambda_2,$$

where λ_1 and λ_2 are the helicities of the two particles and the $d^J_{M\lambda}(\vartheta)$ are Wigner functions. For the incoming CMS state ($\vartheta_0 = \varphi_0 = 0 \Rightarrow \ell_z = 0$) one gets $\langle JM, \lambda_1\lambda_2|\vartheta_0\varphi_0, \lambda_1\lambda_2\rangle \sim \delta_{M\lambda}$, hence $M = \lambda$ and one can drop the index M. By using the parity property (cf. Ref.[8]) $\hat{P}|J,\lambda\rangle = \eta_1\eta_2(-1)^{J-s_1-s_2}|J,-\lambda\rangle$, where η_1, η_2, and s_1, s_2 are the intrinsic parities and spins of the two particles, the construction of states with a well defined parity is straightforward:

$$\hat{P}|J,\lambda;P\pm\rangle := \hat{P}\frac{1}{2}(|J,+\lambda\rangle \pm |J,-\lambda\rangle) = \pm\eta_1\eta_2(-1)^{J-s_1-s_2}|J,\lambda;P\pm\rangle,$$

and we can use them to project out helicity amplitudes with definite parity:

$$\begin{aligned} T^{J\pm}_{\lambda'\lambda} &:= \langle\lambda'|\hat{T}|J\lambda;P\pm\rangle \qquad (2)\\ &= \frac{1}{2}(T^J_{\lambda'\lambda} \pm T^J_{\lambda'-\lambda}) \quad \text{with } T^J_{\lambda'\lambda} = \frac{1}{2}\int T_{\lambda'\lambda}(x) d^J_{\lambda\lambda'}(x)\mathrm{d}x, \; x = \cos\vartheta. \end{aligned}$$

These helicity amplitudes $T^{J\pm}_{\lambda'\lambda}$ have definite, identical J and definite, but opposite P. As is quite obvious this method is valid for any meson-baryon final state combination, even cases as e.g. $\omega N \to \pi\Delta$. In the case of $\pi N \to \pi N$ the $T^{J\pm}_{\lambda'\lambda}$ coincide with the conventional partial wave amplitudes: $T^{J\pm}_{\frac{1}{2}\frac{1}{2}} \equiv T_{\ell\pm}$.

This PWD has been used for calculating pion- and photon-induced ω production on the nucleon. For our first results, we have applied the couplings of set SM95-pt-3 from Ref.[1], i.e. the ω only couples to the nucleon. However, as can be seen in the left panel of Fig. 2 for a reliable calculation of ω production on the nucleon, the inclusion of rescattering is a basic requirement.

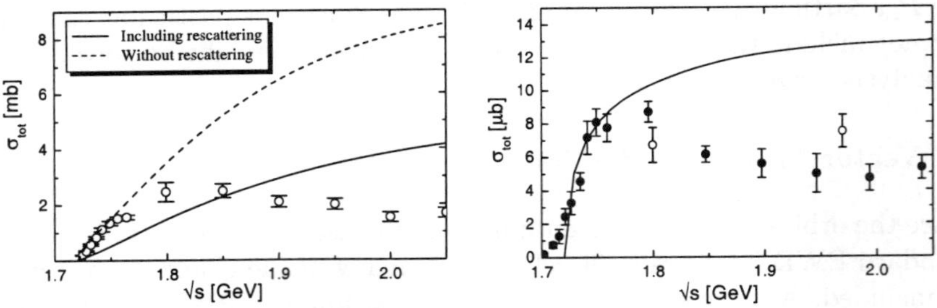

Figure 2. First results for $\pi^- p \to \omega n$ (left) and $\gamma p \to \omega p$ (right).

5 Outlook

The next step of the extension of our coupled channel K matrix model will be the inclusion of ωN data in the determination of resonance properties. Furthermore, since the partial wave formalism is now settled, the inclusion of additional final states, in particular for a more sophisticated description of the $2\pi N$ final state, such as ρN, σN, and $\pi\Delta$ and accounting for their spectral function, is straightforward.

6 Acknowledgments

This work is supported by DFG and BMBF.

References

1. T. Feuster and U. Mosel, *Phys. Rev.* C **58**, 458 (1998), T. Feuster and U. Mosel, *Phys. Rev.* C **59**, 460 (1999).
2. A. Waluyo *et al.*, Los Alamos preprint `nucl-th/0008023`.
3. M.Th. Keil, G. Penner, and U. Mosel, *Phys. Rev.* C **63**, 045202-1 (2001).
4. F.X. Lee and D.B. Leinweber, *Nucl. Phys.* B **73**, 258 (1999) and proceedings to this workshop.
5. S. Capstick and W. Roberts, *Phys. Rev.* D **47**, 2004 (1993) and proceedings to this workshop.
6. H. Höhler, *Landolt-Börnstein* Volume 9, Springer, Berlin 1983.
7. D. Manley *et al.*, *Phys. Rev.* D **30**, 904 (1984).
8. M. Jacob, G.C. Wick, *Ann. Phys.* **7**, 404 (1959).

MULTIPOLE ANALYSIS FOR PION PHOTOPRODUCTION WITH MAID AND A DYNAMICAL MODEL

S. S. KAMALOV[1], D. DRECHSEL[2], L. TIATOR[2] AND S. N. YANG[3]
[1] *Laboratory of Theoretical Physics, JINR, 141980 Dubna, Russia*
[2] *Institut für Kernphysik, Universität Mainz, 55099 Mainz, Germany*
[3] *Department of Physics, National Taiwan University, Taipei, Taiwan*

We present results of new analyses of pion photoproduction data obtained with MAID and a dynamical model

During the last few years we have developed and extended two models for the analysis of pion photo- and electroproduction, the Dynamical Model [1] (hereafter called Dubna-Mainz-Taipei (DMT) model) and the Unitary Isobar Model [2] (hereafter called MAID). The final aim of such an analysis is to shed more light on the dynamics involved in nucleon resonance excitations and to extract N^* resonance properties in an unambiguous way. For this purpose and as a testing ground, we will use benchmark data bases recently created and distributed among different theoretical groups.

The crucial point in a study of N^* resonance properties is the separation of the total amplitude, with partial channels $\alpha = \{l, j\}$,

$$t_{\gamma\pi}^{\alpha} = t_{\gamma\pi}^{B,\alpha} + t_{\gamma\pi}^{R,\alpha} \tag{1}$$

in background $t_{\gamma\pi}^{B,\alpha}$ and resonance $t_{\gamma\pi}^{R,\alpha}$ contributions. In different theoretical approaches this procedure is different, and consequently this can lead to different treatments of the dynamics of N^* resonance excitation. As an example we will consider the two different models DMT and MAID.

In accordance with Ref.[1], in the DMT model the $t_{\gamma\pi}^{B,\alpha}$ amplitude is defined as

$$t_{\gamma\pi}^{B,\alpha}(DMT) = e^{i\delta_\alpha}\cos\delta_\alpha \left[v_{\gamma\pi}^{B,\alpha} + P\int_0^\infty dq' \frac{q'^2\, R_{\pi N}^{(\alpha)}(q_E, q')\, v_{\gamma\pi}^{B,\alpha}(q')}{E - E_{\pi N}(q')} \right], \tag{2}$$

where $\delta_\alpha(q_E)$ and $R_{\pi N}^{(\alpha)}$ are the πN scattering phase shift and full πN scattering reaction matrix in channel α, respectively, and q_E is the pion on-shell momentum. The pion photoproduction potential $v_{\gamma\pi}^{B,\alpha}$ is constructed in the same way as in Ref.[2] and contains contributions from the Born terms with an energy dependent mixing of pseudovector-pseudoscalar (PV-PS) πNN coupling and t-channel vector meson exchanges. In the DMT model $v_{\gamma\pi}^{B,\alpha}$ depends on 7 parameters: The PV-PS mixing parameter Λ_m (see Eq.(12) of Ref.[2]),

4 coupling constants and 2 cut-off parameters for the vector meson exchange contributions.

In the extended version of MAID, the S, P, D and F waves of the background amplitudes $t_{\gamma\pi}^{B,\alpha}$ are defined in accordance with the K-matrix approximation

$$t_{\gamma\pi}^{B,\alpha}(\mathrm{MAID}) = \exp(i\delta_\alpha)\, \cos\delta_\alpha v_{\gamma\pi}^{B,\alpha}(q, W, Q^2)\,, \tag{3}$$

where $W \equiv E$ is the total πN c.m. energy and $Q^2 = -k^2 > 0$ is the square of the virtual photon 4-momentum. Note that in actual calculations, in order to take account of inelastic effects, the factor $\exp(i\delta_\alpha)\cos\delta_\alpha$ in Eqs. (2-3) is replaced by $\frac{1}{2}[\eta_\alpha \exp(2i\delta_\alpha) + 1]$, where both the πN phase shifts δ_α and inelasticity parameters η_α are taken from the analysis of the SAID group[3].

From Eqs. (2) and (3), one finds that the difference between the background terms of MAID and of the DMT model is due to the fact that pion off-shell rescattering contributions (principal value integral) are not included in the background of MAID. From our previous studies of the P wave multipoles in the (3,3) channel [1] it follows that these contributions are effectively included in the resonance sector leading to the dressing of the $\gamma N\Delta$ vertex. However, in the case of S waves the DMT results show that off-shell rescattering contributions are very important for the E_{0+} multipole in the $\pi^0 p$ channel. In this case they have to be taken into account explicitly. Therefore, in the extended version of MAID we have introduced a new phenomenological term in order to improve the description of the π^0 photoproduction at low energies,

$$E_{corr}(MAID) = \frac{\Delta E}{(1 + B^2 q_E^2)^2}\, F_D(Q^2)\,, \tag{4}$$

where F_D is the standard nucleon dipole form factor, $B = 0.71$ fm and ΔE is a free parameter which can be fixed by fitting the low energy π^0 photoproduction data. Thus the background contribution in MAID finally depends on 8 parameters. In both models we also take into account the cusp effect due to unitarity below $\pi^+ n$ threshold for, as described in Ref.[4], i.e.

$$E_{cusp} = -a_{\pi N}\, \omega_c\, ReE_{0+}^{\gamma\pi^+} \sqrt{1 - \frac{\omega^2}{\omega_c}}\,, \tag{5}$$

where ω and $\omega_c = 140$ MeV are the π^+ c.m. energies corresponding to $W = E_p + E_\gamma$ and $W_c = m_n + m_{\pi^+}$, respectively, and $a_{\pi N} = 0.124/m_{\pi^+}$ is the pion charge exchange amplitude.

For the resonance contributions, following Ref.[2], the Breit-Wigner form is assumed in both models , i.e.

$$t_{\gamma\pi}^{R,\alpha}(W,Q^2) = \bar{\mathcal{A}}_\alpha^R(Q^2)\,\frac{f_{\gamma R}(W)\Gamma_R\,M_R\,f_{\pi R}(W)}{M_R^2 - W^2 - iM_R\Gamma_R}\,e^{i\phi_R}\,, \tag{6}$$

where $f_{\pi R}$ is the usual Breit-Wigner factor describing the decay of a resonance R with total width $\Gamma_R(W)$ and physical mass M_R. The phase $\phi_R(W)$ in Eq. (6) is introduced to adjust the phase of the total multipole to equal the corresponding πN phase shift δ_α.

The main subject of our study in the resonance sector is the determination of the strengths of the electromagnetic transitions described by the amplitudes $\bar{\mathcal{A}}_\alpha^R(Q^2)$. In general, these strengths are considered as free parameters which have to be extracted from the analysis of the experimental data. In our two models we have included contributions from the 8 most important resonances, listed in the Table 1. The total number of $\bar{\mathcal{A}}_\alpha^R$ amplitudes is 12 which can also be expressed in terms of the 12 standard helicity elements $A_{1/2}$ and $A_{3/2}$. Thus, to analyze the experimental data we have a total of 19 parameters in

N^*		MAID current	MAID HE fit	DMT HE fit	PDG2000
$P_{33}(1232)$	$A_{1/2}$	-138	-143	—	-135± 6
	$A_{3/2}$	-256	-264	—	-255± 8
$P_{11}(1440)$	$A_{1/2}$	-71	-81	-77	-65± 4
$D_{13}(1520)$	$A_{1/2}$	-17	-6	-7	-24± 9
	$A_{3/2}$	164	160	165	166± 5
$S_{11}(1535)$	$A_{1/2}$	67	81	102	90± 30
$S_{31}(1620)$	$A_{1/2}$	0	86	37	27± 11
$S_{11}(1650)$	$A_{1/2}$	39	32	34	53± 16
$F_{15}(1680)$	$A_{1/2}$	-10	5	10	-15± 6
	$A_{3/2}$	138	137	132	133 ± 12
$D_{33}(1700)$	$A_{1/2}$	86	119	107	104± 15
	$A_{3/2}$	85	82	74	85± 22
PV-PS mixing:	Λ_m	450	406	302	
	ΔE	2.01	1.73	—	
	χ^2/d.o.f.	11.5	6.10	6.10	

Table 1. Proton helicity amplitudes (in $10^{-3}\,GeV^{-1/2}$), values of the PV-PS mixing parameter Λ_m (in MeV) and low-energy correction parameter ΔE (in $10^{-3}/m_{\pi^+}$) obtained after the high-energy (HE) fit.

DMT and 20 parameters in MAID. The final results obtained after the fitting of the high-energy (HE) benchmark data base with 3270 data points in the photon energy range $180 < E_\gamma < 1200$ MeV are given in Table 1.

For the analysis of the low-energy (LE) data base with 1287 data points in the photon energy range $180 < E_\gamma < 450$ MeV in the DMT model we used only 4 parameters: The PV-PS mixing parameter and 3 parameters for the $P_{33}(1232)$ and $P_{11}(1440)$ resonances. In MAID we have one more parameter due to the low energy correction given by Eq. (4). The final results for the helicity elements and the E2/M1 ratio (REM) are given in Table 2. In Table 3 we summarize our results and show the χ^2 ob-

N^*		MAID current	MAID LE fit	DMT LE fit	PDG2000
$P_{33}(1232)$	$A_{1/2}$	-138	-142	—	-135± 6
	$A_{3/2}$	-256	-265	—	-255± 8
$P_{11}(1440)$	$A_{1/2}$	-71	-81	-93	-65± 4
	REM(%)	-2.2	-1.9	-2.1	-2.5± 0.5
	χ^2/d.o.f.	4.76	4.56	3.59	

Table 2. Proton helicity elements (in $10^{-3}\, GeV^{-1/2}$) and REM=E2/M1 ratio (in %) obtained from the LE fit.

tained for different channels and different observables after fitting the LE and HE data bases. Note that we obtain the largest χ^2 in the LE fit for

	LE			HE		
Observables	N	MAID	DMT	N	MAID	DMT
$\frac{d\sigma}{d\Omega}(\gamma,\pi^+)$	317	4.68	3.32	871	6.36	5.95
$\frac{d\sigma}{d\Omega}(\gamma,\pi^0)$	354	7.22	5.74	859	6.87	5.85
$\Sigma(\gamma,\pi^+)$	245	2.79	2.57	546	4.57	6.49
$\Sigma(\gamma,\pi^0)$	192	2.22	1.58	488	7.65	7.65
$T(\gamma,\pi^+)$	107	3.28	2.94	265	3.75	4.17
$T(\gamma,\pi^0)$	72	5.18	4.84	241	5.31	5.65
Total	1287	4.56	3.64	3270	6.10	6.10

Table 3. χ^2/N for the cross sections ($\frac{d\sigma}{d\Omega}$), photon (Σ) and target (T) asymmetries in (γ,π^+) and (γ,π^0) channels obtained after LE and HE fit. N is the number of data points

differential cross sections and target asymmetries in $p(\gamma,\pi^0)p$. Similar results were obtained in practically all other analyses. A detailed comparison with the results of different theoretical groups is given on the website

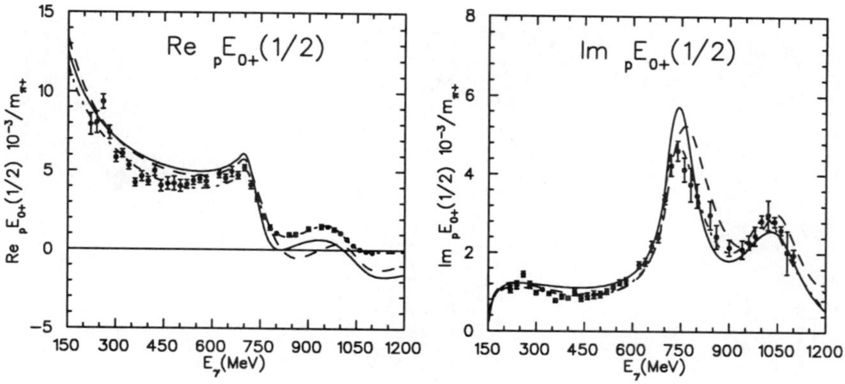

Figure 1. The $_pE_{0+}^{1/2}$ multipole obtained after the HE fit using MAID (solid curves) and DMT (dashed curves). The dash-dotted curves and data points are the results of the global and single-energy fits obtained by the SAID group.

http : //gwdac.phys.gwu.edu/analysis/pr_benchmark.html. Below, in Fig. 1 we show only one interesting example, the E_{0+} multipole in the channel with total isospin 1/2. In this channel contributions from the $S_{11}(1535)$ and $S_{11}(1620)$ resonances are very important. At $E_\gamma > 750$ MeV our values for the real part of the $_pE_{0+}^{1/2}$ amplitude are mostly negative and lower than the results of the SAID multipole analysis. The only possibility to remove such a discrepancy in our two models would be to introduce a third S_{11} resonance. Another interesting result is related to the imaginary part of the $_pE_{0+}^{1/2}$ amplitude and, consequently, to the value of the helicity elements given in Table 1. Within the DMT model for the $S_{11}(1535)$ we obtain $A_{1/2} = 102$ for a total width of 120 MeV, which is more consistent with the results obtained in η photoproduction than with previous pion photoproduction results obtained by the SAID and MAID groups.

References

1. S.S. Kamalov and S.N. Yang, *Phys. Rev. Lett* **83**, 4494 (1999); S.N. Yang, *J. Phys.* G **11**, L205 (1985).
2. D. Drechsel, O. Hanstein, S.S. Kamalov, and L. Tiator, *Nucl. Phys.* A **645**, 145 (1999).
3. R.A. Arndt, I.I. Strakovsky, and R.L. Workman, *Phys. Rev.* C **53**, 430 (1996).
4. J. M. Laget, *Phys. Rep.* **69**, 1 (1981).

Lothar Tiator, Toru Sato, Sabit Kamalov

During a discussion with Latifa Elhouadrhiri as Chair

MODEL DEPENDENCE OF E2/M1

R. M. DAVIDSON
Department of Physics, Applied Physics, and Astronomy
Rensselaer Polytechnic Institute, Troy, NY, 12180, USA
E-mail: davidr@rpi.edu

I present the result of BRAG's preliminary investigation of the model dependence inherent in the extraction of the $N - \Delta$ E2/M1 ratio from the data.

1 Introduction

As was evidenced by the presentations at this meeting, there is both tremendous and diverse activity in the field of baryon physics. The ultimate goal of these investigations is an *understanding* of the nucleon and its excited states. While lattice QCD offers the possibility of first principle calculations of various baryon properties, one still needs QCD-inspired models to develop a physical picture of these complicated objects. There are several competing models on the market, all of which are in fair-to-good agreement with the baryon spectrum. However, it is possible to get the spectrum right with the wrong physics. For example, Bohr's model of the hydrogen atom gives (almost) the correct spectrum, but, lacking the proper physics, has nothing to say about the relative intensities of various transitions.

In baryon physics, as in atomic and nuclear physics, electromagnetic transition amplitudes can be critical tests of the competing models provided the amplitudes can be accurately extracted from the data. The accuracy of this extraction depends not only on the accuracy of the data, but also on the model dependence of the extraction. Since for meson photoproduction a complete set of experiments has not been performed, any analysis of the data requires the use of a model[a] leading to an inherent model dependence of the extracted resonance parameters. The Baryon Resonance Analysis Group (BRAG) has begun investigating the degree of this model dependence.

[a]There are 'model-independent' analyses on the market. These analyses rely on various assumptions such as s- and p-wave dominance. The validity of these assumptions are usually checked using a model.

2 The Participants and their Models

It has been known for some time that the polarized photon asymmetry is particularly sensitive to the E2/M1 ratio. High precision measurements of this asymmetry were done at LEGS[1] and Mainz[2]. Various groups analyzed these data, in conjunction with other data, and the E2/M1 ratios extracted by the different groups ranged from -1.5% to -3.2%, disagreeing within the quoted errors. Subsequently it was discovered[3] that much of the difference between the VPI result[4] and RPI result[5] was due to the fact that different data sets were fitted. If the same data set was fitted, the results from the two different approaches were in excellent agreement. BRAG members decided to broaden this investigation by involving more groups using a wider variety of models to analyze the data. In addition, it was decided to compare not only the resonance parameters, but also the individual multipoles as a function of W. These comparisons can be obtained from either the SAID[6] or BRAG[7] websites.

L. Tiator (Mainz) and I. Strakovsky (GWU) compiled 'bench-mark' data sets to be analyzed by various groups. The low-energy data set, spanning the $\Delta(1232)$ resonance, consisted of 1287 data points comprised of differential cross section data, photon asymmetry data and target asymmetry data for the $p\pi^0$ and $n\pi^+$ channels. These data were sent to a number of groups using different models for analysis. For this present exercise, no attempt was made to account for systematic errors in the data, i.e., only the statistical errors were used in the fits.

Table 1. Results for M1, E2 and E2/M1 obtained by various groups discussed in the text. M1 and E2 are in units of 10^{-3} $\text{GeV}^{-1/2}$ while E2/M1 is given as a percent.

	M1	E2	E2/M1 (%)
RPI	286	-7.2	-2.55
GWU	281	-7.2	-2.57
HA	281	-6.6	-2.35
MAID	275	-5.3	-1.93
KY	280	-6.2	-2.24
AZ	278	-6.3	-2.28
OM	288	-7.8	-2.77
AVG	281.3 ± 4.5	-6.6 ± 0.8	-2.38 ± 0.27

The participating groups and their methods are as follows. The RPI group (denoted by RPI in the Table) used an effective Lagrangian approach[8]. The George Washington group (GWU in the Table) performed a partial wave analysis [9]. The Mainz group performed analyses using both with an isobar model[10] (MAID) and with a dispersion theory model[11] (HA). Aznauryan (AZ) also used a dispersion theory model[12] to analyze the data. Yang and Kamalov (YK) used a dynamical model[13], while Omelaenko (OM) did a multipole analysis. While these models do share some common features, such as unitarity, the dynamics of the models are often quite different. For the two dispersion theory approaches, although the starting point is the same, quite different assumptions were made in order to obtain a manageable system of equations. I refer the reader to the literature for details of these approaches.

3 Results

At the BRAG workshop preceding this meeting, the results of the analyses were presented and compared. In general, the agreement amongst the different multipole sets is good, and where there is large disagreement the origin is understood. Apart from the resonance multipoles, there was also interest in understanding the physics of the $M_{1-}^{1/2}$ multipole. The different analyses tend to agree on the numerical value of this multipole, but differ on its physical interpretation. In some works[8,11,12], it is interpreted as a crossed Δ contribution, while in other works[10] it is interpreted as a Roper contribution.

The results for M1, E2 and E2/M1, defined in terms of K-matrix residues[14], from the various analyses are presented in the table. To give an idea of the spread of the numbers, the simple averages (AVG) and standard deviations are also given. Given the spread of numbers that appear in the previous literature, these results are quite satisfactory and encouraging. These results demonstrate that given high precision data, one can extract from the data a 2% effect with only a modest amount of model dependence. It should be emphasized that it is the spread of numbers in the table that is most important, and not the magnitude of the numbers. The averages in the table should not be understood as the 'preferred' values for these quantities. As noted in Ref.[3], E2 is quite dependent on the data set being fitted. The next step in this investigation is to compile a consistent set of data and reperform the analyses including both the statistical and systematic errors.

Acknowledgments

This research is supported by the U.S. Dept. of Energy grant DE–FG02–88ER40448.

References

1. G. Blanpied *et al.*, *Phys. Rev. Lett.* **69**, 1880 (1992); *ibid.* **79**, 4337 (1997).
2. R. Beck *et al.*, *Phys. Rev. Lett.* **78**, 606 (1997); R. Beck and H. P. Krahn, *ibid.* **79**, 4510 (1997).
3. R. M. Davidson, Nimai C. Mukhopadhyay, M. S. Pierce, R. A. Arndt, I. I. Strakovsky, and R. L. Workman, *Phys. Rev.* C **59**, 1059 (1999).
4. R. A. Arndt, I. I. Strakovsky, and R. L. Workman, *Phys. Rev. Lett.* **79**, 4510 (1997).
5. R.M. Davidson and Nimai C. Mukhopadhyay, *Phys. Rev. Lett.* **79**, 4509 (1997).
6. http://gwdac.phys.gwu.edu
7. http://cnr2.kent.edu/ manley/benchmark.html
8. R.M. Davidson, N.C. Mukhopadhyay, and R.S. Wittman, *Phys. Rev.* D **43**, 71 (1991).
9. R.A. Arndt, I.I. Strakovsky, and R.L. Workman, *Phys. Rev.* C **53**, 430 (1996).
10. D. Drechsel, O. Hanstein, S.S. Kamalov, and L. Tiator, *Nucl. Phys.* A **645**, 145 (1999).
11. O. Hanstein, D. Drechsel, and L. Tiator, *Nucl. Phys.* A **632**, 561 (1998).
12. I.G. Aznauryan, *Phys. Rev.* D **57**, 2727 (1998).
13. S.S. Kamalov and S.N. Yang, *Phys. Rev. Lett.* **83**, 4494 (1999).
14. R.M. Davidson and N.C. Mukhopadhyay, *Phys. Rev.* D **42**, 20 (1990).

ETA PHOTOPRODUCTION IN A COUPLED-CHANNELS APPROACH

A. WALUYO AND C. BENNHOLD

Center for Nuclear Studies, Department of Physics, The George Washington University, Washington, D.C. 20052

G. PENNER AND U. MOSEL

Institut für Theoretische Physik, Universität Giessen, D-35392 Giessen, Germany

Eta photoproduction on the nucleon is studied in a coupled-channels approach based on an effective Lagrangian. We compare to new differential cross section and beam asymmetry data data from the GRAAL collaboration. At energies above the $S_{11}(1535)$ region we find no sign of p-wave nucleon resonance contributions. The forward-peaking of the beam asymmetry data at higher energy is explained in terms of vector meson t-channel contributions. These new polarization data also reveal a novel role played by the $D_{13}(1700)$ state, similar to the role of the $D_{13}(1520)$ at lower energy.

1 Introduction

Among the many possible excitation mechanisms for nucleon resonances, pion scattering and pion photoproduction have supplied most of the information on N^*s over the past fifty years. Recently, however, due to the improvement of experimental facilities, other reactions that may constitute only a small fraction of the total photoabsorption cross section, have begun playing a more important role. Among these is eta photoproduction, which has more recently established itself as a new, powerful tool to selectively probe certain resonances that may be difficult to explore with pions. It is well known that the low-energy behavior of the eta production process is governed by the $S_{11}(1535)$ resonance[1]. A well-known example of the power of the (γ, η) reaction is the extraction of the $A^p_{1/2}$ helicity amplitude of the $S_{11}(1535)$ state. Due to the combined cusp-resonance nature of this resonance, analyses based solely on pion photoproduction consistently underestimate this quantity to be around 0.060 $GeV^{-1/2}$ while extractions from eta photoproduction place the value closer to 0.10 $GeV^{-1/2}$. Recent coupled-channel analyses[2,3] that properly include the cusp dynamics as well as the resonance phenomena have confirmed a range of values consistent with eta photoproduction.

While the very precise threshold measurements of the total and differential cross sections that lead to an improved understanding of the $S_{11}(1535)$ polarization observables, can provide a new doorway[4] to access smaller, non-dominant resonances by relying on the dominant E_{0+} multipole to interfere with a smaller amplitude. Using polarized photon asymmetry data measured at GRAAL it was shown[5] that this interference effect can reveal less well known properties of the $D_{13}(1520)$, such as an ηN branching ratio of less than 0.1% and the helicity amplitude $A^p_{1/2}$.

In order to provide a consistent and complete picture of an individual nucleon resonance, the various possible production and decay channels must be treated in a multichannel framework that preserves unitarity and permits separating resonance from background contributions. Here, we focus our ongoing investigation of N^* properties[6] on the role of selected resonances in the eta photoproduction process. Based originally on the work by Feuster and Mosel[3], we use an expanded version of their model with the asymptotic channels $\gamma N, \pi N, \pi\pi N, \eta N, K\Lambda$, $K\Sigma$, and $\eta' N$ up to the center of mass energy $\sqrt{s}$ of 2.0 GeV.

2 Low-Energy Region: $S_{11}(1535)$ and $D_{13}(1520)$

The threshold region has been discussed in great detail and its dominant features are well understood. The pioneering work of the TAPS collaboration at MAMI B which measured threshold total and differential cross sections with great precision[7] confirmed not only the s-wave dominance but also a very small background, due to a small $g_{\eta NN}$ coupling constant of around 0.1 and the small contributions of the t-channel vector mesons in the threshold region. This can be seen from the absence of p-wave contributions in the threshold differential cross section data. While the TAPS data found the p-wave amplitude to be compatible with zero, they did reveal a small but non-negligible d-wave. With the arrival of the GRAAL photon polarization data[8] and by using the above mentioned inference effect, this d-wave component could be identified as stemming from the $D_{13}(1520)$ state[5,9].

All of the above results are reproduced within our coupled-channels framework, with the results shown in Table 1 and in Figs. 1 and 2. In this second resonance region the data are now of high quality for the $\pi N \to \pi N$, $\gamma N \to \pi N$ and $\gamma N \to \eta N$ reactions. Besides the various two-pion channels the missing link is the $\pi N \to \eta N$ reaction which is still under analysis by the Crystal Ball Collaboration. Including those data should settle remaining questions about the relative size of $S_{11}(1535)$ partial and totals widths. We find an $S_{11}(1535)$ total width of around 250 MeV, towards the upper end of the values from PDG[10], $100-250$ MeV, and considerably larger than the value obtained by another recent coupled-channels analysis[2], 112 MeV, but more in line with Ref.[11], 212 MeV. Our result for the $S_{11}(1535)$ photocoupling is similar to (γ, η) single-channel extractions of this quantity[9]. Our total width for the $D_{13}(1520)$ state is 84 MeV, a bit lower than the values from PDG and the Pitt-ANL group. While our value for the $A^p_{3/2}$ amplitude of the $D_{13}(1520)$ is comparable to other analyses, our value for the $A^p_{1/2}$ is compatible with zero, outside accepted ranges. This finding is especially mysterious in view of the recent work[5] which used the (γ, η) GRAAL data and found a value of $A^p_{1/2} = -0.079$ GeV$^{-1/2}$.

Table 1. Parameters of the nucleon resonances discussed in the text. The first line shows our results while the second line is from PDG.

	$S_{11}(1535)$	$S_{11}(1650)$	$D_{13}(1520)$	$D_{13}(1700)$
M (MeV)	1556 1520-1555	1679 1640-1680	1506 1515-1530	1690 1650-1750
Γ_{tot} (MeV)	252 100-250	219 145-190	84 110-135	202 50-150
Γ_π (MeV)	72 53-83	138 83-135	50 50-60	0.71 5-18
Γ_η (MeV)	157 45-83	26 5-15	0.01 -	1.5 -
$\Gamma_{\pi\pi}$ (MeV)	23 1.5-15	53 15-30	34 48-60	199 85-95
$A^p_{1/2}$ $(10^{-3} GeV^{-1/2})$	111 90±30	46 53±16	2 -24±9	11.2 -18±13

3 Intermediate-Energy Region: $S_{11}(1650)$ and $D_{13}(1700)$

As can be seen in Fig.1, the intermediate-energy region continues to be remarkably absent of p-wave contributions. Thus, while smaller in cross section, the dominant resonance in this region, still interfering with the $S_{11}(1535)$, appears to be the $S_{11}(1650)$. With 26 MeV we find an ηN branching ratio larger than other studies. The flatness of the angular distribution up to a W=1675 MeV appears to rule out any major contribution of the $P_{11}(1710)$ state which had once been hypothesized to have an appreciable decay width into the ηN channel. An unconstrained multichannel fit actually produces a small $P_{11}(1710) \rightarrow \eta N$ width, which however leads to a predicted cross section ruled out by the data. Another feature of interest is the contribution from channel coupling: it is negligible in the threshold region, becomes very

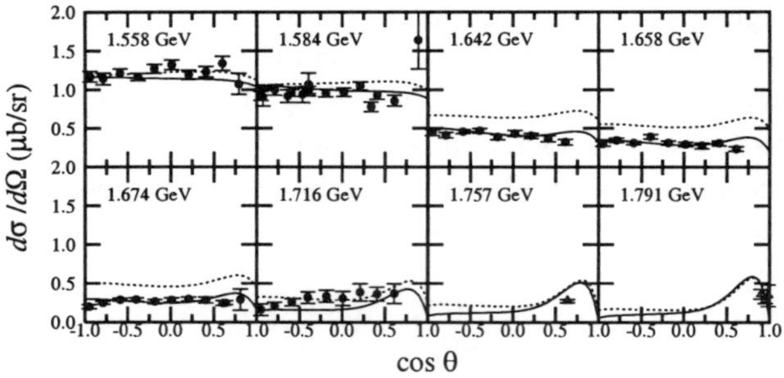

Figure 1. Differential cross section for eta photoproduction. The full curve shows the coupled-channels calculation while the dotted curve has the channel coupling turned off. The data are from MAMI B and GRAAL, except for the panels at W=1.757 and 1.791 GeV which show older data, quoted in previous references.

significant around $W = 1650 MeV$ and then recedes again at higher energies. The nucleon resonance that plays an important role in Σ polarization observable at higher energies is the second D_{13} state 1700 MeV. This is a very inelastic resonance which decouples almost completely from the πN channel. Figure 2 illustrates that the data can not be reproduced if the $D_{13}(1700)$ is excluded. However, the angular distribution of Σ reveals that the situation is more involved compared to the threshold region. The new GRAAL data at higher energies[12] display a forward peaking that varies weakly with W and in our model is naturally explained by vector-meson t-channel contributions. In the differential cross section, the vector mesons show up more prominently above $W = 1700$ MeV where our calculations predict a forward peaking for the cross sections as well. The GRAAL data can neither confirm nor rule out this feature at the present moment, older data at higher energies are sparse with large uncertainties. Final confirmation of the size of the vector meson terms will have to be left to the JLAB data.

Acknowledgments

This work has been supported by US-DOE grant DE-FG02-95ER-40907 and the German BMBF, DFG and GSI Darmstadt.

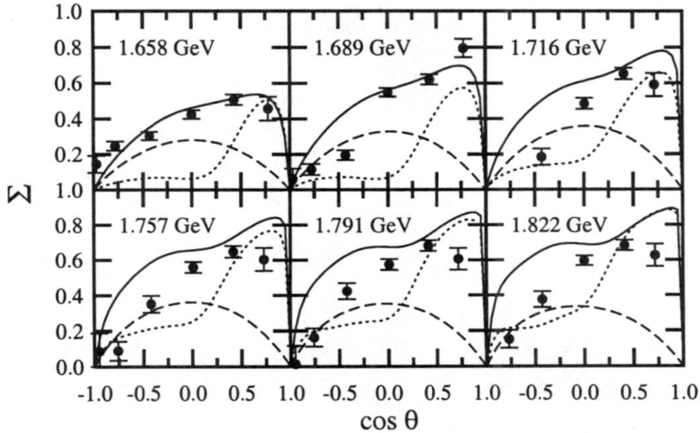

Figure 2. Polarized photon asymmetry for eta photoproduction. The full curve shows our coupled-channels result, while the dashed (dotted) line shows the results without vector mesons ($D_{13}(1700)$). The data are from GRAAL.

References

1. C. Bennhold and H. Tanabe, *Nucl. Phys.* A **530**, 625 (1991); M. Benmerrouche, N.C. Mukhopadhyay and J.-F. Zhang, *Phys. Rev.* D **51**, 3237 (1995).
2. T.P. Vrana, S.A. Dytman, and T.-S. H. Lee *Phys. Rept.* **328**, 181 (2000).
3. T. Feuster and U. Mosel, *Phys. Rev.* C **59**, 460 (1999).
4. L. Tiator, C. Bennhold, and S.S. Kamalov, *Nucl. Phys.* A **580**, 455 (1994).
5. L. Tiator *et al., Phys. Rev.* C **60**, 035210 (1999).
6. C. Bennhold *et al.*, nucl-th/9901066, nucl-th/0008024, and *these proceedings*; A. Waluyo *et al.*, nucl-th/0008023.
7. B. Krusche *et al., Phys. Rev. Lett.* **74**, 3736 (1995).
8. J. Ajaka *et al., Phys. Rev. Lett.* **81**, 1797 (1998).
9. N.C. Mukhopadhyay and N. Mathur, *Phys. Lett.* B **444**, 7 (1998).
10. Particle Data Group, *Eur. Phys. J.* C **15**, 1 (2000).
11. B. Krusche *et al., Phys. Lett.* B **397**, 171 (1997).
12. F. Renard, *et al.*, hep-ex/0011098 and A. D'Angelo, *these proceedings*.

Andreas Thomas

Yongseok Oh

ELECTROWEAK PROPERTIES OF BARYONS IN A COVARIANT CHIRAL QUARK MODEL

S. BOFFI, M. RADICI, L. GLOZMAN

Dipartimento di Fisica Nucleare e Teorica, Università di Pavia and Istituto Nazionale di Fisica Nucleare, Sezione di Pavia, I-27100 Pavia, Italy

R.F. WAGENBRUNN, W. PLESSAS

Institut für Theoretische Physik, Universität Graz, Universitätsplatz 5, A-8010 Graz, Austria

W. KLINK

Department of Physics and Astronomy, University of Iowa, Iowa City, IA 52242, USA

The proton and neutron electromagnetic form factors and the nucleon axial form factor have been calculated in the Goldstone-boson exchange constituent-quark model within the point-form approach to relativistic quantum mechanics. The results, obtained without any adjustable parameter nor quark form factors, are, due to the dramatic effects of the boost required by the covariant treatment in striking agreement with the data.

Constituent quark models (CQM's) provide a useful framework for quantitative calculations of hadron properties in a regime where quantum chromodynamics (QCD) cannot be solved perturbatively. Constituent quarks are effective degrees of freedom described in terms of an Hamiltonian reflecting basic symmetries of QCD. Thus the formalism relies on quantum mechanics with a finite number of degrees of freedom rather than on quantum field theory. However, due to the large value of their kinetic energy constituent quarks are relativistic quasi-particles requiring a relativistic quantum mechanical formulation in terms of unitary representations of the Poincaré group.

Technically, the problem is solved by looking at one of the (unitarily equivalent) forms that are possible when defining the (kinematic) stability subgroup [1]. Here we adopt the point form that has recently attracted some interest in connection with the electromagnetic properties of hadrons [2]. In fact, this form has some advantages. First, the four-momentum operators P^μ containing all the dynamics commute with each other and can be simultaneously diagonalized. Since the Lorentz generators do not contain any interaction terms, the theory is manifestly covariant. Second, the electromagnetic current operator $J^\mu(x)$ can be written in such a way that it transforms as an irreducible tensor operator under the strongly interacting Poincaré group.

Thus the nucleon charge and magnetic form factors can be calculated as reduced matrix elements of such an irreducible tensor operator in the Breit frame.

Here results are presented as a progress report of a more comprehensive programme dealing with electroweak properties of baryons studied within the CQM discussed in Ref. [3] and based on Goldstone-boson-exchange (GBE) dynamics. This type of CQM assumes a pairwise linear confinement potential, as suggested by lattice QCD, with a strength according to the string tension of QCD. The quark-quark interaction is derived from the exchange of pseudoscalar bosons producing the (flavour dependent) hyperfine interaction; in the model only the spin-spin component is utilized, which phenomenologically appears to be the most important in the hyperfine splitting of the baryon spectra. The current operator is a single-particle current operator for point-like constituent quarks. This approach corresponds to a relativistic impulse approximation but specifically in point form. It is called point-form spectator approximation (PFSA).

In Fig. 1 the ratio G_E/G_M of the proton electric to magnetic form factor is shown together with the recent TJNAF data [4]. In Fig. 1 the prediction of the model for the neutron charge form factor is also shown. The solid (dashed) curve is obtained when the theoretical (experimental) value of the proton magnetic moment is used. The difference between the two curves is due to the fact that the calculated proton and neutron charge radii turn out to be in very good agreement with experiment ($r_p^2 = 0.75$ fm^2, $r_n^2 = -0.12$ fm^2) while the magnetic moments are slightly underestimated ($\mu_p = 2.64$ n.m., $\mu_n = -1.67$ n.m.). However, one observes that a very good description of both the proton and neutron e.m. structure is achieved [5]. It is remarkable that no further ingredients beyond the quark model wave functions (such as, e.g., constituent quark form factors) have been employed. What is importantant is that only relativistic boost effects are properly included in point-form relativistic quantum mechanics.

For comparison, results for the neutron charge form factor are also shown when calculated in nonrelativistic impulse approximation, i.e. with the standard nonrelativistic form of the current operator and no Lorentz boosts applied to the nucleon wave functions. Also the case with the confinement potential only has been considered in order to appreciate the role of mixed-symmetry components in the wave functions that are absent without the hyperfine interaction.

A similar approach can be used to study the axial current [6]. According to the PCAC hypothesis this current is not conserved. However, one can always split it into conserved and nonconserved parts with the conserved part

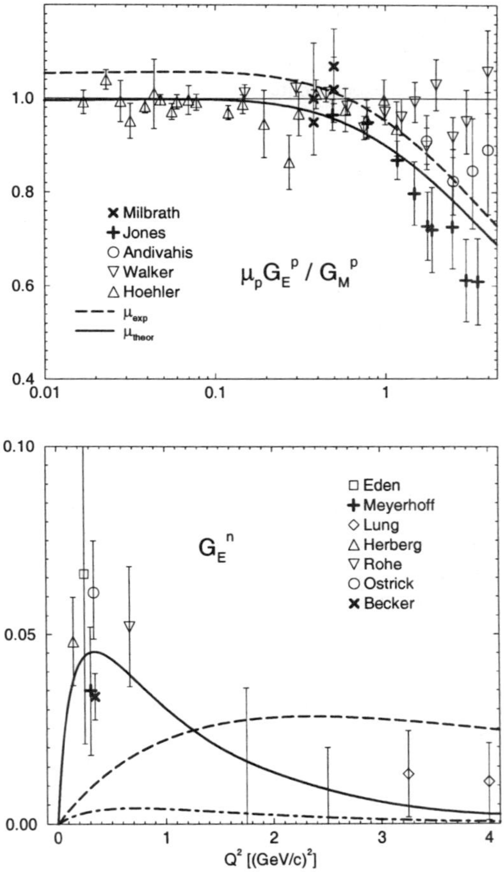

Figure 1. Top: the ratio of the proton electric to magnetic form factor. Solid (dashed) line obtained with the theoretical (experimental) value of the magnetic moment μ. Bottom: the neutron electric form factor. Solid, dashed and dot-dashed lines as predicted by the model in PFSA, the nonrelativistic approximation and the model with the confinement interaction only, respectively.

containing the axial form factor G_A only. Therefore one can calculate G_A in the same point-form approach used for G_E and G_M.

The results are shown in Fig. 2, where comparison is made with data obtained from charged pion electroproduction on protons (see Ref. [7] and references therein). The nonrelativistic result (dashed line) and the calculation without boost (dot-dashed line) are also shown.

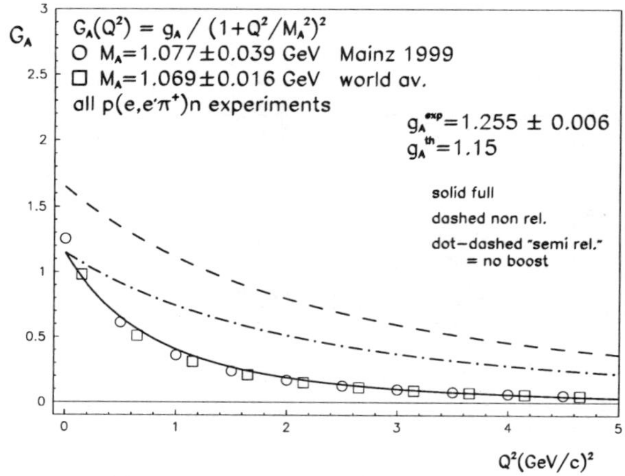

Figure 2. The nucleon axial form factor.

The result is quite good at $Q^2 \neq 0$, while at $Q^2 = 0$ the value of the calculated axial form factor $G_A(0)$ is lower than the value g_A used when fitting the data with a dipole form factor $G_A(Q^2) = g_A/(1 + Q^2/M_A^2)^2$ involving the axial mass M_A. This indicates a deviation of the present results from the usually assumed dipole form. Correspondingly, the axial radius deduced from the slope at $Q^2 = 0$ is lower than the experimental one. With the present model one has $< r_A^2 >^{1/2}= 0.520$ fm to be compared with the experimental value $< r_A^2 >^{1/2}= (0.65 \pm 0.07)$ fm extracted from neutrino experiments and $< r_A^2 >^{1/2}= (0.635 \pm 0.023)$ fm extracted from pion electroproduction [7].

References

1. H. Leutwyler and J. Stern, *Ann. Phys. (N.Y.)* **112**, 94 (1978); B.D. Keister and W.N. Polyzou, *Adv. Nucl. Phys.* **20**, 225 (1991).
2. W.H. Klink, *Phys. Rev.* C **58**, 3587 (1998).
3. L.Ya. Glozman, W. Plessas, K. Varga and R.F. Wagenbrunn, *Phys. Rev.* D **58**, 094030 (1998).
4. M. K. Jones *et al., Phys. Rev. Lett.* **84**, 1398 (2000).
5. R.F. Wagenbrunn, S. Boffi, W. Klink, W. Plessas, and M. Radici, nucl-th/0010048.
6. L.Ya. Glozman, M. Radici, R.F. Wagenbrunn, S. Boffi, W. Klink, and W. Plessas, in preparation.
7. A. Liesenfeld *et al., Phys. Lett.* B **468**, 20 (1999).

LOW LYING $qqqq\bar{q}$ STATES IN THE BARYON SPECTRUM

C. HELMINEN[1], D. O. RISKA[1,2]

[1] *Department of Physics, 00014 University of Helsinki, Finland,*
[2] *Helsinki Institute of Physics, 00014 University of Helsinki, Finland*
E-mail: chelmine@pcu.helsinki.fi, riska@pcu.helsinki.fi

The coupling to light mesons leads to large widths and shifts of the energy of the excited states in the baryon spectrum from the predictions of the constituent quark model with three valence quarks. This coupling may be modelled by admixtures of sea-quark $qqq(q\bar{q})^n$ configurations in the baryon resonances. The structure of $qqqq\bar{q}$ systems and their spectrum in the light and strange baryon sectors are studied with a schematic flavor and spin dependent interaction model, which reproduces the low lying part of the experimental baryon spectrum in the valence quark model. The model reveals that the lowest 5-quark states form a positive parity band in all sectors of the spectrum.

1 Introduction

The constituent quark model with three valence quarks describes the lowest lying excited baryon states as single quark excitations to the excited S-state. The large widths of some of these states indicate strong coupling to the open mesonic channels, which in the quark model would be viewed as sea-quark admixtures. We here compare the energy of $qqqq\bar{q}$ states with the energy of low lying baryon resonances with corresponding quantum numbers.

2 The $qqqq\bar{q}$ System

A translationally invariant Hamiltonian model for a $qqqq\bar{q}$ system, the simplest version of which would be the harmonic oscillator Hamiltonian, may be written as

$$H = \sum_{i=1}^{5} \frac{\vec{p}_i^{\,2}}{2m_i} - \frac{\vec{P}^2}{2M} + \frac{1}{2}\sum_{i<j}^{5}(C[r_i - r_j]^2 + V_0) + \sum_{i=1}^{5} m_i \,. \tag{1}$$

The confining interaction, the color dependence of which is taken to be $\sim \boldsymbol{\lambda}_i^C \cdot \boldsymbol{\lambda}_j^C$, will be equal for qq- and $q\bar{q}$-pairs due to the color symmetry structure of the $qqqq\bar{q}$ system[1]. If the constituent masses of the quarks and the antiquark are taken to be equal, Eq. (1) may, by a suitable change of variables, be rewritten as a Hamiltonian describing four uncoupled harmonic oscillators. The value of the confining constant C may be determined from the oscillator parameter in the three quark baryon system, which in turn is determined by the empirical N(1440)-N splitting. The excited spectrum of the oscillator Hamiltonian is organized in shells, the ground state shell of which is formed of states with negative parity because of the negative parity of the antiquark. The lowest positive parity states occur in the first excited P-shell ($L = 1$) and have excitation energies $\omega \approx 228$ MeV, with a quark or an antiquark in a p-state.

For the hyperfine interaction between quarks we use the same schematic flavor spin interaction, which in the chiral constituent quark model is known to yield a spectrum for the qqq states which agrees well with the empirical baryon spectrum up to $\sim$ 1700 MeV [2,3],

$$H_\chi = -C_\chi \sum_{i<j}^{N} \boldsymbol{\lambda}_i^F \cdot \boldsymbol{\lambda}_j^F \boldsymbol{\sigma}_i \cdot \boldsymbol{\sigma}_j \,. \tag{2}$$

The hyperfine interaction between quarks and antiquarks is taken to be negligible, and thus $N = 4$. When using the same value for C_χ as in the $SU(3)$ flavor symmetric version of the chiral constituent quark model, this hyperfine interaction is strong enough to bring the lowest $L = 1$ states below the lowest states with $L = 0$, and therefore the lowest 5-quark states form a band with positive parity states. The N(1440), $\frac{1}{2}^+$ state has a complex structure and is expected to have large sea-quark and possible more exotic components[4,5]. We take the empirical value for the real part of the pole position of this resonance as the reference point for the energy of the positive parity ground state of the 5-quark system.

The flavor decuplet baryons can be derived from the $[4]_F$ and $[31]_F$ four-quark subsystems by adding an antiquark, the flavor octet states are derived from the $[31]_F$, $[22]_F$ and $[211]_F$ subsystems, and the flavor singlet baryon state from the $[211]_F$ subsystem. To derive some of the states, e.g. nucleon-like states from the four-quark symmetry $[211]_F$, it is necessary to assume the presence of "hidden" strangeness ($s\bar{s}$-pairs). The energies of these 5-quark configurations will, due to $SU(3)_F$ breaking, be shifted upwards by $\sim 2\delta m$, where δm is the difference between the constituent masses of the strange and light flavor quarks. The calculated spectrum up to 1900 MeV of N-like 5-quark states without and with hidden strangeness is shown in Fig. 1 along with the empirically known N resonances. In Fig. 2 the corresponding situation for Δ-like 5-quark states is shown. The 5-quark states with strangeness -1 contain at least one strange quark, and accordingly their spectrum begins at a level that is shifted about δm above the corresponding level for non-strange 5-quark

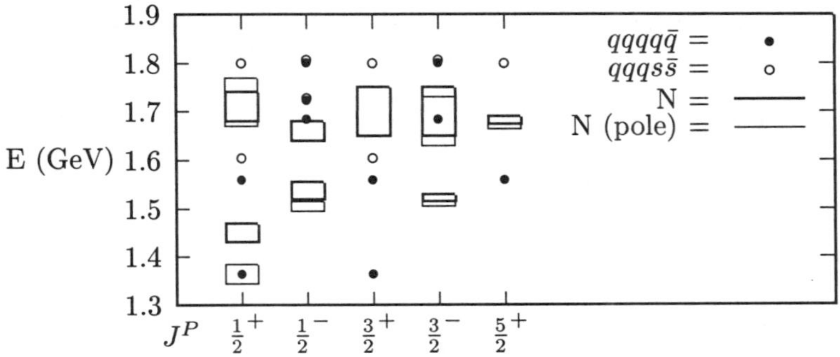

Figure 1. N(exp) and N-like 5-quark states.

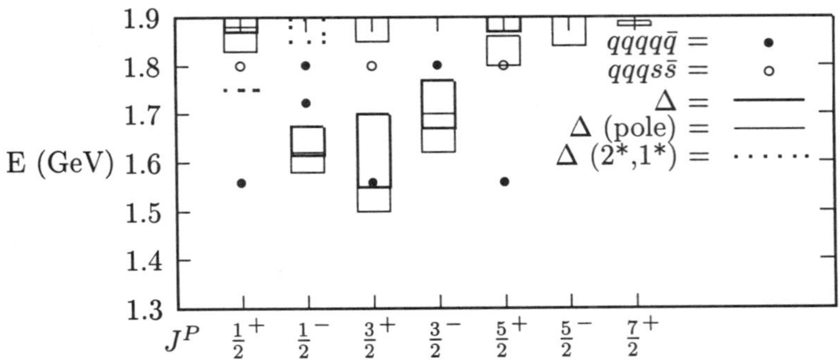

Figure 2. Δ(exp) and Δ-like 5-quark states.

states. The spectra of the Λ- and Σ-like 5-quark states up to 2000 MeV are shown in Fig. 3 and Fig. 4, respectively.

3 Summary

There are strong empirical and phenomenological indications for large sea-quark admixtures in the low lying negative and positive parity states of the baryon resonances. The main result of the present study is that in the chiral constituent quark model, a quark model with a flavor-spin dependent hyperfine interaction, the lowest $qqqq\bar{q}$ states in the baryon spectrum form bands of positive parity states, while there are also higher lying negative parity states.

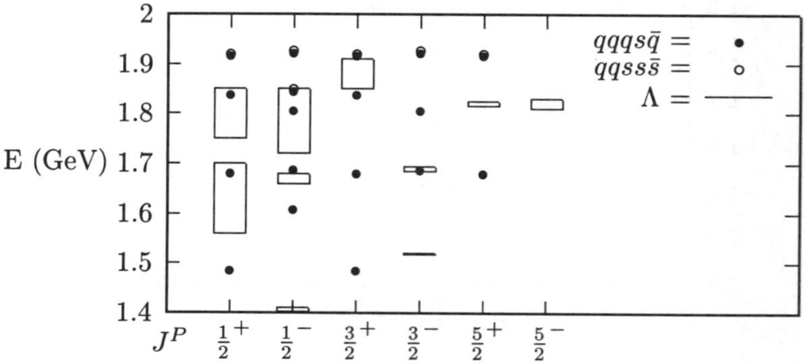

Figure 3. Λ(exp) and Λ-like 5-quark states.

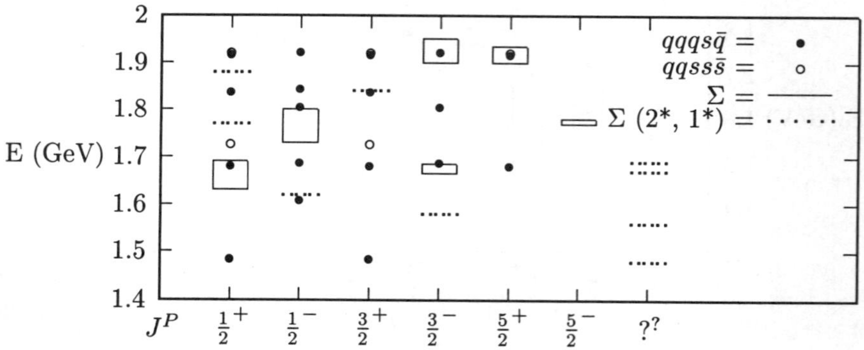

Figure 4. Σ(exp) and Σ-like 5-quark states.

References

1. M. Genovese, J.-M. Richard, Fl. Stancu, and S. Pepin, *Phys. Lett.* B **425**, 171 (1998).
2. L.Ya. Glozman and D.O. Riska, *Phys. Reports* **268**, 263 (1996).
3. L.Ya. Glozman, W. Plessas, K. Varga, and R.F. Wagenbrunn, *Phys. Rev.* D **58**, 094030 (1998).
4. H.P. Morsch and P. Zupranski, *Phys. Rev.* C **61**, 024002 (2000).
5. V.D. Burkert, *Probing the Structure of Nucleons in the Resonance Region*, nucl-th/0005033.

PARITY DOUBLETS FROM A RELATIVISTIC QUARK MODEL

U. LÖRING, B. METSCH
Institut für Theoretische Kernphysik der Universität Bonn,
Nußallee 14-16, D-53115 Bonn, GERMANY
E-mail: metsch@itkp.uni-bonn.de

The $N-$ and the Λ–excitation spectrum exhibit parity doublets, *i.e.* states of the same spin but with opposite parity being almost degenerate in mass. It is shown that in a relativistic quark model with instantaneous interaction kernels, where confinement is implemented by a linearly rising potential and the major mass splittings are generated from an interaction based on instanton effects, this degeneracy occurs quite naturally: Once the parameters of the instanton induced interaction are fixed to reproduce the ground state octet-decuplet splittings, some states are selectively lowered to a position which is degenerate with states of opposite parity. Some observable consequences are briefly discussed.

A glance at the nucleon– and $\Lambda-$ excitation spectrum[1] reveals a conspicuous degeneracy of some states with the same spin and opposite parity. Prominent examples are $N\frac{5}{2}^+(1680)$–$N\frac{5}{2}^-(1675)$, $N\frac{9}{2}^+(2220)$–$N\frac{9}{2}^-(2250)$, $\Lambda\frac{5}{2}^+(1820)$–$\Lambda\frac{5}{2}^-(1830)$. Although one might also regard $\Delta\frac{5}{2}^+(1905)$–$\Delta\frac{5}{2}^-(1930)$ and $\Delta\frac{9}{2}^+(2300)$–$\Delta\frac{9}{2}^-(2400)$ as parity partners, the situation seems less clear: for the first because of the nearby $\Delta\frac{5}{2}^+(2000)$-resonance and for the second because of the relatively large splitting. In the Σ–spectrum no clear indications of parity doublets is found. In the literature these observations have been related to a phase transition from the Nambu–Goldstone mode of chiral symmetry to the Wigner–Weyl mode in the upper part of the baryon spectrum[2,3,4]. In the present contribution we will show how this feature can be understood in the context of a (relativistic) constituent quark model on the basis of the quark *dynamics*, where the major spin-dependent mass splittings are induced by instanton effects ('t Hooft's force).

Our relativistically covariant constituent quark model is based on the Bethe-Salpeter equation for three-quark bound states with instantaneous interaction kernels. The details of the model are described elsewhere[5,6,7]. Quarks are assumed to possess an effective constituent mass and confinement is implemented by a linearly rising 3-body string potential with a Dirac structure, which is a combination of scalar and time-like vector structures chosen such that unwanted spin-orbit effects are minimized. The major spin-dependent mass splittings are generated by a flavor dependent 2-particle interaction, which is motivated by instanton effects. This force affects flavor-antisymmetric qq-pairs only, and consequently this interaction does not act on flavor symmetric states, such as the Δ-resonances, which are thus determined by the dynamics of the confinement potential alone. Accordingly, the constituent quark masses and the confinement parameters were determined by a fit to the spectrum in this sector[6].

The residual instanton induced interaction does act on particular flavor octet states. Once the strengths of this interaction have been adjusted to account for the ground state nucleon and Λ-mass we find that in fact one can describe the major spin-dependent mass splittings in the nucleon spectrum quite well[6], see Fig. 1. In particular one finds that in this manner the Roper

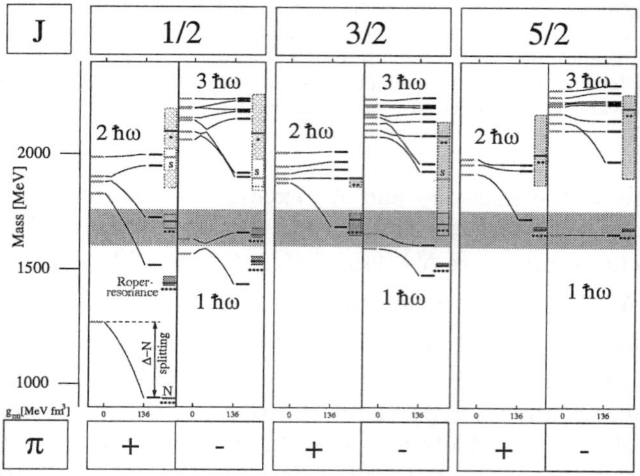

Figure 1. Parity doublets for low lying nucleon resonances. The left side of each column shows the spectrum calculated with the confinement interaction alone. The curve shows the effect of the instanton induced interaction with increasing coupling g_{nn} of the nonstrange-nonstrange quark interaction until the N–Δ splitting is reproduced. This result is compared to the experimental spectrum[1], right part of each column, where uncertainties in the resonance position are indicated by boxes.

resonance can be accounted for quite naturally. Moreover one finds a selective lowering of those substates of a major oscillator shell (which in spite of the linear confinement adopted here still provides an adequate classification of states with confinement alone) that contain so called scalar diquarks, *i.e.* quark pairs with trivial spin and angular momentum. This is found in particular for the highest spin states in a given oscillator shell N, see Fig. 2. For given $N\hbar\omega$ the maximum total angular momentum for a state containing such a scalar diquark is $J = (L_{\max} = N) + \frac{1}{2}$. 't Hooft's force lowers this

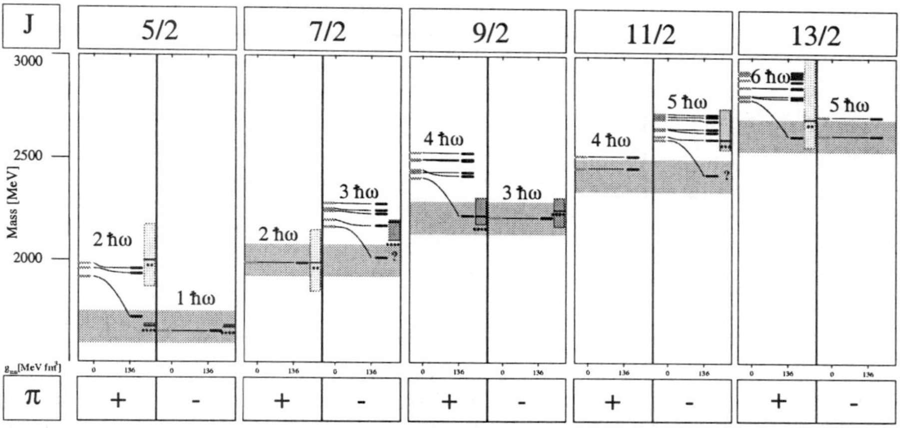

Figure 2. Parity doublets for higher lying nucleon resonances. See also caption to Fig. 1.

state enough to become almost degenerate with the unaffected spin-quartet state of the oscillator shell with $N-1$, which has opposite parity but the same total angular momentum: $J = (L_{\text{max}} = N-1) + \frac{3}{2}$. In this way patterns of approximate parity doublets for *all* lowest excitations in the sectors $J = \frac{5}{2}$ to $J = \frac{13}{2}$ are formed *systematically*. In the $N\frac{5}{2}^{\pm}$ and $N\frac{9}{2}^{\pm}$ sectors this scenario is nicely confirmed experimentally by the well-established parity doublets $N\frac{5}{2}^{+}(1680)$–$N\frac{5}{2}^{-}(1675)$ and $N\frac{9}{2}^{+}(2220)$–$N\frac{9}{2}^{-}(2250)$. In the $N\frac{7}{2}$ sector, however, the present experimental findings seem to deviate from such a parity doubling structure due to the rather highly determined resonance position of the $N\frac{7}{2}^{-}(2190)$. Although this state is given a four-star rating[1] an investigation of this sector with new experimental facilities such as the CLAS detector at CEBAF (JLab) or the Crystal Barrel detector at ELSA (Bonn) would be highly desirable. The same mechanism explains approximate parity doublet structures also for states with lower angular momentum as *e.g.* the N^* doublets in the second resonance region around ~ 1700 MeV with spins $J^\pi = \frac{1}{2}^{\pm}$, $\frac{3}{2}^{\pm}$, and $\frac{5}{2}^{\pm}$ (see Fig. 1).

Observable consequences of this parity doubling scenario should manifest in a different shape of electromagnetic $p\gamma^* \to N^*$ transition form factors of both members of a doublet due to their significantly different internal strutures. Fig. 3 shows as an example the magnetic multipole $\gamma^* p \to N\frac{5}{2}^{+}(1680)$ and $\gamma^* p \to N\frac{5}{2}^{-}(1675)$ transition form factors: That member of the doublet, which is affected by 't Hooft's force ($N\frac{5}{2}^{+}$), exhibits a rather strong scalar diquark correlation and thus its structure should be more compact compared to its unaffected doublet partner ($N\frac{5}{2}^{-}$) whose structure is expected to be

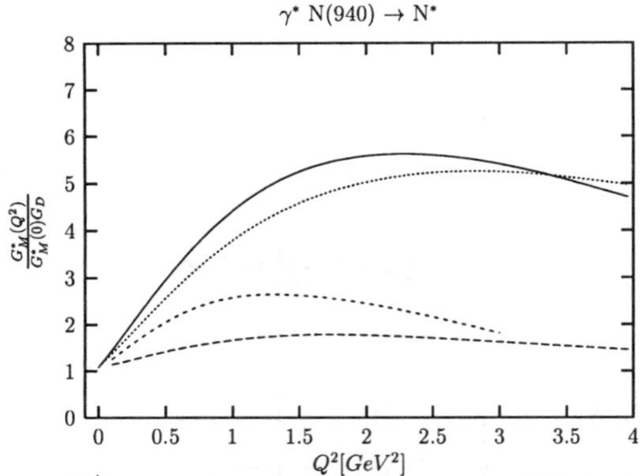

Figure 3. The $\gamma^* p \to N\frac{5}{2}^+(1680)$ (dotted line) and $\gamma^* p \to N\frac{5}{2}^-(1675)$ (short dashed line) transition form factors $G_M^*(Q^2)$ divided by the dipole form $G_D(Q^2)$ and normalized to their threshold values $G_M^*(0)$. For comparison also the $\gamma^* p \to N\frac{1}{2}^-(1535)$ (solid line) and $\gamma^* p \to \Delta\frac{3}{2}^+(1232)$ (dashed line) are shown.

rather soft. Consequently, the transition form factor to the latter resonance decreases faster than that to its doublet partner with the scalar diquark contribution.

In the strange sector[7] 't Hooft's force accounts in a similar way for the prominent doublets of the Λ-spectrum. At the same time instanton-induced effects are found to be significantly weaker in the Σ-spectrum, thus explaining the fact that no clear experimental indications of parity doublets are observed in this sector.

The contributions of K. Kretzschmar and H. R. Petry are gratefully acknowledged.

References

1. Particle Data Group, *Eur. Phys. J.* C **15**, 1 (2000).
2. L. Y. Glozman, *Phys. Lett.* B **475**, 329 (2000)
3. T. D. Cohen and L. Y. Glozman, hep-ph/0102206.
4. M. Kirchbach, *Nucl. Phys.* A **689**, 157 (2001).
5. U. Löring, K. Kretzschmar, B. Ch. Metsch, and H. R. Petry, hep-ph/0103287, accepted for publication in *Eur. Phys. J.* **A**.
6. U. Löring, B. Ch. Metsch, and H. R. Petry, hep-ph/0103289, accepted for publication in *Eur. Phys. J.* **A**.
7. U. Löring, B. Ch. Metsch, and H. R. Petry, hep-ph/0103290, accepted for publication in *Eur. Phys. J.* **A**.

A DISPERSION THEORETICAL APPROACH TO VIRTUAL COMPTON SCATTERING OFF THE PROTON

B. PASQUINI[1], D. DRECHSEL[2], M. GORCHTEIN[2], A. METZ[3], M. VANDERHAEGHEN[2]

[1] *ECT*, I-38050 Villazzano (Trento), and INFN, Trento, Italy*
[2] *Institut für Kernphysik, J. Gutenberg-Universität, D-55099 Mainz, Germany*
[3] *Division of Physics and Astronomy, Faculty of Science, Vrije Universiteit, 1081 HV Amsterdam, The Netherlands*

We present a dispersion relation formalism for virtual Compton scattering (VCS) off the proton, which provides a new tool to extract the generalized polarizabilities of the proton from VCS observables over a large energy range.

Virtual Compton scattering (VCS) off the proton has recently seen an increased interest both theoretically and experimentally [1]. In VCS off a proton target, a virtual photon interacts with the proton and a real photon is emitted in the process. At low energy of the outgoing real photon, the VCS reaction amounts to a generalization of real Compton scattering (RCS) in which both energy and momentum of the virtual photon can be varied independently, and it allows us to extract response functions, parametrized by the generalized polarizabilities (GPs) of the proton. The first experimental results for GPs of the proton have recently been obtained at MAMI [2], and further experimental programs are underway at JLab [3] and MIT-Bates [4].

Until now, VCS experiments at low outgoing photon energies were analyzed in terms of low-energy expansions (LEXs). In the LEX, only the leading term (in the energy of the real photon) of the response to the quasi-constant electromagnetic field is taken into account. This leading term depends linearly on the GPs. When increasing the photon energy, the sensitivity of the VCS cross section to the GPs becomes larger, and it is therefore promising to extract GPs with enhanced precision from VCS observables at larger energies. To this end, a reliable estimate of higher order terms beyond the LEX is of utmost importance. The situation can be compared to RCS, for which one uses a dispersion relation (DR) formalism to extract the polarizabilities at energies above pion threshold, with generally larger effects on the observables [5,6]. In recent investigations [7] we extended such a dispersion analysis for the VCS process, and here we report the main features of this dispersive approach.

To calculate the VCS process, we expand the VCS tensor into a basis of 12 independent amplitudes F_i ($i = 1, ..., 12$), which are functions of 3 invariants: Q^2, $\nu = (s-u)/(4M_N)$, and t. The nucleon structure information is contained in the non-Born contributions F_i^{NB} for which one can write unsubtracted DRs with respect to the variable ν at fixed t and fixed Q^2 :

$$\text{Re}F_i^{NB}(Q^2, \nu, t) = \frac{2}{\pi}\mathcal{P}\int_{\nu_{thr}}^{+\infty} d\nu' \frac{\nu' \, \text{Im}_s F_i(Q^2, \nu', t)}{\nu'^2 - \nu^2}, \qquad (1)$$

with $\text{Im}_s F_i$ the discontinuities across the s-channel cuts of the VCS process. However, due to the high energy behavior of the amplitudes F_i, the unsubtracted dispersion integral in Eq. (1) converges only for 10 of the 12 amplitudes. To construct the remaining two amplitudes, denoted by F_1 and F_5, in an unsubtracted dispersion framework, we proceed in an analogous way as in the case of RCS [5]. This amounts to closing the contour of the integral in Eq. (1) by a semi-circle of finite radius ν_{max} (instead of the usually assumed infinite radius) in the complex plane, i.e. the unsubtracted dispersion integrals for F_1 and F_5 are evaluated along the real ν-axis in a finite range $(-\nu_{max} \leq \nu \leq +\nu_{max})$, and the remaining contribution from the finite semi-circle of radius ν_{max} in the complex plane is described by an "asymptotic contribution". The imaginary parts of $\text{Im}_s F_i$ are calculated through unitarity taking into account the dominant contribution from πN intermediate states. For the pion photo- and electroproduction helicity amplitudes we use the MAID2000 analysis [8]. The asymptotic contribution to the amplitude F_5 results from t-channel π^0-exchange, while the asymptotic contribution to the amplitude F_1 originates predominantly from t-channel $\pi\pi$ intermediate states. In addition, it turns out that higher-energy dispersive contributions $(\pi\pi N, ...)$ mainly affect the F_1 and F_2 amplitudes. Since the dispersive terms beyond πN are very poorly known, we parametrize these contributions to F_1 and F_2 by energy independent constants, fixed at arbitrary Q^2, $\nu = 0$ and $t = -Q^2$. This parametrization involves two free parameters which can be expressed in terms of the electric, $\alpha(Q^2)$, and magnetic, $\beta(Q^2)$, polarizabilities, which have to be fitted to experimental VCS data at each fixed value of Q^2. However, in order to provide predictions for VCS observables at different values of Q^2, we take the following parametrization for the Q^2 dependence of the scalar GPs

$$\beta(Q^2) - \beta^{\pi N}(Q^2) = \frac{(\beta - \beta^{\pi N})}{(1 + Q^2/\Lambda_\beta^2)^2}, \;\; \alpha(Q^2) - \alpha^{\pi N}(Q^2) = \frac{(\alpha - \alpha^{\pi N})}{(1 + Q^2/\Lambda_\alpha^2)^2}, \qquad (2)$$

where the RCS values $(\beta - \beta^{\pi N})$ and $(\alpha - \alpha^{\pi N})$ are fitted to RCS data [9].

In Fig. 1, we show the results for the structure functions which can be extracted within the LEX formalism from unpolarized VCS cross sections. The VCS response function P_{LT} reduces to the magnetic polarizability β at

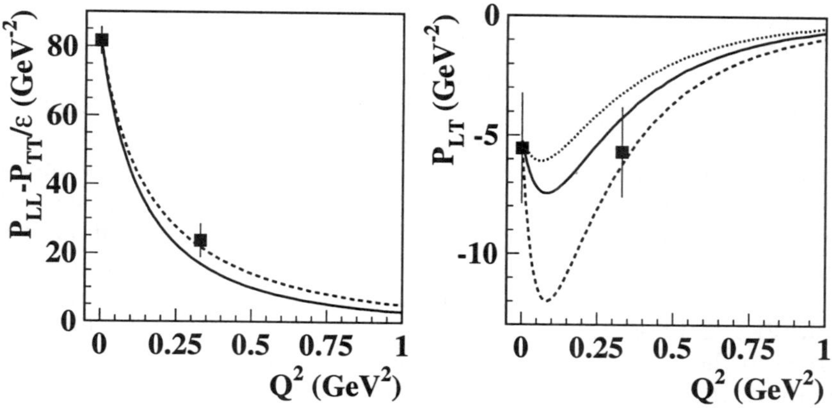

Figure 1. Left panel: result for the structure function $P_{LL} - P_{TT}/\varepsilon$ for $\varepsilon = 0.62$. Full curve: calculation with the asymptotic part of $\alpha(Q^2)$ according to Eq. (2) with $\Lambda_\alpha = 1$ GeV (full curve) and $\Lambda_\alpha = 1.4$ GeV (dashed curve). Right panel: Result for the structure function P_{LT} with the asymptotic contribution of $\beta(Q^2)$ according to Eq. (2) with $\Lambda_\beta = 0.7$ GeV (dotted curve), $\Lambda_\beta = 0.6$ GeV (full curve), and $\Lambda_\beta = 0.4$ GeV (dashed curve). The RCS data are from Ref.[9], and the VCS data at $Q^2 = 0.33$ GeV^2 from Ref.[2].

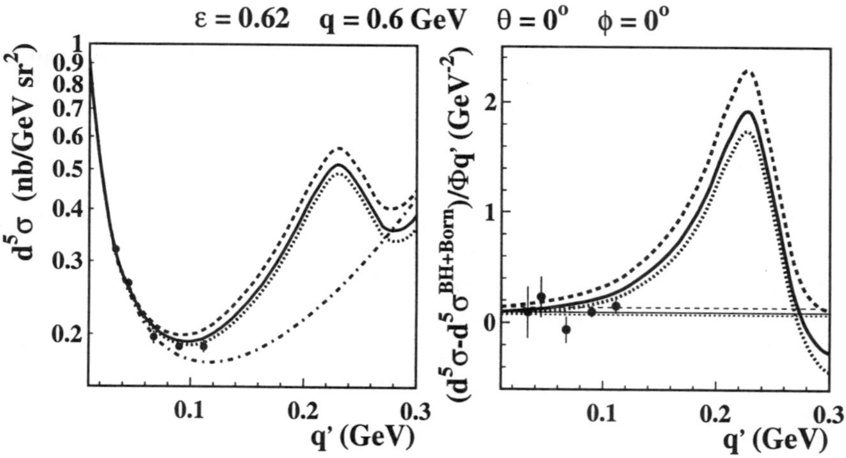

Figure 2. Left panel: VCS cross section as function of the outgoing-photon energy q′ in MAMI kinematics. Dashed-dotted curve: BH + Born contribution. The total DR results are calculated using a fixed value of $\Lambda_\alpha = 1$ GeV and three values of Λ_β : $\Lambda_\beta = 0.7$ GeV (dotted curve), $\Lambda_\beta = 0.6$ GeV (solid curve), and $\Lambda_\beta = 0.4$ GeV (dashed curve). Right panel: Results for $(d^5\sigma - d^5\sigma^{\mathrm{BH+Born}})/\Phi \mathrm{q}'$ as function of q′, where Φ is a phase space factor. The thick curves show the DR calculation with the full q′ dependence and the thin horizontal curves are the DR results within the LEX formalism. The curves correspond to the same values of Λ_α and Λ_β as in the left panel. The data are from Ref.[2].

the real photon point, while at finite Q^2, it contains the contribution from $\beta(Q^2)$ and a spin-dependent GP. The total result for P_{LT} results from a large dispersive πN contribution and a large asymptotic contribution to $\beta(Q^2)$ with opposite sign. The large cancellation between these contributions leads to an interesting structure in the Q^2 region around 0.1 GeV^2, where forthcoming data are expected from an experiment at MIT-Bates[4]. The response function shown in the left panel of Fig. 1 is given by a combination of P_{LL}, proportional to $\alpha(Q^2)$, and P_{TT}, containing only spin GPs. For this combination $P_{LL} - 1/\varepsilon P_{TT}$, the πN dispersive contributions are quite small, and the evolution in Q^2 is mainly determined by the asymptotic contribution from $\alpha(Q^2)$.

In Fig. 2 we show the DR predictions of VCS observables for photon energies in the $\Delta(1232)$-resonance region. It is seen that the region between pion threshold and the Δ-resonance peak displays an enhanced sensitivity to the GPs through the interference with the rising Compton amplitude due to Δ-resonance excitation. For example, at q$' \simeq 0.2$ GeV/c, the predictions for P_{LT} in Fig. 1 for $\Lambda_\beta = 0.4$ GeV and $\Lambda_\beta = 0.6$ GeV give a difference of about 20 % in the non-Born squared amplitude. In contrast, the LEX prescription results in a relative effect of about 10% or less for the same two values of P_{LT}.

VCS data both below and above threshold and at the same value of Q^2 (in the range $Q^2 = 1 - 2$ GeV^2) are under analysis at JLAB [10]. These investigations will provide an interesting check on whether the present DR formalism can extract a consistent value of the GPs by describing the data in both energy regions.

References

1. M. Vanderhaeghen, *Eur. Phys. J.* A **8**, 455 (2000), and references therein.
2. J. Roche *et al.*, *Phys. Rev. Lett.* **85**, 708 (2000).
3. G. Audit *et al.*, CEBAF Report No. PR 93-050, 1993.
4. J. Shaw *et al.*, MIT-Bates proposal No. 97-03, 1997.
5. A. L'vov, V.A. Petrun'kin, M. Schumacher, *Phys. Rev.* C **55**, 359 (1997).
6. D. Drechsel, M. Gorchtein, B. Pasquini, and M. Vanderhaeghen, *Phys. Rev.* C **61**, 015204 (1999).
7. B. Pasquini, D. Drechsel, M. Gorchtein, A. Metz, and M. Vanderhaeghen, *Phys. Rev.* C **62**, 052201 (2000); hep-ph/0102335, submitted to *Eur. Phys. J.*
8. D. Drechsel, O. Hanstein, S. Kamalov, and L. Tiator, *Nucl. Phys.* A **645**, 145 (1999).
9. V. Olmos de León, *et al., Eur. Phys. J.* A **10**, 207 (2001).
10. G. Laveissiere, *these proceedings*; S. Jaminion, *these proceedings*.

NEUTRON CHARGE FORM FACTOR AND QUADRUPOLE DEFORMATION OF THE NUCLEON

A. J. BUCHMANN

Institut für Theoretische Physik, Eberhard Karls Universität Tübingen, Auf der Morgenstelle 14, D-72076 Tübingen, Germany

E-mail: alfons.buchmann@.uni-tuebingen.de

A quark model relation between the neutron charge form factor and the $N \to \Delta$ charge quadrupole form factor is used to predict the $C2/M1$ ratio in the $N \to \Delta$ transition from the elastic neutron form factor data. Excellent agreement with the pion electroproduction data is found, indicating the validity of the suggested relation. The implication of the negative $C2/M1$ ratio for the intrinsic deformation of the nucleon is discussed.

The geometrical shape of the proton can be determined from its *intrinsic* quadrupole moment [1]

$$Q_0^p = \int d^3r \, \rho^p(\mathbf{r}) \, (3z^2 - r^2), \tag{1}$$

where $\rho^p(\mathbf{r})$ is the not necessarily spherically symmetric charge density of the proton. If $Q_0^p > 0$, the proton is prolate (cigar-shaped); if $Q_0^p < 0$, it is oblate (pancake-shaped); and if $Q_0^p = 0$, the charge density inside the proton is spherically symmetric. The intrinsic quadrupole moment, which is defined with respect to the body-fixed frame, must be distinguished from the *spectroscopic* quadrupole moment measured in the laboratory. The latter is zero due to angular momentum selection rules. Q_0^p can be inferred by measuring electromagnetic quadrupole transitions between the nucleon ground and its low-lying excited states, or by measuring the quadrupole moment of an excited state, e.g. the $\Delta(1232)$ with $J > 1/2$.

Another possible way to obtain information on the shape of the nucleon has recently been suggested [2,3,4]. In the constituent quark model with two-body exchange currents a connection between the neutron charge form factor $G_{C0}^n(\mathbf{q}^2)$ and the $N \to \Delta$ charge quadrupole form factor $G_{C2}^{p\to\Delta^+}(\mathbf{q}^2)$ has been found [3]

$$G_{C2}^{p\to\Delta^+}(\mathbf{q}^2) = -\frac{3\sqrt{2}}{\mathbf{q}^2} \, G_{C0}^n(\mathbf{q}^2), \tag{2}$$

where $\mathbf{q}$ is the three-momentum transfer of the virtual photon. Together with the known SU(6) relation $G_{M1}^{p\to\Delta}(\mathbf{q}^2) = -\sqrt{2} G_{M1}^n(\mathbf{q}^2)$ (which remains intact after adding exchange currents [3]) this provides a determination of the C2/M1 ratio through the *elastic* charge and magnetic neutron form factors which agrees well with C2/M1 data from pion electroproduction experiments (see Fig.1). By comparing the low $\mathbf{q}$ expansion of the left and right hand side of Eq.(2), the $N \to \Delta$ quadrupole moment $Q_{p\to\Delta}$ can be expressed in terms of the known neutron charge radius r_n^2, and the $N \to \Delta$ quadrupole transition radius $r^2_{p\to\Delta^+}$ is obtained from the fourth moment r_n^4 of $\rho^n(r)$

$$Q_{p\to\Delta^+} = \frac{1}{\sqrt{2}} r_n^2, \quad r^2_{p\to\Delta^+} = \frac{3}{10}\frac{r_n^4}{r_n^2}. \tag{3}$$

Experimentally [5] one finds $r_n^2 = -0.113(3)$ fm^2 and $r_n^4 = -0.32(8)$ fm^4, from which we predict $Q_{p\to\Delta^+} = -0.080(2)$ fm^2 and $r^2_{p\to\Delta^+} = -0.84(21)$ fm^2.

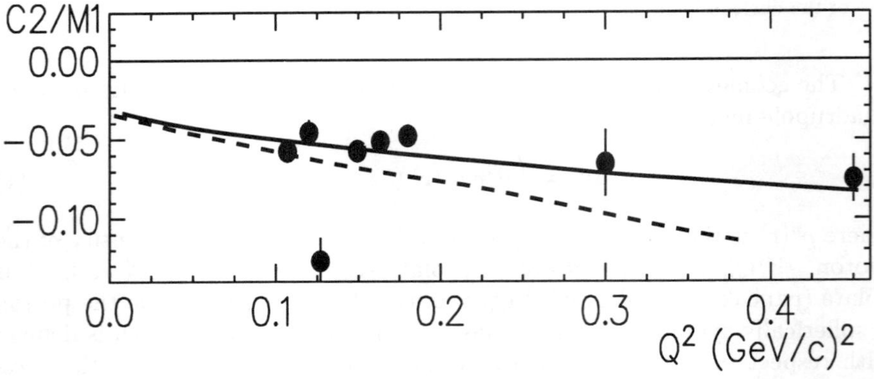

Figure 1. The C2/M1 ratio in the electromagnetic $N \to \Delta$ transition: (i) explicit quark model calculation (dashed curve) [3]; (ii) as obtained from the *elastic* neutron form factor data according to Eq.(2) (solid curve) [5]; (iii) experimental pion electroproduction data [7].

Quite generally, the baryon's charge density consists of a sum of one-, two-, and three-quark pieces: $\rho(\mathbf{q}) = \rho_{[1]}(\mathbf{q}) + \rho_{[2]}(\mathbf{q}) + \rho_{[3]}(\mathbf{q})$. The two- and three-quark terms describe the nonvalence quark degrees of freedom (e.g. $q\bar{q}$ pairs) seen by the electromagnetic probe. In deriving Eq.(2) we have made the following assumptions: (i) the baryon wave functions with orbital angular momentum $L = 0$ contain only valence quark degrees of freedom; (ii) one- and three-body operators are suppressed in comparison to the two-body term $\rho_{[2]}$. Assumption (i) is not a restriction provided the nonvalence degrees of freedom are included in the form of many-body operators [8], and assumption (ii) is supported by several investigations [2,4,6]. Eq.(2) is then a result of the dominance of $\rho_{[2]}$ for both observables, the spin-flavor structure of $\rho_{[2]}$, and the SU(6) spin-flavor symmetry of the N and Δ wave functions.

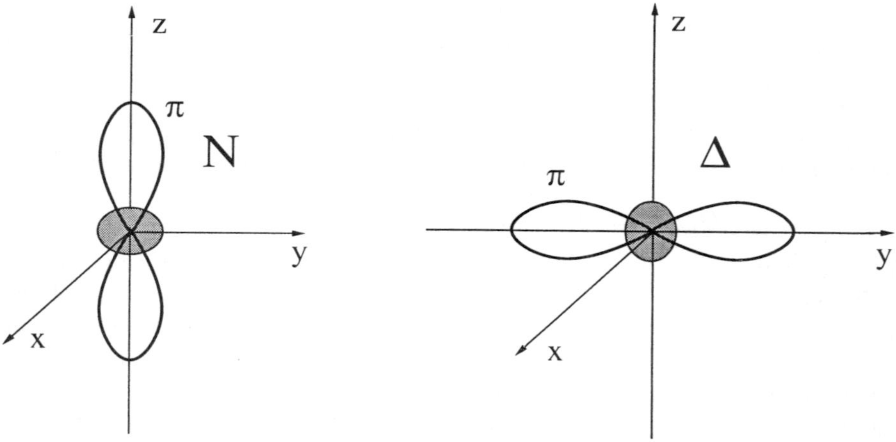

Figure 2. Intrinsic quadrupole deformation of the nucleon (left) and Δ (right) in the pion cloud model. In the N the p-wave pion cloud is concentrated along the polar (symmetry) axis, with maximum probability of finding the pion at the poles. This leads to a prolate deformation. In the Δ, the pion cloud is concentrated in the equatorial plane producing an oblate intrinsic deformation.

This can be seen as follows. A multipole expansion of, e.g., the gluon exchange charge operator $\rho_{[2]}$ up to quadrupole terms gives (with $\mathbf{q} = q\mathbf{e}_z$) the following decomposition in spin-isospin space

$$\rho_{[2]} = -B \sum_{i<j} e_i \left(2\boldsymbol{\sigma}_i \cdot \boldsymbol{\sigma}_j - (3\boldsymbol{\sigma}_{i\,z} \boldsymbol{\sigma}_{j\,z} - \boldsymbol{\sigma}_i \cdot \boldsymbol{\sigma}_j) \right) + (i \leftrightarrow j) \, , \tag{4}$$

where $\boldsymbol{\sigma}_{i\,z}$ is the z component of the spin operator of quark i, and e_i is the quark charge. B contains the orbital and color part common to the spin-scalar (C0) and spin-tensor (C2) part of the operator. Note that there is a *fixed* relative strength between the C0 term and the C2 term. Evaluating $\rho_{[2]}$ between SU(6) spin-flavor N and Δ wave functions, one obtains $r_n^2 = 4B$ and $\sqrt{2}Q_{p\to\Delta^+} = 4B$. This leads to Eq.(3). A generalization of this derivation to finite momentum transfers is straightforward and leads to Eq.(2).

In order to calculate Q_0^p from the observable $Q_{p\to\Delta^+}$ we need a model. We have calculated Q_0^p using three different nucleon models [4]. In the quark model we find that Eq.(3) implies

$$Q_0^p = -Q_0^{\Delta^+} = -r_n^2 \, , \tag{5}$$

i.e., a prolate intrinsic deformation of the proton and an oblate intrinsic deformation of the Δ^+. We also see that the neutron charge radius r_n^2 and the quadrupole deformation of the nucleon are intimately related phenomena, which reflect the $q\bar{q}$ degrees of freedom in the nucleon.

Also in the pion cloud model [9], a relation between r_n^2 and $Q_{p\to\Delta^+}$ and between the intrinsic quadrupole moments of the N and Δ is obtained [4]

$$Q_{p\to\Delta^+} = Q_{\Delta^+} = r_n^2, \qquad Q_0^p = -Q_0^{\Delta^+} = -r_n^2. \tag{6}$$

Even though the valence quark core of the nucleon may have a small oblate deformation (as one would obtain from the small D-state admixture of the valence quarks), the major contribution to the intrinsic quadrupole moment comes from the p-wave coupling of the pion cloud to the valence quark core. From angular momentum coupling the pion cloud in the proton is oriented along the polar axis (see Fig.2). This leads to an overall prolate deformation of the nucleon. Similarly, for the Δ with spin 3/2, angular momentum conservation implies that the pion cloud lies in the equatorial plane characteristic of an oblate intrinsic deformation. In addition, we have calculated Q_0^p and $Q_0^{\Delta^+}$ in the Bohr-Mottelson collective model [4], with the same qualitative results, namely a prolate shape of the N and an oblate shape of the Δ.

References

1. H. Frauenfelder and E. M. Henley, Subatomic Physics, Englewood Cliffs, New Jersey, 1974.
2. A. J. Buchmann, E. Hernández, and A. Faessler, *Phys. Rev.* C **55**, 448 (1997).
3. A. J. Buchmann, *Nucl. Phys.* A **670**, 174c (2000).
4. A. J. Buchmann and E. M. Henley, *Phys. Rev.* C **63**, 015202 (2001).
5. P. Grabmayr and A. J. Buchmann, *Phys. Rev. Lett.* **86**, 2237 (2001).
6. A. J. Buchmann and R. F. Lebed, *Phys. Rev.* D **62**, 096005 (2000).
7. R. Gothe, Proc. of N*2000, Newport News, World Scientific, Singapore, 2001, p. 153; R. Siddle *et al., Nucl. Phys.* B **35**, 93 (1971); J.C. Alder *et al., Nucl. Phys.* B **46**, 573 (1972).
8. G. Morpurgo, *Phys. Rev.* D **40**, 2997 (1989); G. Dillon and G. Morpurgo, *Phys. Lett.* B **448**, 107 (1999).
9. E. M. Henley and W. Thirring, Elementary Quantum Field Theory, McGraw-Hill, New-York, 1962.

THE GROUND AND RADIAL EXCITED STATES OF THE NUCLEON IN A RELATIVISTIC SCHRÖDINGER-TYPE MODEL

S. B. GERASIMOV

Bogoliubov Laboratory of Theoretical Physics,
Joint Institute for Nuclear Research, 141980 Dubna, Russia
E-mail: gerasb@thsun1.jinr.ru

Using the relativistic potential Schrödinger-type equations earlier discussed for two-body (*i.e.* quark-antiquark) systems, we calculate masses of the ground and higher radial excitations of the nucleon with the emphasis on the possibility to obtain, within the standard 3q-constituency of baryons and the potential interaction of quarks, the first radial-excited state around the mass of the Roper resonance $\sim$ 1440 MeV.

In this report, we present evaluation and qualitative discussion of the nucleon mass and masses of several low radial excitations following from a generalization of our approach[1] to an adopted particular form of the three-quark confinement interaction suggested by the lattice QCD simulation[2], taken, however, with one qualitatively important difference concerning the sign and magnitude of the constant term entering into the analytically fitted form of the lattice measurement of the 3q-potential.

Unlike the Salpeter-type equations with the relativistic kinetic energies in the usual "square-root" form and potential terms included as they stand in the Schrödinger equation, *e.g.*[3], our approach is based on the Schrödinger-type equation, constructed from the squared forms of the one-particle Dirac equation for each quark in the given system. Following the elaborated scheme, discussed in some detail elsewhere, we write down the three-quark equation with the particular three-body world-scalar potential $V_s^{conf}(1,2,3)$ and the standard pairwise (strong) "Coulomb" terms $\sum_{i,j} V_v(r_{ij}); i,j = 1,2,3$

$$\frac{W^2 - 9m_q^2}{2W}\Psi(1,2,3) = [\sum_{i=1}^{3}\frac{3\vec{p}^2}{2W} + \frac{3m_q}{W}V_s^{conf}(1,2,3)$$
$$+\frac{1}{2}\sum_{i<j}V_{ij}^c + \frac{3}{2W}V_s^{conf^2}(1,2,3) - \mathcal{O}(V_{ij}^c)^2) - \mathcal{O}(V_{ij}^c V_{ik}^c)]\Psi(1,2,3), \quad (1)$$

where all spin-dependent terms are relegated for the subsequent perturbative treatment, and the explicit form of terms arising from squaring the sum of pairwise Coulomb potentials has not been given. Now, after introducing the

Jacobi coordinates $\vec{R} = (1/3)(\sum_i \vec{r_i}), \eta = (1/\sqrt{2})(\vec{r_1} - \vec{r_2}), \xi = (\sqrt{2/3})[(\vec{r_1} + \vec{r_2})/2 - \vec{r_3}]$ and separating the c.m. motion, we introduce the hyperspherical coordinates $\rho = \sqrt{\eta^2 + \xi^2}$, $\eta = \rho cos\theta, \xi = \rho sin\theta$, and can proceed in full analogy with known methods of solving the nonrelativistic Schrödinger three-body equations.

In the hypercentral approximation with the minimal hypermomentum $K = K_{min} = 0$ [4,6], we get an analog of the one-dimensional Schrödinger equation with "effective" power potentials having the (quasi)harmonic oscillator-, linear- and Coulomb- parts, and the effective centrifugal part including terms arising from angular averaging the squared Coulomb potentials. This equation has the following form

$$\frac{d^2\chi_0}{d\rho^2} + \frac{2}{3}W[\frac{W^2 - 9m_q^2}{2W} - \frac{3m_q}{W}\tilde{V_s}(\rho) - V^c(\rho) - \frac{3}{2}WV_s^{conf^2}(\rho) - \frac{l'(l'+1)}{\rho^2}]\chi_0(\rho) = 0 \;, \tag{2}$$

where $\tilde{V_s}(\rho) = 2m_q V_s^{conf}$, and the effective centrifugal term $\sim l'(l'+1)$ includes terms of order $\sim \rho^{-2}$ arising from averaging the Coulomb potentials squared (l' would equal $3/2$ if the terms $\sim \mathcal{O}(\alpha_s^2)$ are neglected).

Further, we use a simple and very convenient approach[5] to deal approximately with sums of power potentials. Namely, combining known "kinetic potentials" [5] for the Coulomb, linear, and oscillator potentials, one obtains a kind of "effective" auxiliary spectrum where, as evident from Eq.(2), the known functions of W play the role of nonrelativistic energy and reduced mass, while the same spectral parameter W enters also into coefficients defining the strength of "effective" power potentials. We obtain the spectrum of our relativistic problem via the numerical solution of the algebraic equation for W, having the meaning of the consistency condition for all W's entering in different ways the obtained auxiliary quasi-nonrelativistic spectrum.

As a model representation of the world-scalar confinement potential in baryons, one can use the picture of flux-tubes or strings attached to the quarks and to a junction, combined with the adiabatic approximation, where the strings are assumed to adjust rapidly to the quark motion. A potential is generated by minimizing the string length for each set of quark positions and has the form

$$V_s^{conf} = a_{3q}L_{min} + C_{3q} \;, \tag{3}$$

where the relation of L_{min} with the Jacobi coordinates $\vec{\eta}$ and $\vec{\xi}$, or with $\vec{r}_{ij}, i,j = 1,2,3$ is given, e.g., in Ref. [2]. For a_{3q}, we have taken the value $a_{3q} =$

Table 1. Calculated masses of the ground state and radial excited states in GeV

$N(L,n_r)$	$m_N(\alpha_s = 0.6; m_q = 0.30)$	$m_N(\alpha_s = 0.5; m_q = 0.25)$
$N(0,0)$	1.087	1.165
$N(0,1)$	1.64	1.71
$N(0,2)$	2.12	2.17
$N(0,3)$	2.54	2.55

.16 GeV^2 and for an overall constant energy shift C_{3q} we take the *negative* value -.8 GeV which is in accord with other studies of the hadron spectrum [3,6], but is in sharp distinction with the proposed fit to lattice calculation [2], where this quantity is found to be substantial and *positive.* For the Coulomb potential, we use the standard form

$$V_v^c = -\frac{1}{2}\sum_{i<j}\frac{\kappa}{r_{ij}} , \tag{4}$$

with $\kappa = (4/3)\alpha_s$, and $\alpha_s = .5 \div .6$. Those potentials are functions of the angle ϑ between the two Jacobi coordinates, we expand, following the approximation proposed in Ref.[7], in Legendre polynomials of $cos\vartheta$ and keep only the term with $l = 0$ for subsequent averaging over the hyperspherical angles. In this way, the masses of the ground and several radial excited states of the nucleon have been calculated for two sets of some typical values of parameters listed in Table 1. Concerning the inclusion of the very important short-range spin-spin interaction between constituent quarks, we mention the qualitative effects of two types of the contact spin-dependent potential which will be considered perturbatively: the flavour-independent one originating from the one-gluon exchange (V_{OGE}) and the flavour-dependent potential, acting mainly between quark pairs with zero spin and isospin, which is derived in the instanton model (V_{INS}). For light relativistic quarks, we use both potentials keeping the simple-minded energy dependence, collected from an expression for the interaction of two free quarks each with the energy assumed to be equal to that of the off-mass-shell quark in a given bound state. Schematically, we have

$$\Delta m^*_{OGE} = C_{OGE}(\frac{1}{2\varepsilon(N^*)})^2|\Psi_{N^*}(0)|^2 , \tag{5}$$

$$\Delta m^*_{INS} = C_{INS}(\frac{m_q + \varepsilon(N^*)}{2\varepsilon(N^*)})^2|\Psi_{N^*}(0)|^2 . \tag{6}$$

At last, using an analog of the Schwinger relation between the mean value of $\langle\delta(\vec{r}_{ij})\rangle$ and the mean value of the derivative of the potential terms in the same quantum state, and calculating the last one directly from the known spectrum with the help of the Feynman-Hellmann theorem, we come to the estimation of the ratio $|\Psi_{N^*_{n_r}}(0)|^2/|\Psi_N(0)|^2$, which for $n_r = 1$ is larger than unity by 30-40 % owing to the presence of the confinement potential squared, i.e., the "effective" oscillator potential. The value of $|\Psi_N(0)|^2$ is connected with the mass difference of the $\Delta(1240)$-resonance and the nucleon or, equivalently, with the mass shift of the nucleon when the contact spin-spin interaction is switched on, when the nature of the spin-dependent interaction is fixed. Hence we are able to estimate the expected shift of the "zeroth-order" mass of the first radial excited state while either the OGE-type or the instanton-type spin-spin interaction is switched on and to conclude that only with parameters of the second column in Table 1 and with the instanton-type force, the corrected mass $m(n_r = 1, L = 0)$ turns out close to the mass of the Roper resonance $\simeq$ 1440 MeV. Alternatively, if the instanton dominance turns to be physically irrelevant, the mass of the first radial state lies higher than the Roper resonance, and an additional argument would appear for a near-lying hybrid-baryon state.

In addition to this main conclusion of our work, we would like to stress the mystery of the difference between the phenomenological and lattice-calculated constant terms in the analytic approximation of the $3q$-confinement potential which has still to be unraveled.

In conclusion, the author would like to thank the Organizers of the "NSTAR-2001" Workshop for warm hospitality and support. The author is also grateful to Prof. M.Müller-Preussker of the Humboldt-University, Berlin, and members of his group for helpful discussions and comments.

References

1. S.B.Gerasimov, in *Prog. Part. Nucl. Phys.* **8**, Ed. Sir D. Wilkinson, Pergamon Press, 1982, p.207; *Multiquark interactions and QCD*, JINR-D1,2-81-728, Dubna, 1981, p.51; hep-ph/9812509;hep-ph/0001068.
2. T.T.Takahashi, *et al. Phys. Rev. Lett.* **86**, (2001) 18.
3. S.Capstick and N.Isgur, *Phys. Rev.* D **34**, 2809 (1986).
4. J.M.Richard, *Phys. Lett.* B **100**, 515 (1981).
5. R.L. Hall, *J. Math. Phys.* **25**, 2708 (1984).
6. A.M. Badalyan, D.I. Kitoroage, and D.S. Pariysky, *Yad. Fiz.* **46**, 226 (1987); *ibid* **47**, 807 (1988).
7. M. Fabre de la Ripelle and Yu.A. Simonov, *Ann. Phys.(NY)* **212**, 235 (1991).

VECTOR MESON PHOTOPRODUCTION IN THE QUARK MODEL

Q. ZHAO
Department of Physics, University of Surrey, Guildford, GU2 7XH, UK
E-mail: qiang.zhao@surrey.ac.uk

We present a quark model study of the ω meson photoproduction near threshold. With a limited number of parameters, all the data in history are reproduced. The roles played by the s- and u-channel processes (resonance excitations and nucleon pole terms), as well as the t-channel *natural* (Pomeron) and *unnatural* parity (pion) exchanges are clarified. This approach provides a framework for systematic studies of vector meson photoproduction near threshold.

1 Introduction

For a long time, the experimental study of the neutral vector meson (ω, ρ^0 and ϕ) photoproduction was concentrated on high energy regions, where the diffractive process played a dominant role and could be accounted for by a soft Pomeron exchange model. Recently, the availability of high intensive electron and photon beams at JLAB, ELSA, ESRF, and SPring-8 gives accesses to excite nucleons with clean electromagnetic probes. Thus, vector meson production via resonance excitations provides an ideal tool to study the non-diffractive mechanisms in vector meson photoproduction near threshold. Concerning the resonance excitations in this reaction, the other essential motivation is to search for "missing resonances", which were predicted by the nonrelativistic constituent quark model (NRCQM) [1], but not found in πN scattering. Vector meson photoproduction near threshold might provide supplementary knowledge of those missing resonances and their couplings to vector mesons.

In this contribution, a quark model approach to vector meson photoproduction near threshold is applied to ω meson photoproduction. Our purpose is to provide a framework on which a systematic study of resonance excitations becomes possible.

Our model consists of three processes: (i) s- and u-channel vector meson production with an effective Lagrangian (S+U); (ii) t-channel Pomeron exchange ($\mathcal{P}$) for ω, ρ^0 and ϕ production [2]; (iii) t-channel light meson exchange. In particualar, in ω meson photoproduction the π^0 exchange is taken into account.

In the $SU(6)\otimes O(3)$ symmetry limit, the constituent quark ψ couples to a vector meson ϕ_m^μ via an effective Lagrangian [3,4]:

$$L_{eff} = \overline{\psi}(a\gamma_\mu + \frac{ib\sigma_{\mu\nu}q^\nu}{2m_q})\phi_m^\mu\psi, \tag{1}$$

where a and b are overall parameters introduced at quark the level for baryon states. In this way, all the s- and u-channel resonances and nucleon pole terms can be included. The t-channel vector meson exchange and contact term from the Lagrangian will only contribute in charged vector meson photoproduction. In $\gamma p \to \omega p$, eight low-lying resonances: $P_{11}(1440)$, $S_{11}(1535)$, $D_{13}(1520)$, $P_{13}(1720)$, $F_{15}(1680)$, $P_{11}(1710)$, $P_{13}(1900)$, and $F_{15}(2000)$, with quark harmonic oscillator shell $n \leq 2$ are explicitly included [5], while higher mass states are treated degenerate with n. We refer the readers to Refs. [3,4] for details of this approach.

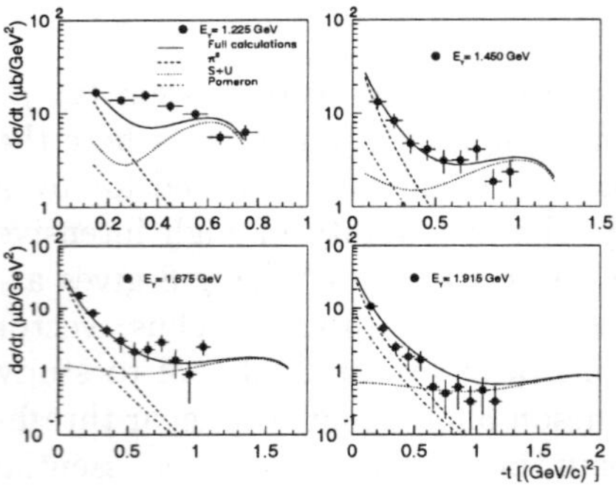

Figure 1. Differential cross section for $\gamma p \to \omega p$.

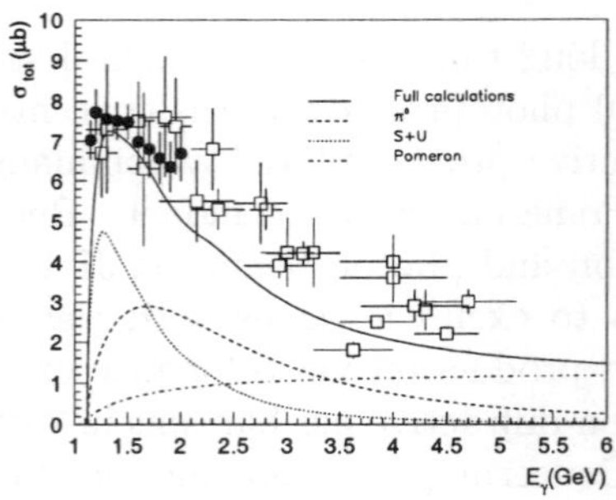

Figure 2. Total cross section for $\gamma p \to \omega p$.

2 Analysis

In Fig. 1, differential cross sections for four energy bins, E_γ =1.225, 1.450, 1.675 and 1.915 GeV, are presented and compared with the SAPHIR data [6]. Results for exclusive processes are also presented. They show that near threshold the π^0 exchange plays a dominant role over the other two processes, in particular at small angles. With increasing energy the *natural* parity Pomeron exchange becomes more and more important which will produce interesting interfering effects in polarization observables. In Ref. [4] we show that the *natural* and *unnatural* parity exchanges can be constrained well by the measurement of forward angle parity asymmetries [7]. The dotted curves in Fig. 1

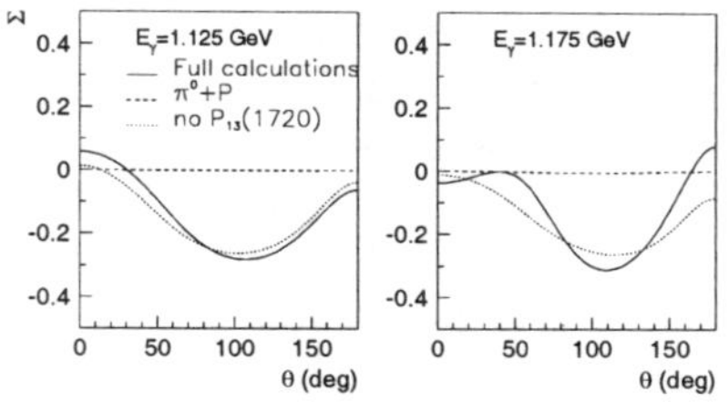

Figure 3. Polarized beam asymmetry. See the text for notations.

represent the exclusive calculations of the *s*- and *u*-channel processes. They account for the large angle behavior in the differential cross sections, but have only a small impact on the small-angle cross sections. This feature justifies the method by which we constrain the *t*-channel processes.

In Fig. 2, the total cross section for $\gamma p \rightarrow \omega p$ is given. Interestingly, the figure shows that the dominant process near threshold is the S+U, although it falls down rapidly with increasing energy. The Pomeron exchange becomes dominant over the pion exchange above $E_\gamma \sim 3.5$ GeV, which is consistent with the experimental results [7]. Note that in Ref. [9], a phenomenological model with parameters predicted by the $^3P_0$ quark-pair-creation model [10] obtains quite different results for exclusive processes.

For the purpose of investigating individual resonance excitations, we have to turn to polarization observables in which small effects from individual resonances might be picked up. Following the convention of Ref. [11], we present predictions for $\Sigma \equiv (2\rho^1_{11} + \rho^1_{00})/(2\rho^0_{11} + \rho^0_{00})$, in Fig. 3, where ρ^1 and ρ^0 are density matrix elements in helicity space [12]. One of the most important features of Σ is that large asymmetries cannot be produced by the $\mathcal{P}$ or $\mathcal{P}+\pi^0$. As shown by the solid curves, large asymmetries are produced by the interferences between the $\mathcal{P}+\pi^0$ and S+U processes.

In this study, we find that the $P_{13}(1720)$ and $F_{15}(1680)$, which are classified by the $[56,^2 8, 2, 2, J]$ quark model representation, play a strong role in this reaction. The $S_{11}(1535)$ and $D_{13}(1520)$ of $[70,^2 8, 1, 1, J]$ have also relatively large effects. In Fig. 3, we show that without the $P_{13}(1720)$, the asymmetry will be significantly changed (see the dotted curves). The predictions can be compared to the preliminary data from GRAAL collaboration [13].

Another interesting observable with a polarized photon beam is, $\Sigma_A \equiv (\rho^1_{11} + \rho^1_{1-1})/(\rho^0_{11} + \rho^0_{1-1}) = (\sigma_\parallel - \sigma_\perp)/(\sigma_\parallel + \sigma_\perp)$, where $\sigma_\parallel$ and $\sigma_\perp$ represent the pion decay cross sections of the ω meson with the pions submerged in or perpendicular to the photon polarization plane. As found in Ref. [4], this observable is more sensitive to small contributions from individual resonances.

In summary, we present a quark model study of vector meson photoproduction near threshold by applying it to $\gamma p \to \omega p$. It provides a framework on which a systematic investigation of resonance excitations in vector meson production can be built. Our predictions should be confronted with the forthcoming data from GRAAL and JLAB in the near future.

Acknowledgments

Fruitful discussions with E. Hourany concerning the GRAAL experiment are acknowledged.

References

1. N. Isgur and G. Karl, *Phys. Lett.* B **72**, 109 (1977); *Phys. Rev.* D **23**, 817 (1981); R. Koniuk and N. Isgur, *Phys. Rev.* D **21**, 1868 (1980).
2. A. Donnachie and P.V. Landshoff, *Phys. Lett.* B **185**, 403 (1987); *Nucl. Phys.* **311**, 509 (1989).
3. Q. Zhao, Z.-P. Li and C. Bennhold, *Phys. Lett.* B **436**, 42 (1998); *Phys. Rev.* C **58**, 2393 (1998).
4. Q. Zhao, *Phys. Rev.* C **63**, 025203 (2001).
5. Particle Data Group, D.E. Groom *et al.*, *Euro. Phys. J.* C**15**, 1 (2000).
6. F.J. Klein, Ph.D. thesis, Univ. of Bonn, Bonn-IR-96-008 (1996); πN *Newslett.* **14**, 141 (1998).
7. J. Ballam *et al.*, *Phys. Rev.* D **7**, 3150 (1973).
8. R. Erbe *et al.*, *Phys. Rev.* D **175**, 1669 (1968); H. R. Crouch *et al.*, *Phys. Rev.* **155**, 1468 (1967); Y. Eisenberg *et al.*, *Phys. Rev.* D **5**, 15 (1972); Y. Eisenberg *et al.*, *Phys. Rev. Lett.* **22**, 669 (1969); D. P. Barber *et al.*, *Z. Phys.* C **26**, 343 (1984); W. Struczinski *et al.*, *Nucl. Phys.* B **108**, 45 (1976).
9. Y. Oh, A.I. Titov, and T.-S.H. Lee, *Phys. Rev.* C **63**, 025201 (2001); *these proceeding.*
10. See e.g. S. Capstick and W. Roberts, Los Alamos preprint nucl-th/0008028, for a recent review; S. Capstick, *these proceedings.*
11. M. Pichowsky, Ç. Şavklı, and F. Tabakin, *Phys. Rev.* C **53**, 593 (1996).
12. K. Schilling, P. Seyboth, and G. Wolf, *Nucl. Phys.* B **15**, 397 (1970).
13. J. Ajaka *et al.*, Proceedings of the 14th International Spin Physics Symposium, Osaka, Japan, October 16-21, 2000.

STATUS OF NUCLEON RESONANCES WITH MASSES $M < M_N + M_\pi$

R. BECK[1], S. N. CHEREPNYA[2], L. V. FIL'KOV[2], V. L. KASHEVAROV[2], M. ROST[1], AND TH. WALCHER[1]

[1] *Institut für Kernphysik, Johannes Gutenberg-Universität, Mainz, Germany*

[2] *Lebedev Physical Institute, Leninsky Prospect 53, 117924 Moscow, Russia*
E-mail: filkov@x4u.lpi.ruhep.ru

We discuss different interpretations of the peaks observed a few years ago by Tatischeff *et al.* [1] in the missing mass spectra of the reaction $pp \to \pi^+ pX$, which were declared as new exited nucleon states with small masses. A study of the possible production of such states in the process $\gamma p \to \pi^+ N^* \to \pi^+ + \gamma\gamma n$ by analyzing the invariant mass spectrum of $\gamma\gamma n$ is proposed. It is shown that the experimental data obtained at MAMI-B allows one to analyze this process and to get information about the existence of excited nucleon states with small masses.

A few years ago three narrow bumps have been observed in missing mass spectra of the reaction $pp \to p\pi^+ X$ by Tatischeff *et al.* [1] at $M_X = 1004$, 1044, and 1094 MeV. These bumps were interpreted as new nucleon resonances N^*. The values of the masses $M_X = 1004$ and 1044 MeV are below $m_N + m_\pi$ and so these states can decay with an emission of photons only. If they decay into γN, then these resonances have to contribute to Compton scattering on the nucleon. However, the analysis [2] of the existing experimental data on this process completely excluded such N^* as intermediate states in the Compton scattering on the nucleon.

In Ref. [3] it was assumed that these states belong to the totally antisymmetric $\underline{20}$-plet of the spin-flavor $SU(6)_{FS}$. Such an N^* can transit into a nucleon only if two quarks from N^* participate in the interaction. Then the simplest decay of N^* with the masses 1004 and 1044 MeV is $N^* \to \gamma\gamma N$. This assumption could be checked, in particular, by investigating the reaction $\gamma p \to \gamma X$ or $\gamma p \to \pi X$ in the photon energy region of about 800 MeV.

Another interpretation of the states found in work [1] was suggested in Refs. [4,5]. In these works the reaction $pd \to p + pX_1$ has been studied with the aim of searching for supernarrow dibaryons (SND), the decay of which into two nucleon is forbidden by the Pauli exclusion principle. Three peaks have been observed in invariant mass spectra of pX_1 states at $M_{pX_1} = 1904 \pm 2, 1926 \pm 2$, and 1942 ± 2 MeV (see Fig.1a). The analysis of the angular distribution of the protons from the decay of pX_1 states showed that the peaks found can be explained as manifestation of the SNDs. Additional information about

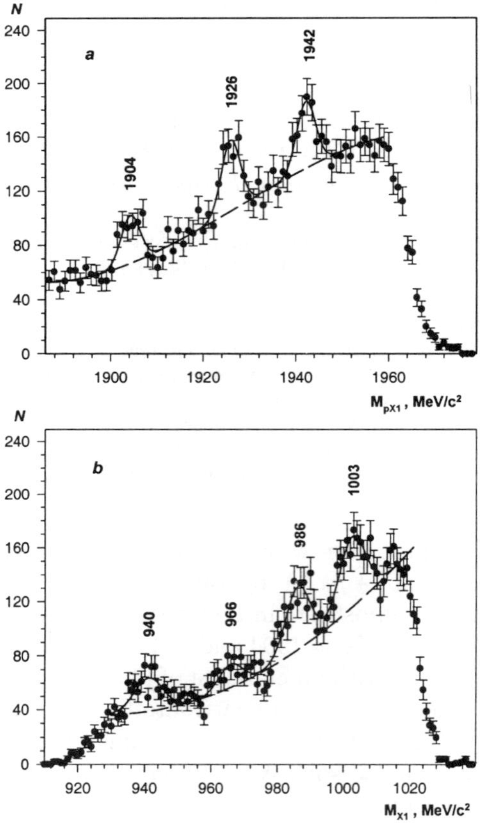

Figure 1. The invariant mass M_{pX_1} (a) and missing mass M_{X_1} (b) spectra.

the nature of these states have been obtained by analysing the missing mass M_{X_1} spectra. If the observed state is a dibaryon decaying mainly into two nucleons, then X_1 is a neutron and M_{X_1} has to be equal to the neutron mass m_n. If the value of M_{X_1}, obtained from experiment, differs essentially from m_n, then $X_1 = \gamma + n$ and the found state is SND.

Fig. 1b demonstrates the missing mass M_{X_1} spectrum obtained in Refs [4,5]. As is seen from this figure, besides the peak at the neutron mass caused by the process $pd \to p + pn$, a resonance like behavior of the spectrum is observed at 966 ± 2, 986 ± 2, and 1003 ± 2 MeV. These values of M_{X_1} coincide with the simulated ones and differ essentially from the neutron mass. Hence, the dibaryons found are really SNDs.

It should be noted that the peak at $M_{X_1} = 1003 \pm 2$ MeV corresponds to the bump observed in Ref. [1]. Taking into account the found connection between SNDs and resonancelike states X_1, authors of the works [4,5] assumed that the peaks from Ref. [1] at 1004 and 1044 MeV are not exited nucleons but resonance like states $X_1 = \gamma + n$ caused by the possible existence and decay of the SNDs with the masses 1942 and 1982 MeV. Such X_1 are not real resonances and cannot give contribution to Compton scattering on the nucleon.

However, a SND can also decay into NN^*. Unfortunately, the experiment [4,5] could not unambiguously discriminate an exited nucleon from an X_1 state. To clarify this question we propose to study the exited nucleon production in the process

$$\gamma + p \to \pi^+ + N^* \to \pi^+ + \gamma\gamma n \tag{1}$$

by analysing the invariant mass spectrum of the $\gamma\gamma n$ at the incident photon energy from 537 up to 817 MeV. The data on this process can be obtained from the experiment on radiative π^+ meson photoproduction from the proton, which has been carried out at MAMI-B. In this experiment γ, π^+, and n were detected, so we have enough data to reconstruct the invariant mass of the $\gamma\gamma n$ from 985 up to 1075 MeV. The lowest value of $M_{\gamma\gamma n}$ is due to the threshold of the final photon energy (15 MeV) and the highest one corresponds to the sum of masses of neutron and π^0 meson.

The main background processes are $\gamma p \to \pi^0\pi^+ n$ and $\gamma p \to \pi^0\pi^0\pi^+ n$. To suppress the contribution of double pion photoproduction, we will consider the invariant mass of two photons $M_{\gamma\gamma} < 110$ MeV. As the π^+ meson and N^* must fly in the same plane, we have an additional condition on the difference of pion and N^* azimuthal angles: $160° < |\phi_{\pi^+} - \phi_{N^*}| < 200°$. This condition allows one to suppress the contribution of triple pion photoproduction.

The results of the simulation of the N^* production in process (1), at the conditions of radiative π^+ meson photoproduction from proton experiment and for an exposition time of 100 hours, are shown in Fig.2. The calculations obtained without any cuts are presented in Fig.2a. Fig.2b demonstrates the final result after both cuts. As is seen from this figure, the background can be well suppressed and, if N^* states exist, they would be well recognizable.

In the experiment on radiative π^+ meson photoproduction we had about 1000 hours of exposition time. As result, we expect to get about 600, 3500, and 12000 events for the exited nucleon states with the masses 986, 1004, and

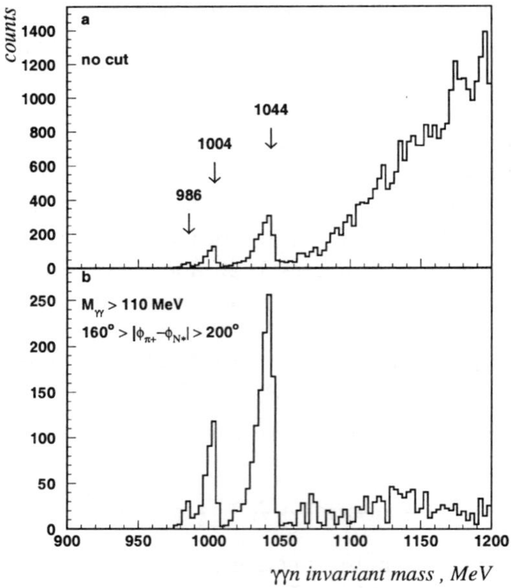

Figure 2. Invariant mass spectra of $\gamma\gamma n$. (a) – without any cuts; (b) – taking into account conditions $M_{\gamma\gamma} < 110$ MeV and $160° < |\phi_{\pi^+} - \phi_{N^*}| < 200°$.

1044 MeV, respectively. So this experiment can give important information about the possible of existence of excited nucleon states with small masses.

This work was supported in part by RFBR, grant No 00-02-04014 NNIO_a, and Deutsche Forschungsgemeinschaft (SFB 443).

References

1. B. Tatischeff, J.Yonnet, and N. Willis *et al.*, *Phys. Rev. Lett.* **79**, 601 (1997).
2. A.I. L'vov and R.L. Workman, *Phys. Rev. Lett.* **81**, 1936 (1998).
3. A.P. Kobushkin, nucl-th/9804069.
4. L.V. Fil'kov, V.L. Kashevarov, E.S. Konobeevski *et al.*, VII Conf. CIPANP2000, Quebec City, Canada, 22-28 May 2000, AIP Conf. Proceed. v.549, p.267; nucl-th/0009044.
5. L.V. Fil'kov, V.L. Kashevarov, and E.S. Konobeevski *et al.*, nucl-th/0101021, Proc. XV Intern. Seminar on High Energy Physics "Nuclear Physics and Quantum Chromodynamics", Dubna, 25-29 September, 2000.

SPIN STRUCTURE OF THE $\Delta(1232)$ AND INELASTIC COMPTON SCATTERING

A. I. L'VOV

Lebedev Physical Institute, Leninsky Prospect 53, Moscow 117924, Russia
E-mail: lvov@x4u.lebedev.ru

Radiative transitions $\gamma\Delta(1232) \to N^*$ are discussed in the nonrelativistic quark model with spin-orbit corrections for the 70-plet $L^P = 1^-$ nucleon resonances N^*. The reaction $\gamma N \to \gamma\Delta$ is considered as a tool to measure some of these transitions. A particular sensitivity to photoexcitations of $S_{11}(1535)$, $D_{13}(1700)$, and $D_{15}(1675)$ is predicted.

1 Introduction: The GDH Sum Rule for the $\Delta(1232)$

My motivation to analyze photocouplings of the $\Delta(1232)$ was initially related to a problem of saturation of the Gerasimov–Drell–Hearn sum rule [1] for the Δ [2]. Generally, the GDH integral

$$I = \int_{\rm thr}^{\infty} \frac{d\omega}{\omega}\,[\sigma_{1+s}(\omega) - \sigma_{1-s}(\omega)] = \pi s \left(\frac{\mu}{s} - \frac{eZ}{M}\right)^2 \quad (1)$$

$$= \sum_{N^* \neq B} \frac{4\pi M^* M}{(M^{*2} - M^2)^2} \left\{|A(1,s)|^2 - |A(1,-s)|^2\right\}, \quad (2)$$

must give the magnetic moment μ of the baryon B of mass M, spin s and electric charge eZ through the total photoabsorption cross sections $\sigma_{1\pm s}$ for the target spin parallel or anti-parallel to the photon helicity $\lambda = +1$. The second line re-expresses the integral (1) through the electromagnetic transition amplitudes $A(\lambda,\sigma)$ of $\gamma B \to N^*$ in the zero-width approximation for the resonances N^*. Usually photocouplings A_Λ are defined instead of $A(\lambda,\sigma)$,

$$A(\lambda,\sigma) \equiv \langle N^*(\vec p + \vec k), \lambda+\sigma \mid -\vec\epsilon_\lambda \cdot \vec J_{\rm e.m.}(\vec k) \mid B(\vec p), \sigma\rangle = \sqrt{\frac{M^{*2} - M^2}{M^*}}\,\xi A_\Lambda\,, \quad (3)$$

where $\Lambda = \lambda + \sigma$ is the total helicity, $\lambda = +1$, and $\xi = \pm 1$ is a phase factor determined by the pion decay amplitude of the resonance N^* [3]. Collinear kinematics $\vec p \parallel \vec k$ is assumed in Eq. (3).

In the framework of the nonrelativistic quark model (NQM), there is a fundamental difference between the GDH sum rule for the nucleon and for the Δ. The radiative $M1$ transitions $N \to \Delta$ with the helicities $1 + \frac{1}{2} \to \frac{3}{2}$ and $1 + (-\frac{1}{2}) \to \frac{1}{2}$ nearly saturate the sum rule for the nucleon giving $I = 216\ \mu$b in close agreement with the experimental values $I_p = 205\ \mu$b and $I_n = 233\ \mu$b inferred from the nucleon anomalous magnetic moments. Meanwhile the same $N \leftarrow \Delta$ transition gives a big *negative* contribution $I = -323\ \mu$b to the GDH sum rule for the Δ [note the helicities $1+(-\frac{3}{2}) \to -\frac{1}{2}$] in drastic disagreement with the manifestly positive r.h.s. of Eq. (1). Assuming that the GDH sum rule is valid for all hadrons including the Δ, we conclude that the Δ must have strong photocouplings to nucleon resonances of spin $J \geq \frac{5}{2}$. These resonances must compensate $\Delta \to N, N^*$ contributions from spin $J \leq \frac{3}{2}$ states, all of which are negative.

2 Photocouplings in NQM: The Role of Spin-Orbit Corrections

It was found long ago [4] that nonrelativistic constituent models do not obey the GDH constraint (1). Relativistic corrections are needed in order to restore the validity of the GDH sum rule. They are especially important for high-lying states. For a target having all constituents (quarks) in the s-shell and thus having the angular and magnetic moments completely composed of the spins of the constituents, it is sufficient to include only spin-orbit corrections,

$$H^{\mathrm{SO+NA}}_{\mathrm{e.m.}} = \sum_i \frac{e_i \vec{\sigma}_i}{8m_i^2} \cdot (\vec{p}_i \times \vec{E}_i - \vec{E}_i \times \vec{p}_i) + \sum_{i \neq j} \frac{e_i}{4M_{\mathrm{tot}}} \left(\frac{\vec{\sigma}_i}{m_i} - \frac{\vec{\sigma}_j}{m_j} \right) \cdot \vec{p}_j \times \vec{E}_i \,. \quad (4)$$

Apart from ordinary additive spin-orbit terms (SO), this equation contains important non-additive (NA) two-body pieces which appear due to the Wigner rotation of quark spins into the center-of-mass frame of the baryon [4,5].

One can illustrate the role of the SO+NA terms taking the radiative transition $\gamma N \to \Delta$ as example. Equation (3) defines the photocoupling A_Λ of the nucleon as a Lorentz-invariant quantity. However, when nonrelativistic wave functions and the nonrelativistic electromagnetic current are used in Eq. (3), the result does depend on the frame used. For example, the photocoupling $A_{3/2}$ found in the c.m. frame ($\vec{p} = -\vec{k}$), in the Breit frame ($\vec{p} = -\frac{1}{2}\vec{k}$) and in the lab frame ($\vec{p} = 0$) is equal to -175, -194 and -219×10^{-3} GeV$^{-1/2}$, respectively [oscillator wave functions were used in this calculation with the oscillator parameter $a = 0.41$ GeV and $m_q = 336$ MeV]. When the spin-orbit terms (4) are included, the result turns out to be nearly frame-independent: -198, -194 and -182×10^{-3} GeV$^{-1/2}$, respectively. In this example SO and NA corrections are equally important and they are seen to nearly restore the Lorentz invariance of the transition amplitude.

Using the Karl–Isgur quark model (for a recent review see Ref. [3]) with the spin-orbit correction (4), we (re)calculated the photocouplings of the nucleon to negative-parity baryons $|70,\ L^P = 1^-\rangle$. With three exceptions, we found a full agreement with the previous work [5] [those exceptions are the SO contribution to the amplitude $A_{3/2}$ for $\gamma N \to D_{33}(1700)$ and NA contributions to the amplitudes $A_{1/2}$, $A_{3/2}$ for $\gamma N \to S_{13}(1620)$].

Then we calculated photocouplings of the Δ. Our results for Δ^+ are given in the Table 1 as example, separately for the cases when a pure nonrelativistic approximation is used (columns NR) and when the spin-orbit corrections (4) are included, too (columns SO). Also (some of) the predictions of Carlson–Carone [6] are shown (columns CC1) which have been obtained through a fit to experimental data on *nucleon* photocouplings using one-body quark operators dominating in the large N_c limit of QCD. For *large* amplitudes, there is a qualitative agreement between our results (when SO is included) and the results of Carlson–Carone with the exception of a few signs which are probably related with different phase conventions used for the baryon wave functions; our convention follows Ref. [3].

Our calculation suggests a strong photocoupling of the $\Delta(1232)$ to the lowest $J = \frac{5}{2}$ state, which is $D_{15}(1675)$. The $\Delta \to D_{15}$ transition contributes much to the GDH sum rule for the Δ and helps to reduce a gap between the negative value Eq. (2) evaluated through the lowest 56 and 70-plet baryons N^* and the manifestly positive r.h.s. of Eq. (1). See Ref. [2] for further detail.

Table 1. $\Delta^+(1232)$ photocouplings to $|70,\ L^P = 1^-\rangle$ baryons in units of 10^{-3} GeV$^{-1/2}$. The phase factor of ξ, Eq. (3), is included into A_Λ.

N^*	$A_{-1/2}$			$A_{1/2}$			$A_{3/2}$			$A_{5/2}$		
	NR	SO	CC1	NR	SO	CC1	NR	SO	CC1	NR	SO	CC1
$S_{11}(1535)$	−88	−75	108	−94	−86	4						
$D_{13}(1520)$	61	20	62	48	1	−16	22	−19	−90			
$S_{11}(1650)$	−196	−115	−63	−107	−60	−128						
$D_{13}(1700)$	−87	−64	43	−164	−138	123	−198	−175	169			
$D_{15}(1675)$	9	−12	24	−26	−62	−19	−97	−147	−113	−201	−267	−258
$S_{31}(1620)$	46	46	−42	−27	−27	73						
$D_{33}(1700)$	81	81	−54	47	47	0	0	0	55			

3 Differential Cross Section of $\gamma N \to \gamma\Delta$

Using the above photo-amplitudes as functions of the photon energy, we estimated differential cross section of inelastic Compton scattering $\gamma N \to \gamma\Delta$ in the resonance region. Both s and u channel resonances, with experimental masses and widths, were included as well as seagull contributions, cf.

Ref. [7]. Dominating contributions to the reaction off the proton come from the $S_{11}(1535)$ and $D_{13}(1700)$ resonances, see Fig. 1. For the neutron target, the $D_{11}(1675)$ resonance dominates. Note that $D_{11}(1675)$ is decoupled from the proton [i.e., the Moorhouse selection rule remains valid for D_{15} even after the NA corrections are included], so that the reaction off the neutron, via $\gamma d \to \gamma p_s \Delta^0$, turns out to be the only efficient way to study the $\gamma \Delta D_{15}$ transition.

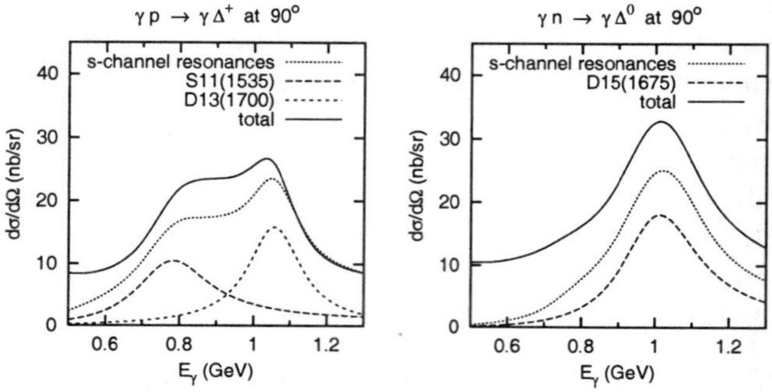

Figure 1. Differential cross sections (c.m.) of inelastic Compton scattering in the NQM including spin-orbit corrections. Contributions of s-channel resonances (without u-channel and seagull contributions) are shown separately.

The differential cross section of inelastic Compton scattering is seen to be compatible with that of elastic γN scattering. Therefore we may conclude that the original idea of Carlson–Carone [6] to study photocouplings of Δ in the reaction $\gamma N \to \gamma \Delta$ can probably be realized in practice.

References

1. S.B. Gerasimov, *Yad. Fiz.* **2**, 598 (1965); S.D. Drell and A.C. Hearn, *Phys. Rev. Lett.* **16**, 908 (1966).
2. A.I. L'vov, Proc. Symposium on the Gerasimov-Drell-Hearn sum rule and the nucleon spin structure in the resonance region, World Scientific, Singapore (2001), eds. D. Drechsel and L. Tiator, p. 365 [nucl-th/0010060].
3. S. Capstick and W. Roberts, *Prog. Part. Nucl. Phys.* **45**, 241 (2000).
4. F.E. Close and L.A. Copley, *Nucl. Phys.* B **19**, 477 (1970); R.A. Krajcik and L.L. Foldy, *Phys. Rev. Lett.* **24**, 545 (1970).
5. F.E. Close and Z. Li, *Phys. Rev.* D **42**, 2194 (1990).
6. C.E. Carlson and C.D. Carone, *Phys. Lett.* B **441**, 363 (1998).
7. S. Capstick and B.D. Keister, *Phys. Rev.* D **46**, 84 (1992).

STRUCTURE OF THE ROPER RESONANCE FROM ALPHA-PROTON AND PI-NUCLEON SCATTERING

H.-P. MORSCH

Institut für Kernphysik, Forschungszentrum Jülich, D-52425 Jülich,
E-mail: morsch@fz-juelich.de

P. ZUPRANSKI

Soltan Institute for Nuclear Physics, Pl-00681 Warsaw, Poland
E-mail: Pawel.Zupranski@fuw.edu.pl

Existing data on the Roper resonance N*(1440) excitation in hadronic reactions were reanalyzed in a consistent way, which requires the assumption of two P_{11} structures, a scalar N* resonance and a second order excitation of the Δ resonance. In this way, all data on α-p and π-N scattering are well described. This also yields a consistent description of $2\pi^o$-photoproduction, where the second structure is observed only. The fact that the first N* resonance (which is strongly excited in α-p) is not seen with photons, supports the conjecture of a pure scalar N* excitation.

1 Introduction

Since the discovery of the lowest N* resonance, $P_{11}(1440)$, by D. Roper, this resonance has been the subject of intense discussions, because its phenomenological description in π-N scattering as well as its theoretical interpretation caused severe problems. Resonance parameters extracted from the different observables are not consistent. From a fit of the resonance in the total π-N cross section[1] resonance parameters M and Γ are obtained which are about 1470 MeV and 350 MeV, respectively. A lower mass and a smaller width is suggested from the speed[2] (derivative of the scattering amplitude T with respect to the mass $|dT/dm|$) which shows a pronounced peak structure centered at about 1370 MeV. In addition, this resonance has an anomalous behaviour in Im T^{in}, which describes the absorption from the elastic channel: instead of a peak as observed for other resonances, only a fall-off to lower absorption is found. The P_{11} resonance is dominant in the elastic π-N channel; the inelasticities into 2π(s)-N (2π coupled to isospin=0) and π-Δ, studied in the π-N$\rightarrow$ $\pi\pi$N reaction[3], are of the order of 10-20 %.

More recently, a resonance in the same mass region has been observed[4] in α-p scattering (Saturne resonance). This excitation exhausts a large fraction of the corresponding monopole sum rule; therefore, it has been interpreted as a compression or breathing mode of the nucleon, from which the nucleon compressibility has been deduced.

Theoretically, a $N^*(1/2,1/2)^+$ resonance at low excitation energy is of particular interest. In the constituent quark model[5] such a resonance corresponds to a 1s→2s quark excitation which should lie at a much higher excitation energy. Assuming meson-exchange instead of gluon exchange[6] the $N^*(1/2,1/2)^+$ comes down to the experimental value. Interestingly, in the Skyrmion model[7] (which treats the baryon structure as a mesonic field), a monopole mode is the lowest excitation of the baryon. Finally, it has been speculated that this resonance might correspond to a hybrid structure.

We report on an attempt to understand the data from π-N and α-p scattering in a consistent way and to resolve the above mentioned problems in π-N.

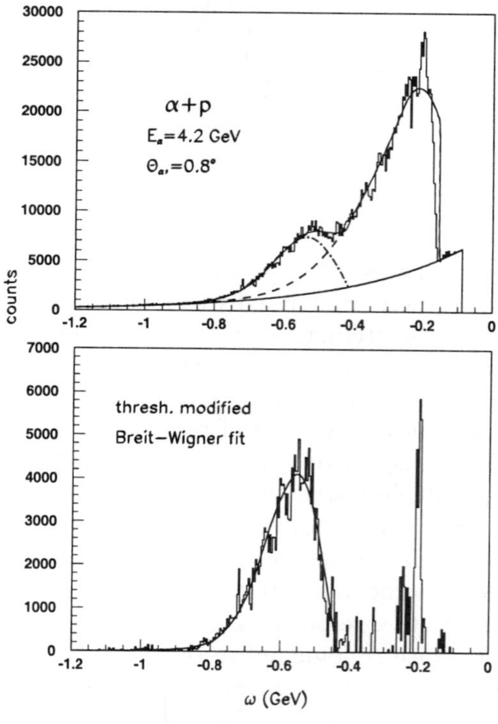

Figure 1. Missing energy spectrum from inelastic α-p scattering. Dashed line: contribution from projectile Δ excitation, dot-dashed line: N^* excitation. In the lower part the instrumental background as well as the projectile excitation is subtracted.

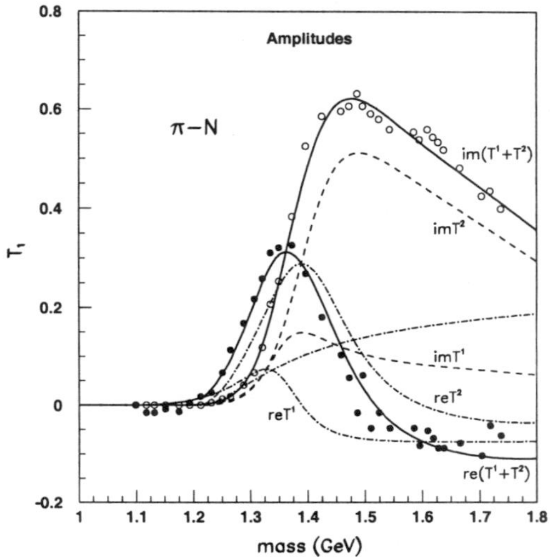

Figure 2. π-N amplitudes T_1 (real and imaginary parts given by solid and open points, respectively) in comparison with our fits. The separate amplitudes for the two resonances are given by the broken lines.

2 T-Matrix Description of π-N and α-p Scattering

For a partial wave with angular momentum l the π-N scattering cross section is given by $\sigma_l = (2l+1)\ 4\pi/k^2 \cdot |T_l|^2$, where the mass dependence of the π-N amplitude T_l may be given by resonances and a background $T_l(m) = \sum_i T_l^i(m)+B(m)$. In a realistic description $T_l^i(\mathrm{m})$ may be given by momentum dependent Breit-Wigner forms with threshold cut-off functions:

$$T_l^i(m) = thres_i^{n\pi}(m)\ \frac{(\Gamma_i^{el}/\Gamma_i)\cdot\Gamma_i(m)/2}{M_i - m - i\Gamma_i(m)/2} \tag{1}$$

with

$$\Gamma_i(m) = \left(\frac{q_\pi(m)}{q_\pi(M_i)}\right)^{(2l+1)m_i^{cut}} \Gamma_i, \tag{2}$$

where $q_\pi(m)$ is the pion c.m. momentum. The $q^{(2l+1)}$ dependence of the width is due to the centrifugal barrier. With increasing mass the barrier influence decreases and we need a cut-off m^{cut}; this is important to obtain the observed resonance fall-off to larger masses. Details of the used forms for $thres_i^{n\pi}(m)$ and m_i^{cut} are given in Ref.[8]. Using this form, a good description of the α-p spectra in Fig. 1 is obtained with the N* resonance parameters M=1390±20 MeV and Γ=190±30 MeV. In addition, the excitation of the Δ resonance in the projectile[8] has to be taken into account.

Assuming one resonance with parameters deduced from α-p, a good description of only part of the π-N data, speed plot and partial cross sections into the the 2π(s)-N channel, is obtained. Other data, e.g total cross sections and inelasticities are not described. A good description of **all** π-N scattering data can be obtained only if a second resonance structure in the region of the Roper resonance is assumed (see Fig. 2). This structure lies at higher excitation and has a large width of the order of 350 MeV. It can be understood as a second order excitation of the Δ resonance[8].

3 Comparison with γ-N

In photoinduced reactions the Δ resonance is strongly excited, therefore it is of interest to investigate the importance of the 2nd resonance (second order excitation of the Δ) in these systems. However, in 2π-photoproduction non-resonant $\Delta\pi$ contributions are generally stronger than the resonance effects in question. Therefore, we looked into $2\pi^o$ production data[9], where non-resonant $\Delta\pi$ contributions are strongly reduced. The results of our calculations are shown in Fig. 3. Indeed, a large part of the observed cross section is described by the second resonance.

It is important to verify, whether the first N* (Saturne resonance) could be observed in 2π-photoproduction. Arbitrarily normalized cross sections are given in Fig. 3 by the dashed lines. The comparison with the data shows clearly that this resonance is not seen. This supports a picture of a pure scalar excitation.

The present results can be tested further in exclusive α-p→ α' N* experiments. As in α-p the lower N* component is excited only, branching ratios for the N* decay are expected to be very different from those observed in π-N→N*. First results from exclusive α-p scattering exist[10] which seem to support the above conclusions.

References

1. R.A. Arndt, I.I. Strakovsky, R.L. Workman, and M.M. Pavan, *Phys. Rev.* C **52**, 2120 (1995), and refs. therein;
   R.A. Arndt, *et al.*, new results from SAID 1999, unpublished.

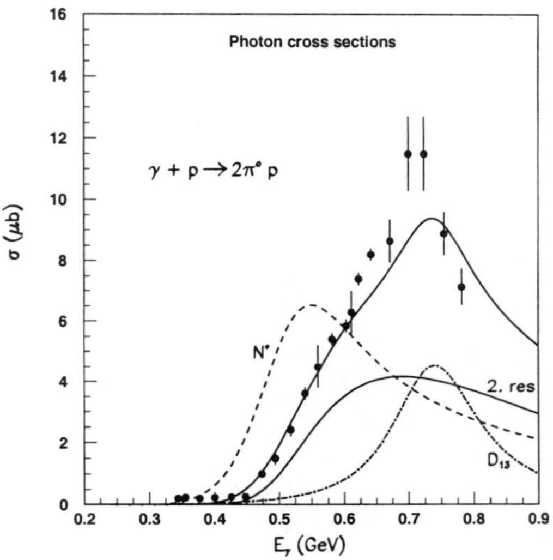

Figure 3. Cross sections for neutral 2π photoproduction in comparison with our calculations, in which the contribution from the $D_{13}(1520)$ is included (dot-dashed line). The shape of the first N* resonance is given by the dashed line.

2. G. Höhler and A. Schulte, πN *Newslett.* **7**, 94 (1992); G. Höhler, πN *Newslett.* **9**, 1 (1993).
3. D.M. Manley, R.A. Arndt, Y. Goradia, and V.L. Teplitz, *Phys. Rev.* D **30**, 904 (1984).
4. H.P. Morsch *et al., Phys. Rev. Lett.* **69**, 1336 (1992).
5. See e.g. N. Isgur and G. Karl, *Phys. Rev.* D **18**, 4187 (1978) S. Capstick and N. Isgur, *Phys. Rev.* D **34**, 2809 (1986).
6. L.Y. Glozman and D.O. Riska, *Phys. Rep.* **268**, 263 (1996).
7. Hajduk and B. Schwesinger, *Phys. Lett.* B **140**, 172 (1984); B. Schwesinger, *Nucl. Phys.* A **537**, 253 (1992).
8. H.P. Morsch and P. Zupranski, *Phys. Rev.* C **61**, 024002 (1999).
9. F. Härter, *et al., Phys. Lett* B **401**, 229 (1997); M. Wolf. *et al., Eur. J.* A **9**, 5 (2000).
10. G.D. Alkhazov, *et al.*, Proceedings of the COSY workshop on “Baryon Excitations”, May 2000, eds. T. Barnes and H.P. Morsch, Forschungszentrum Jülich, p. 53.

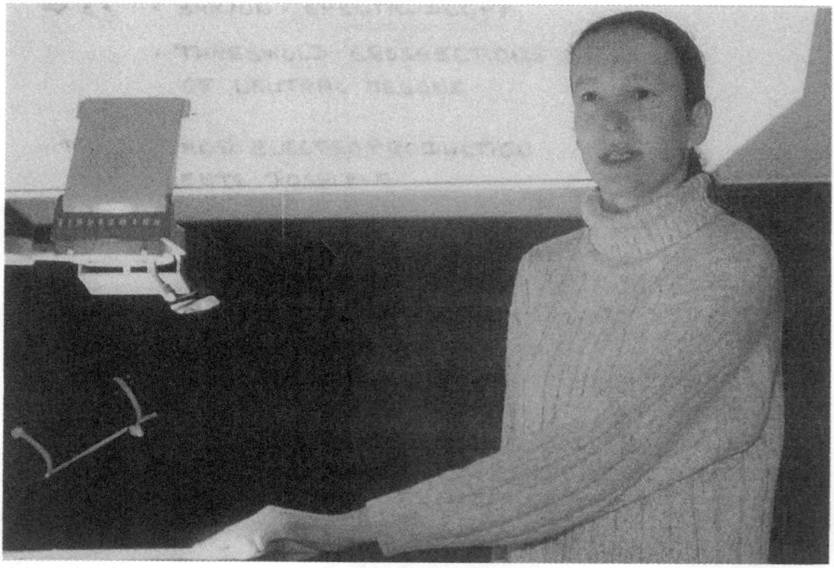

Ulrike Thoma

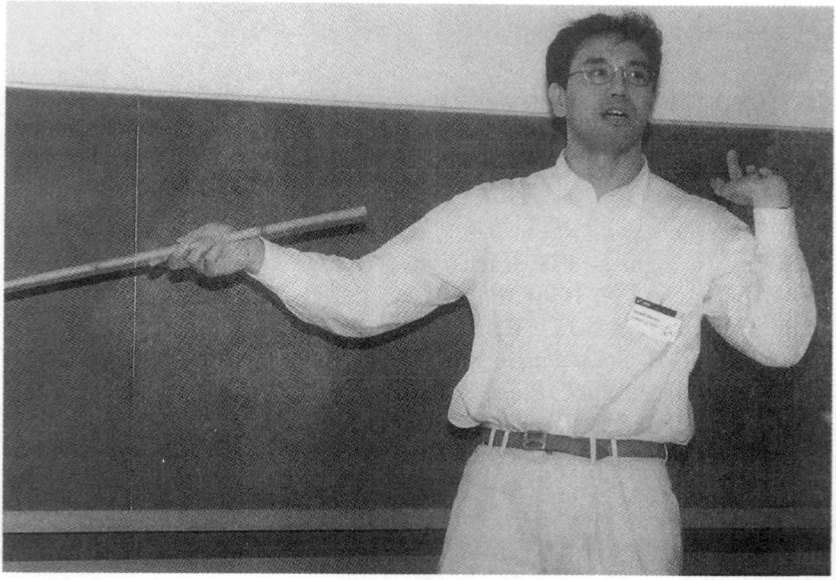

Takashi Nakano

STRUCTURE OF THE NUCLEON INVESTIGATED BY COMPTON SCATTERING

G. GALLER, S. WOLF, M. CAMEN, K. KOSSERT, S. PROFF,
M. SCHUMACHER, F. WISSMANN
Zweites Physikalisches Institut, Universität Göttingen, D-37073 Göttingen, Germany
E-mail: schumacher@physik2.uni-goettingen.de

J. AHRENS, H.-J. ARENDS, R. BECK, F. HÄRTER, P. JENNEWEIN,
J. PEISE, I. PREOBRAJENSKI, M. SCHMITZ
Institut für Kernphysik, Universität Mainz, D-55099 Mainz, Germany

A.M. MASSONE, C. MOLINARI, P. OTTONELLO, A. ROBBIANO,
M. SANZONE
Dipartimento di Fisica dell'Universita di Genova and INFT-Sezione di Genova, I-16146 Genova, Italy

A.I. L'VOV
P.N. Lebedev Physical Institute, Moscow 117924, Russia

V. LISIN, R. KONDRATIEV
Institute for Nuclear Research, Moscow 117312, Russia

P. GRABMAYR, T. HEHL
Physikalisches Institut, Universität Tübingen, D-72076 Tübingen, Germany

Experimental differential cross sections for Compton scattering by the proton measured with the **LAR**ge **A**cceptance arrangement at the tagged-photon facility MAMI (Mainz) are interpreted in terms of the nonsubtracted dispersion theory based on the SAID-SM99K parameterization of photo-meson amplitudes. Using the new global average for the difference of electric and magnetic polarizabilities of the proton $\alpha - \beta = (10.5 \pm 0.9_{\rm stat+syst} \pm 0.7_{\rm model}) \times 10^{-4}{\rm fm}^3$ a backward spin-polarizability of $\gamma_\pi = (-37.1 \pm 0.6_{\rm stat+syst} \pm 3.5_{\rm model}) \times 10^{-4}{\rm fm}^4$ and a E2/M1 ratio of EMR(340 MeV)$=(-1.7 \pm 0.4_{\rm stat+syst} \pm 0.2_{\rm model})$ % have been obtained.

Considerable progress has been made in Compton scattering of real photons due to the first successful operation of a **LAR**ge **A**cceptance experiment at **MAMI** (Mainz)[1], covering the angular range of photon scattering-angles from $\theta_\gamma^{\rm lab} = 30°$ to $150°$ and photon energies from $E_\gamma = 250$ MeV to 800 MeV. The **LARA** arrangement makes use of a photon arm consisting of 150 Pb glass detectors to detect photons from the (γ, γ) and (γ, π^0) reactions and a proton arm registering the complete kinematics of recoil protons by two wire chambers and an array of time-of-flight detectors. Furthermore, the dispersion theory[2,3] of Compton scattering by the nucleon which formerly was restricted to the first resonance, has been extended to cover also the second resonance region[2]. Compton scattering is described by six invariant amplitudes $A_i(\nu,t)$ where $\nu = E_\gamma + t/4m$, $t = (k-k')^2$ with E_γ, k being the laboratory energy and four vector of the incoming photon and k' the four vector of the scattered photon. The real parts of the invariant amplitudes are written in the form[2]

$$\mathrm{Re}A_i(\nu,t) = A_i^{\mathrm{Born}}(\nu,t) + A_i^{\mathrm{int}}(\nu,t) + A_i^{\mathrm{as}}(\nu,t) \tag{1}$$

with

$$A_i^{\mathrm{Born}}(\nu,t) = \frac{a_i(t)}{\nu^2 - t^2/16m^2},$$
$$A_i^{\mathrm{int}}(\nu,t) = \frac{2}{\pi}\mathcal{P}\int_{\nu_{\mathrm{thr}}(t)}^{\nu_{\mathrm{max}}(t)} \mathrm{Im}A_i(\nu',t)\frac{\nu' d\nu'}{\nu'^2-\nu^2}, \tag{2}$$

and $E_{max} = \nu_{max}(t) - t/4m = 1.5\,\mathrm{GeV}$. The quantities $\mathrm{Im}A_i(\nu,t)$ entering into the integral parts are calculated from the pion photoproduction multipoles $E_{l\pm}, M_{l\pm}$ in the parameterizations SAID-SM99K[4] or MAID2K[5]. In the backward direction Compton scattering is mainly given by the spin-dependent amplitude A_2 and spin-independent amplitude A_1 which are dominated by their asymptotic contributions. The asymptotic contribution of A_2 is given by the Low amplitude of π^0 exchange

$$A_2^{\mathrm{as}}(t) \simeq A_2^{\pi^0}(t) = \frac{g_{\pi NN}F_{\pi^0\gamma\gamma}}{t-m_{\pi^0}^2}\tau_3 F_\pi(t), \tag{3}$$

which can be calculated on an absolute scale. The asymptotic contribution of A_1 is constructed in the "σ"-pole approximation

$$A_1^{\mathrm{as}}(t) \simeq A_1^{\sigma}(t) = \frac{g_{\sigma NN}F_{\sigma\gamma\gamma}}{t-m_\sigma^2}, \tag{4}$$

where $g_{\sigma NN}F_{\sigma\gamma\gamma}$ is given by the difference of the electric and magnetic polarizabilities $\alpha - \beta$ through

$$2\pi(\alpha-\beta) + A_1^{\mathrm{int}}(0,0) = -A_1^{\mathrm{as}}(0,0) = \frac{g_{\sigma NN}F_{\sigma\gamma\gamma}}{m_\sigma^2}, \tag{5}$$

with the integral part being a minor contribution. For $(\alpha-\beta)$ the new global average for the proton[6] $(\alpha-\beta) = (10.5 \pm 0.9_{\mathrm{stat+syst}} \pm 0.7_{\mathrm{model}}) \times 10^{-4}\mathrm{fm}^3$ has been adopted. The quantity m_σ has a strong impact on the differential cross section in the second resonance region and at backward angles. Cor-

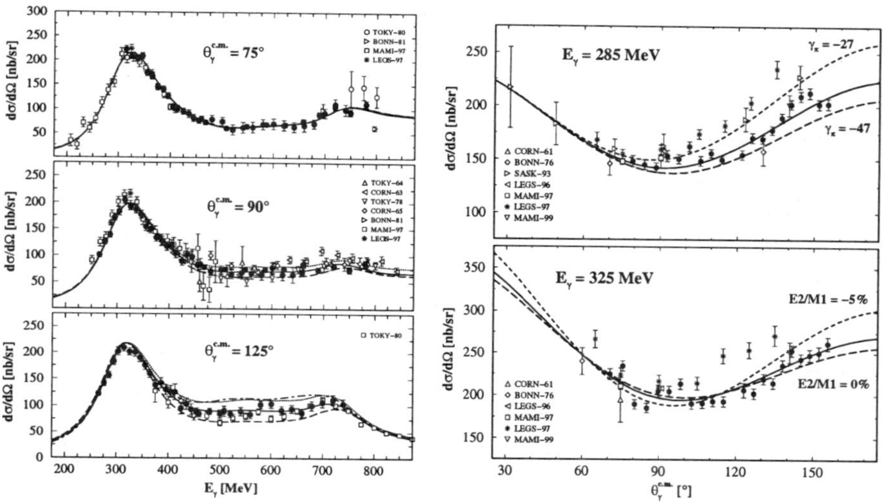

(a) Energy distribution

(b) Angular distribution

Figure 1. Energy and angular distributions of the present experiment (filled circles) compared with previous data and predictions. The solid curves are calculated using the structure parameters SAID+COMPTON of Table 1 and $m_\sigma = 600$ MeV. Subfigures (a): Upper dotted curves $m_\sigma = 800$ MeV, lower dashed curves $m_\sigma = 400$ MeV. The dash-dotted curve in the lower left subfigure is calculated from the structure parameters MAID+COMPTON of Table 1 and $m_\sigma = 600$ MeV. Subfigures (b): Variation of γ_π and E2/M1 at energies of largest sensitivity to these parameters.

respondingly, the data obtained in that range mainly serve as a test of the σ-pole ansatz and for the determination of the quantity m_σ. A further structure parameter which can be determined form the present experiment is the backward spin polarizability

$$\gamma_\pi = -\frac{1}{2\pi m}[A_2^{\text{non-Born}}(0,0) + A_5^{\text{non-Born}}(0,0)] \tag{6}$$

with m being the nucleon mass.

In a first step the validity of the σ-pole parameterization has been tested. This is illustrated in Fig. 1 (a). It is apparent that a good fit to the data is obtained with the SAID-SM99K parameterization if a parameter of $m_\sigma = 600$ MeV is applied. The dash-dotted curve at $\theta_\gamma^{c.m.} = 125°$ which almost coincides with the upper dotted curve has been calculated using $m_\sigma = 600$ MeV but with the MAID2K parameterization. The analysis showed that with this latter parameterization it was not possible to fit the data in the second reso-

Table 1. Electromagnetic structure parameters of the proton as determined from fits to the present Compton scattering data using the SAID-SM99K (lines 1-3) and MAID2K (lines 7-9) parameterizations. The same quantities as contained in the SAID-SM99K (lines 4-6) and MAID2K (lines 10-12) parameterizations.

Method	Structure parameter		
SAID+ COMPTON	$\lvert M_{1+}^{(3/2)}(320\,\text{MeV})\rvert$	=	$(39.7 \pm 0.3_{\text{stat+syst}} \pm 0.03_{\text{model}})^{a)}$
	EMR(340 MeV)	=	$(-1.7 \pm 0.4_{\text{stat+syst}} \pm 0.2_{\text{model}})\%$
	γ_π	=	$(-37.1 \pm 0.6_{\text{stat+syst}} \pm 3.5_{\text{model}})^{b)}$
SAID	$\lvert M_{1+}^{(3/2)}(320\text{MeV})\rvert$	=	$39.74^{a)}$
	EMR(340 MeV)	=	-1.68%
	γ_π	=	$-38.20^{b)}$
MAID+ COMPTON	$\lvert M_{1+}^{(3/2)}(320\,\text{MeV})\rvert$	=	$(39.8 \pm 0.3_{\text{stat+syst}} \pm 0.03_{\text{model}})^{a)}$
	EMR(340 MeV)	=	$(-2.0 \pm 0.4_{\text{stat+syst}} \pm 0.2_{\text{model}})\%$
	γ_π	=	$(-40.9 \pm 0.4_{\text{stat+syst}} \pm 2.5_{\text{model}})^{b)}$
MAID	$\lvert M_{1+}^{(3/2)}(320\,\text{MeV})\rvert$	=	$39.92^{a)}$
	EMR(340 MeV)	=	-2.19%
	γ_π	=	$-38.18^{b)}$

a) $\times 10^{-3}/m_{\pi^+}$ b) $\times 10^{-4}\text{fm}^4$

nance region without destroying the agreement in the first resonance region. Therefore, it appeared reasonable to base the determination of γ_π and of the E2/M1 ratio on the SAID-SM99K parameterization as illustrated in Fig.1 (b), leading to the results labeled SAID+COMPTON in Table 1. Our result for γ_π is in disagreement with the one of the LEGS group[7] which gave a smaller value of $\gamma_\pi^{\text{LEGS}} = (-27.1 \pm 2.2_{\text{stat+syst}}{}^{+2.8}_{-2.4}{}_{\text{model}}) \times 10^{-4}\text{fm}^4$. The difference can be traced back to a difference in the measured differential cross sections.

References

1. G. Galler *et al., Phys. Lett.* B (in press); arXiv:nucl-ex/0102003.
2. A.I. L'vov, V.A. Petrun'kin, and M. Schumacher, *Phys. Rev.* C **55**, 359 (1997).
3. D. Drechsel, M. Gorchtein, B. Pasquini, and M. Vanderhaeghen, *Phys. Rev.* C **61**, 015204 (2000).
4. R.A. Arndt, I.I. Strokovsky, and R.L. Workman, *Phys. Rev.* C **53**, 430 (1996).
5. D. Drechsel *et al., Nucl. Phys.* A **645**, 145 (1999).
6. V. Olmos de León *et al. Eur. Phys. J.* (in press).
7. J. Tonnison *et al., Phys. Rev. Lett.* **80**, 4382 (1998).

VIRTUAL COMPTON SCATTERING AT JEFFERSON LAB: PRELIMINARY RESULTS IN THE POLARIZABILITY DOMAIN AT Q^2=1 AND 1.9 GEV^2.

S. JAMINION
FOR THE JEFFERSON LAB HALL A COLLABORATION
Laboratoire de Physique Corpusculaire, Université BLAISE PASCAL/CNRS-IN2P3, F63177 Aubière CEDEX, France
E-mail : jaminion@clermont.in2p3.fr

Virtual Compton Scattering (VCS) off the proton below pion threshold allows the measurement of the Generalized Polarizabilities (GPs) of the proton. These new observables put stringent constraints on models of nucleon structure. The E93050 experiment performed at Jefferson Lab, investigated the VCS process at Q^2=1 and 1.9 GeV^2. Preliminary values of $(ep \to ep\gamma)$ cross sections are presented. The extraction of structure functions (containing the GPs) from these data is presently under study.

1 Virtual Compton Scattering

Virtual Compton Scattering (VCS) off the proton below pion threshold allows the measurement of the Generalized Polarizabilities (GPs) of the proton. VCS is accessible via photon electro-production, which is the coherent sum of two processes (see Fig. 1):

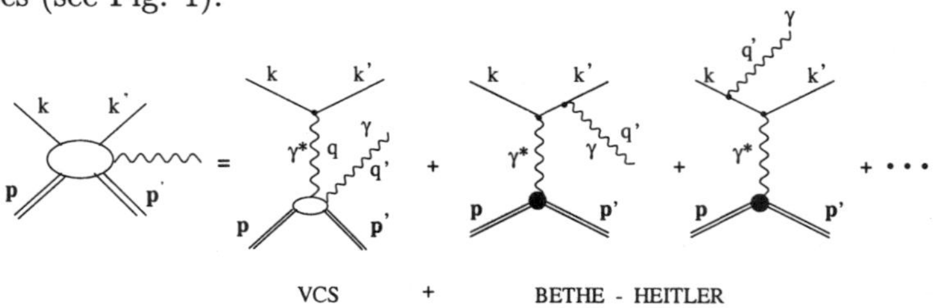

Figure 1. Exclusive photon electroproduction on the proton.

- Bethe-Heitler: the photon is emitted by either the incoming or the outgoing electron.
- Virtual Compton scattering: the photon is emitted by the nucleon (Born term) or by an excited nucleon. This latter amplitude is called Non-Born and contains the unknown physics. It is parametrized by the GPs, which are related to the deformability of the proton in an applied EM field.

The Bethe-Heitler process and the Born terms are entirely calculable within Quantum Electrodynamics, using the EM form factors of the proton. The fivefold ($ep \to ep\gamma$) cross section can be written as [1]:

$$d^5\sigma_{exp} = d^5\sigma_{BH+B} + \left[\mathcal{M}_0 - \mathcal{M}_0^{BH+B}\right] * \left[\Phi q'_{cm}\right] + \mathcal{O}(q'_{cm}) \; , \qquad (1)$$

where $d^5\sigma_{BH+B}$ is the cross section for the Bethe-Heitler process and the Born terms, $[\Phi q'_{cm}]$ is a phase space factor, and $[\mathcal{M}_0 - \mathcal{M}_0^{BH+B}]$ is the term containing Generalized Polarizabilities at lowest order in q'_{cm}, the outgoing photon momentum in the (γp) center of mass. From this term, two structure functions can be extracted: $P_{LL} - \frac{1}{\epsilon}P_{TT}$ and P_{LT} (in an unpolarized experiment at fixed Q^2 and ϵ) which are combinations of 5 GPs.

The first VCS measurement was performed at Mainz [2] at $Q^2 = 0.33$ GeV^2. The second VCS experiment E93050 [3] was performed at Jefferson Lab at $Q^2 =$ 1 and 1.9 GeV^2, including an investigation in the resonance region [4].

2 Virtual Compton Scattering at Jefferson Laboratory.

We used the JLab 4 GeV continuous electron beam and the two High Resolution Spectrometers of Hall A, which allow the detection of the scattered electron in coincidence with the recoiling proton. The outgoing photon is the "missing particle".

To determine the GPs, we want a large phase space in the (γp) center of mass. We have defined 17 (for Q^2=1 GeV^2) and 14 (for Q^2=1.9 GeV^2) different hadron arm settings to cover the entire phase space. Our first goal is the precise measurement of absolute 5-fold differential cross sections $d^5\sigma(ep \to ep\gamma)$ over a wide range of photon angles and momenta.

Cuts were made to select photon electroproduction events and allow subtraction of background from elastic scattering.

The missing mass squared spectrum of Fig. 2 shows the γ and π^0 peaks after applying cuts. The data analysis relies also on an accurate Monte Carlo simulation of the solid angle, including resolution effects and a realistic cross section (Bethe-Heitler+Born).

3 Preliminary Results for Cross Sections

Figure 3 shows a selection of our experimental cross sections $d^5\sigma(ep \to ep\gamma)/d\Omega_{e'lab}dE_{e'lab}d\Omega_{\gamma\gamma^* cm}$ at Q^2=1 [5] and 1.9 GeV^2 [6]. They are determined in the leptonic plane (ϕ=0 and 180 °) at given values of q'_{cm} and $\theta_{\gamma\gamma^*\ cm}$, the polar angle between the real and the virtual photon. The solid curve represents the Bethe-Heitler+Born calculation, i.e. our reference

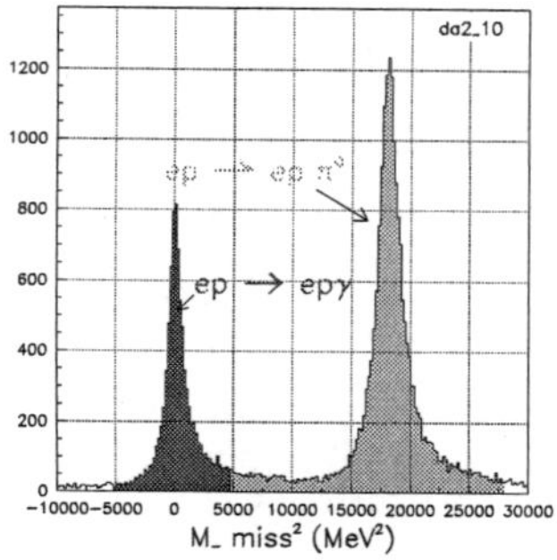

Figure 2. Missing mass squared spectrum for coincidence events. The window for the VCS analysis is [±5000] MeV^2.

cross section for the process with no polarizability effect (rigid proton). This calculation is done with the following choice of proton EM form factors: Bosted's parametrization of G_M^P [7] and the JLab measurement of $\frac{\mu G_E^P}{G_M^P}$ [8].

At Q^2=1 GeV^2, the data of Fig. 3 (left) below pion threshold show a clear trend, similar to what is expected from a polarizability effect, i.e. a departure from Bethe-Heitler+Born (solid curve) increasing with q'_{cm} and enhanced at forward $\theta_{\gamma\gamma^* \ cm}$. An iterative procedure involving the Monte Carlo is presently under investigation in order to extract the structure functions $P_{LL} - \frac{1}{\epsilon}P_{TT}$ and P_{LT} from these cross section data.

At Q^2=1.9 GeV^2, only the data in the region of backward angles are shown on Fig. 3 (right), below and above π^0 threshold. When the outgoing photon angle is restricted to the vicinity of the leptonic plane, the cross section increases. This is consistent with expected enhancement due to polarizabilities. The dashed curve is the effect predicted by a Dispersion Relation calculation[9].

To conclude, the VCS data at Q^2=1 and 1.9 GeV^2 are promising and should allow the extraction of information about the proton Generalized Polarizabilities in a more complete stage of the analysis.

We thank the Hall A technical staff and the JLab Accelerator division for their support of this experiment and B.Pasquini and M.Vanderhaeghen for fruitful discussions. This work was supported by DOE contract DE-AC05-84ER40150 under which the Southeastern Universities Research Association (SURA) operate the Thomas Jefferson National Accelerator Facility.

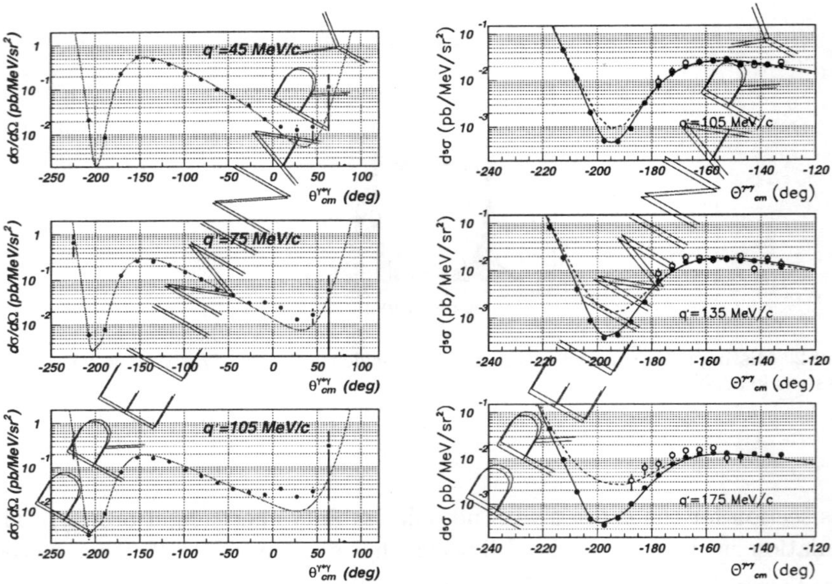

Figure 3. The cross section for the $(ep \to ep\gamma)$ reaction from JLab data (see text) at Q^2=1 GeV^2 (left) and Q^2=1.9 GeV^2 (right). The abscissa is the angle between the real and the virtual photon measured in the leptonic plane: [-180,0]° for ϕ=180° and the complementary range for ϕ=0°. Black dots: data integrated over all out-of-plane angles, open circles (right only): data restricted to ± 9° around the leptonic plane.

References

1. P.A.M Guichon and M. Vanderhaeghen, Virtual Compton Scattering off the Nucleon, *Prog. Part. Nucl. Phys.* **41**, 125 (1998).
2. J.Roche *et al., Phys. Rev. Lett.* **85**, 708 (2000).
3. P.Y.Bertin, P.A.M. Guichon and C.Hyde-Wright, *CEBAF proposal PR93-050* (1993).
4. G. Laveissiere, *these proceedings.*
5. Ph.D. thesis of N. Degrande, Gent University, Belgium (2001).
6. Ph.D. thesis of S. Jaminion, Clermont-Ferrand, Univ. Blaise Pascal (2000).
7. P.E. Bosted, *Phys. Rev.* C **51**, 409 (1995).
8. M.K. Jones *et al., Phys. Rev. Lett.* **84**, 1398 (2000).
9. B. Pasquini, D. Drechsel, M. Gorchtein, A. Metz, and M. Vanderhaeghen, hep-ph/0102335 (Feb 2001).

MEASUREMENT OF THE CROSS SECTION ASYMMETRY Σ FOR $\gamma p \to \pi^\circ p$ OVER THE RANGE $E_\gamma = 0.5 - 1.1$ GEV

F. V. ADAMIAN, A. YU. BUNIATIAN, G. S. FRANGULIAN, P. I. GALUMIAN, V. H. GRABSKI, A. V. HAIRAPETIAN, H. H. HAKOBIAN, V. K. HOKTANIAN, J. V. MANUKIAN, A. M. SIRUNIAN, A. H. VARTAPETIAN, H. H. VARTAPETIAN, AND V. G. VOLCHINSKY

Yerevan Physics Institute, 2 Alikhanian Brothers St., Yerevan, 375036 Armenia

R. A. ARNDT, I. I. STRAKOVSKY, AND R. L. WORKMAN

Center for Nuclear Studies, Department of Physics
The George Washington University, Washington, DC 20052, USA

The cross section asymmetry Σ has been measured for the photoproduction of π°-mesons off protons, using the polarized photon beam of the Yerevan 4.5 GeV synchrotron, over the second and third resonance regions. We compare with existing experimental data and recent phenomenological analyses. The influence of these measurements on such analyses is also considered.

1 Introduction

Single-pion photoproduction has been used extensively to explore the electromagnetic properties of nucleon resonances, and most determinations of the $N\gamma$ resonance couplings[1] have been obtained through multipole analyses of this reaction. A survey of existing data in the 1 GeV region shows most polarization measurements to have only one or two angular points at a given beam energy, often with rather large uncertainties. This allows a great deal of freedom in multipole analyses, and is clearly insufficient to pick out more than the strongest resonance signals.

In this experiment, we have performed systematic measurements of the cross section asymmetry Σ for the reaction $\gamma p \to \pi^\circ p$ over the energy range $E_\gamma = 0.5 - 1.1$ GeV and at pion CM angles $\theta_\pi^* = 85^\circ - 125^\circ$. The experimental data (158 data points) from π°-meson photoproduction constitute the first systematic and high statistics measurement of Σ for this reaction channel over the present kinematic region. Final results of the measurements will published soon in Ref.[2].

2 Experiment and Data Analysis

Recoil protons were detected by a magnetic spectrometer in coincidence with one photon of π° decay registered in the Čerenkov spectrometer. The incident photon energy and the CM scattering angle were reconstructed on average with an accuracy of $\sigma_{E_\gamma}/E_\gamma \approx 1\%$ and $\sigma_{\theta^*} \approx 0.6°$, with corresponding acceptances of $\approx 18\%$ and $\approx 7.5°$, respectively.

The background $p\gamma$ rate was dominated by double pion production processes generated by the high energy tail from coherent bremsstrahlung spectrum and did not exceed 5%. The systematic errors of the measurements were determined mainly by the accuracy of the linear polarization calculations, which stayed within $\approx 2.3\%$ over the full energy range. Overall systematic uncertainties for each energy bin did not exceed 3% and were included in the error bars of the experimental data. Further details are available in Ref.[2].

3 Results and Discussion

Our results are compared to existing data[3,4,5,6,7] and the predictions of phenomenological analyses[8,9] in Fig. 1. At $\theta^*_\pi = 90°$, where the most systematic measurements exist, one sees good consistency between prior measurements and our present results. Near 900 MeV, our data disagree with data from MIT[4] and some early SLAC[3] measurements, but are in good agreement with later measurements from another SLAC group[5]. Our values are also consistent with the results of this group for $\theta^*_\pi = 110°$. Results for the angular distributions at $E_\gamma = 700$, 750, and 800 MeV are in agreement with the existing data as well.

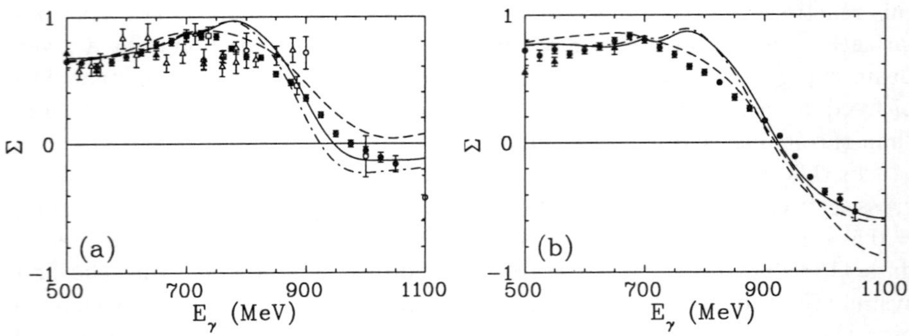

Figure 1. Energy dependence of π° photoproduction asymmetry Σ at $\theta^* = 90°$ (a) and 120° (b), respectively. Experimental data are from Yerevan, present experiment (filled circles), SLAC[3] (open triangles), MIT[4] (open circles), SLAC[5] (open squares), Kharkov[6] (filled triangles), and Yerevan, previous measurements[7] (filled squares). Solid (dash-dotted) curves give WI00 (FA00) results by GW[9] versus MAID2000 results by the Mainz group[8] (dashed curves).

While both the MAID[8] and SAID[9] analyses reproduce the qualitative behavior of Σ, a shape discrepancy is noticeable in the GW fit. The main difference between multipole analyses and experiment is observed above $E_\gamma =$ 700 MeV where there have been few previous measurements. Inclusion of our data in the GW fit results in an improved description, but shape differences remain. The Mainz fit appears to have a shape more consistent with the data. (It should be noted that some of the older fits, in particular the fit of Ref.[10], predict a shape consistent with the Mainz result.)

It is interesting to compare these results to those recently published by the GRAAL collaboration[11]. In that work, a similar comparison was made for Σ measurements in the reaction $\gamma p \to \pi^+ n$. There a deviation from GW predictions was noted in the 800 − 1000 MeV range, at backward angles. It was suggested that this discrepancy could be removed by a change in the N(1650) photo-decay amplitude ($A_{1/2}$). We have examined the GRAAL angular distribution at 950 MeV (W_{CM} = 1630 MeV) and agree that a change (reduction) in the $E_{0+}^{1/2}$ multipole can account for the shape difference. Here too, for neutral pion production and energies near 900 MeV, a reduction of this multipole in FA00 by about 20% (in modulus) results in an improved description. A reduction of the same amount is found in comparing the energy-dependent and single-energy solutions. The effect is displayed in Fig. 2. Note that this modification to the GW fit, while improving the agreement with data, actually worsens the agreement with the Mainz prediction at the most backward angles. It will be important to see how the Mainz fit adjusts to fit these new

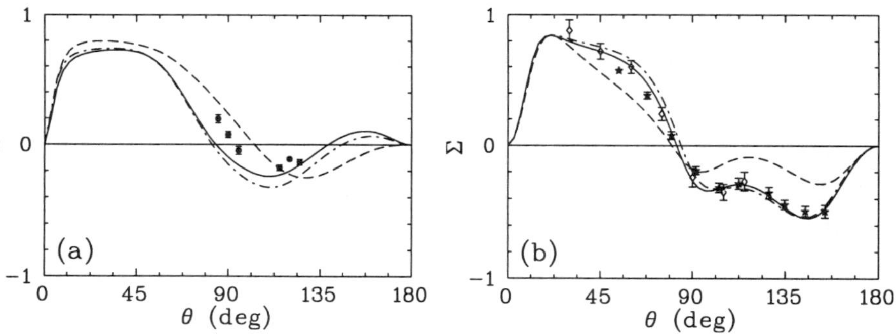

Figure 2. Angular dependence of π° (a) and π^+ (b) photoproduction asymmetry Σ at $E_\gamma = 950\,MeV$. π° data are present measurements. π^+ data are from GRAAL[11] (filled asterisk) and DNPL[12] (open diamond). Plotted are fits FA00 (dash-dotted), FA00 with a modified $E_{0+}^{1/2}$ multipole (see text) (solid), and the MAID2000 result[8] (dashed).

data. Fits employing identical data sets would also be useful in determining whether differences are due to the chosen formalism or data constraints.

Acknowledgments

The YERPHI group is indebted to the synchrotron staff and cryogenic service for reliable operation during the experiment. This work was supported in part by the Armenian Ministry of Science (Grant-933) and the U. S. Department of Energy Grant DE–FG02–99ER41110. The GW group gratefully acknowledges a contract from Jefferson Lab under which this work was done. Jefferson Lab is operated by the Southeastern Universities Research Association under the U. S. Department of Energy Contract DE–AC05–84ER40150.

References

1. D. E. Groom *et al.*, *Eur. Phys. J.* C **15**, 1 (2000).
2. F. V. Adamian *et al.*, to be published in *Phys. Rev.* C (2001).
3. R. Zdarko and E. Dally, *Nuovo Cim.* A **10**, 10 (1972).
4. J. Alspector *et al.*, *Phys. Rev. Lett.* **28**, 1403 (1972).
5. G. Knies *et al.*, *Phys. Rev.* D **10**, 2778 (1974).
6. V. Ganenko *et al.*, *Phys. At. Nucl.* (former *Sov. J. Nucl. Phys.)* **23**, 162 (1976); A. A. Belyaev *et al.*, *Nucl. Phys.* B **213**, 201 (1983) and references therein.
7. R. O. Avakyan *et al.*, *Phys. At. Nucl.* (former *Sov. J. Nucl. Phys.)* **40**, 588 (1984) and references therein.
8. D. Drechsel, O. Hanstein, S. S. Kamalov, and L. Tiator, *Nucl. Phys.* A **645**, 145 (1999); L. Tiator *et al.*, in preparation.
9. R. A. Arndt *et al.*, *Phys. Rev.* C **53**, 430 (1996); R. A. Arndt *et al.*, in preparation.
10. I. M. Barbour *et al.*, *Nucl. Phys.* B **141**, 253 (1978).
11. J. Ajaka *et al.*, *Phys. Lett.* B **475**, 372 (2000).
12. P. J. Bussey *et al.*, *Nucl. Phys.* B **154**, 205 (1979).

POSITIVE PION PHOTOPRODUCTION AND COMPTON SCATTERING AT GRAAL

V. KOUZNETSOV FOR THE GRAAL COLLABORATION,
University "Tor Vergata", Physics Department, Via della Ricerca Scientifica 1, I-00133, Rome, Italy,
and
Institute for Nuclear Research, 117312, Moscow, Russia.
E-mails: Slava@roma2.infn.it, Slava@cpc.inr.ac.ru

The GRAAL facility was designed to measure polarization observables, in particular the Σ beam asymmetry in photon-induced reactions. The polarized and tagged photon beam is produced by backscattering of the laser light on the 6.04 GeV electrons circulating in the storage ring of the European Synchrotron Radiation Facility (Grenoble, France). With the green 514 nm laser light, the tagged spectrum covers the energy range of 0.55-1.1 GeV, for the UV line this range is 0.8-1.5 GeV. The linear beam polarization varies from ~0.45 at the lower energy limits to 0.98 at the upper ones.

The GRAAL detection system provides the detection and the identification of all types of final state particles in an almost 4π solid angle, and has a cylindrical symmetry which makes it suitable for measurements of the Σ beam asymmetry. The detailed description of the GRAAL facility and the procedure to derive the Σ beam asymmetry from experimental data is in [1].

Experimental data have been collected at GRAAL for ηp, $\pi^+ n$, $\pi^0 p$, $K^+\Lambda$, ωp, and γp channels. In this communication, the current state-of-art for $\pi^+ n$ photoproduction and Compton scattering is reported.

For $\pi^+ n$ photoproduction, results were first obtained with the green laser. This data set includes 92 points measured at the energies from 0.6 to 1.05 GeV [1]. In Fig.1, these results are shown together with the most accurate previous results [4,5,6]. Our results are in good agreement with the old experiments while have essentially smaller error bars. The data points were produced in the almost unmeasured region of backward angles up to 160° where only few previous points of low accuracy were available. The beam asymmetries are compared with two solutions of the GWU-SAID group: the published SM95, and the most recent WI00 [2]. The latter was developed using the fit to the data base with our data included in. At energies below 0.75 GeV and at 1.05 GeV, the SM95 solution is confirmed, while between 0.8 and 1 GeV it is rather low at backward angles. The WI00 solution is a significant step toward our results. However, it still differs in the region of 0.85-1 GeV at backward angles and seems to be slightly worse at forward angles at the same energies.

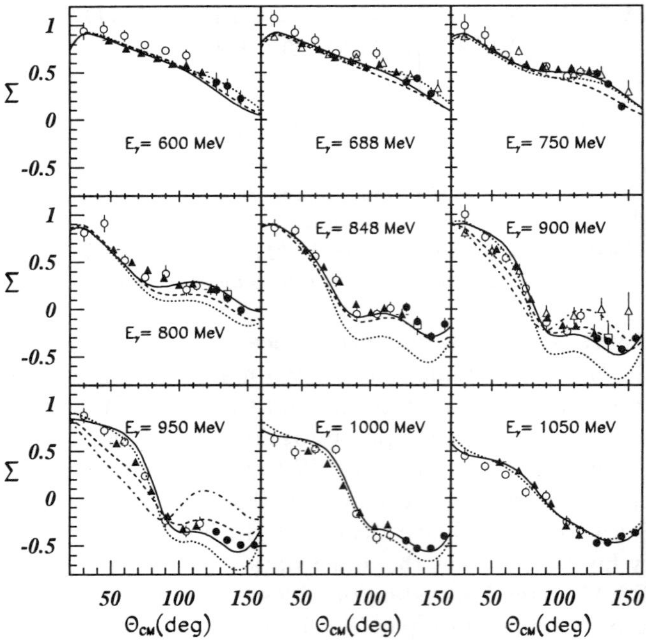

Figure 1. Σ-asymmetry for positive pion photoproduction at different incident γ energies (green laser data). Black triangles and circles are our results; open circles are the results of the Daresbury group [4], open triangles and squares are from SLAC [5,6]. The solid and dotted lines are the WI00 and SM95 solutions of the SAID partial-wave analysis respectively [2], and the dashed line is the prediction of MAID2000 [3].

The results are also compared with predictions of a unitary isobar model of Drechel et al. [3], often known as MAID2000. This model seems to reproduce reasonably our data up to 950 MeV, while above this region it differs from them.

Recently, new Σ results have been produced using data collected with the UV laser. New data set includes 237 Σ beam asymmetry points covering the wide energy and angular ranges of 0.8 - 1.5 GeV and 40 – 160° [8]. In Fig.2, the samples of points from the new and preceding data sets are shown at overlapping energies. The new data have been measured using the UV laser which produces the different beam spectrum and the different dependence of the beam polarization on the beam energy. Essential modifications in the apparatus and the analysis procedure have been implemented. Nevertheless, the reproducibility of our results is excellent and proves the quality of the

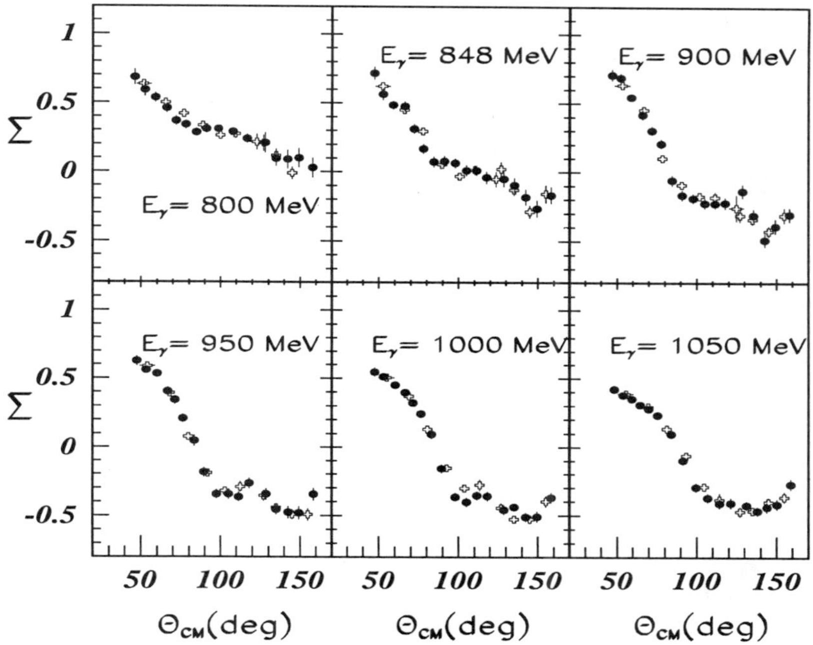

Figure 2. Comparison of new and published data sets. Black circles are the new points obtained with the UV laser, crosses are the results of the previous measurement with the green laser [1].

data.

The main difficulty of the experimental study of Compton scattering is the separation of this low-yield channel from the kinematically close background of π^0 photoproduction. The almost 4π acceptance of the GRAAL detector for photons makes it possible to reject efficiently π^0 events detecting both photons from $\pi^0 \to 2\gamma$ decays [7]. Fig.3 shows the spectrum of missing energy calculated as a difference between measured and expected energy of the scattered photon. The clear peak near 0, corresponding to Compton events, stands on the top of the flat background originating from the π^0 photoproduction. First Compton results, expected soon, will be beam asymmetry data at backward angles $\Theta_{cm} = 150°$ in the energy range from 0.6 to 1.5 GeV. Furthermore the experiments will be extended to the wide angular range of $60 - 150°$. The possibility to measure Compton scattering on the neutron is also offered by the powerful GRAAL detection system.

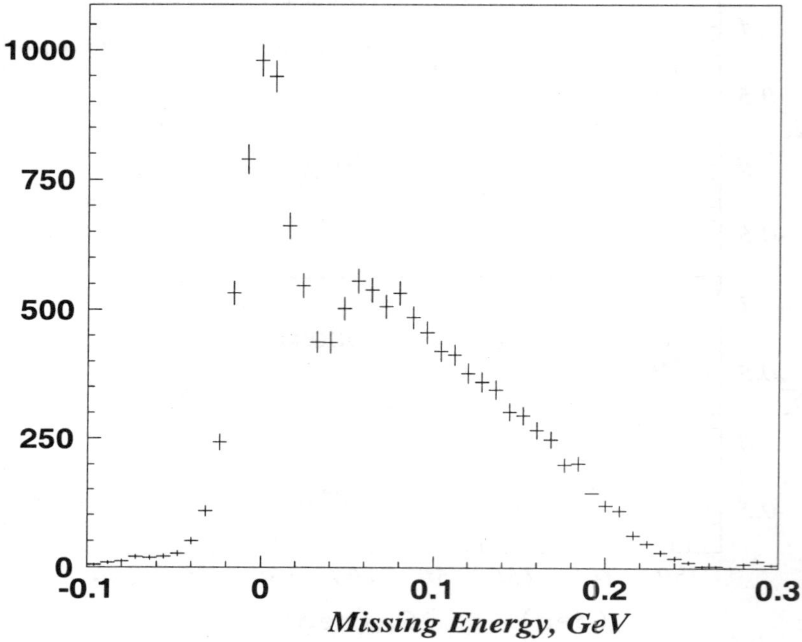

Figure 3. Spectrum of missing energy at $E_{beam} = 0.8$ GeV and $\Theta_{cm} = 150°$. The peak at 0 corresponds to Compton scattering.

References

1. J.Ajaka *et al., Phys. Lett.* B **475**, 372-377 (2000).
2. R.Arndt, I.Strakovsky, and R.Workman, *Phys. Rev.* C **53** 430 (1996); the WI00, and SM95 solutions are available via telnet said.phys.vt.edu, user said.
3. D.Drechsel, O.Hanstein, S.Kamalov, and L.Tiator, *Nucl. Phys.* A **645**, 145 (1999), and S.Kamalov, private communication. Predictions are available at URL http://www.kph.uni-mainz.de/MAID/.
4. P.J.Bussey *et al., Nucl. Phys.* B **154**, 205 (1979).
5. G.Knies *et al., Phys. Rev.* D **10**, 2778 (1974).
6. R.Zdarko and E.Dolly, *Nuovo Cimento* A **10**, 10 (1972).
7. V.Kouznetsov, Proc. of Workshop on Virtual Compton Scattering, Clermont-Ferrand, France, 25-29 June 1996, 208-218.
8. Publication in preparation.

VIRTUAL COMPTON SCATTERING AND PION ELECTRO-PRODUCTION IN THE NUCLEON RESONANCE REGION

G. LAVEISSIERE
FOR THE JEFFERSON LAB HALL A COLLABORATION
Laboratoire de Physique Corpusculaire IN2P3-CNRS, Université Blaise Pascal, 24 Av. des Landais, 63177 Aubière, France
E-mail: geraud@clermont.in2p3.fr

Preliminary results of a p(e,e'p)X experiment in the nucleon resonance region are presented: the π^0 and γ electro-production on hydrogen at backward angle (120 degree to 180 degree), 4-momentum transfer of 1.0 GeV^2, and center of mass energy W ranging from 0.95 GeV to 2.0 GeV.
For the π^0 electro-production, the transverse and longitudinal components of the cross section are extracted. Comparisons with the MAID2000 prediction (for π^0) and with results from the Dispersion Relation formalism (for VCS) are discussed.

1 Introduction

Study of the nucleon resonance region through poorly investigated channels as $p(e,e'p)\gamma$ or $p(e,e'p)\pi^0$ at intermediate Q^2 (1.0 GeV^2) improve our understanding of the nucleon resonances and provide accurate constraints on the theoretical models and partial-wave analyses for the nucleon multipoles.

2 Experimental Considerations

The E93050 experiment [1] was performed in the Hall A of Jefferson Lab in 1998 and intended to measure the p(e,e'p)X reaction. A 100% duty cycle beam of electrons is scattered off a liquid hydrogen target, and the twin high resolution spectrometers of Hall A are used to detect the outgoing electron and proton. The third particle (π^0 or γ) is identified using a missing mass separation technique.

The resonance region was covered at Q^2=1.0 GeV^2 and the total center of mass energy W ranges from W=0.95 GeV to 2.0 GeV. As the beam energy had to remain constant throughout the entire period, the photon polarisation rate ϵ ranges from 0.95 to 0.75 along with W. The missing particle was detected at backward angles in the $\gamma * p$ center of mass, typically from 120 to 180°. The spectrometers acceptance allows to detect the full azimuthal distribution of the missing particle.

The Hall A spectrometers, which are equipped with a standard set of detectors (wire chambers, scintillators and Cerenkov counters), are used to detect the final state of the reaction. The vertex is reconstructed using an optical database, and the final momentum resolution is 10^{-4}.

3 Pion Electroproduction

A Monte-Carlo simulation has been used to calculate the acceptance. The cross section used in the simulation stems from the MAID2000 program [2]. All radiative corrections [3] and experimental resolution effects have been taken into account.

First, the 5-fold differential cross section is evaluated in each bin in the 5-dimensional phase space [W,Q^2,ϵ,cos$\theta^*_{\gamma\gamma}$,ϕ]. The result is then divided by the conventional flux factor (Hand convention) to extract the 2-fold differential cross section. The angular azimuthal distribution is then considered and a second order polynomial fit is applied to the data. These parameters allow to extract the transverse and longitudinal terms of the cross section. The resulting $\sigma_T + \epsilon.\sigma_L$, σ_{TL} and σ_{TT} response functions are shown in Fig. 1.

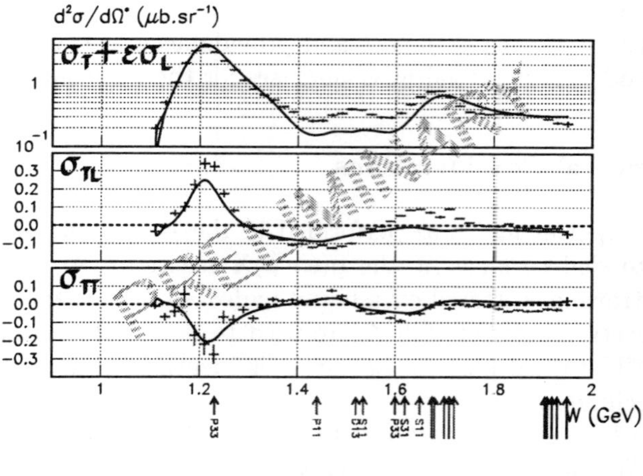

Figure 1. Extraction of the transverse and longitudinal terms of the π^0 electro-production cross section.

The relative Q^2 derivative of the $\sigma_T + \epsilon.\sigma_L$ term was also extracted and the result is shown in Fig. 2a.

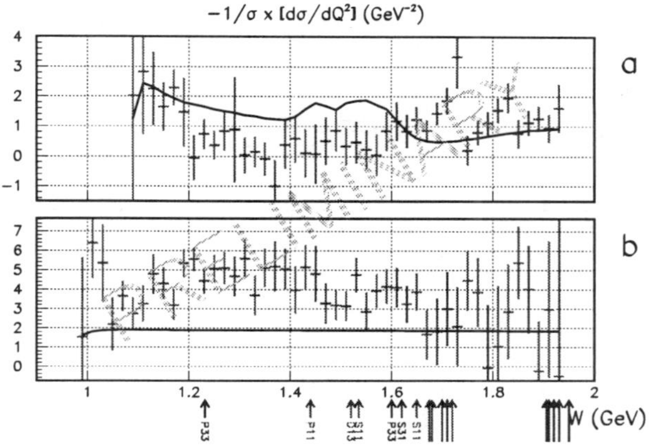

Figure 2. Evolution in Q^2 of the π^0 electro-production (a) and VCS cross section (b). The solid curve represents the theoretical prediction, MAID for the π^0 channel and Bethe-Heitler+Born for the photon channel.

4 Virtual Compton Scattering (VCS)

Detailed procedures of extraction of the Virtual Compton Scattering cross sections have been discussed in S. Jaminion's presentation [4] at this workshop.

In the resonance region, the VCS cross section has been extracted at two points in the azimuthal angle ϕ: 0 and 180° for $\theta^*_{\gamma\gamma} = 167°$ (see Fig. 3), using data in a $\pm 90^o$ interval in ϕ around the considered point. The Bethe-Heitler + Born theoretical cross section has been introduced in the Monte-Carlo simulation to compute the acceptance. This process introduces some uncertainties in the cross section extraction. Nevertheless, these uncertainties should stay reasonably small because the shape of the angular distributions is close to the Bethe-Heitler+Born case. Finally, a comparison with the new dispersion relation formalism [5] is shown for $\phi = 180^o$.

A linear fit has also been performed to the data in order to extract the relative Q^2 evolution of the cross section at Q^2=1.0 GeV^2 (see Fig. 2b).

5 Conclusion

This experiment has allowed the first extraction of the excitation curve for the $p(e,e'p)\gamma$ process at backward center-of-mass angle. The π^0 electro-

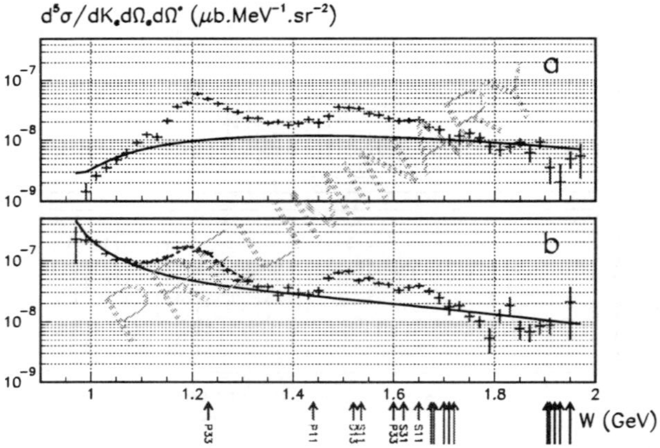

Figure 3. Virtual Compton Scattering W excitation curve at Q^2=1.0 GeV^2, $cos\theta^*_{\gamma\gamma} = -0.975$ and $\phi = 0^o$ (a), $\phi = 180^o$ (b). The solid curve represents the Bethe-Heitler + Born contribution. The dashed curve shows the prediction of the dispersion relation formalism for a particular set of parameters.

production excitation curve shows some important discrepancies with respect to the MAID 2000 predictions. In both cases, no evidence of any missing resonance has been found.

We wish to thank the Jefferson Lab accelerator staff for all the precious help they provided, as well as B. Pasquini, M. Vanderhaeghen and L. Tiator for their numerous advice. This work was supported by DOE contract DE-AC05-84ER40150 under which the Southeastern Universities Research Association (SURA) operates the Thomas Jefferson National Accelerator Facility.

References

1. P.Y. Bertin, C. Hyde-Wright, and P.A.M. Guichon, CEBAF proposal **PR93050** (1993).
2. D. Drechsel, O. Hanstein, S.S. Kamalov and L. Tiator, *Nucl. Phys.* A **645**, 145 (1999), http://www.kph.uni-mainz.de/MAID.
3. M. Vanderhaeghen, J.M. Friedrich, D. Lhuillier, D. Marchand, L. Van Hoorebeke, and J. Van de Wiele, *Phys. Rev.* C **62**, 025501 (2000).
4. S. Jaminion, *these proceedings.*
5. B. Pasquini, D. Drechsel, M. Gorchtein, A. Metz and M. Vanderhaeghen, *Phys. Rev.* C **62**, 052201(R).

SINGLE SPIN BEAM ASYMMETRY MEASUREMENTS FROM SINGLE π^{o} ELECTROPRODUCTION IN THE $\Delta(1232)$ RESONANCE REGION

K. JOO

Thomas Jefferson National Accelerator Facility, 12000 Jefferson Avenue, Newport News, VA 23606, USA
E-mail: kjoo@jlab.org

New measurements of the electroproduction of the $\Delta(1232)$ resonance through $p(\vec{e},e^{'}p)\pi^{o}$ reaction have been performed. The data were taken with the CEBAF Large Acceptance Spectrometer (CLAS) at Jefferson Lab. Single spin beam asymmetries were measured over a range of four-momentum transfer Q^2=(0.3 - 0.9 GeV^2). The high statistical accuracy of this data set provides new and unique information on the interference between resonant and non-resonant amplitudes, and is expected to provide strong constraints on dynamical models of the $N \to \Delta$ transition form factors. A preliminary multipole analysis will be presented.

1 Introduction

Single pion electroproduction in the $\Delta(1232)$ resonance has been studied as a means of exploring the physics underlying the structure of the nucleon. Most previous measurements have focused on unpolarized cross section measurements. A new experiment using CLAS at Jefferson Lab/Hall B has been performed to measure single spin beam asymmetries over a large kinematic range. These will provide new and unique information on the interference between resonant and non-resonant amplitudes which are not available from unpolarized cross section measurements.

When polarized electrons are used, the differential cross section for single π^{o} electroproduction is given by:

$$\frac{d\sigma}{d\Omega} = \sigma_T + \epsilon\sigma_L + \epsilon\sigma_{TT}\sin^2\theta\cos(2\phi) + \sqrt{\epsilon(\epsilon+1)}\sigma_{LT}\sin\theta\cos\phi + h\sqrt{\epsilon(1-\epsilon)}\sigma'_{LT}\sin\theta\sin\phi \quad (1)$$

where h is the helicity of polarized electrons. σ_{LT} and σ'_{LT} are the real and imaginary parts of the same transverse-longitudinal interference amplitudes. For the case of the $\Delta(1232)$ resonance, it is not unreasonable to restrict oneself to S- and P-wave amplitudes. Then one can expand σ'_{LT} as:

$$\sigma'_{LT} = D'_0 P_0(\cos\theta) + D'_1 P_1(\cos\theta) \quad (2)$$

where

$$D_0' = -Im(S_{0+}^*(M_{1-} - M_{1+} + 3E_{1+}) + E_{0+}^*(S_{1-} - 2S_{1+}))$$
$$D_1' = -6Im(S_{1+}^*(M_{1-} - M_{1+} + 3E_{1+}) + E_{1+}^*(S_{1-} - 2S_{1+})). \qquad (3)$$

If there would be no background amplitudes, D_0' and D_1' would be identically zero. Any non-zero D_0' and D_1' contributions therefore have to come from interferences from resonant amplitudes with non-resonant ones.

2 Data Analysis and Results

The data were taken using a polarized electron beam of 1.515 GeV at 100 % duty factor incident on a liquid hydrogen target. Scattered electrons and protons were detected in CLAS. The electron beam polarization was approximately 0.7.

For this analysis, we performed an energy independent multipole analysis. The electron helicity dependent response function, σ'_{LT} is isolated by measuring the single spin beam asymmetry, $A_{LT'}$ which is defined as

$$A_{LT'} = (\sigma^+ - \sigma^-)/(\sigma^+ + \sigma^-) \qquad (4)$$

For an absolute normalization, we used measured unpolarized cross sections. D_0' and D_1' in equation (3) were extracted by fits to the polar angle distributions of σ'_{LT} up to the D-wave contributions. Figure 1 shows the results for σ'_{LT} compared with MAID98 [1] and Sato-Lee [2] calculations and Figure 2 shows our results for D_0' and D_1' which are also compared with MAID98 [1], MAID2000 [1] and Sato-Lee [2] calculations. Error bars are statistical error and results are very preliminary. A strong non-zero σ'_{LT} over the entire $\Delta(1232)$ resonance region indicates significant contributions from non-resonant amplitudes. Both MAID [1] and Sato-Lee [2] calculations qualitatively describe our results. More detail studies on systematic errors are underway using GEANT based Monte Carlo simulation.

Analysis for single π^+ electroproduction is also underway. Results from both $p\pi^0$ and $n\pi^+$ channels will provide strong constraints on dynamical models for the multipoles in the $\Delta(1232)$ resonance region.

References

1. D.Drechsel, O.Hanstein, S.S. Kamalov, and L. Tiator, private communication.
2. T.Sato and T.-S.H. Lee, *Phys. Rev.* C **54**, 2660 (1996), and T.-S.H. Lee, private communication, 2000.

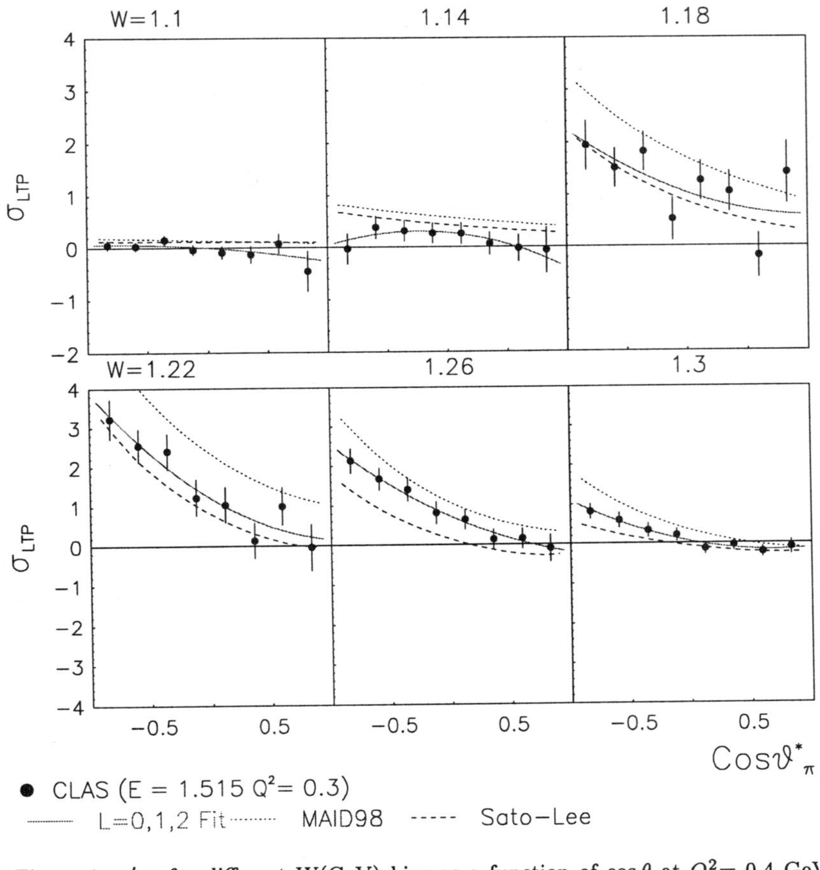

Figure 1. σ'_{LT} for different W(GeV) bins as a function of $\cos\theta$ at $Q^2 = 0.4$ GeV^2. Error bars are statistical only. Results are very preliminary.

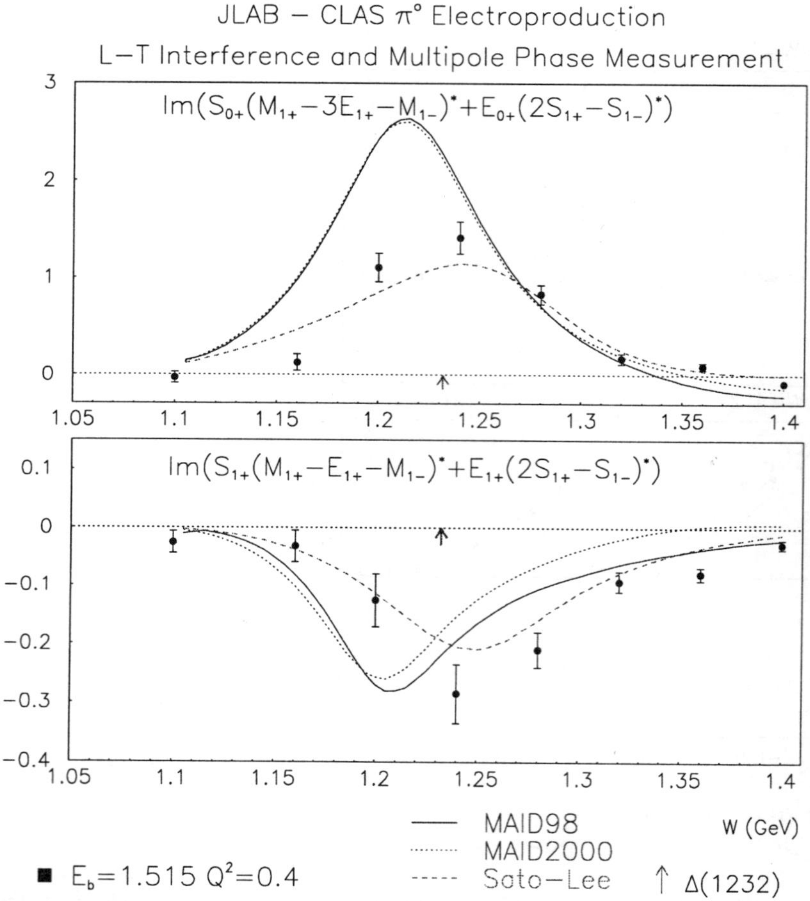

Figure 2. Top figure: D'_0 in μb as a function of W (GeV). Bottom figure: D'_1 in μb as a function of W (GeV) at Q^2= 0.4 GeV^2. Error bars are statistical only. Results are very preliminary.

THE NEW CRYSTAL BALL EXPERIMENTAL PROGRAM

W. J. BRISCOE

Department of Physics and Center for Nuclear Studies
The George Washington University, Washington, DC 20052, USA
E-mail: briscoe@gwu.edu

The Crystal Ball Spectrometer is being used at Brookhaven National Laboratory in a series of experiments which study all neutral final states of π^-p and K^-p induced reactions. We report about the experimental set up and progress in obtaining new results for the charge exchange $\pi^- p \to \pi^\circ$n, two π° production $\pi^- p \to \pi^\circ\pi^\circ$n, η production $\pi^- p \to \eta$n, and radiative capture reaction $\pi^- p \to \gamma$n reactions.

1 Introduction

The major goal of Nuclear Physics is to understand the strong interaction. The best candidate theory is Quantum Chromodynamics, QCD, which attempts to explain the strong interaction in terms of underlying quark and gluon degrees of freedom. The study of the structure of baryons and their excitations in terms of the elementary quark and gluon constituents is thus pivotal to our understanding of nuclear matter within QCD. Within this goal our motivations for the particular reactions of interest are briefly described in the following paragraphs.

The radiative decay of a resonance provides the ideal laboratory for testing theories of the strong interaction, gives us insight into the fundamental interactions between mesons and nucleons, and allows us to probe into the structure of the nucleon itself. In particular, these data are important in the study of the radiative decay of the neutral Roper resonance. They can be combined with recent JLab Hall B data for the reactions $\gamma p \to \pi^+$n and $\gamma p \to \pi^\circ$p which study the mesonic decays of the charged Roper in the incident photon energy region from 400 to 700 MeV. In addition, comparison of our data to the new JLab data taken on the inverse reaction $\gamma n \to \pi^-$p, using a deuteron target, tests extrapolation techniques for the deuteron correction and study medium effects within the deuteron.

The elusive charge exchange process has been the weakest link in partial-wave and coupled-channel studies. The accurate data that we obtained in this momentum region will help in improving the determinations of the isospin-odd s-wave scattering length, the πNN coupling constant, and the π-N σ term. In addition, better charge exchange data helps in evaluating the mass splitting of the Δ and the charge splitting of the P_{33} resonance and may result in new values for the $P_{11}(1440)$ mass and width.

Two pion production provides a means of studying sequential pion resonant decays; with neutral pions we study the $\pi\pi$ interaction in the absence of final-state Coulomb effects and owing to isospin considerations there is no contribution of ρ decay.

Near threshold η-production measurements provide data useful in verifying models of η-meson production and are also necessary for extraction of the η-N scattering length. Precise η production data are necessary to resolve ambiguities in the resonance properties of the $S_{11}(1535)$ and in the η photoproduction helicity amplitudes.

2 Experimental Considerations

Figures and diagrams are found on the web site: http://bmkn8.physics.ucla.edu/Crystalball/.

2.1 The Crystal Ball

We used the SLAC Crystal Ball to make these measurements at the C6 line at the Brookhaven National Laboratory, BNL, Alternating Gradient Synchrotron, AGS, with pion momenta from 147 MeV/c to 760 MeV/c. Data are taken simultaneously on all reactions which helps ensure that background events are accurately subtracted. Data taking using the Crystal Ball began in July 1998 and continued until late November 1998.

The Crystal Ball is a segmented, electromagnetic calorimetric spectrometer, covering 94% of 4π steradians. It was built at SLAC and used for meson spectroscopy measurements there for three years. It was then used at DESY for five years of experiments and put in storage at SLAC from 1987 until 1996 when it was moved to BNL by our collaboration.

The Crystal Ball is constructed of 672 hygroscopic NaI crystals, hermetically sealed inside two mechanically separate stainless steel hemispheres. The crystals are viewed by photomultipliers, PMT. There is an entrance and exit tunnel for the beam, LH_2 target plumbing, and veto counters.

The crystal arrangement is based on the geometry of an icosahedron (20 triangular faces or "major-triangles" arranged to form a spherical shape). Each "major-triangle" is subdivided into four "minor-triangles", which in turn consist of nine individual crystals. Each crystal is shaped like a truncated triangular pyramid, points towards the interaction point, is optically isolated, and is viewed by a PMT which is separated from the crystal by a glass window. The beam pipe is surrounded by 4 scintillators covering 98% of the target tunnel (these scintillators form the veto-barrel).

This high degree of segmentation provides excellent resolution. Electromagnetic showers in the ball are measured with an energy resolution of $\sigma/E = 2.7\%/E[GeV]^{1/4}$. Shower directions are measured with a resolution in θ of $\sigma = 2^\circ$–3° for energies in the range 50–500 MeV; the resolution in ϕ is $2^\circ/\sin\theta$. Typically, 98% of the deposited energy of each photon is contained in a cluster of thirteen crystals (a crystal with its twelve nearest neighbors). The thickness of the NaI amounts to nearly one hadron interaction length resulting in two-thirds of the charged pions interacting in the detector. The minimum ionization energy deposited is 197 MeV; the length of the counters corresponds to the stopping range of 233 MeV for $\mu^\pm$, 240 MeV for $\pi^\pm$, 341 MeV for $K^\pm$, and 425 MeV for protons. The preliminary energy calibration is performed using the 0.661 MeV γ's from a ^{137}Cs source. The final energy calibration is done using three reactions: i) $\pi^- p \to \gamma n$ at rest, yielding an isotropic, monochromatic γ flux of 129.4 MeV; ii) $\pi^- p \to \pi^\circ n$ at rest, yielding a pair of photons in the energy range 54.3—80 MeV, almost back to back; and iii) $\pi^- p \to \eta n$ at threshold, yielding two photons, about 300 MeV each, in coincidence almost back to back. The PMT analog pulses are sent to ADCs for digitization. Analog sums of the signals from each minor-triangle are available for trigger purposes.

2.2 Beam

The final stages of the C6 beamline consist of four quadrupoles and a dipole that form a beam momentum spectrometer. Wire chambers are located on both sides of the dipole to track the particles through the spectrometer. The momentum resolution is 0.3%. The scintillators located up and down stream of the dipole provide TOF information and the coincidence trigger for the beam. Scintillators surround the LH2 target to provide a charged particle veto. Two columns of scintillator neutron counters are located downstream of the Crystal Ball. A beam veto scintillator is located further downstream. A concrete shield wall located upstream of the beam stop shields the Crystal Ball from low energy photons from the stop. A Cerenkov counter is located just after this wall to monitor electron contamination in the beam.

The usual trigger consists of: a beam coincidence trigger, no downstream beam veto, and a total energy-over-threshold signal from the Crystal Ball. The Crystal Ball trigger is normally a total energy trigger. A trigger based on the distribution of the energy in different regions of the Crystal Ball was also used to provide a more restrictive trigger in certain cases.

2.3 Simulations

The primary reason so few data are available for the radiative capture reaction is the difficulty in separating its contribution from other reactions. This is mainly due to the significant background from $\pi^- p \to \pi^\circ n$ whose cross-section is about 50 times larger. The geometry of the Crystal Ball provides the capability of discriminating against multiple γ-rays that arise from the decay of π^o,s and η,s. However, because of the large entrance and exit tunnels there is a 20% chance that one of the two γ's from say π^o decay is missed, resulting in a fake one-photon event.

The separation of signal for $\pi^- p \to \gamma n$ from 'background' was investigated with GEANT. A full-fledged Monte-Carlo simulation of the Crystal Ball including all 672 NaI crystals, the hydrogen target, its mechanical support and the down-stream neutron counters was done. Electromagnetic showers are propagated by EGS within GEANT. Our Monte-Carlo includes such subtleties as secondaries from photon and pion breakup of a nucleus in the NaI, photon split-offs (a single photon cluster split into two) and backward Compton scattering. It reproduces and improves upon the photon energy and angular resolution that were measured in the course of the Crystal Ball's eight-year tenure at SLAC and DESY.

The large solid angle acceptance of the Crystal Ball and the additional use of the forward neutron detector wall lead to a rejection factor of 40-150 for the background events from $\pi^- p \to \pi^\circ n$. Since the total cross-section ratio is approximately 50 at 700 MeV/c, we expect a signal to background of about 2 to 1 for the measurement of $\pi^- p \to \gamma n$ within the angular range of about 70° to about 140°. This background and thus the angular range of the radiative capture measurement can be further reduced by using endcaps to increase the solid angle.

3 Results

Results of the 1998 experimental runs are being reviewed by the collaboration and will be available in the near future. See the Crystal Ball Collaboration given above web site for the most recent information.

4 Acknowledgments

The author acknowledges the support of the United States Department of Energy and the George Washington University Research Enhancement Fund.

OBSERVATION OF ETA-MESIC NUCLEI IN PHOTOREACTIONS: RESULTS AND PERSPECTIVES

G. A. SOKOL, A. I. L'VOV AND L. N. PAVLYUCHENKO
Lebedev Physical Institute, Leninsky Prospect 53, Moscow 117924, Russia
E-mail: gsokol@x4u.lebedev.ru

Recent results from the LPI experiment on searching for η-mesic nuclei in photoreactions are discussed and further perspectives are summarized.

This talk concerns experimental investigations of new objects of the nuclear physics, η-mesic nuclei ($_{\eta}A$) which are bound systems of the η-meson and a nucleus. Discovery and learning properties of these objects are of fundamental significance for understanding interactions of η-meson with nucleons and nucleon resonances, and for understanding the behavior of hadrons in nuclear matter.

A mechanism for the creation of η-mesic nuclei in photoreactions and their decay into a πN pair is shown in Fig. 1. At the first stage of the reaction, an incoming photon produces a slow η-meson and a fast nucleon N_1 escaping the nuclear target. Then the η-meson propagates in nuclear matter and undergoes multiple rescattering off nucleons. At last, the η annihilates producing an πN_2 pair escaping the nucleus. The $S_{11}(1535)$ nucleon resonance plays a fundamental role in all that dynamics. It mediates creation and annihilation of the η and it binds the η-meson in the nucleus due to an $S_{11}(1535)$-induced ηN interaction (attraction). Knowing the energies and momenta of the ingoing (γ) and outgoing (N_1, πN_2) particles, one can determine the energies and momenta of the η and $S_{11}(1535)$ propagating in nuclear matter. The width and binding energy of the η in the nucleus can be determined as well.

Theoretical estimates given in Refs. [1,2] show that the binding effects lead to a full dominance of the reaction mechanism related with the multiple rescattering of η and a formation of the intermediate η-nucleus over a mechanism of nonresonance background production of the πN pairs in the subthreshold invariant-mass region $\sqrt{s_{\pi N}} < m_\eta + m_N$ of the subprocess $\eta + N \to \pi + N$. A peak in the mass distribution of πN is theoretically expected in this region. This is illustrated in Fig. 2 where the spectral function $S(E)$ of the (kinetic) energy E of η in the nucleus is shown which is proportional to the number of ηN collisions the η experiences when traveling through the nucleus. Another spectral function, $S(E,q)$, shows a distribution of the produced πN pairs over their total energy $(E + m_\eta + m_N)$ and momentum q. The presence of the ηN

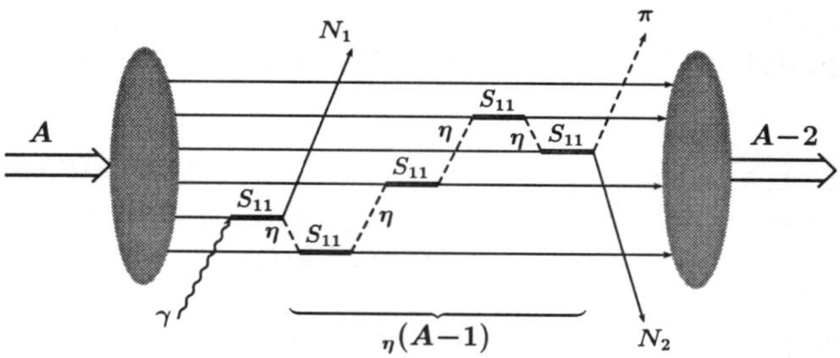

Figure 1. Mechanism of creation and decay of an η-mesic nucleus and its decay into a πN pair.

attractive potential produces strong enhancements in the spectral functions in the energy-momentum region corresponding to the bound ηA states.

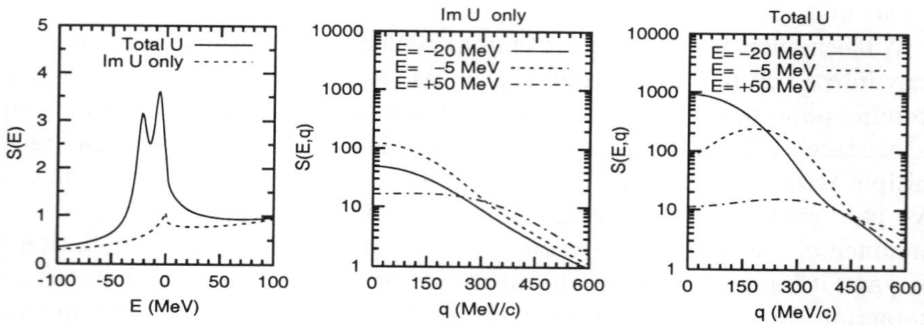

Figure 2. Spectral functions $S(E)$ and $S(E,q)$ (in arbitrary units) of the (kinetic) energy E and momentum q of η in the nucleus. They are found with a rectangular-well optical potential simulating the nucleus ^{12}C. For a comparison, results obtained with dropping the attractive (i.e. real) part of the ηA potential are also shown.[2]

The first experimental evidence for formation of η-nuclei in photoreactions was recently obtained in Ref. [2]. The experiment was performed at the bremsstrahlung photon beam of the 1 GeV electron synchrotron at Lebedev Physical Institute. The end-point photon energies $E_{\gamma max} = 650$ and 850 MeV were used which are below and above η-photoproduction threshold off the free nucleon. The reaction studied was

$$\gamma + {}^{12}\mathrm{C} \to p(n) + {}^{11}_{\eta}\mathrm{B}({}^{11}_{\eta}\mathrm{C}) \to \pi^+ + n + X,$$

in which energy-momentum correlations of π^+ and n were studied. Two time-of-flight scintillation spectrometers for detection of π^+ and n were placed in opposite directions at 90° with respect to the photon beam. Note that the $\pi^+ n$ pairs flying transversely to the photon beam cannot be produced via the one-step reaction $\gamma p \to \pi^+ n$ in the nucleus, whereas they naturally appear through an intermediate production and annihilation of a slow η-meson. Such transversely flying pairs have indeed been observed in the experiment. Moreover, a resonance peak in the total energy of the $\pi^+ n$ pairs has been found which appeared when the photon energy $E_{\gamma max} = 850$ MeV exceeded the η-production threshold (this is the kinematics of "effect+background"[2]) and did not appear when $E_{\gamma max} = 650$ MeV (the kinematics of "background"[2]), see Fig. 3. This peak was interpreted as a manifestation of decays of the bound η's in the nucleus, i.e. a formation and decay of η-mesic nuclei.

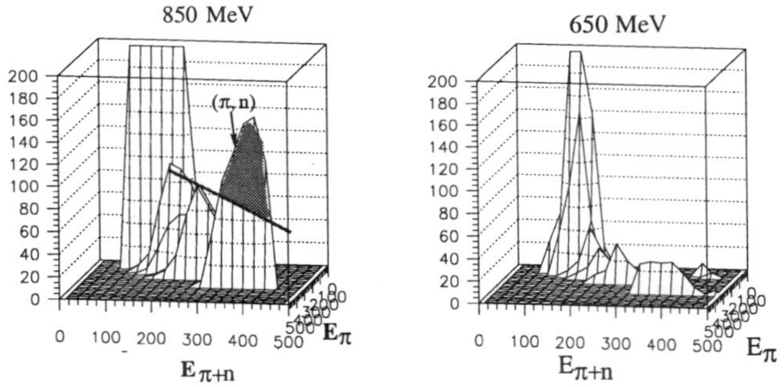

Figure 3. Distribution over the total kinetic energy of the $\pi^+ n$ pairs for the "effect + background" run ($E_{\gamma\,\max} = 850$ MeV, the left panel) and for the "background" run ($E_{\gamma\,\max} = 650$ MeV, the right panel) obtained after unfolding the raw spectra.

Subtracting a smooth background and unfolding the measured (raw) spectra with experimental resolutions, the 1-dimensional energy distribution over the total energy of the $\pi^+ n$ pairs has been found (Fig. 4). It develops a peak near 90 MeV lying below the mass of the $S_{11}(1535)$ resonance and even below the threshold energy $m_\eta + m_N = 1486$ MeV, thus indicating the presence of binding effects. The width of the peak is about 150 MeV which is compatible with the experimental resolution and broadening effects due to the Fermi motion of the nucleons and the bound η in the intermediate η-nucleus.

In prospect, studies of the η-mesic nuclei lying at the intersection of the nuclear physics and the physics of hadrons promise to bring new information important for both the fields. The η-mesic nuclei provide a unique possibility to learn interactions of η-mesons with nucleons and nucleon resonances, both

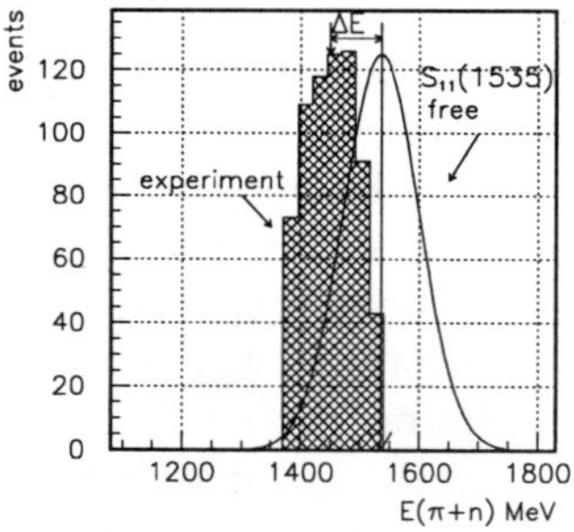

Figure 4. Distribution over the total energy of the π^+n pairs after a subtraction of the background.

free and in the nuclear matter. Detecting and measuring the energy of the nucleon knocked out in the process of quasi-free η-production on nucleons in the nucleus, one can tag the energy of the η staying in the nucleus. Used with the tagged photons technique, this opens the possibility to study the energy dependence of interactions between the η and nuclear constituents.

We hope to perform further photoproduction experiments in order to measure binding energies of the η in different nuclei in the $A = 3{-}16$ mass range. Since the η energy levels and widths in nuclei depend on many important characteristics like, e.g., the ηN potential and the self-energy of the $S_{11}(1535)$ in the nuclear matter, the expected data might be very useful for further progress in understanding exotic nuclear systems and the behaviour of hadrons in nuclei.

Acknowledgment

This work was supported by the Russian Foundation for Basic Research, grant 99-02-18224.

References

1. A.I. L'vov, Mesons and Light Nuclei '98, World Scientific, Eds. J. Adam *et al.*, p. 469 [nucl-th/9810054].
2. G.A. Sokol *et al., Fizika* B **8**, 81 (1998) [nucl-ex/9905006]; nucl-ex/0011005.

BARYON SPECTROSCOPY: EXPERIMENTS AT PNPI

I. LOPATIN

Petersburg Nuclear Physics Institute, Gatchina, Leningrad district, 188350 Russia
E-mail: lopatin@mail.pnpi.spb.ru

Differential cross sections of $\pi^- p$ charge exchange scattering are measured in the region of low-lying πN resonances. The experiment is carried out at the pion channel of the PNPI synchrocyclotron by detecting the recoil neutron in coincidence with one gamma from the decay $\pi^0 \to 2\gamma$. Results obtained for the backward scattering are presented in comparison with the predictions of different partial-wave analyses. Measurements of the spin rotation parameters in $\pi^+ p$ elastic scattering are performed at the pion channel of the ITEP synchrotron at incident pions momenta of 1.43 and 1.62 MeV/c. Obtained values of these parameters agree with the predictions of the partial-wave analysis VPI and obviously contradict the analysis KH–80 which was used till now by the Particle Data Group as a basis for extracting characteristics of πN resonances.

1 Study of $\pi^- p$ Charge Exchange Scattering in the Region of Low-Lying Pion-Nucleon Resonances

Most of the information on characteristic features of nonstrange excited baryons comes till now from πN data, mainly elastic scattering, consisting of cross sections and polarization measurements. The data are synthesized by a partial-wave analysis (PWA). The baryon spectrum can be extracted from PWA results using Breit-Wigner fits or speed plots.

At present, an accuracy in determining the characteristics of πN resonances is limited mostly by a lack of high quality experimental data on $\pi^- p$ charge exchange scattering. To improve the situation and fill the gap in the data-base, measurements of differential cross sections (DCS) for the reaction $\pi^- p \to \pi^0 n$ are now underway at PNPI in the energy range from 300 to 600 MeV (corresponding values of momenta are 417 to 725 MeV/c). Physicists of the University of California at Los Angeles (head of group Prof. B.M.K.Nefkens) and Abilene Christian University (head of group Prof. M.E.Sadler) provided the experiment with some equipment and participated in the measurements at initial steps.

At the first stage, measurements were performed for scattering into the backward hemisphere [1]. The recoil neutrons were detected in coincidence with one of the gammas from the decay $\pi^0 \to 2\gamma$. The energy of the neutrons was measured using the time-of-flight technique, with a 5 m base. Gamma detectors were placed at angles kinematically conjugated with the neutron detectors. Two kinds of total absorption electromagnetic calorimeters were used for detecting gammas. One was made of eight Čerenkov lead glass SF-5 blocks, another is a four-to-four array of CsI(Na) crystals. An energy calibration of both types of gamma-detectors was performed using electrons with energies from 70 to 500 MeV available at the pion channel of the PNPI synchrocyclotron.

Results obtained for three centre-of-mass scattering angles are presented in Fig. 1. the new data excel in accuracy all previous experiments (values

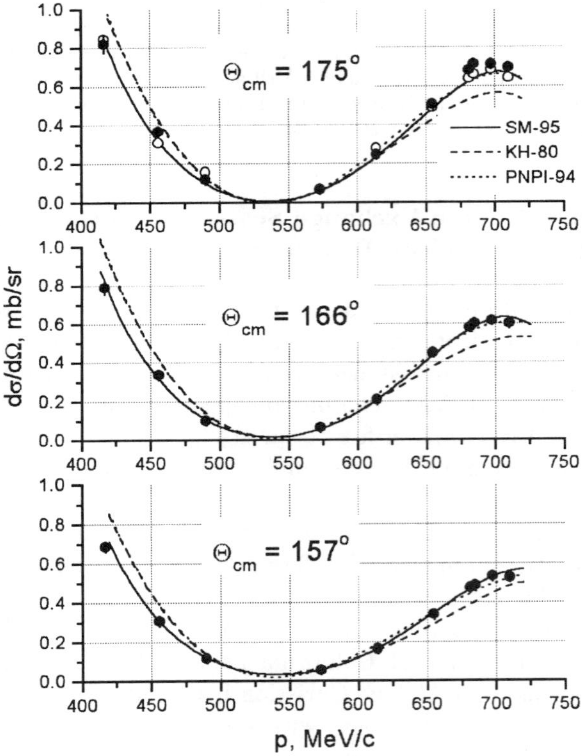

Figure 1. Results of measuring differential cross sections of $\pi^- p$ charge exchange scattering to backward angles. Results obtained with Čerenkov spectrometers are shown by solid circles, those obtained with CsI(Na) crystals by open circles. Shown by the curves are the predictions of partial-wave analyses SM–95, PNPI–94 and KH–80.

of statistical errors are typically within circles); moreover, for momenta below 600 MeV/c no experimental data existed at all prior to this work. For the angle $\Theta_{cm} = 175°$ two sets of data are presented: one was obtained using Čerenkov lead glass blocks as gamma-detector and another with CsI(Na) crystals. Since two gamma-detectors have essentially different values of angular acceptance, a comparison of DCS values obtained for these two cases will allow one to estimate a systematic error in calculating the acceptance. This error does not exceed 10%.

Shown by the curves in Fig. 1 are the predictions of three different partial-wave analyses: KH–80, PNPI–94 and VPI (solution SM–95). Our experimental results agree satisfactorily with the SM–95 PWA and seem to contradict the predictions of KH–80 PWA. This can be considered as an indication that the results of KH–80 PWA may be incorrect in some energy ranges and need to be revised.

2 Measurement of the Spin Rotation Parameters for π^+p Elastic Scattering in the Second Resonance Region

The above described experiments cover the region of low-lying pion-nucleon resonances. But most of the πN resonances have masses more than 1750 MeV which corresponds to $T_\pi > 1000$ MeV. As it was mentioned earlier, our knowledge or the characteristics of these resonances comes from partial-wave analyses of experimental πN data. And till now this knowledge is rather poor – even for the well established resonances. There exist essential discrepancies between values of the resonance mass and width given by the three global PWAs: KH, CMB and VPI. In some cases the πN resonances found in the analyses KH and CMB – and having even the three-star rating in notations of the Particle Data Group – were not observed in the analysis VPI.

A closer inspection of the above situation shows that the only way to remove such ambiguities is to measure the spin rotation parameters. Such measurements were performed recently by the PNPI–ITEP collaboration. The experiment was carried out at the pion channel of the ITEP accelerator at the incident pion momenta of 1.43 and 1.62 GeV/c [2]. This experiment is rather complicated since it requires a polarized proton target of special type – with the polarization vector lying in the horizontal plane. To extract the spin rotation parameters, it is necessary to measure an asymmetry of the secondary scattering of the recoil protons on nuclei with the known analyzing power (usually carbon). Two different types of proton polarimeters were used at different stages of the experiment. One was a multi-plate polarimeter made of optical spark chambers with graphite electrodes; a special television system

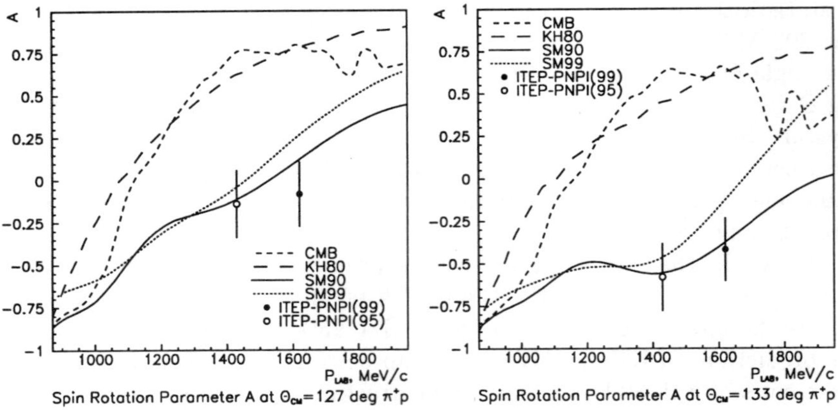

Figure 2. Results of measuring the spin rotation parameter A in π^+p elastic scattering. Shown by curves are the predictions of different partial-wave analyses.

was developed for filmless read-out in this case. Another type of polarimeter consists of one thick graphite block (36.5 g/cm^2) with two sets of spark chambers – in front and behind of this block – to detect the recoil proton before and after the secondary scattering; the analyzing power of this polarimeter was determined experimentally using the beam of polarized protons available at ITEP.

The obtained results are shown in Fig. 2 in comparison with the predictions of different PWAs. One can see that the experimental data confirm the predictions of the analysis VPI (solutions SM–90 and SM–99) and obviously contradict the predictions of the analyses KH and CMB. This conclusion is very important since all characteristics of πN resonances presented in listings of the Particle Data Group are based on the PWAs KH and CMB. The fact that the experimental data do not confirm the predictions of these analyses leads to the conclusion that these Listings need to be revised.

References

1. I.V.Lopatin *et al.*, Preprint PNPI–2363, Gatchina, 32p. (2000).
2. I.G.Alekseev *et al.*, *Phys. Lett.* B **351**, 585 (1995); *Phys. Lett.* B **485**, 32 (2000).

MEASUREMENT OF THE CROSS SECTION ASYMMETRY IN DEUTERON PHOTODISINTEGRATION BY LINEARLY POLARIZED PHOTONS IN THE ENERGY RANGE $E_\gamma = 0.8 - 1.6$ GEV

F. ADAMIAN, A. AGANIANTS, N. DEMEKHINA, G. FRANGULIAN, V. GRABSKI, A. HAIRAPETIAN, H. HAKOBIAN, I. KEROPIAN, A. LEBEDEV, ZH. MANUKIAN, G. MOVSESIAN, E. MURADIAN, A. OGANESIAN, R. OGANEZOV, A. SIRUNIAN, H. TOROSIAN, H. VARTAPETIAN, V. VOLCHINSKI

Yerevan Physics Institute, Armenia

YU. BORZUNOV, S. CHUMAKOV, L.GOLOVANOV, N. MOROZ, YU. PANEBRATSEV, S. SHIMANSKI, A. TSVENEV

Joint Institute for Nuclear Research, Dubna, Russia

M. REKALO

Kharkov Institute of Physics and Technology, Kharkov, Ukraine

The first measurements of the cross section asymmetry Σ for the deuteron photodisintegration process at cm angle of 90° up to 1.6 GeV were performed at Yerevan Electron Synchrotron. These results are in reasonable agreement with previous measurements at lower energy. Our data show agreement with the asymptotic meson exchange model predictions in the energy range 0.8 –1.6 GeV.

1 Introduction

The exclusive deuteron photodisintegration reaction has been one of the important processes to study the problems of nuclear and particle physics.

At the low energy region $E_\gamma = 0.1-0.6$ GeV the process $\gamma d \rightarrow pn$ has been investigated over many years. Beginning from 1980 we study this process with polarized photons, measuring the asymmetry Σ [1] as well as in double polarization experiments [2] up to the photon energy $E_\gamma = 1.0$ GeV. The results were compared with various theoretical models, based on the meson, nucleon and isobar degrees of freedom [1].

During the last years the interest to study the process $\gamma d \rightarrow pn$ at energies above E_γ= 1.0 GeV is growing. This is mainly explained by the possible contribution of new degrees of freedom (quark, gluon) expected already at energies as small as $E_\gamma = 1.4$ GeV.

Such assumption follows from the results of measurements performed at SLAC [3] and TJNAF [4] in the energy range E_γ = 1.4 - 4.0 GeV where at $\theta_p^* =$

90° a scaling - like behavior for the cross section of the process $\gamma d \to pn$ was observed. This result is in agreement with the prediction of the constituent quark counting rules ($d\sigma/dt \sim S^{-11}$) [5].

Theoretical models have been developed also for energies $E_\gamma > 1.0$ GeV, trying to describe the existence of early scaling in the reaction $\gamma d \to pn$ within the framework of models based on meson-baryons [6,7,8] or quark-gluon degrees of freedom [9,10,11]. Among those models only two models, in the energy range up to $E_\gamma = 4.0$ GeV and the angular interval $\theta_p^* = 36° - 90°$, seem to explain the behavior of the differential cross section of the process $\gamma d \to pn$ [3,4], the asymptotic meson-baryon exchange model and its modification [7,8], and the QCD rescattering model [11].

The authors of these two models note the necessity of the measurement of spin observables, which can give the possibility to discriminate between the single quark counting rules and the theoretical models.

2 Experimental Setup

The experiment was carried out at the linearly polarized photon beam generated by coherent bremsstrahlung of 4.5 GeV electrons at the Yerevan Synchrotron, on a internal 100 μm diamond crystal target [1]. The experimental setup is shown in Fig. 1. The external beam is shaped by a system of collimators and sweeping magnets and it is (10×10) mm^2 wide at the target position. The beam monitoring system is based on the thirty-channel pair spectrometer (PS-30), which consists of a bending magnet, a set of removable converters and eleven telescopes of scintillation counters. During the experiment particular emphasis has been placed on maintaining the stability of the coherent peak position (checks of the coherent peak were carried out every 40-50 sec).

In this experiment the new liquid deuterium target of 300 mm length has been used, that allows one to increase the detection yield as compared to the 90 mm target, used in previous experiments [1]. The protons are detected by the magnetic spectrometer (MS) consisting of a double focusing system, a bending magnet, a telescope of four trigger counters and three scintillation momentum hodoscope counters (H_1, H_2, H_3). The angular and momentum acceptances and corresponding resolutions of the MS are $\Delta\Omega$=7× 10^{-3} sr, $\Delta P/P$=17% and $\sigma_\theta = 1.3°$, and $\sigma P/P$ ~1.8%, respectively.

The neutrons are detected by a time-of-flight hodoscope detector (NS-18). Six anticoincidence counters (for the detection of charged particles) are mounted in front of the NS. For the γ-quanta rejection a lead converter of the 2 cm thickness is placed in front of the veto counters.

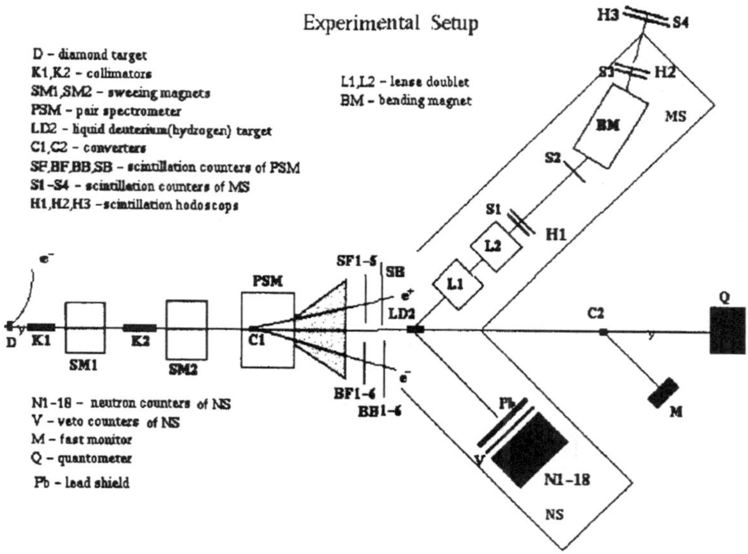

Figure 1. Experimental setup. In the frame, the neutron spectrometer NS-18 from the front.

3 Results and Summary

The data obtained for the asymmetry Σ in the kinematic range E_γ=0.8 – 1.6 GeV and $\theta_p^* = 90°$ are presented in Fig. 2. The energy resolution RMS (E_γ) for the photons varies from 40 to 140 MeV depending on E_γ. The error bars include the statistical uncertainties and the uncertainty of the calculated photon polarization (5%).

For Σ we observe a better agreement with the asymptotic meson exchange model predictions [7] in the measured energy range. We note a possible increase for Σ at E_γ >1.4 GeV. This increase, if confirmed in future measurements, may indicate a change in the production mechanism at these energies.

Our results in the energy range E_γ = 1.0 - 1.6 GeV show values $\Sigma > 0$, in strong disagreement with the HHC hypothesis ($\Sigma = -1$). This result is the first test of HHC for exclusive photoreactions in the regime of scaling (E_γ >1.4 GeV, for $\gamma d \to pn$).

In future work, we plan to reduce the statistical errors of Σ for the energies higher than $E_\gamma = 1.4\ GeV$ and to make a measurement of Σ at E_γ = 2 GeV.

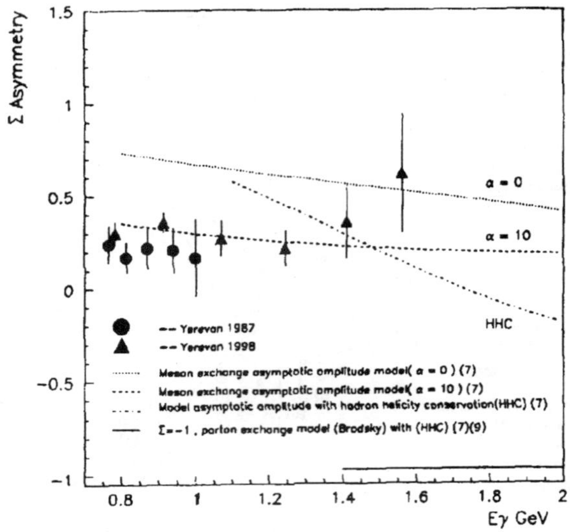

Figure 2. The energy dependence of the cross-section asymmetry Σ for $\theta_p = 90^\circ$ in the cms.

References

1. F. V. Adamian *et al., JETP Lett.* **39**, 239 (1984); *J. Phys. G. Nucl. Part. Phys.* **17**, 1189 (1991).
2. F. V. Adamian *et al., J. Phys. G. Nucl. Part. Phys.* **14**, 831 (1988); R. O. Avakian *et al., Jad. Fiz.* **52**, 313 and 618 (1990).
3. J. Napolitano *et al., Phys. Rev. Lett.* **61**, 2530 (1988); J. E. Beltz *et al., Phys. Rev. Lett.* **74**, 646 (1995).
4. C. Bochna *et al., Phys. Rev. Lett.* **81**, 4576 (1998).
5. V. A. Matveev *et al., Lett. Nuovo Cimento* **7**, 719 (1973); S. J. Brodsky *et al., Phys. Rev. Lett.* **31**, 1153 (1973).
6. T. S. H. Lee, ANL report PHY-5253-TH (1988).
7. S. I. Nagorny *et al., Sov. J. Nucl. Phys.* **55**, 189 (1992); Proc. 14th AIP Conference "Few Body Problems in Physics" Williamsburg 1994, p.757.
8. A. E. L. Dieperink and S. I. Nagorny, *Phys. Lett.* B **456**, 9 (1999).
9. S. J. Brodsky and J. R. Hiller, *Phys. Rev.* C **28**, 475 (1983).
10. L. A. Kondratjuk *et al., Phys. Rev.* C **48**, 2491 (1993).
11. L. L. Frankfurt *et al., Phys. Rev. Lett.* **84**, 3045 (1999).

NUCLEON RESONANCES IN LATTICE QCD

F. X. LEE

Center for Nuclear Studies, Department of Physics,
The George Washington University, Washington, DC 20052, USA and
Jefferson Lab, 12000 Jefferson Avenue, Newport News, VA 23606, USA

Results are presented for N* masses in the 1/2 and 3/2 sectors using an anisotropic lattice QCD action. Clear splittings from the nucleon ground state are observed with smeared operators and 500 configurations. The first excited states in the 1/2+ and 1/2- channels are further isolated, using a correlation matrix and the variational method. The importance of spin projection in the 3/2 sector is demonstrated. The basic pattern of these splittings is consistent with experimental observations at the quark masses explored, although further studies at smaller quark masses are needed to explain the details of the splittings.

1 Introduction

There is increasing experimental information on the excited states of the nucleon (N*s) from JLab and other accelerators, as evidenced by this N* workshop and others. Associated with the experimental effort is the desire to understand N* structure from first principles. The N* spectrum is the emblem of QCD, just like the hydrogen atom spectrum is the emblem of quantum mechanics. Given that state-of-the-art lattice QCD simulations have produced a ground-state spectrum that is very close to the observed values [1], it is important to extend beyond the ground states. The rich structure of the N* spectrum [2] provides a fertile ground for exploring how the internal degrees of freedom in the nucleon are excited and how QCD works in a wider context. One outstanding example is the splitting pattern between the observed 1/2+ spectrum and its parity partner: the 1/2− spectrum, as shown in Fig. 1. The splittings are a direct manifestation of spontaneous chiral symmetry breaking of QCD, because without it QCD predicts parity doubling in the baryon spectrum.

Several lattice studies of the N* spectrum have appeared recently, employing different versions of putting the quarks on the lattice. In [3] the $1/2\pm$ spectrum was first explored with a $\mathcal{O}(a^2)$ improved next-nearest-neighbor action. In [4] the splitting $N_{1/2-} - N_{1/2+}$ was carefully examined with the domain-wall fermion action. In [5] results are reported for the $N_{1/2\pm}$ and the $\Delta_{3/2\pm}$ based on the non-perturbative $\mathcal{O}(a)$ improved clover action. All the calculations show a clear splitting of the $N_{1/2-}$ from the ground-state nucleon. Here, we report further calculations in the $1/2\pm$ sectors using a highly-improved action

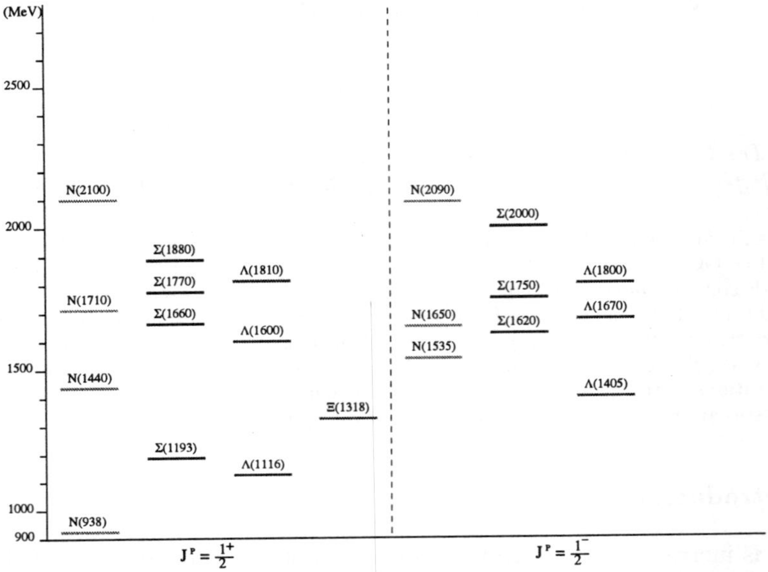

Figure 1. Some of the observed N*s in the spin 1/2 sector.

on a anisotropic lattice.

2 Method

In the 1/2 sector, two independent, local interpolating fields coupling to the nucleon are considered:

$$\chi_1(x) = \epsilon^{abc}\left(u^{Ta}(x)C\gamma_5 d^b(x)\right)u^c(x)\,, \tag{1}$$

$$\chi_2(x) = \epsilon^{abc}\left(u^{Ta}(x)Cd^b(x)\right)\gamma_5 u^c(x)\,. \tag{2}$$

The interpolating fields for other members of the octet can be found by appropriate substitutions of quark fields [6]. Despite having explicit positive-parity by construction, these interpolating fields couple to both positive and negative parity states. A parity projection is needed to separate the two. In the large Euclidean time limit, the correlator with Dirichlet boundary condition in the time direction and zero spatial momentum becomes

$$G(t) = \sum_{\mathbf{x}} < 0|\chi(x)\,\bar{\chi(0)}|0 > \tag{3}$$

$$= \lambda_+^2 \frac{\gamma_4 + 1}{2} e^{-M_+ t} + \lambda_-^2 \frac{-\gamma_4 + 1}{2} e^{-M_- t} \tag{4}$$

The relative sign in front of γ_4 provides the solution: taking the trace of G(t) with $(1 \pm \gamma_4)/4$ respectively. A method on how to do the projection at finite momentum was also proposed in [3] which involves only diagonal elements. Three types of correlators are considered in this work: two diagonal ones $N_1 =< \chi_1 \bar{\chi_1} >$ and $N_2 =< \chi_2 \bar{\chi_2} >$, and the crossed $N_3 =< \chi_2 \bar{\chi_1} > + < \chi_1 \bar{\chi_2} >$.

In the 3/2 sector, the local interpolating field considered is :

$$\chi_\mu(x) = \epsilon^{abc} \left(u^{Ta}(x) C \gamma_5 \gamma_\mu d^b(x) \right) \gamma_5 u^c(x) \tag{5}$$

which has negative-parity by construction but also couples to positive parity. The operator has overlap with both spin 3/2 and 1/2 components, so a spin projection is needed to separate them out. This is done by applying the projection operator

$$P_{\mu\nu} = g_{\mu\nu} - \frac{1}{3}\gamma_\mu\gamma_\nu - \frac{1}{p^2}(\gamma \cdot p \gamma_\mu p_\nu + p_\mu \gamma_\nu \gamma \cdot p) \tag{6}$$

to the total correlation function $G_{\mu\nu}$:

$$G_{\mu\nu}^{3/2} = \sum_\lambda G_{\mu\lambda} P_{\lambda\nu}. \tag{7}$$

The 1/2 component can be isolated by

$$G_{\mu\nu}^{1/2} = G_{\mu\nu} - G_{\mu\nu}^{3/2}. \tag{8}$$

The parity projection is done in the same manner as in the 1/2 sector.

The anisotropic gauge action of [7], and the anisotropic D234 quark action of [8] are used. Both have tadpole-improved tree-level coefficients. Similar anisotropic actions have been used to study systems with heavy quarks [9,10,11]. One advantage of anisotropic lattices is that with modest lattice sizes one can access large spatial volumes while having a fine temporal resolution, important when the states are extended and heavy. A $10^3 \times 30$ lattice with $a_s \approx 0.24$ fm and anisotropy $\xi = a_s/a_t = 3$ is used. The tadpole factor from the plaquette mean link is $u_s = 0.81$ while u_t is set to 1. In all, 500 configurations are analyzed. On each configuration 9 quark propagators are computed using a multi-mass solver, with quark masses ranging from approximately 780 to 90 MeV. The corresponding mass ratio π/ρ is from 0.93 to 0.60. The source is located at the 2nd time slice. A gaussian-shaped, gauge-invariant smearing function [12] was applied at both the source and the sink to increase the overlap with the states in question. Statistical errors are derived from a bootstrap procedure.

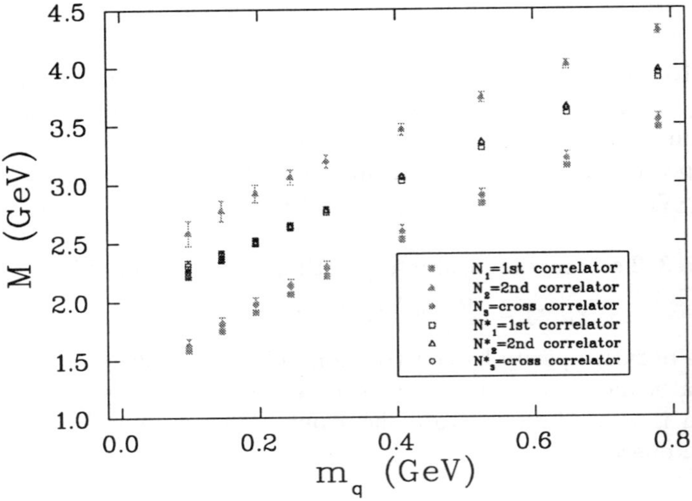

Figure 2. Masses of $N_{1/2\pm}$ as a function of quark mass for all three types of correlators.

3 Results and Discussion

Figure 2 shows the extracted masses as a function of the quark mass. The N_1 and N_3 give roughly the same mass, while N_2 yields a consistently higher mass. It turns out that χ_2 has very little overlap with the ground-state nucleon and couples mostly to the first excited state, the Roper $N_{1/2+}(1440)$. This point will be clarified below. All three negative-parity correlators give consistent masses. The ordering of $N_{1/2+}(1440)$ and $N_{1/2-}(1535)$ at the quark masses considered is inverted compared to that from experiment. These results are consistent with those observed in [4]. It would be interesting to see if the ordering is reversed at smaller quark masses where meson cloud becomes more important. The crossing can be expected even in the quenched approximation where quarks wiggling backward in time provide part of the meson cloud. Of course quenched QCD is sick at small quark masses, so a correct description must include the sea quarks. There is already evidence of such crossing at very small quark masses in the case of a_0 and a_1 mesons [13,14]. Preliminary results with overlap quarks seem to a cross over to the correct orderings [15]. The effects of the meson cloud show up as non-analytic behavior in the quark mass. Some curvature is already apparent in the data, and so a linear extrapolation is undesirable.

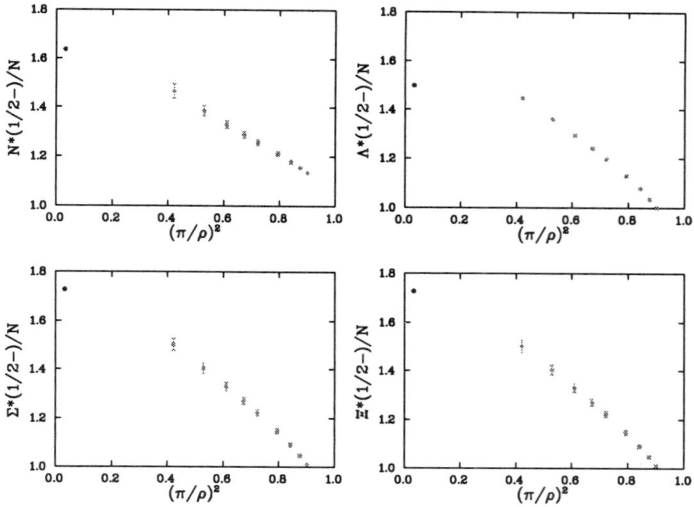

Figure 3. Mass ratio N*(1/2-)/N as a function of mass ratio $(\pi/\rho)^2$. The experiment points are indicated on the left.

In order to make contact with experiment, Fig. 3 presents mass ratios of the 1/2− states to the nucleon ground state as a function of $(\pi/\rho)^2$. This way of presenting data has minimal dependence on the uncertainties in determining the scale and the quark masses. These ratios appear headed in the right direction compared to experiment. The $\Lambda_{1/2-}(1405)$ deserves special attention since it requires a larger curvature. At present it is almost degenerate with the $N_{1/2-}(1535)$, while in the physical spectrum its mass is lower despite having a heavier strange quark. This has been a long-standing puzzle. The origin of the mass suppression of the $\Lambda_{1/2-}(1405)$ lies in strong resonance channels such as the KN and $\pi\Sigma$ which would present themselves as nonlinearities in the quark-mass extrapolation. In the language of QCD sum rules, it is mainly due to the relative reduction of the strange quark condenstate [16]. It would be extremely interesting to see how it pans out on the lattice.

To extract information on the first excited states in each sector, the 2 by 2 correlator matrix from the two interpolating fields is considered, using the variational method originally proposed in [17]. Figure 4 shows the results in the nucleon channel. It confirms that χ_2 couples mostly to the the $N_{1/2+}(1440)$. The splitting of the Roper $N_{1/2+}(1440)$ from the nucleon is roughly 5 times that of the $N_{1/2-}(1620)$ from the $N_{1/2-}(1535)$, consistent with experimental observations. This may be traced back to the structure of the interpolating

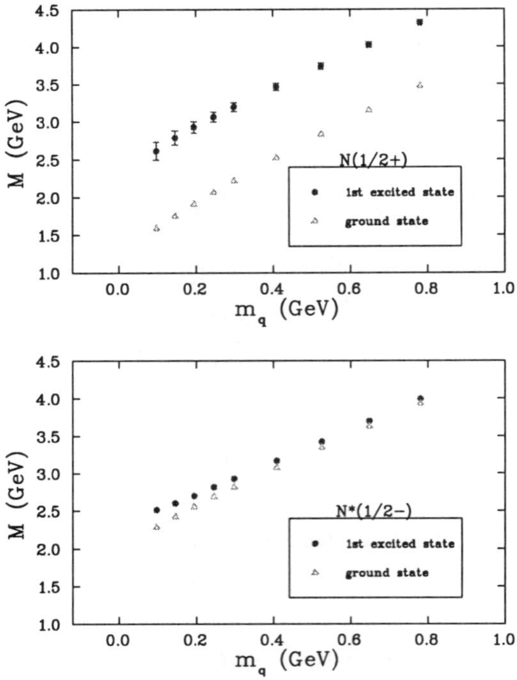

Figure 4. Ground states and 1st excited states in the 1/2 sector from the correlator matrix analysis.

fields used to excite the states. The χ_1 and χ_2 have upper spinor components of the order 1 and $(p/E)^2$ respectively, whereas the interpolating fields for 1/2- states, $\chi_1^- = \gamma_5\chi_1$ and $\chi_2^- = \gamma_5\chi_2$, both have upper spinor components of the order p/E which also introduce derivatives of the quark field operators. In the quark model, the two nearby low-lying $N_{1/2-}$ states of the physical spectrum are described by coupling three quarks to a spin-1/2 or spin-3/2 spin-flavor wave function coupled in turn to one unit of orbital angular momentum. In a relativistic field theory one does not expect χ_1^- and χ_2^- to isolate individual states. However, it is expected that χ_1^- will predominantly excite the lower-lying $j = 1/2$ state associated with $\ell = 1$ and $s = 1/2$ while χ_2^- will predominantly excite the higher-lying state associated with $s = 3/2$. This expectation is borne out in the variational analysis. The results should improve with a larger basis of interpolating fields. An independent method based on maximum entropy [18] is being applied to verify the results.

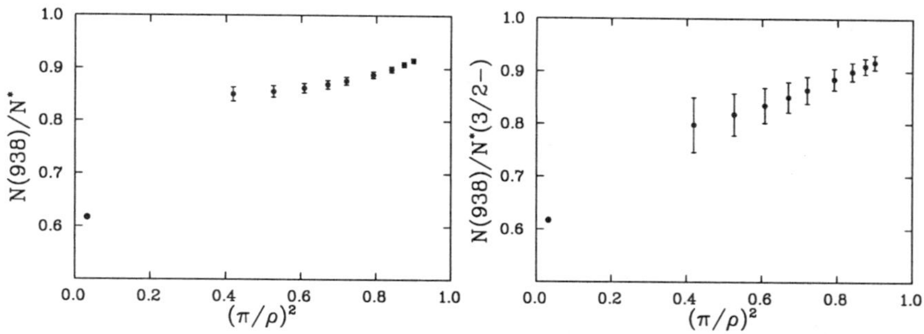

Figure 5. Mass ratio N/N*(3/2-) as a function of mass ratio $(\pi/\rho)^2$. The left graph shows the result extracted from the total correlation function. The right graph shows the result after spin projection. The experiment points are indicated on the left.

Fig. 5 shows the results in the 3/2- sector. The spin projection significantly improves the trend towards the experimental point, but at a price of increased statistical noise. A signal for 1/2- state can also be extracted (not shown), but with even larger error bars. The parity projection was done in the same manner as in the 1/2 sector. The 3/2- state in the nucleon channel is lower than the 3/2+ state, consistent with experiment.

4 Conclusion

It appears that the low-lying N*s in the 1/2 and 3/2 sectors are now accessible on the lattice. In this work, the potential of anisotropic lattices, smeared operators and correlator matrix, and the importance of spin projection in the 3/2 sector is demonstrated in probing the N*s. These N* states have proven elusive on the lattice and require good operators and statistics to isolate. The pattern of the splittings is mostly consistent with experiment. More definite comparisons should address the systematics: finite a, finite volume, and most importantly, the chiral limit. There are still unresolved issues with regard to the details of the splittings, such as the inverted ordering of the $N_{1/2+}(1440)$ and the $N_{1/2-}(1535)$, and the degeneracy of the $\Lambda_{1/2-}(1405)$ and the $N_{1/2-}(1535)$ at present quark masses, that depend on this limit. One way to proceed is to use ChPT to extrapolate the lattice data at quark masses where the quenched approximation is benign, like that proposed in [19]. The more desirable way is to be able to simulate directly at or near the physical quark masses. Toward this end, good chiral properties are essential, and the domain-wall [4] and the overlap quarks [15,20,21] seem to hold the promise.

5 Acknowledgment

This work is supported in part by U.S. DOE under grant DE-FG02-95ER40907. I thank D.B. Leinweber and K.F. Liu for helpful discussions. I also thank the organizers of the workshop for a rewarding experience.

References

1. S. Aoki, *et al., Phys. Rev. Lett.* **84**, 238 (2000).
2. Particle Data Group, *Eur. Phys. J.* C **15**, 1 (2000).
3. F.X. Lee and D.B. Leinweber, *Nucl. Phys.* B (Proc. Suppl.) **73**, 258 (1999).
4. S. Sasaki, *Nucl. Phys.* B (Proc. Suppl.) **83**, 206 (2000); hep-ph/0004252; T. Blum and S. Sasaki, hep-lat/0002019; and these proceedings.
5. D. Richards, *Nucl. Phys.* B (Proc. Suppl.) **94**, 269 (2001)
6. D.B. Leinweber, R.M. Woloshyn, and T. Draper, *Phys. Rev.* D **43**, 1659 (1991).
7. C.J. Morningstar and M. Peardon, *Phys. Rev.* D **56**, 4043 (1997).
8. M. Alford, T.R. Klassen, and G.P. Lepage, *Phys. Rev.* D **58**, 034503 (1998); *Nucl. Phys.* **B496**, 377 (1997).
9. R. Lewis, *Nucl. Phys.* B (Proc. Suppl.) **94**, 359 (2001)
10. R.M. Woloshyn, *Phys. Lett.* B **476**, 309 (2000).
11. P. Chen, X. Liao, and T. Manke *Nucl. Phys.* B (Proc. Suppl.) **94**, 342 (2001).
12. S. Güsken, *Nucl. Phys.* B (Proc. Suppl.) **17**, 361 (1990).
13. H. Thacker, private communication.
14. S.J. Dong, F.X. Lee, K.F. Liu, and J.B. Zhang, *Nucl. Phys.* B (Proc. Suppl.) **94**, 752 (2001).
15. F.X. Lee, K.F. Liu, S.J. Dong, and J.B. Zhang, unpublished.
16. D. Jido and M. Oka, hep-ph/9611322.
17. M. Lüscher and U. Wolff, *Nucl. Phys.* B **339**, 222 (1990).
18. M. Asakawa, T. Hatsuda, and Y. Nakahara, hep-lat/9909137, hep-lat/0011040.
19. D.B. Leinweber, A.W. Thomas, K. Tsushima, and S.V. Wright, *Phys. Rev.* D **61**, 074502 (2000).
20. H. Neuberger, *Nucl. Phys.* B (Proc. Suppl.) **83-84**, 67 (2000); *Phys. Lett.* B **417**, 141 (1998).
21. S.J. Dong, F.X. Lee, K.F. Liu, and J.B. Zhang, *Phys. Rev. Lett.* **85**, 5051 (2000).

LATTICE STUDY OF NUCLEON PROPERTIES WITH DOMAIN WALL FERMIONS

S. SASAKI

Department of Physics, University of Tokyo,
7-3-1 Hongo, Tokyo 113-0033, Japan
E-mail: ssasaki@phys.s.u-tokyo.ac.jp

Domain wall fermions (DWF) are a new fermion discretization scheme with greatly improved chiral symmetry. Our final goal is to study the nucleon spin structure through lattice simulation using DWF. In this paper, we present our current progress on two topics toward this goal: 1) the mass spectrum of the nucleon excited states and 2) the iso-vector vector and axial charges, g_V and g_A, of the nucleon.

1 Introduction

The RIKEN-BNL-Columbia-KEK QCD Collaboration has been pursuing the domain-wall fermion (DWF) method[1] in lattice quantum chromodynamics (QCD). In DWF an extra fifth dimension is added to the lattice. By manipulating the *domain-wall* structure of the fermion mass in this fifth dimension, we control the number of light fermion species in the other four space-time dimensions. These light fermions possess exact chiral symmetry in the limit of an infinite fifth dimension. In particular: 1) fermion near-zero mode effects are well understood[2], 2) explicit chiral symmetry breaking induced by a finite extra dimension is described by a single residual mass parameter, which is very small in the present calculation, in the low-energy effective lagrangian[2], and 3) non-perturbative renormalization works well[3].

DWF is a promising new approach for treating fermions on the lattice. However, we need several tests of DWF in the baryon sector to reach our final goal of establishing the spin structure of the nucleon from first principles. Here we report our recent studies of the mass spectrum of the nucleon and its excited states and the nucleon matrix elements of the iso-vector vector and axial charges in quenched lattice QCD with DWF. Although most of the latter results are preliminary, the conclusive results in the former subject have been reported in Ref.[4].

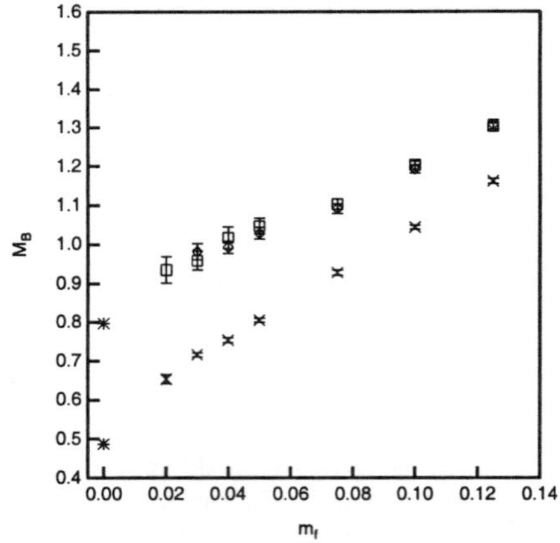

Figure 1. N and N^* (square and diamond) masses versus the quark mass m_f in lattice units. Note the large N-N^* mass splitting which is within 10% (in the chiral limit) of the experimental value (bursts).

2 Nucleon Excited States

First, we discuss the mass spectrum of the nucleon N and its excited states (the negative-parity nucleon N^* and the positive-parity first excited nucleon N') by means of a systematic investigation utilizing two distinct interpolating operators $B_1^{\pm}$ and $B_2^{\pm}$. For an explanation of those operators, see Ref.[4]. Our quenched DWF calculation was employed on a lattice with size $16^3 \times 32 \times 16$, gauge coupling $\beta = 6/g^2 = 6.0$, and domain wall height $M_5 = 1.8$. Additional details of our simulation can be found in Ref.[4].

In Fig.1 we show the low-lying nucleon spectrum as a function of the quark mass, m_f in lattice units ($a^{-1} \approx 1.9$ GeV set from aM_ρ=0.400(8) in the chiral limit). B_1^+ gives the ground-state nucleon mass N (cross). The N^* mass estimates (square and diamond) are extracted from both B_1^- and B_2^-. The corresponding experimental values for N and N^* are marked with lower and upper stars. Both N^* mass estimates extracted from two distinct operators agree with each other. The large N-N^* mass splitting is clearly evident.

In contrast to the negative parity operators, we find that the mass estimates from a second operator, B_2^+, are considerably larger than the ground state obtained from B_1^+. This suggests that B_2^+ has negligible overlap with the nucleon ground state and provides a signal for the positive-parity *excited*

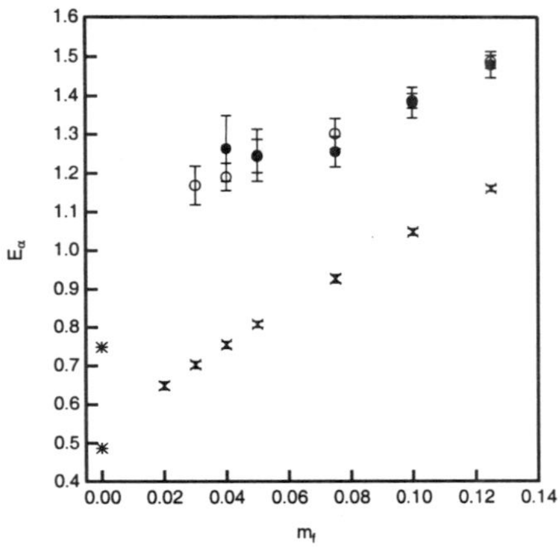

Figure 2. The mass of the positive-parity excited state (circles) is too high compared to the nucleon ground state (cross) to account for the observed splitting.

nucleon N'. To justify this possibility, we employ a sophisticated approach which utilizes the transfer matrix of a 2×2 correlation function[5] constructed from both B_1^+ and B_2^+. The diagonalization of the transfer matrix yields the excited state[5]. Figure 2 shows a comparison of the fitted mass from $\langle\langle B_2^+(t)\bar{B}_2^+(0)\rangle\rangle$ (circle) and the estimated mass from the average effective mass given by the smaller eigenvalue of the transfer matrix (bullet). The cross symbol corresponds to the nucleon ground state mass evaluated from the larger eigenvalue of the transfer matrix, which is quite consistent with the fitted mass from $\langle\langle B_1^+(t)\bar{B}_1^+(0)\rangle\rangle$

Another important conclusion can be drawn from Fig.1 and Fig.2. In the heavy quark mass region, the ordering of the negative-parity nucleon (N^*) and the positive-parity excited nucleon (N') is inverted relative to experiment. This remarkable result was originally reported in our early paper[4] and subsequently confirmed in Ref.[6]. Further systematic calculation is required to determine whether this ordering switches to the observed ordering as one approaches the chiral limit.

3 Nucleon Matrix Elements

The nucleon (iso-vector) axial charge g_A is a particularly interesting quantity. We know precisely the experimental value $g_A = 1.2670(35)$ from neutron beta decay. Why does g_A deviate from unity in contrast to the vector charge, $g_V = 1$? The simple explanation is given by the fact that the axial current is only partially conserved in the strong interaction while the vector current is exactly conserved. Thus, the calculation of g_A is an especially relevant test of the chiral properties of DWF in the baryon sector. In addition, the calculation of g_A is an important first step in studying polarized nucleon structure functions since $g_A = \Delta u - \Delta d$ where $\langle p, s|\bar{q}_f\gamma_5\gamma_\mu q_f|p, s\rangle = 2s_\mu \Delta q_f$ with $s^2 = -1$ and $s \cdot p = 0$.

We follow the standard practice[8,9] for the calculation of g_V and g_A. The three-point functions for the local vector current $J_V^f = \bar{q}_f\gamma_4 q_f$ and axial current $J_A^f = \bar{q}_f\gamma_5\gamma_i q_f$ $(i = 1, 2, 3)$ are defined by

$$G_\Gamma^{u,d}(t, t') = \mathrm{Tr}[P_\Gamma \sum_{\vec{x},\vec{x}'} \langle TB_1(\vec{x}, t) J_\Gamma^{u,d}(\vec{x}', t')\bar{B}_1(0, 0)\rangle] \tag{1}$$

with $\Gamma = V$ (vector) or A (axial) where $P_V = P_+ = (1 + \gamma_4)/2$ and $P_A = P_+\gamma_i\gamma_5$. For the axial current, the three-point function is averaged over $i = 1, 2, 3$. The lattice estimates of vector and axial charges can be derived from the ratio between two- and three-point functions

$$g_\Gamma^{\mathrm{lattice}} = \frac{G_\Gamma^u(t, t') - G_\Gamma^d(t, t')}{G_N(t)}\,, \tag{2}$$

where $G_N(t) = \mathrm{Tr}[P_+ \sum_{\vec{x}} \langle TB_1(\vec{x}, t)\bar{B}_1(0, 0)\rangle]$.

Recall that in general lattice operators $\mathcal{O}_{\mathrm{lat}}$ and continuum operator $\mathcal{O}_{\mathrm{con}}$ are regularized in different schemes. The operators are related by a renormalization factor $Z_\mathcal{O}$: $\mathcal{O}_{\mathrm{con}}(\mu) = Z_\mathcal{O}(a\mu)\mathcal{O}_{\mathrm{lat}}(a)$. This implies that the continuum value of vector and axial charges are given by $g_\Gamma = Z_\Gamma g_\Gamma^{\mathrm{lattice}}$. In the case of conventional Wilson fermions, the renormalization factor Z_A is usually estimated in perturbation theory (Z_A differs from unity because of explicit symmetry breaking). For DWF, the conserved axial current receives no renormalization. This is not true for the lattice local current. An important advantage with DWF, however, is that the lattice renormalizations, Z_V and Z_A, of the local currents are the same[3] so that the ratio $(g_A/g_V)^{\mathrm{lattice}}$ directly yields the continuum value g_A.

Our preliminary results are analyzed on 200 quenched gauge configurations at $\beta = 6.0$ on a $16^3 \times 32 \times 16$ lattice with $M_5 = 1.8$[7]. We choose a fixed separation in time of the nucleon interpolating operators, $t = t_{\mathrm{source}} - t_{\mathrm{sink}}$ and $t' < t$, with currents inserted in between. We take $t_{\mathrm{source}} = 5$ and $t_{\mathrm{sink}} = 21$[7].

In Fig.3 we show the dependence of the vector renormalization, $Z_V = 1/g_V^{\mathrm{lattice}}$ on the location of current insertions. A good plateau is observed

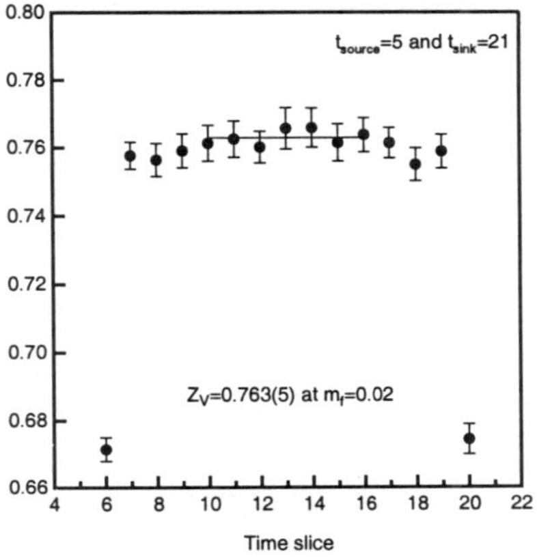

Figure 3. Dependence of vector renormalization, $Z_V = 1/g_V^{\text{lattice}}$, on t', at $m_f = 0.02$. A good plateau is observed.

so that Z_V is certainly well behaved. The value 0.763(5) at $m_f = 0.02$ (obtained by averaging over time slices denoted by the solid line in Fig.3) agrees well with $Z_A = 0.7555(3)$, which was obtained from a completely different

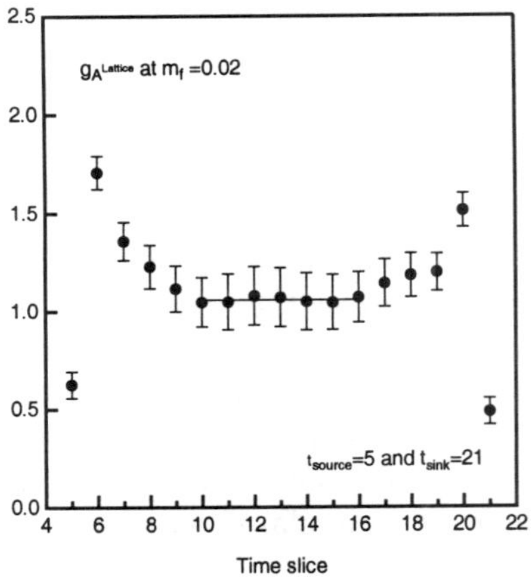

Figure 4. The lattice axial charge, g_A^{lattice}, at $m_f = 0.02$. A good plateau is seen in the range $10 \leq t \leq 16$.

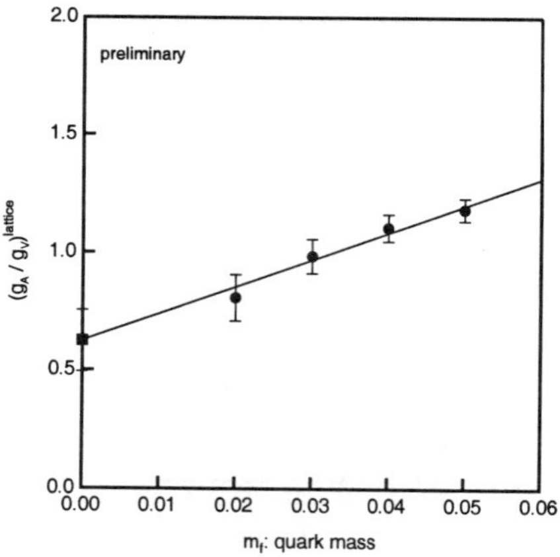

Figure 5. Dependence of $(g_A/g_V)^{\text{lattice}}$ on m_f.

calculation of meson two-point correlation functions based on the relation $\langle A_\mu^{\text{conserved}}(t)\bar{q}\gamma_5 q(0)\rangle = Z_A \langle A_\mu^{\text{local}}(t)\bar{q}\gamma_5 q(0)\rangle^2$.

For the axial charge, g_A^{lattice}, plateaus are seen for $10 \le t \le 16$ in Fig.4, so the charge ratios $(g_A/g_V)^{\text{lattice}}$ at each m_f are averaged in this time slice range. We find that there is a fairly strong dependence of $(g_A/g_V)^{\text{lattice}}$ on m_f as shown in Fig.5. A simple linear extrapolation to $m_f = 0$ yields $g_A = 0.62(13)$, which is roughly 1/2 the experimental value[7]. However, a simple linear ansatz may not describe the data, in which case the result in the chiral limit may be even smaller.

To compare to results using Wilson fermions, we plot g_A versus the square of the π-ρ mass ratio in Fig.6. Our result is given by the bullets. We also include two heavier mass points (with large errors) from an earlier simulation using a larger separation between the source and the sink. The triangles and diamonds are from Wilson simulations at relatively strong[8] and weak coupling[9]. At first glance, the DWF and Wilson fermions seem to be in rough agreement except for the lightest point. However, our DWF results have a strong mass dependence. This may be a finite volume effect; the Wilson results at strong coupling were simulated on a lattice with a volume which is twice ours.

A couple of comments on the mass dependence of our data are in order. First, it is interesting to note that our results appear consistent with the value 5/3 (marked as star) in the heavy quark limit, while the others seem

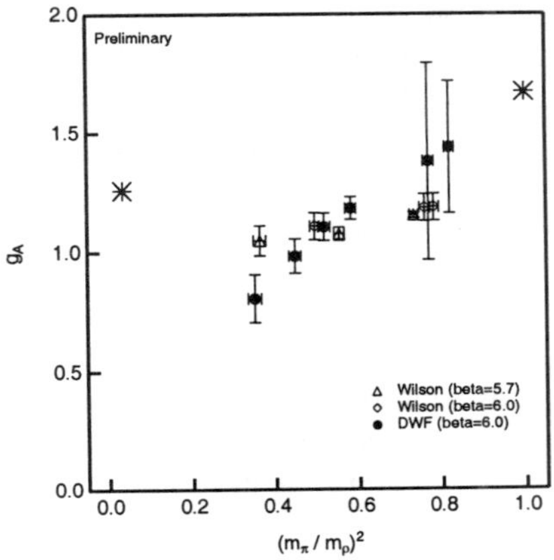

Figure 6. g_A versus $(m_\pi/m_\rho)^2$.

inconsistent with this limit. Second, our results may also be consistent with vanishing in the chiral limit. This can be explained through the axial Ward-Takahashi identity which governs g_A. If the PCAC relation $m_\pi^2 \propto m_f$ is modified, for example by chiral logarithms, the nucleon matrix element of the pseudoscalar density does not develop a pole as $m_f \to 0$, and the r.h.s of the identity vanishes in the chiral limit. Thus, g_A must also vanish as $m_f \to 0$. Indeed, we already know that the PCAC relation for the pion mass is modified in the quenched approximation by two effects: zero-modes of the Dirac operator and the quenched chiral logarithm[2]. Further investigation of the Ward-Takahashi identity is under way.

4 Conclusions

We have explored several nucleon properties in quenched lattice QCD using domain wall fermions toward our final goal of studying the spin structure of the nucleon from first principles.

Our quenched DWF calculation reproduces very well the large mass splitting between the nucleon $N(939)$ and its parity partner $N^*(1535)$[4]. We have also calculated the mass of the first positive-parity excited state $N'(1440)$ by the diagonalization of a 2×2 matrix correlator and confirmed that it is heavier than the negative-parity excited state $N^*(1535)$[4]. A remaining puzzle is whether or not a switching of N^* and N' occurs close to the chiral limit.

A preliminary calculation of iso-vector vector and axial charges shows that all the relevant three-point functions are well behaved. Z_V determined from the nucleon matrix element of the vector current agrees closely with that from an NPR study of quark bilinears[3] and a direct calculation using meson correlation functions[2]. This indicates that $g_V = 1$ and $Z_V = Z_A$ are mutually satisfied in our quenched DWF calculation. However, a linear extrapolation of g_A to the chiral limit yields a value which is a factor of two smaller than the experimental value. We are currently investigating the Ward-Takahashi identity that governs g_A to shed light on this behavior. We also plan to check related systematic effects arising from finite lattice volume and quenching (for example quenched chiral logarithms, zero modes, and the absence of the full pion cloud), especially in the lighter quark mass region.

Acknowledgments

The author would like to thank the organizers of NSTAR2001 for an invitation. This work was done in collaboration with Tom Blum and Shigemi Ohta as a part of the RIKEN-BNL-Columbia-KEK QCD Collaboration. We thank RIKEN, Brookhaven National Laboratory, and the U.S. Department of Energy for providing the facilities essential for the completion of this work.

References

1. An early review of domain wall fermions is given in T. Blum, *Nucl. Phys.* B (Proc. Suppl.) **73**, 167 (1999) ; For a recent review see P. Vranas, *Nucl. Phys.* B (Proc. Suppl.) **94** 177 (2001).
2. T. Blum, *et al.*, hep-lat/0007038.
3. T. Blum, *et al.*, hep-lat/0102005.
4. S. Sasaki, T. Blum, and S. Ohta, hep-lat/0102010; S. Sasaki, *Nucl. Phys.* B (Proc. Suppl.) **83**, 206 (2000) and hep-ph/0004252.
5. M. Lüscher and U. Wolff, *Nucl. Phys.* B **339**, 222 (1990).
6. F.X. Lee, hep-lat/0011060; D. Richards, hep-lat/0011025.
7. T. Blum, S. Ohta, and S. Sasaki, *Nucl. Phys.* B (Proc. Suppl.) **94**, 295 (2001).
8. M. Fukugita *et al., Phys. Rev. Lett.* **75**, 2092 (1995).
9. M. Göckeler *et al., Phys. Rev.* D **53**, 2317 (1996).

QUARK-HADRON DUALITY: RESONANCES AND THE ONSET OF SCALING

W. MELNITCHOUK
Special Research Centre for the Subatomic Structure of Matter,
Adelaide University, 5005, Australia, and
Jefferson Lab, 12000 Jefferson Avenue, Newport News, VA 23606, USA

We discuss the origin of Bloom-Gilman duality and the relationship between resonances and scaling in deep-inelastic scattering. We present a simple quantum mechanical model which reproduces the essential features of Bloom-Gilman duality at low Q^2, and describe applications of local duality relating structure functions at $x \sim 1$ and elastic electromagnetic form factors.

1 Introduction

Quark-hadron duality addresses some of the most fundamental issues in strong interaction physics, in particular the nature of the transition from the perturbative to the nonperturbative region of QCD. In its broadest form, it postulates that calculations of physical quantities performed in either a quark-gluon or hadronic basis should yield identical results. Although duality is in practice almost never realized exactly, there are rare cases where the average of hadronic observables can be described to good accuracy within perturbative QCD.

The correspondence between hadronic and partonic descriptions has in fact been observed in a number of processes, such as e^+e^- annihilation into hadrons, heavy quark decays, and inclusive electron-hadron scattering. It also forms the basis for theoretical approaches such as QCD sum rules. However, despite many years of studying duality, a deep understanding of its origins remains elusive. In this review, I shall discuss some recent progress made in unraveling the origin of duality in electron scattering, first observed some 30 years ago by Bloom and Gilman.

2 Bloom-Gilman Duality

In studying inelastic electron scattering in the resonance region and the onset of scaling, Bloom and Gilman found[1] that the inclusive F_2 structure function at low W generally follows a global scaling curve which describes high W data, to which the resonance structure function averages. More recently, this behavior was confirmed in high precision measurements of the F_2 structure function in the resonance region at Jefferson Lab[2]. The similarity between the averaged resonance and scaling structure functions implies that the lowest moment of F_2 is approximately independent of Q^2. Furthermore, the data clearly demonstrate that duality works remarkably well for each of the low-lying resonances, including the elastic, to rather low values of Q^2 (~ 0.5 GeV^2), as Fig. 1 illustrates.

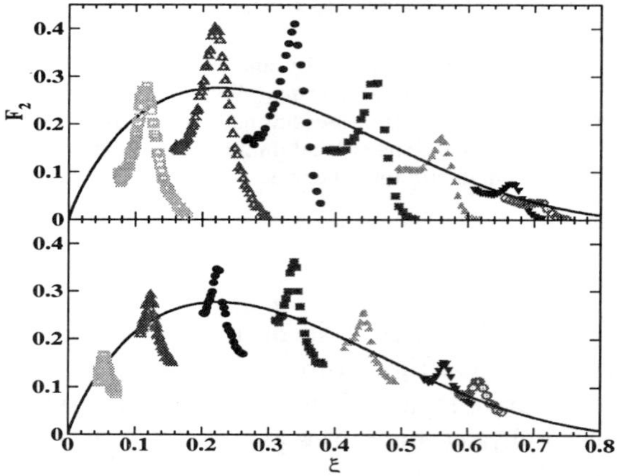

Figure 1. Proton F_2 structure function versus the Nachtmann scaling variable, ξ, in the first (upper panel) and second (lower panel) resonance regions, for various Q^2 between 0.07 GeV^2 (smallest ξ) and 3 GeV^2 (largest ξ). The solid line is the global scaling curve determined from all nucleon resonance data[3].

Before the advent of QCD, Bloom-Gilman duality was interpreted in the context of finite-energy sum rules[4], in analogy with the s and t channel duality observed in hadron-hadron scattering. In QCD, Bloom-Gilman duality can be reformulated[5] in the language of the operator product expansion, in which moments of structure functions are organized in powers of $1/Q^2$. The leading terms are associated with free quark scattering, and are responsible for the scaling, while the $1/Q^2$ terms involve interactions between quarks and gluons. The weak Q^2 dependence of the low moments of the structure function is then interpreted as indicating that the non-leading, $1/Q^2$-suppressed, interaction terms do not play a major role even at low Q^2.

An important consequence of duality is that the strict distinction between the resonance and deep-inelastic regions becomes entirely artificial. To illustrate this, consider that at $Q^2 = 1$ GeV^2 around 2/3 of the total cross section comes from the resonance region, $W < 2$ GeV. However, the resonances and the deep-inelastic continuum conspire to produce only about a 10% correction to the lowest moment of the scaling F_2 structure function at the same Q^2. Even though each resonance is built up from a multitude of twists, when combined the resonances interfere in such a way that they resemble the leading twist component.

This can be even more dramatically illustrated by considering QCD in the large N_c limit, where the hadron (or more specifically, meson) spectrum consists of infinitely narrow, noninteracting resonances. Since the quark level calculation still yields a smooth scaling curve, one sees that an averaging over hadrons must be invoked even in the scaling limit[6]. In the next section we demonstrate this duality explicitly in a simple model of large N_c QCD.

3 Understanding the Origin of Duality

The essential features of the dynamics behind Bloom-Gilman duality can be exposed with the help of a simple model in which the hadronic spectrum is dominated by infinitely narrow resonances made of valence quarks. To strip away irrelevant details which may complicate the illustration, we consider scattering from a relativistic scalar quark confined to an infinitely massive core by an oscillator-like potential. Although a model with these features will not give a realistic description of data, it will allow us to study the critical questions of when and why duality holds. Furthermore, it affords exact solutions for the complete spectrum of excited states.

If all the excited states are infinitely narrow resonances, the structure function is given entirely by a sum of squares of transition form factors[6],

$$\mathcal{S}(\nu, \vec{q}^{\,2}) = \sum_{N=0}^{N_{\max}} \frac{|\vec{q}|}{4E_0 E_N} \, |F_{0N}(\vec{q})|^2 \, \delta(E_N - E_0 - \nu) \,, \tag{1}$$

where E_0 and E_N are the energies of the ground state and N-th excited state, respectively, and F_{0N} is the transition form factor. The sum over N is restricted to a maximum value $N_{\max}$ allowed by the available energy transfer, ν. Note that for a scalar probe, the structure function has dimensions of $(\text{mass})^{-2}$. The corresponding scaling variable[6],

$$u = \frac{1}{2m} \left(\sqrt{\nu^2 + Q^2} - \nu \right) \left(1 + \sqrt{1 + \frac{4m^2}{Q^2}} \right) \,, \tag{2}$$

is defined with respect to the quark mass, m (rather than the target mass, M, which is infinite), and takes into account nonzero mass effects at finite Q^2. In the limit $Q^2 \to \infty$, the variable $u \to (m/M)x$.

In Fig. 2 we show the onset of scaling for the structure function $\mathcal{S}$ as a function of u (the energy-dependent δ-function has been smoothed out by a Breit-Wigner shape with some width in order to display the u dependence). With increasing Q^2 the resonances are seen to move out towards higher u, as observed in the physical spectrum in Fig. 1. Furthermore, the area under the curves remains approximately constant, indicating that global duality is reproduced by the model. Remarkably, the resonance "spikes" at lower Q^2 also tend to oscillate (at least qualitatively) around the scaling curve, reminiscent of the local duality observed in the proton F_2 data. The curve at $Q^2 = 5$ GeV^2 is already close to the asymptotic scaling function[6],

$$\mathcal{S}(u) = \frac{E_0}{\sqrt{\pi}\beta} \exp\{(E_0 - mu)^2/\beta^2\}, \tag{3}$$

which is obtained by taking the continuum limit for the energy, and taking the sum over all excited states (β is related to the relativistic string constant).

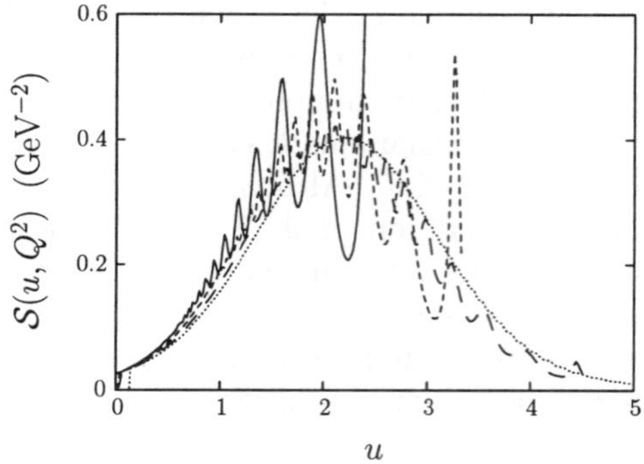

Figure 2. Scalar structure function versus the scaling variable u, for $Q^2 = 0.5$ (solid), 1 (short-dashed), 2 (long-dashed) and 5 GeV^2 (dotted).

Note that the curves in Fig. 2 are at fixed Q^2, but sweep over all ν. Since $\vec{q}^{\,2} = Q^2 + \nu^2$, as ν is increased, more and more thresholds for creating excited states are crossed, so that the density of states per $\vec{q}^{\,2}$ interval increases correspondingly. In fact, the correct density of states is crucial to compensate for the falling off with $\vec{q}^{\,2}$ of each individual form factor, $F_{0N} \sim \vec{q}^{\,N} \exp(-\vec{q}^{\,2}/4\,\beta^2)$. Had the spacing between the excited states not been Regge-like, $E_N \sim \sqrt{N}$, but linear as for the nonrelativistic case, the density of states would not have been sufficient to produce duality. While it remains an interesting exercise to see whether potentials other than a harmonic oscillator give rise to similar behavior, this simple example illustrates that Bloom-Gilman duality appears as a natural consequence of even the simplest quantum mechanical systems, with the only requirements being confinement and a correct treatment of kinematics.

4 Applications of Local Duality

If the inclusive–exclusive connection via local duality is taken seriously, one can relate structure functions measured in the resonance region to electromagnetic transition form factors[1,5]. Isolating an individual resonance contribution to the inclusive structure function is problematic, however, since the separation of the resonance from the nonresonant background is model-dependent. For the extreme case of elastic scattering, on the other hand, there is no background below the pion production threshold, so the extraction of the elastic form factors from the inclusive structure function data (and vice versa) is free of this ambiguity.

The elastic magnetic form factor of the proton has in fact been extracted[2] from the recent Jefferson Lab data, and found to agree to within 30% with the experimental form factor for $Q^2 < 5$ GeV^2. Conversely, empirical electromagnetic form factors at large Q^2 can be used to predict the $x \to 1$ behavior of deep-inelastic structure functions[1]. Knowledge of structure functions at large x is vital for several reasons — the $x \to 1$ behavior, for instance, is very sensitive to mechanisms for spin-flavor SU(6) symmetry breaking, for which there are nonperturbative and perturbative QCD predictions[7,8].

Of particular interest is the polarization asymmetry, A_1, which at large Q^2 is given by the ratio of spin-dependent to spin-averaged structure functions, $A_1 \approx g_1/F_1$. Assuming that the area under the elastic peak is the same as the area under the scaling function (at much larger Q^2) when integrated from the pion threshold to the elastic point[1], the polarization asymmetry at threshold can be written as[9]:

$$A_1(x_{\rm th}) = \left\{ \frac{G_M\,(G_M - G_E)}{4M^2(1+\tau)^2} + \frac{1}{1+\tau}\left(\frac{d(G_E G_M)}{dQ^2} + \tau \frac{dG_M^2}{dQ^2} \right) \right\} \Big/ \frac{dG_M^2}{dQ^2} \quad (4)$$

where $x_{\rm th} = Q^2/(2m_\pi M + m_\pi^2 + Q^2)$, and $\tau = Q^2/4M^2$. In the limit $x_{\rm th} \to 1$, corresponding to $Q^2 \to \infty$, both F_1 and $g_1 \sim dG_M^2/dQ^2$, so that $A_1^{p,n} \to 1$. Note that SU(6) symmetry predicts that for valence quarks $A_1 = 5/9$ for the proton and 0 for the neutron, while the perturbative QCD expectation based on helicity conservation is $A_1^{p,n} \to 1$ as $x \to 1$.

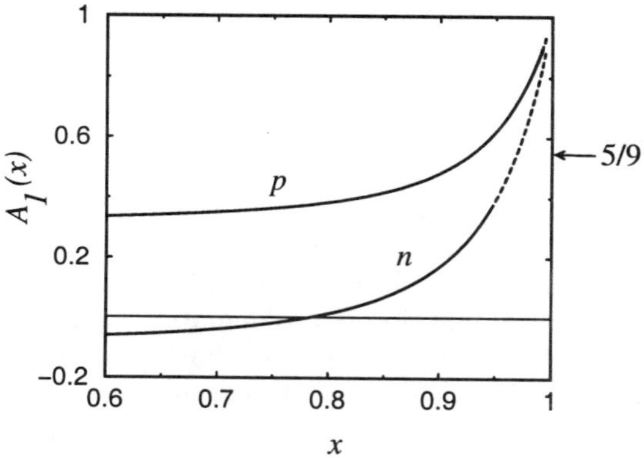

Figure 3. Polarization asymmetries A_1 for the proton and neutron at large x.

Using parameterizations of global form factor data, we show in Fig. 3 the proton and neutron asymmetries A_1 as a function of $x = x_{\rm th}$. The solid curves represent the asymmetry calculated from actual form factor data, while the dashed extensions at larger x illustrate the extrapolation of the form factors beyond the currently measured regions of Q^2. Unfortunately the current data on A_1 extend only out to an average $\langle x \rangle \sim 0.5$, and are inconclusive about the $x \to 1$ behavior. While the proton A_1 data do indicate a rise at $x \sim 0.5 - 0.6$, the neutron asymmetry is, within errors, consistent with zero over the measured range. It will be of great interest in future to observe whether, and at which x and Q^2, the A_1 asymmetries start to approach unity.

Expressions similar to Eq. (4) can be derived also for other structure functions[9]. The ratios of the neutron to proton F_1, F_2 and g_1 structure functions are shown in Fig. 4 as a function of x, with x again evaluated at $x_{\rm th}$. Asymptotically, each of the structure functions approaches dG_M^2/dQ^2, so that in the dipole approximation the n/p ratios $\sim \mu_n^2/\mu_p^2$. Also indicated in Fig. 4 are some leading twist model predictions for the ratio F_2^n/F_2^p, namely 2/3 from SU(6) symmetry, 3/7 from broken SU(6) with helicity conservation, and 1/4 from broken SU(6) with scalar diquark dominance[8]. Note, however, that the structure functions predicted from the duality relations contain both leading twist and higher twist contributions (for a discussion of the conditions under which the form factors can yield leading twist structure functions see Close & Isgur[10]).

The reliability of the duality predictions is of course only as good as the quality of the empirical data on the electromagnetic form factors and resonance structure functions. While the duality relations are expected to

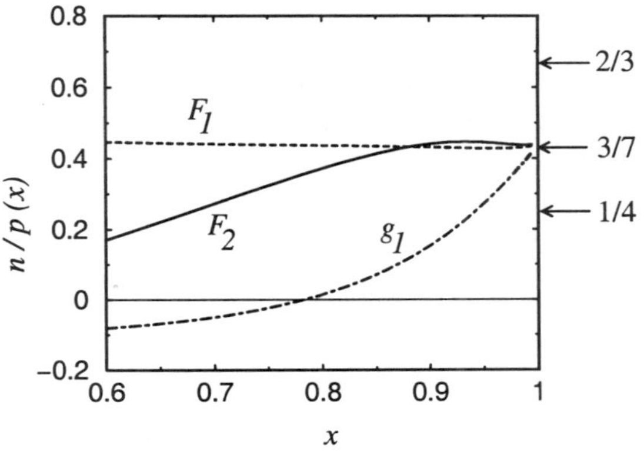

Figure 4. Neutron to proton ratio for the F_1 (dashed), F_2 (solid) and g_1 (dot-dashed) structure functions at large x.

be progressively more accurate with increasing Q^2, the difficulty in measuring form factors at large Q^2 also increases. More data on form factors at larger Q^2 would allow more accurate predictions for the $x \to 1$ structure functions, and new experiments at Jefferson Lab, Mainz and elsewhere will provide valuable constraints.

5 Conclusion

Quark-hadron duality offers the prospect of addressing the physics of the transition from the strong to weak coupling limits of QCD, where neither perturbative QCD nor effective descriptions such as chiral perturbation theory are applicable. While considerable insight into quark-hadron duality has been gained from recent theoretical studies, it will be important in future to understand more quantitatively the features of the electron scattering data in the resonance region and the phenomenological N^* spectrum in terms of realistic models of QCD.

On the experimental side, the spin and flavor dependence of duality can be most readily accessed through semi-inclusive scattering, which requires both high luminosity and a high duty factor. An energy upgraded Jefferson Lab would be an ideal facility to study meson production in the current fragmentation region at moderate Q^2, allowing the onset of scaling to be tracked in the pre-asymptotic regime. This would shed considerable light on the relationship between incoherent (single quark) and coherent (multi-quark) processes, and on the nature of the quark $\to$ hadron transition in QCD.

References

1. E.D. Bloom and F.J. Gilman, *Phys. Rev. Lett.* **16**, 1140 (1970).
2. I. Niculescu *et al.*, *Phys. Rev. Lett.* **85**, 1182 (2000).
3. C.S. Armstrong *et al.*, *Phys. Rev.* D **63**, 094008 (2001).
4. R. Dolen, D. Horn, and C. Schmid, *Phys. Rev. Lett.* **19**, 402 (1967).
5. A. de Rújula, H. Georgi, and H.D. Politzer, *Ann. Phys.* **103**, 315 (1975).
6. N. Isgur, S. Jeschonnek, W. Melnitchouk, and J.W. Van Orden, *Phys. Rev.* D, in print, hep-ph/0104022.
7. W. Melnitchouk and A.W. Thomas, *Phys. Lett.* B **377**, 11 (1996).
8. N. Isgur, *Phys. Rev.* D **59**, 034013 (1999).
9. W. Melnitchouk, *Phys. Rev. Lett.* **86**, 35 (2001).
10. F.E. Close and N. Isgur, hep-ph/0102067.

Wally Melnitchouk

GENERALIZED GDH SUM RULE AND SPIN-DEPENDENT ELECTROPRODUCTION IN THE RESONANCE REGION

J. P. CHEN*

Jefferson Lab, Newport News, VA 23606, USA

E-mail: jpchen@jlab.org

We have measured the spin-parallel and spin-perpendicular asymmetries and cross sections for inclusive $^3\vec{H}e(\vec{e},e')$ reaction from the quasielastic to beyond the resonance region. The Q^2 range covered is from 0.1 to 1 GeV^2. From these data, the ^{3}He and the neutron spin structure and the Q^2 evolution of the generalized Gerasimov-Drell-Hearn (GDH) sum are extracted. Preliminary results are presented. Plans for future studies on this subject are discussed. Also discussed is a new approved experiment in the resonance region at higher Q^2 range to study spin duality.

1 Introduction

The nucleon spin structure has been of central interest ever since the EMC experiment found that at small distances the quarks carry only a fraction of the nucleon spin. Going from shorter to larger distances the quarks are dressed with gluons and $q\bar{q}$ pairs and acquire more and more of the nucleon spin. How is this process evolving with the distance scale (or 4-momentum transfer Q^2)? At the two extreme kinematic regions we have two fundamental sum rules: the Bjorken sum rule[1] for $Q^2 \to \infty$ and the Gerasimov-Drell-Hearn (GDH) sum rule[2,3] at $Q^2 = 0$. A study of the connection between the two sum rules, in particular the evolution of the GDH sum rule in the low Q^2 region, will help us understand the transition from the incoherent processes of deep inelastic scattering (DIS) off the partons (quark-gluon picture) to the resonance dominated coherent processes (hadronic picture).

2 Generalized GDH Sum Rules

2.1 Bjorken Sum Rule

From 20 years of spin structure experiments in the deep inelastic region, one of the most important outcomes is the experimental test of the Bjorken sum rule to better than 10%. The Bjorken sum rule relates the spin structure over the entire energy range to a static property of the nucleon, the axial coupling

**FOR THE JEFFERSON LAB E94-010 COLLABORATION*

constant g_A:

$$\int_0^1 (g_1^p - g_1^n)dx = \frac{1}{6}g_A \ . \tag{1}$$

The Bjorken sum rule is based on very general principles (QCD) and is valid at the Bjorken limit: $Q^2 \to \infty$, $\nu \to \infty$ while $x = Q^2/2m\nu$ is finite. For finite Q^2, there is a QCD correction factor.

2.2 GDH Sum Rule

Another fundamental sum rule, the Gerasimov-Drell-Hearn (GDH) sum rule, holds at the other extreme of the kinematics, $Q^2 = 0$:

$$\int_{thr}^{\infty} (\frac{\sigma_{3/2} - \sigma_{1/2}}{\nu})d\nu = \frac{2\pi^2\alpha}{M^2}\kappa^2 \ . \tag{2}$$

Here again the sum rule relates the helicity (doubly polarized) cross section difference over the whole energy range to a static property of the nucleon, the anomalous magnetic moment of the nucleon, κ.

The GDH sum rule is derived using the dispersion relation (without subtraction) on the forward Compton scattering amplitude, combined with the Optical and the Low Energy Theorems. The input assumptions are very general. The only assumption, which could be open to "reasonable" questions, is the no-subtraction condition in the dispersion relation. Initial partial wave analysis[4] based on unpolarized and singly polarized data, provided an estimate of the double polarization cross section, and therefore the indirect 'experimental' sum, which indicated a discrepancy from the GDH sum rule. The direct experimental test became possible only recently, with the availability of high luminosity polarized beams and polarized targets and advanced detection technology.

The first direct measurement was performed on the proton at MAMI over an energy range from 200 MeV to 800 MeV[5]. Experiments extending the proton measurement to higher energies are on-going at Bonn or planned at several other facilities. An analysis by the Mainz group[6] shows that the GDH sum rule for the proton is reasonably satisfied. However, a similar analysis shows a significant discrepancy for the neutron GDH sum rule. Measurements of the GDH sum for the neutron are also planned[7].

2.3 Generalized GDH Sum Rule

A number of models have been proposed to extend the GDH integral for the proton and neutron to finite Q^2 and connect it to the Bjorken sum rule in the deep inelastic regime [8,9]. Recently, Ji and Osborne[10] made a rigorous generalization of the GDH sum rule to the entire region of Q^2. Under the same assumptions as the GDH sum rule, Ji and Osborne derived the generalized GDH sum rule:

$$4 \int_{el}^{\infty} \frac{G_1(\nu, Q^2)}{\nu} d\nu = S_1(Q^2)$$

where $S_1(Q^2)$ is the forward virtual Compton Scattering amplitude. The GDH sum rule and Bjorken sum rule are the two limiting cases ($Q^2 = 0$ and $Q^2 = \infty$) of the generalized GDH sum rule. Other than the two limiting cases, $S_1(Q^2)$ can also be calculated at small Q^2, where hadrons are the relevant degree of freedom, with Chiral Perturbation theory and at large Q^2, where quarks and gluons(partons) are the relevant degree of freedom, with a higher order QCD expansion (twist expansion). At small Q^2, it was calculated to leading order[11,10] and to next-to-leading order[12] using Chiral Perturbation theory in the Heavy Baryon approximation. Efforts are underway to calculate the generalized GDH sum rule to next-to-leading order without the Heavy Baryon approximation[13]. At large Q^2, twist-2 and twist-4 terms have been calculated. Efforts are underway to calculate to higher orders.

An important question in this connection is how low in Q^2 the Bjorken sum rule can be evolved using the high twist expansion. Recent estimates[10] suggest a Q^2 value as low as 0.5 GeV^2. Also at the other end, chiral perturbation theory is applicable at small Q^2, and may allow the evolution of the GDH sum rule to Q^2 up to about 0.2 GeV^2. Theoretical efforts (such as lattice calculations) are needed to bridge the remaining gap.

This would be the first time that hadronic structure is described by a fundamental theory in the entire kinematical regime, from $Q^2 = 0$ to ∞.

3 JLab E94-010 Experiment

Experiment JLab E94-010[14] has recently been completed with a polarized ^{3}He target to study the Q^2 evolution of the GDH integral for ^{3}He and the neutron. The experiment covers a range of Q^2 from 0.1 to 1 GeV^2 and from the quasi-elastic to beyond the resonance region. Figure 1 shows the kinematic coverage of the experiment.

3.1 The Experimental Setup

The experiment was carried out in Hall A of Jefferson Lab. A highly polarized (> 70%) electron beam with a current of up to 15 μA was scattered off a

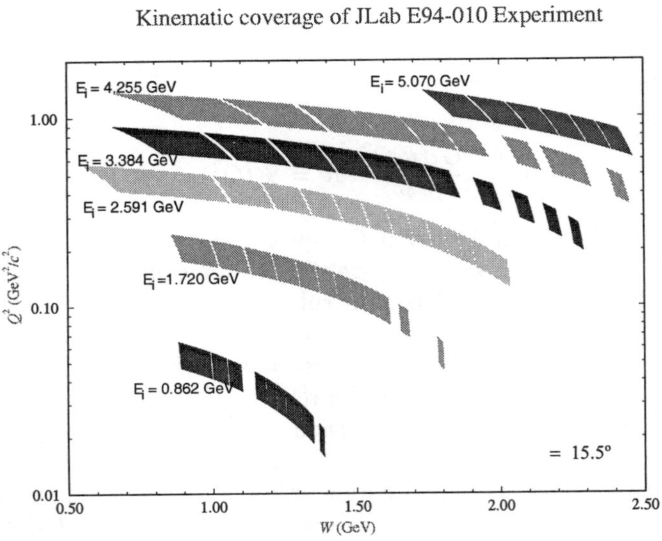

Figure 1. The kinematics of the E94-010 experiment.

high density (10 atm gaseous in a 40 cm long glass cell) polarized ^{3}He target (30-40% polarization in-beam). The polarized ^{3}He target utilized the spin-exchange principle. Rubidium atoms were polarized by optical pumping, and the polarization was transferred to ^{3}He nuclei via spin-exchange collisions. The target was polarized either parallel or perpendicular to the beam direction and the polarization was measured with two independent methods: NMR with Adiabatic Fast Passage, and EPR (Electron Paramagnetic Resonance). Both methods have an uncertainty of about 4%, and the two methods agree with each other within the uncertainty. An additional check is obtained by measuring the elastic asymmetry, which gives a measurement of the product of the beam and target polarization.

Two nearly identical spectrometers, set at the same scattering angle of 15.5°, were used to detect the scattered electrons. Both spectrometers were equipped with the standard electron detector packages. The comparison of the data from the two spectrometers provides a check of the data quality and minimizes some systematic uncertainties.

3.2 Preliminary Results

The data analysis has almost been completed. The spectrometer acceptance was carefully studied with a Monte-Carlo simulation compared to elastic data measured on a 7-foil ^{12}C target with sieve slits. Detector efficiencies were thoroughly studied. The measured ^{3}He elastic asymmetries were checked against the world elastic data weighted by the beam and target polarizations. The measured ^{3}He elastic cross section was checked against the world elastic data. Radiative corrections were performed using the formalism of Kuchto, Shumeiko amd Akushevich[15]. The spin structure functions g_1 and g_2 were extracted using the parallel cross section difference and the perpendicular cross section difference. The quantity $\sigma'_{TT} = \frac{1}{2}(\sigma_{1/2} - \sigma_{3/2})$ was extracted from g_1 and g_2:

$$\sigma'_{TT} = \frac{4\pi^2\alpha}{MK}(g_1 - \frac{Q^2}{\nu^2}g_2)$$

where $K = \nu - Q^2/2M$ is the virtual photon flux. To compute the generalized GDH sum, σ'_{TT} is needed at constant Q^2 values. We used an interpolation to extract σ'_{TT} for 10 constant Q^2 values from 0.1 to 1 GeV^2. The generalized GDH integral was computed using σ'_{TT}, at constant Q^2, integrating from the pion threshold to an invariant mass W at 2 GeV. The results of the ^{3}He GDH integral are shown in Fig. 2 as filled circles. (For the lowest Q^2 of 0.1 GeV^2, the interpolated σ'_{TT} covered only the dominant Δ peak and a little beyond. To reach $W = 2\ GeV$, extrapolation was used).

The neutron GDH integrals were extracted using the PWIA model of Ciofi degli Atti and Scopetta[16]. The extraction of the neutron GDH sum will be discussed in more detail in the next subsection. Also plotted in fig. 2 are the high Q^2 results from HERMES. For comparison, the predictions of Drechsel, Kamalov and Tiator[6] are plotted. Finally, the real photon GDH sum rule prediction for the neutron is also shown.

These data from the first precision measurement of the generalized GDH integral at low Q^2 ($0.1\ GeV^2 < Q^2 < 1\ GeV^2$) show a dramatic change in the value of the integral from what is observed at high Q^2 ($> 1\ GeV^2$). While not unexpected from theoretical models, our data illustrate the large sensitivity of the GDH integral to the transition from incoherent partonic behavior to the coherent hadronic behavior. Understanding the transition region will provide a bridge to connect the quark-gluon picture to the hadronic picture.

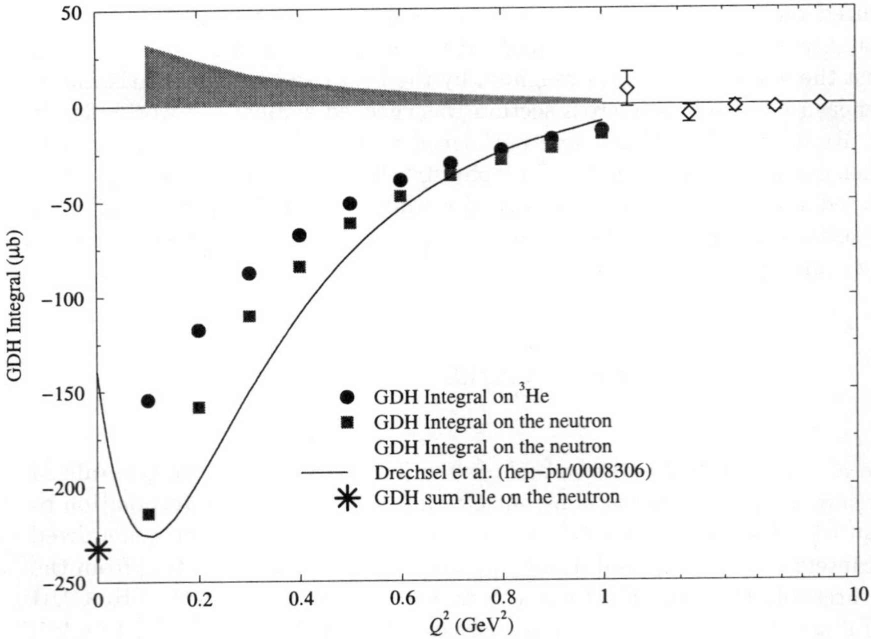

Figure 2. Preliminary results of the GDH integral on ^{3}He and the neutron. The integration was performed from the pion threshold to $W = 2\ GeV$. HERMES results at high Q^2 and the Mainz model predictions are shown for comparison. The real photon GDH sum rule value is also plotted.

3.3 Extraction of the Neutron Results

Due to the small components of S' and D waves in the ^{3}He ground state, the polarized ^{3}He target is only approximately a polarized neutron target. J. Friar et al.[17] calculated the effective polarization of the neutron and proton in a polarized ^{3}He target. The results are that the effective neutron polarization is about 87% and the effective proton polarization is about -3%. Recently, Ciofi degli Atti and Scopetta[16] studied the extraction of the neutron spin structure functions and the generalized GDH sum from a polarized ^{3}He target. Although at the resonance region at low Q^2, the extracted neutron spin structure function g_1 could differ significantly from g_1 of the free neutron, the extracted neutron GDH sum does not differ as much from the free neutron GDH sum. The reason is that the Fermi motion, being the main effect, does not affect the integrated results. Other theory groups[18] will also study this problem in the future.

4 GDH with Nearly Real Photons: JLab E97-110

The preliminary results connect well with the high Q^2 data from HERMES. However, at the low Q^2 end, the experiment reached its limit at $Q^2 = 0.1\ GeV^2$, where the uncertainty is relatively large. The region below $Q^2 = 0.2\ GeV^2$ is of special interest, since it is the region where Chiral Perturbation Theory is expected to be valid. Moreover, by measuring the slope near $Q^2 = 0$, one could extrapolate to the real photon $Q^2 = 0$ to test the fundamental real photon GDH sum rule.

JLab experiment E97-110[19], which is planned to take data next year, will measure the generalized GDH with polarized ^{3}He in the very low Q^2 region ($0.01\ GeV^2 < Q^2 < 0.5\ GeV^2$) and with higher electron energy (up to 6 GeV). It uses new septum magnets to reach small scattering angles for the Hall A spectrometers to enable us to reach the very low Q^2 region. The experiment will enable us to extrapolate to the real photon point and also study the convergence of the sum rule.

5 Outlook and Summary

The spin structure and the GDH sum rule study are one of the major efforts at Jefferson Lab. Another experiment measuring the resonance spin structure at higher Q^2 ($1\ GeV^2 < Q^2 < 5\ GeV^2$) was recently approved to study quark-hadron duality in the spin structure function[20]. JLab is planning to upgrade the accelerator to 12 GeV, which would allow a test of the high energy (Regge) behavior of the GDH sum rule.

JLab E94-010 is the first completed experiment measuring the spin structure and the generalized GDH sum rule for ^{3}He and the neutron in the low Q^2 region. Preliminary results were shown. A future extension to nearly real photon kinematics and higher energies is planned. A new experiment in the resonance region at higher Q^2 values was recently approved. Future GDH studies will include a high energy test with the JLab 12 GeV energy upgrade. Fruitful results from the GDH sum rule and spin structure study are expected in the next few years.

Acknowledgments

This work was supported by the US Department of Energy (DOE) and the US national Science Foundation. The Southeastern Universities Research Association operates the Thomas Jefferson National Accelerator Facility for the DOE under contract DE-AC05-84ER40150.

References

1. J. D. Bjorken, *Phys. Rev.* **148**, 1467 (1966).
2. S. B. Gerasimov, *Sov. J. Nucl. Phys.* **2**, 430 (1966).
3. S. D. Drell and A. C. Hearn, *Phys. Rev. Lett.* **16**, 908 (1966).
4. I. Karliner, *Phys. Rev.* D **7**, 2717 (1973); R.L.Workman and R. Arndt, *Phys. Rev.* D **45**, 1789 (1992); V. Burkert and Zh. Li, *Phys. Rev.* D **47**, 46 (1993); A. Sandorfi, C. S. Whisnant, and M. Khandaker, *Phys. Rev.* D **50**, 6681 (1994).
5. Mainz GDH Collaboration, J. Ahrens *et al., Phys. Rev. Lett.* **84**, 5950 (2000).
6. D. Drechsel, S.S. Kamalov and L. Tiator, *Phys. Rev.* D **63**, 114010 (2001).
7. J. Ahrens *et al.*, MAMI proposal; JLab E94-117, spokespersons, J. P. Chen, S. Gilad, and S. Whisnant.
8. M. Anselmino, B. Ioffe, and E. Leader, *Sov. J. Nucl. Phys.* **49**, 136 (1989).
9. V. Burkert and B. Ioffe, *JETP* **105**, 1153 (1994).
10. X. Ji and J. Osborne, hep-ph/9905010 (1999).
11. V. Bernard, N. Kaiser and U. Meissner, *Phys. Rev.* D **48**, 3062 (1993).
12. X. Ji, C. Kao and J. Osborne, *Phys. Lett.* B **472**, 1 (2000).
13. U. Meissner, Proceedings of GDH 2000, eds. D. Drechsel and L. Tiator, World Scientific, Singapore (2001), p. 47.
14. JLab E94-010, Spokespersons Z. E. Meziani, G. Cates, and J. P. Chen.
15. T. V. Kuchto and N. M. Shumeiko, *Nucl. Phys.* B **219**, 412 (1983); I. V. Akushevich and N. M. Shumeiko, *J. Phys.* G **20**, 513 (1994).
16. C. Ciofi degli Atti and S. Scopetta, *Phys. Lett.* B **44**, 223 (1997).
17. J. Friar *et al., Phys. Rev.* C **42**, 2310 (1990).
18. W. Glöckle and G. Salme, private communication.
19. JLab E97-110, Spokespersons J. P. Chen, G. Cates, and F. Garibaldi.
20. JLab E01-012, Spokespersons N. Liyanage, J. P. Chen, and S. Choi.

DOUBLE POLARIZATION MEASUREMENTS USING THE CLAS AT JLAB

R. C. MINEHART
Physics Dept., U. Virginia
Charlottesville, VA 22906, USA
E-mail: minehart@virginia.edu

for the CLAS collaboration

Preliminary results for measurements of inclusive and exclusive scattering of polarized electrons on a polarized $^{15}NH_3$ and $^{15}ND_3$ targets using the CLAS detector at JLAB are presented.

1 Motivation and Theoretical Description

The CLAS collaboration at Jefferson Laboratory is studying the scattering of polarized electrons from dynamically polarized $^{15}NH_3$ and $^{15}ND_3$ targets. Here we present preliminary results of the analysis of the first data run (EG1) carried out in the fall of 1998. A second run began in Sept., 2000 and finished in mid-April, 2001. Two beam energies, 2.5 GeV and 4.3 GeV were used in the first run.

The spin-dependence of inclusive electron scattering on the proton and neutron has been of great interest ever since the discovery from deep inelastic scattering (DIS) measurements that only a fraction of the nucleon spin could be attributed to the intrinsic spin of the quarks. Although the interpretation of spin dependence of inclusive scattering at low Q^2 is not as straightforward, its measurement is nevertheless important, particularly in the nucleon resonance region ($W < 3$ GeV), where the spin dependence is sensitive to the structure of individual resonances.

The main topic of this paper is the analysis of inclusive scattering of longitudinally polarized electrons from longitudinally polarized protons. The differential cross section for electron polarization, P_e, and nucleon polarization, P_t, is

$$\frac{d\sigma}{d\Omega_e dE'_e} = \Gamma_T \left[\sigma_T + \epsilon\sigma_L + P_e P_t \sigma_{et}\right] \quad (1)$$

where Γ_T is the transverse flux factor, ϵ is the virtual photon polarization. $\sigma_T(W, Q^2)$ and $\sigma_L(W, Q^2)$ are the absorption cross sections for transverse and longitudinal virtual photons. The double polarization absorption cross section is given by

$$\sigma_{et} = \sqrt{1-\epsilon^2}\cos\theta_\gamma A_1\sigma_T + \sqrt{2\epsilon(1-\epsilon)}\sin\theta_\gamma A_2\sigma_T \tag{2}$$

where θ_γ is the angle between the virtual photon and proton, A_1 is the asymmetry for the helicity $\frac{1}{2}$ and $\frac{3}{2}$ states of the virtual photon and proton, and $A_2(Q^2, W)$ arises from transverse-longitudinal interference. Explicitly,

$$A_1(Q^2,W) = \frac{\sigma^T_{1/2} - \sigma^T_{3/2}}{2\sigma_T}, \qquad A_2(Q^2,W) = \frac{\sigma_{LT}}{\sigma_T} \tag{3}$$

With the full data set from both EG1 runs we expect to carry out a kinematical separation of A_1 and A_2, but for now we are using a model for the relatively small contributions of A_2 and $R \equiv \sigma_L/\sigma_T$.

A_1 and A_2 are linearly related to the spin structure functions, g_1 and g_2,

$$g_1(x,Q^2) = \frac{\tau}{1+\tau}\left(A_1 + \frac{1}{\sqrt{\tau}}A_2\right)F_1(x,Q^2) \tag{4}$$

$$g_2(x,Q^2) = \frac{\tau}{1+\tau}\left(\sqrt{\tau}A_2 - A_1\right)F_1(x,Q^2) \tag{5}$$

where $\tau \equiv \frac{\nu^2}{Q^2}$, F_1 is a structure function known from unpolarized electron scattering measurements, and $x = \frac{Q^2}{2M\nu}$ is the Bjorken scaling variable.

At low Q^2, the first moment of g_1, defined as $\Gamma_1 \equiv \int_0^1 g_1(x,Q^2)dx$, is dominated by the resonance region. At high Q^2, Γ_1 is known from DIS experiments to be positive. At $Q^2 = 0$, if the contribution from elastic scattering ($x = 1$) is excluded, the GDH[1] sum rule sets the slope of Γ_1 at Q^2 to be negative, which implies that Γ_1 must change sign at some positive Q^2.

The determination of the spin structure functions from experiment can be carried out by measuring the difference in the normalized scattering rate for opposite signs of P_eP_t. The difference in the two rates, except for a small correction due to a polarization of the ^{15}N, is due to the cross section difference for hydrogen. The normalization is obtained from the beam flux measured with the Faraday Cup and the target thickness. This measurement requires an accurate knowledge of the experimental acceptance for each (Q^2, W) bin. Knowledge of acceptance is not needed to measure the experimental asymmetry ratio, defined by

$$A_{exp} = \frac{N(+) - N(-)}{N(+) + N(-)} \times \frac{1}{f_{dil}} \tag{6}$$

but a dilution factor, f_{dil}, is required to convert the total yield from NH_3 to the yield from the hydrogen. The experimental asymmetry can be expressed in terms of the photon asymmetries as

$$A_{exp} = P_e \; P_t \sqrt{1-\epsilon^2} \; \cos\theta_\gamma \left[\frac{A_1 + \eta A_2}{1 + \epsilon \sigma_L / \sigma_T} \right] \tag{7}$$

where η is a kinematical quantity.

In the resonance region, A_1 is sensitive to the structure of the transition cross sections. For example, for excitation of the $\Delta(1232)$, $\sigma^T_{3/2} = 3\sigma^T_{1/2}$ which, in the absence of other amplitudes, leads to $A_1 = -0.5$. On the other hand, pure excitation of the Roper resonance at 1440 MeV, which is a spin 1/2 state with $\sigma_{3/2} = 0$, would yield the value $A_1 = 1$. A rise of A_1 from -0.5 in the $\Delta(1232)$ region towards positive values in the Roper region could imply a substantial contribution from the Roper.

The CLAS detector is well-suited to the study of exclusive reactions involving multi-particle final states, and we are analyzing data for the single pion production channels, $\vec{p}(\vec{e}, e'p)\pi^0$, $\vec{p}(\vec{e}, e'\pi^+)n$, and $\vec{n}(\vec{e}, e'p)\pi^-$. The target neutron is, of course, bound inside the deuteron. For all of these reactions, the missing mass technique is used to identify the undetected particle. The cross section for polarized single pion production can be written in terms of virtual photon absorption cross sections as,

$$d\sigma = \Gamma_T \left(\sigma_u + P_e \sigma_e + P_t \sigma_t + P_e P_t \sigma_{et}\right) \tag{8}$$

The polarization cross sections depend on combinations of amplitudes that are not accessible in unpolarized measurements, and typically are sensitive to interferences between large and small amplitudes.

2 Description of the Experiment

The CLAS detector[2] was used to study reactions initiated by longitudinally polarized electron beams with energies of 2.6 and 4.3 GeV incident on polarized NH_3 and ND_3 targets. The beam polarization, which was approximately 70%, was measured frequently with a Moller polarimeter, and was reversed once per second. The beam current, which was monitored with a Faraday cup, was typically in the range of a 1-4 nA, corresponding to a luminosity of about $4 - 15 \times 10^{33}$ cm^{-2}.

The target consisted of NH_3 or ND_3 pellets in a 1 cm thick cell immersed in liquid Helium at a temperature of 1K. A pair of superconducting Helmholtz coils produced a 5 T polarizing magnetic field oriented along the beam direction. The polarization was driven by a 140 GHz RF field. The 50° opening angle of the Helmholtz coils is well matched to the acceptance of the CLAS Cerenkov detector and calorimeter. The electron beam was rastered over the target to expose the target material to a uniform electron flux.

Altogether we obtained 283 M electron scattering events from NH_3 at 2.6 GeV and 174 M at 4.3 GeV. We obtained 112 M from ND_3 at 2.5 GeV.

3 Data Analysis and Results

Electrons are selected from negative tracks associated with a signal in the Cerenkov counter, and with momentum matching the energy measured in the calorimeter. For absolute cross section measurements, the tracks are restricted to a fiducial volume in which the detection efficiency is essentially 100%. For asymmetry measurements no fiducial cuts are applied. A small correction, dependent on momentum and production angles, is made to the electron momentum to account for tracking errors. This correction was chosen to minimize the width of the elastic peak in the W spectrum for hydrogen.

Measurements with NH_3 and ND_3 were interspersed with measurements with a ^{12}C target and an empty target. A target model was used to simulate the ^{15}N and NH_3 spectra from the carbon measurements. For W>1.4 GeV, the simulated NH_3 spectra were compared to the measured spectra to obtain the effective density of the NH_3 target. Using this density the simulated ^{15}N data was subtracted from the measured NH_3 data to obtain the contribution of electron scattering from protons. Direct measurements in our second run with an ^{15}N target agree well with the simulated spectra.

The product $P_B P_T$ for beam and target polarization was obtained directly from the data, using two different techniques. Using the inclusive electron spectra for several Q^2 bins, the asymmetry in the region of the elastic e-p peak was compared to the known elastic asymmetry to obtain the polarization product. The consistency of $P_B P_T$ for the different Q^2 bins served as a check on the data. A second independent measurement was obtained from exclusive elastic ep data. For a limited range of kinematics both the scattered electron and recoil proton were in the angle region allowed by the target's Helmholtz coils. The strict correlation in angle and momentum makes it easy to separate the elastic ep events from the quasi-elastic scattering from nitrogen. With this technique there is no need to use the carbon data to subtract the nitrogen contribution. The average target polarization was in the range of 50-60% for NH_3 and 15-20% for ND_3. Much higher average polarizations were achieved in the second data run.

The electrons for each sign of $P_e P_t$ were sorted into Q^2 and W bins. Experimental asymmetries, defined in Eq. 6, were used to obtain the combination

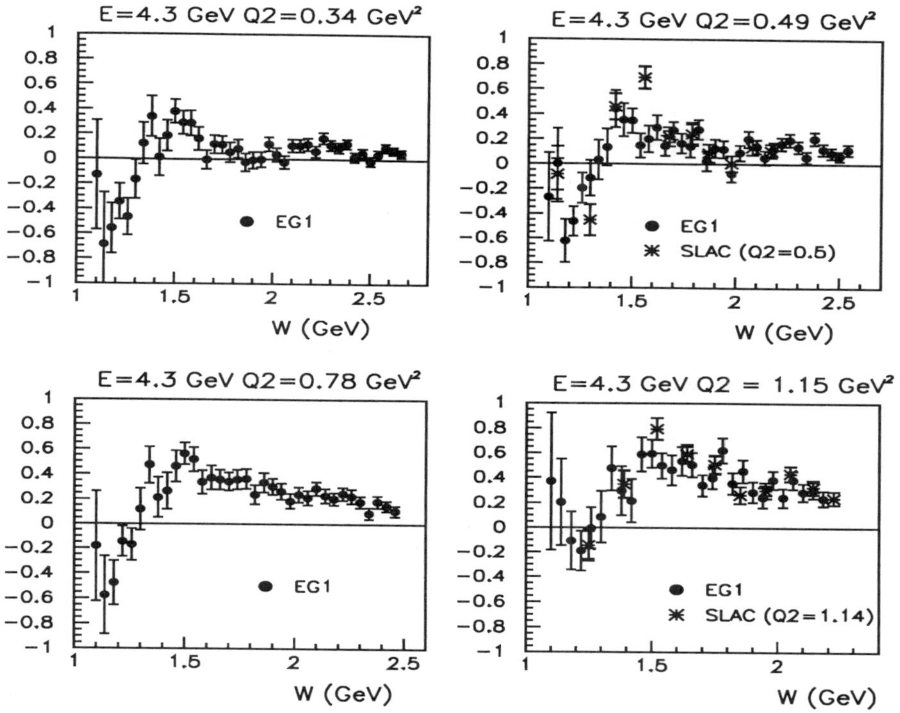

Figure 1. $A_1 + \eta A_2$ vs. W(GeV) for the proton (preliminary). Our data are shown as circles. SLAC data (10 points in each of the right side plots) are shown as crosses.

$A_1 + \eta A_2$ for the proton. Radiatively corrected results for four representative Q^2 bins are shown in Fig. 1 for an incident beam energy of 4.3 GeV. Data from SLAC[4] are included for comparison. Using a model calculation of F_1, A_1, A_2 and R, the dominant A_1, and g_1, can be extracted from our measurement of $A_1 + \eta A_2$. The model makes use of all the world's ep data along with reasonable assumptions about the Q^2 and x dependence of the structure functions. Results for g_1 for $Q^2 = 1$ GeV^2 bin are shown in Fig. 2. The result for the same quantity from our model is shown as a solid line. We can integrate the $g_1(x, Q^2)$ obtained in this way to calculate the measured contribution to the integral Γ_1. (As mentioned before, we exclude elastic scattering from this calculation.) The result is shown in Fig. 3(a). Also shown in this figure is

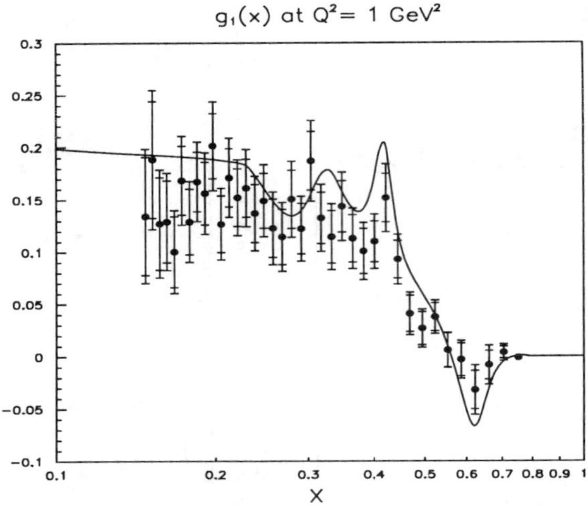

Figure 2. $g_1(x)$ vs. x for the proton, for $Q^2 = 1$ GeV^2 (Preliminary). The solid line is a calculation using the model described in the text. Double bars indicate statistical and (statistical + systematic) errors.

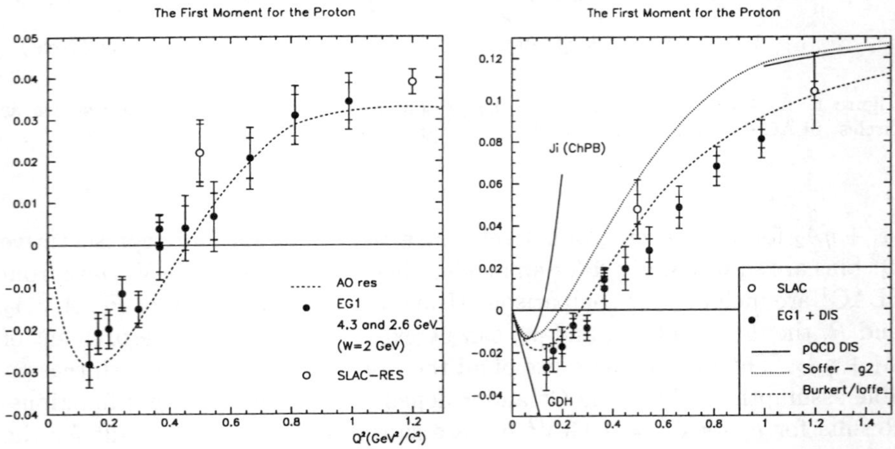

Figure 3. Γ_1 vs. Q^2 for the proton (Preliminary). Double bars indicate statistical and (statistical + systematic) errors. (a) Solid circles show the integral using only our measurements. Data from SLAC are shown with open circles. The dashed line is a calculation using AO, (b) Our model is used to extend the integral to $x = 0$. See the text for an explanation of the curves.

a comparison to a CEBAF AO calculation using only resonance production. The integral can also be extended down to $x = 0$ by using the model calculation for $x < x_{thr}$. The results are also shown in Fig. 3. The slope predicted by the GDH sum rule and the chiral perturbation calculation of Ji et al.[7] at low Q^2 are also indicated. The DIS trend is shown at large Q^2. The calculations of Soffer et al.[8] and Burkert and Ioffe[9] are shown. The expected reversal in sign of Γ_1 at $Q^2 \approx 0.3$ GeV^2 is observed. We have also extracted $A_1 + \eta A_2$ for the deuteron. The results clearly show resonance structure. However, the deuterium data taken in the first EG1 run period were too limited to draw any detailed conclusions. The second EG1 run contains extensive data on the deuteron, so that we expect high quality results from its analysis.

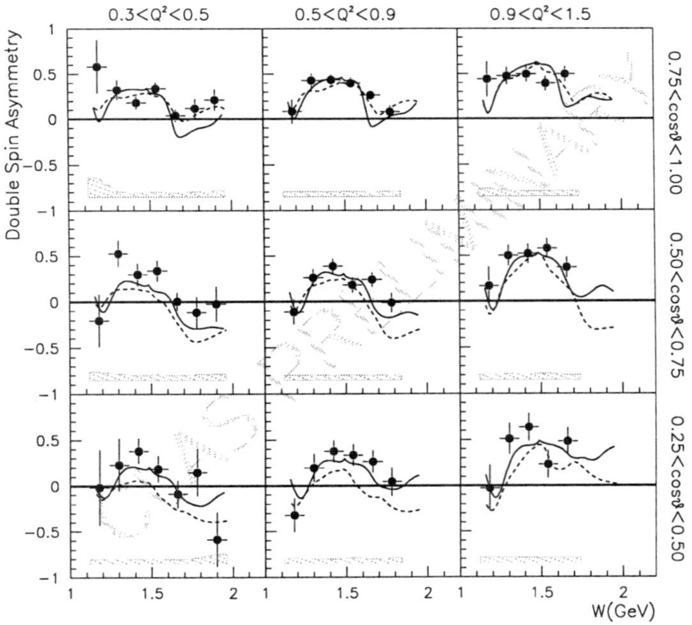

Figure 4. Double asymmetry for $\vec{p}(\vec{e}, e'\pi^+)p$. The horizontal bins are from left to right, $.3 < Q^2 < .5$, $.5 < Q^2 < .9$, and $.9 < Q^2 < 1.5$. The vertical bins are from bottom to top, $.25 < \cos\theta < .5$, $.5 < \cos\theta < .75$, and $.75 < \cos\theta < 1$. The solid lines were obtained using MAID2000, and the dashed lines were obtained using the AO program.

To illustrate the status of our analysis of exclusive reactions, examples of the results for the radiatively corrected double asymmetry vs. W in single π^+ production at 2.5 GeV are shown in Fig. 4 for nine $(Q^2, \cos\theta)$ bins. We have taken advantage of the characteristic angle dependence and integrated the asymmetry over the hadronic azimuthal angle, ϕ^*. The figures also show the results of calculations using the MAID2000[10] program and the AO[3] program. Both calculations reproduce the general features of the data and the dependence on resonance structure.

Acknowledgments

This report represents the work of the entire CLAS collaboration at Jefferson Lab.

References

1. S. Gerasimov, *Yad. Fiz.* **2**, 598 (1965); S.D. Drell and A.C. Hearn, *Phys. Rev. Lett.* **16**, 908 (1966).
2. See previous presentations, e.g. R. C. Minehart, p. 308, R. DeVita, p. 336, Proceedings of NSTAR2000, eds. V. D. Burkert *et al.*, World Scientific, Singapore (2001).
3. AO Program, V.D. Burkert, unpublished.
4. K. Abe *et al.*, *Phys. Rev.* D **58**, 112003 (1998).
5. J. Edelmann, G.Piller, N. Kaiser, and W. Weise, *Nucl. Phys.* A **665**, 125 (2000).
6. T. Gehrmann and W.J. Stirling, *Phys. Rev.* D **53**, 6100 (1996).
7. Xiang-Dong Ji, Chung-Wen Kao, and J. Osborne, *Phys. Lett.* B **472**, 1 (2000).
8. J. Soffer and O.V. Teryaev, *Phys. Rev. Lett.* **70**, 3372 (1993), *Phys. Rev.* D **51**, 25 (1995).
9. V.D. Burkert and B.L. Ioffe, *Phys. Lett.* B **296**, 223 (1992), *J. Exp. Theor. Phys.* **78**, 619 (1994).
10. D. Drechsel, O. Hanstein, S.S. Kamalov, and L. Tiator, *Nucl. Phys.* A **645**, 145 (1999).

THE HELICITY DEPENDENT EXCITATION SPECTRUM OF THE NUCLEON AND THE GDH SUM RULE

A. THOMAS FOR THE GDH- AND A2-COLLABORATIONS

Institut für Kernphysik, Universität Mainz,
J.-J. Becherweg 45, D-55099 Mainz, Germany
E-mail: thomas@kph.uni-mainz.de

The GDH collaboration is measuring the photoabsorption cross sections of circularly polarized photons on longitudinally polarized protons to determine, for the first time, a double polarization observable in a large kinematical range, which will provide new information about the excitation spectrum of the nucleon via an enhancement of small multipole amplitudes in interference terms.
The Gerasimov-Drell-Hearn (GDH) sum rule was derived in 1965 [1,2] and connects the helicity dependent photoabsorption cross sections with the anomalous magnetic moment of the nucleon. With our new data it is now possible to check the behaviour of the GDH sum rule integral experimentally[3].
The experiment has been started using the polarized electron beam of the Mainz accelerator MAMI in the energy range 140 - 800 MeV and at the moment is being continued at the Bonn accelerator ELSA up to an energy of approximately 3 GeV. We have finished the data taking period at MAMI for the proton in 1998 and we plan to do the measurement on the polarized neutron in 2002.

1 Experimental Setup

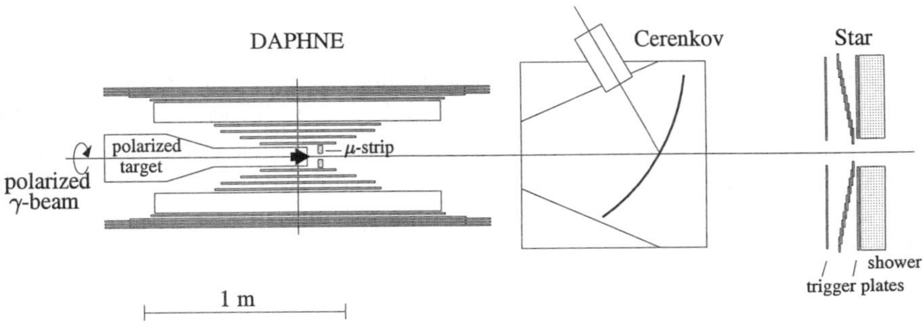

Figure 1. Experimental setup for the Mainz GDH-apparatus

The MAMI accelerator is using a strained GaAs crystal source[4] to deliver polarized electron beams with a maximum energy of 855 MeV. The photons were produced in the A2-Glasgow-Mainz tagging facility[5], which first determines the bremsstrahlungs photon energy and second monitors continuously

the degree of polarization of the electrons by detecting the asymmetry in the Møller process. The helicity transfer to the photon can be calculated reliably [6] and is energy dependent. Our maximum tagged photon energy was 800MeV, the degree of electron polarization was typically about 75%.
A solid state 'frozen spin' polarized target[7] which had been integrated into the 4π-detector DAPHNE was used. The target material butanol (C_4H_9OH), which had been chemically doped with paramagnetic radicals to allow the process of 'Dynamic Nuclear Polarization', was cooled in a $^3He/^4He$ dilution refrigerator. After 4 hours values for the degree of proton polarization of 80% - 85% were reached by irradiation with microwaves. The polarization was maintained with a relaxation time of 200 hours for the proton spins in the 'frozen spin' mode at 50 mK and a holding field of 0.4 Tesla produced by a thin superconducting solenoid[8] that was integrated into the refrigerator and operates at 1.2K. The target material was stored in a 2 cm long 2 cm diameter teflon container, leading to a total thickness of $9 \times 10^{22} \frac{protons}{cm^{-2}}$.
The cylindrical detector DAPHNE [9] was especially designed for handling multi-particle final states by provision of a large solid angle (94% of 4π), particle identification and moderate efficiency for neutral particles. It consists of three multi-wire proportional chambers and six layers of plastic scintillators.

2 Results

We have published our results for photon energies upto 450 MeV for the single pion production channels on the proton [10] and recently submitted a second paper on the total photoabsorption cross section for energies up to 800MeV [3]. Beside the importance of this data to check the GDH sum rule experimentally and to measure the forward spin polarizability, new information about the nucleon's excitation spectrum can be extracted. In this paper we compare the data with the MAID[11] and the SAID[12] predictions. These partial wave analyses are mainly based on data stemming from unpolarized and single polarized measurements.

2.1 The $\Delta(1232)$ resonance region

There has been an extended program at different laboratories to increase our knowledge about the EMR ratio for the $\Delta(1232)$ resonance[13,14]. The determination of the double polarization observable E will provide new and complementary information to clarify this question. In Fig. 2 we have compared our results[15] with predictions from the multipole analysis MAID2000 using EMR ratios of -2.5%, 0% and +2.5%. A more complete analysis of our data in collaboration with the theoretical groups from Mainz (MAID) and Washington (SAID) is planned.

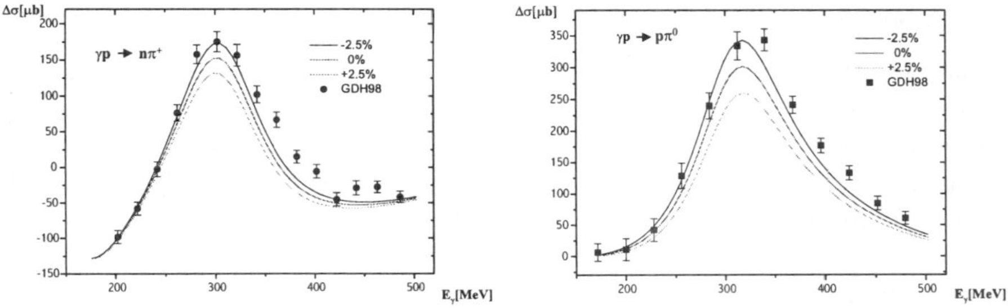

Figure 2. Sensitivity of the GDH observable $\sigma_{3/2} - \sigma_{1/2}$ for the EMR ratio of the $\Delta(1232)$ resonance for single π^+ and π^0 production. The curves were produced using MAID2000.

2.2 *The* 2nd *resonance region*

In the 2nd resonance region we can distinguish in our data all the contributing partial reaction channels. Figure 3 shows our preliminary results for single π^0 production. The agreement with the predictions from MAID2000 can be improved by changing by 30% the parametrisation of the multipoles E_{2-} and M_{2-} that drive the D_{13}-excitation. Combining the data from our double polarized experiment with the new beam asymmetry data from Grenoble[16] and

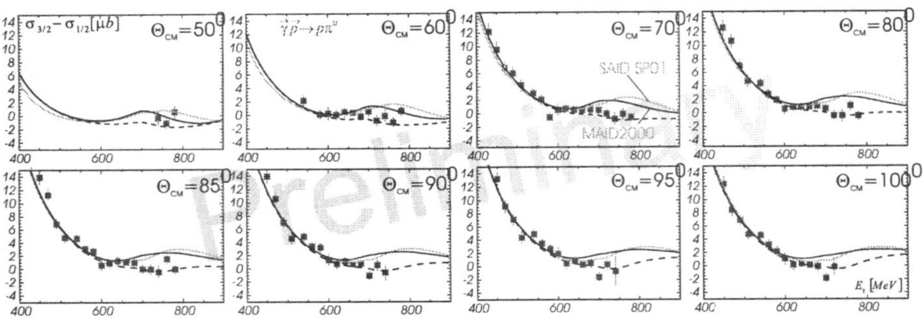

Figure 3. The Asymmetry $\sigma_{3/2} - \sigma_{1/2}$ for selected angles compared to the MAID2000 (black) and SAID SP01 (grey) analyses. The dashed curve was produced using MAID2000 with a modified parametrisation for the $D_{13}(1520)$ resonance.

Yerevan[17] that were presented at this conference will improve the knowledge on the higher, strongly overlapping resonances.

2.3 Double pion production

In the MAMI B energy range up to 800MeV we are analyzing our data for the asymmetry $\sigma_{3/2} - \sigma_{1/2}$ for the double pion production channels[18]. Comparing our results with predictions from different theoretical models[19,20] will give a new insight into the double pion production mechanism, especially the role of the $D_{13}(1520)$ and the $\Delta(1700)$ is under discussion.

Acknowledgments

This work was supported by the Deutsche Forschungsgemeinschaft (SFB 443).

References

1. S.B.Gerasimov, *Yad. Fiz.* **2**, 598 (1965), (*Sov. J. Nucl. Phys.* **2**, 430 (1966)).
2. S.D.Drell and A.C.Hearn, *Phys. Rev. Lett.* **16**, 908 (1966).
3. J.Ahrens *et al.*, submitted to *Phys. Rev. Lett.*
4. K.Aulenbacher *et al., NIM* A **391**, 498 (1997).
5. S.Hall *et al., NIM* A **368**, 698 (1996).
6. H.Olsen and L.C.Maximon, *Phys. Rev.* **114**, 887 (1959).
7. Ch.Bradtke *et al., NIM* A **436**, 430 (1999).
8. H.Dutz *et al., NIM* A **340**, 272 (1994).
9. G.Audit *et al., NIM* A **301**, 473 (1991).
10. J.Ahrens *et al., Phys. Rev. Lett.* **84**, 5950 (2000).
11. Drechsel, Hanstein, Kamalov, and Tiator, *Nucl. Phys.* A **645**, 145-174 (1999); http://www.kph.uni-mainz.de/MAID/.
12. R.A.Arndt *et al., Phys.Rev.* C **53**, 430 (1996); http://gwdac.phys.gwu.edu.
13. R.Beck and H.P.Krahn *et al., Phys. Rev. Lett.* **78**, 606 (1997).
14. G.S.Blanpied *et al., Phys. Rev. Lett.* **79**, 4337 (1997).
15. I.Preobrajensky, Ph.D. thesis in preparation, Mainz 2001.
16. A.D'Angelo, *these proceedings.*
17. H.H.Hakobyan, *these proceedings.*
18. M.Lang, Ph.D. thesis in preparation, Mainz 2001.
19. J.Nacher, *these proceedings.*
20. M.Vanderhaeghen and H.Holvoet, private communication, Mainz 2001.

STATIC MAGNETIC MOMENT OF THE $\Delta(1232)$

M. KOTULLA FOR THE TAPS AND A2 COLLABORATIONS
II. Physikalisches Institut Universitaet Giessen, Heinrich-Buff-Ring 16, 35392 Giessen, Germany

The reaction $\gamma\, p \rightarrow \pi^{\circ}\gamma'\, p$ has been measured with the TAPS photon spectrometer at the Mainz microtron facility. This reaction channel provides access to the static magnetic moment of the $\Delta^{+}(1232)$ resonance. Preliminary energy and angular differential cross section are presented and compared to recent calculations of the $\gamma\, p \rightarrow \pi^{\circ}\gamma'\, p$ reaction.

1 Introduction

The static properties of baryons like magnetic moments or polarizabilities carry valuable information about the baryonic structure. In particular, they provide an important testing ground for QCD based calculations in the confinement region. It is generally assumed that the $\Delta^{+}(1232)$ resonance has a similar quark structure as the proton, except that the spins couple to J=3/2 instead of J=1/2 as for the proton.

However, due to its short lifetime it is experimentally very difficult to investigate the internal structure of the Δ resonance. In general very little experimental information is available outside the ground state SU(3) octet. Table 1 shows predictions of different quark models for μ_{Δ} in comparison to the experimental status.

Kondratyuk and Ponomarev[7] proposed a method to investigate the static electromagnetic structure of the Δ isobar. Figure 1 shows an energy level diagram with the proton (nucleon) as the ground state and the Δ as the first excited state. The Δ structure can be probed by exciting the proton with a photon to a Δ, which subsequently emits a real photon followed by the decay into a nucleon and a pion. Spin and parity conservation requires that the electromagnetic transition is magnetic dipole (M1) radiation. This $\Delta \rightarrow \Delta\gamma$ amplitude is proportional to $\mu_{\Delta^{+}}$ and was recently investigated in Ref.[8,9]. The next allowed multipole is the higher order electric quadrupole (E2) transition, but this contribution is generally assumed to be small (e.g. compare lattice calculation[3] and E2/M1 mixing in the $\gamma p \rightarrow \Delta$ transition[10]).

Therefore, the measurement of the reaction $\gamma\, p \rightarrow \pi^{\circ}\gamma'\, p$ provides access to $\mu_{\Delta^{+}}$. Unfortunately, the final state $\pi^{\circ}\gamma'\, p$ can result from several reaction processes (compare Fig. 2). The advantage of the reaction $\gamma\, p \rightarrow \pi^{\circ}\gamma'\, p$ is that there are only heavy particles, Δ and proton, contributing to the

	$\mu_{\Delta^{++}}/\mu_N$	$\mu_{\Delta^{+}}/\mu_N$	$\mu_{\Delta^{0}}/\mu_N$	$\mu_{\Delta^{-}}/\mu_N$
PDG2000[1]	3.7–7.5	–	–	–
SU(3): $\mu_\Delta = Q_\Delta \cdot \mu_p$	5.58	2.79	0	-2.79
RQM[2]	4.76	2.38	0	-2.38
lattice QCD[3]	4.9±0.6	2.5±0.3	0	-2.5±0.3
χPT[4]	4.0±0.4	2.1±0.2	-0.17±0.04	-2.25±0.25
χQSM[5]	4.73	2.19	-0.35	-2.90
LCQSR[6]	4.4±0.8	2.2±0.4	0.0	-2.2±0.4

Table 1. Predictions of different quark models for μ_Δ in comparison to the experimental status (PDG2000).

bremsstrahlung radiation. Consequently the bremsstrahlung contributions are of the same order as the interesting $\Delta \to \Delta\gamma$ transition. In addition the dominance of the resonant reaction process of the reaction $\gamma\, p \to \pi^\circ p$ leads to the assumption that the background Born contributions are playing a minor role which makes the extraction of μ_{Δ^+} easier. The reaction channel $\gamma\, p \to \pi^+\gamma'\, n$ is in that sense less favorable for extracting the magnetic moment of the Δ^+ isobar. In summary, a consistent theoretical description of all contributing processes is crucial for extracting μ_{Δ^+}.

The magnetic moment of the Δ^{++} isobar was extracted in a similar way from the reaction $\pi^+\, p \to \pi^+\gamma'\, p$. Two measurements at the University of California (UCLA)[11] and Schweizerisches Institut für Nuklearforschung (SIN, todays name PSI)[12] have been performed and as a result of many theoretical analyses of these data the Particle Data Group[1] quotes a range of 3.7–7.5 μ_N for $\mu_{\Delta^{++}}$.

2 Experimental Setup and Analysis Methods

The reaction $\gamma\, p \to \pi^\circ\gamma'\, p$ was measured at the electron accelerator Mainz microtron (MAMI)[13,14] using the Glasgow tagged photon facility[15] and the photon spectrometer TAPS[16]. A quasi monochromatic photon beam was produced via bremsstrahlung tagging. The photon energy range covered was 205–820 MeV with an average energy resolution of 2 MeV. The TAPS detector consisted of six blocks each with 62 hexagonally shaped BaF_2 crystals arranged in an 8×8 matrix and a forward wall with 138 BaF_2 crystals arranged in a 11×14 rectangle. The six blocks were located in a horizontal plane around the target at angles of ±54°, ±103° and ±153° with respect to the beam axis. Their distance to the target was 55 cm and the distance of

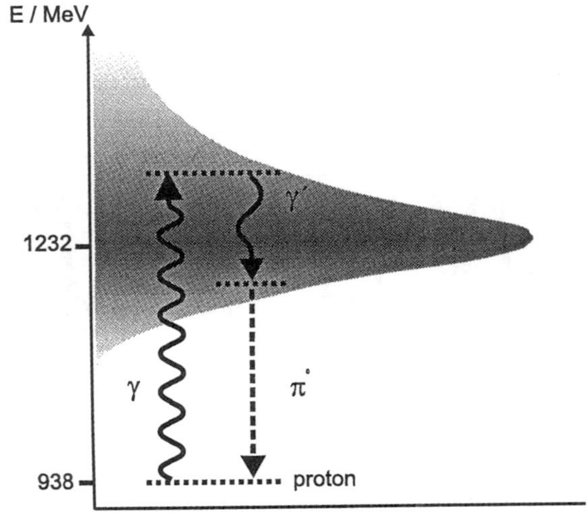

Figure 1. Method to study the static electromagnetic properties of the $\Delta^+(1232)$ isobar. The γ' transitions carries the desired information.

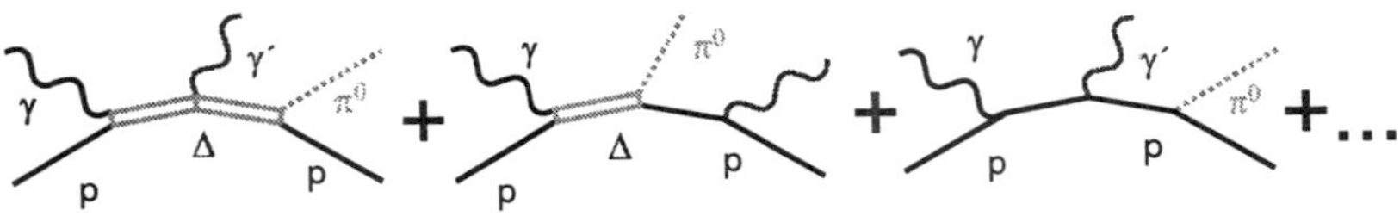

Figure 2. *Left:* diagram with an amplitude sensitive to μ_{Δ^+}. *Middle:* a *Delta*-resonant bremsstrahlung diagram and *on the right* a Born diagram as an example for other possible processes which also lead to the $\pi^\circ\gamma'\,p$ final state.

the forward wall was 60 cm. This setup covered ≈38% of the full solid angle. All BaF_2 modules were equipped with 5 mm thick plastic detectors for the identification of charged particles. The liquid hydrogen target was 10 cm long with a diameter of 3 cm.

The reaction was exclusively measured, i.e. the four-momenta of all particles in the final state were measured. The π^0 mesons were detected via their two-photon decay channel and identified in a standard invariant mass analysis using the measured photon energies and angles as input. For the data shown in the talk, the two π^0 decay photons and the third photon in the final state were distinguished by using the π^0 invariant mass as a selection criterion. The two photons with an invariant mass closest to the π^0 mass were assigned to be

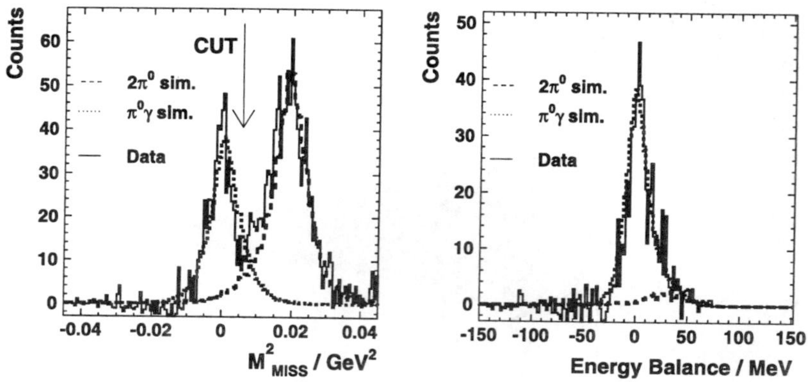

Figure 3. *Left:* Missing energy of the π^0 p in the final state. The peak near 0.02 GeV^2 originated from $2\pi^0$ production and was cut away. For the peak at 0 GeV^2 the energy balance of the reaction was checked (*right hand side*). *Right:* Energy balance for the reaction $\gamma\ p \rightarrow \pi^{\circ}\gamma'\ p$. The dashed and dotted lines show the corresponding simulated lineshapes using GEANT3.

the decay photons. The protons were identified using the excellent time resolution of the TAPS detector. The characteristic time of flight dependence on the energy of the proton and a pulse shape analysis were sufficient to obtain a very clean proton signal.

Exploiting the kinematic overdetermination of the reaction, further kinematic checks were performed. Special attention had to be paid to the $2\pi^0$ production as a possible background channel. This arises from the limited coverage of the full solid angle since one of the four $2\pi^0$ decay photons might have escaped undetected. In a first step, the conservation of the total momentum in the three Cartesian directions was checked respectively. After that a missing mass analysis was performed to discriminate a possible $2\pi^0$ contamination. The following missing mass was calculated:

$$M^2_{MISS} = ((E_{\pi^0} + E_p) - (E_{beam} + m_p))^2 - ((\vec{p}_{\pi^0} + \vec{p}_p) - (\vec{p}_{beam}))^2 \qquad (1)$$

where $E_{\pi^0}, \vec{p}_{\pi^0}, E_p, \vec{p}_p$ denote the energy and momenta of the π^0 and proton in the final state and m_p the proton mass. The resulting distribution (Fig. 3 left hand side) allowed an efficient discrimination of the $2\pi^0$ background. A Monte Carlo simulation using GEANT3 [17] of the $2\pi^0$ and $\pi^0\gamma'$ reactions reproduced the lineshape of the measured data. As the final kinematic check the energy balance was calculated to test energy conservation:

$$E_{BAL} = (E_{beam} + m_p) - (E_{\pi^0} + E_p + E_{\gamma'}) \qquad (2)$$

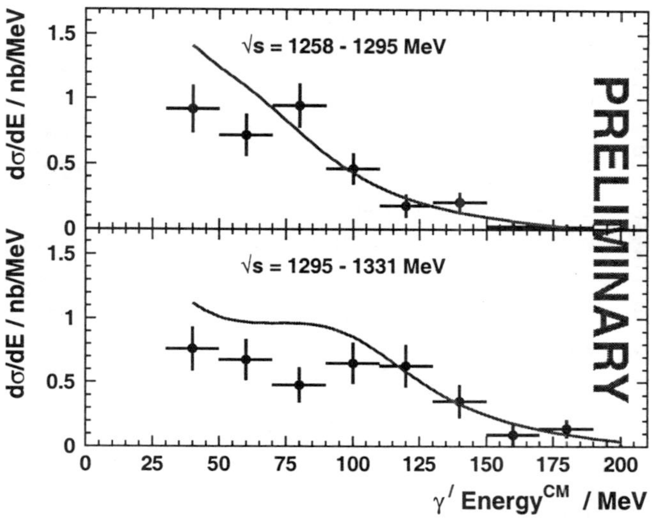

Figure 4. Preliminary energy differential cross section for two incident photon beam energies. The solid line shows the Drechsel, Vanderhaeghen calulation which includes the bremsstrahlung diagrams (Δ resonant and non-resonant Born terms). The calculation is shown for a value of $\mu_{\Delta^+} = 3\mu_N$.

the notation is the same as in Eq. (1). Figure 3 (right hand side) shows the resulting clean identification of the $\gamma\, p \to \pi^{\circ}\gamma'\, p$ reaction channel.

The cross section can be deduced from the rate of the $\pi^0\gamma' p$ events divided by the number of hydrogen atoms per cm^2, the photon beam flux, the branching ratio of π^0 into two photons and the detector and analysis efficiency. The detector and analysis efficiency were calculated performing a Monte Carlo simulation using the GEANT3 code.

3 Preliminary Results and Conclusion

The measured differential cross sections for the reaction $\gamma\, p \to \pi^{\circ}\gamma'\, p$ are shown in Fig. 4 and 5 for two different incident photon beam energies. Since the data analysis is not finished yet, only preliminary cross sections are presented.

The first series of calculations, only including the resonant $\Delta \to \Delta\gamma$ process, were done by Machavariani, Faessler, Buchmann[8] and Drechsel, Vanderhaeghen et al.[9]. Both groups use the effective Lagrangian formalism and the

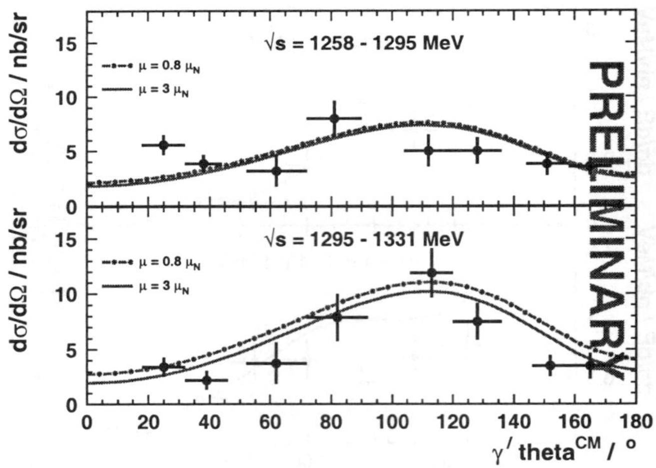

Figure 5. Preliminary angular differential cross section for two incident photon beam energies. The solid line shows the Drechsel, Vanderhaeghen calulation which includes the bremsstrahlung diagrams (Δ resonant and non-resonant Born terms) for a value of $\mu_{\Delta^+} = 3\mu_N$. The dashed dotted line shows the same calculation for a value of $\mu_{\Delta^+} = 0.8\mu_N$. The cross sections are integrated over photon energies $E_{\gamma'} > 40$ MeV.

latter one in addition a quark model approach to describe the reaction. Since these calculations are incomplete, they can not reproduce the measured cross sections.

Recently Drechsel and Vanderhaeghen[18] extended their calculation and included bremsstrahlung diagrams (resonant Δ and non-resonant Born diagrams) as indicated in Fig. 2. This calculation is shown in comparison to the preliminary data and a favorable agreement can be stated.

Figure 5 shows the sensitivity of the cross section $\frac{d\sigma}{d\Omega_\gamma}$ to a variation of μ_{Δ^+}. For an incident photon energy $E_\gamma = 450$ MeV the difference between $\mu_{\Delta^+} = 0.8\mu_N$ and $\mu_{\Delta^+} = 3.0\mu_N$ is at the 10–15% level. For a given incident excitation energy of the Δ, the transition energy of the photon γ' is given by the Δ spectral function (compare Fig. 1). The sensitivity can therefore be further increased by integrating the angular dependence for photon energies above e.g. $E_{\gamma'} > 80$ MeV.

In conclusion, the possibility of measuring the reaction $\gamma\ p \rightarrow \pi^\circ \gamma'\ p$ has been demonstrated. Further investigations have to be made in order to explore the accuracy of the Δ^+ magnetic moment that can be extracted from the present data set.

As an outlook, the current studies call for a dedicated experiment using a 4π detector with a high luminosity photon beam in order to measure the reaction $\gamma\ p \rightarrow \pi^{\circ}\gamma'\ p$ with high statistical precision. Special kinematical regions for the cross section $\frac{d\sigma}{dE_{\gamma}d\Omega_{\gamma}d\Omega_{\pi}}$ promise a higher sensitivity to μ_{Δ^+}. The same holds for measuring the photon asymmetry using a linearly polarized photon beam. Such an experiment is being prepared for the Crystal Ball detector at the Mainz Microtron accelerator facility[19].

Acknowledgments

It is a pleasure to acknowledge inspiring discussions with M. Vanderhaeghen, D. Drechsel, A. Machavariani and A. Faessler. I would like to thank the accelerator group of MAMI as well as many other scientists and technicians of the Institut fuer Kernphysik at the University of Mainz for the outstanding support. This work is supported by DFG Schwerpunktprogramm "Untersuchung der hadronischen Struktur von Nukleonen und Kernen mit elektromagnetischen Sonden".

References

1. Review of Particle Physics (Part. Data Group), *Eur. Phys. J.* C (2000).
2. F. Schlumpf, *Phys. Rev.* D **48**, 4478 (1993).
3. D.B. Leinweber *et al., Phys. Rev.* D **46**, 3067 (1992).
4. M.N. Butler *et al., Phys. Rev.* D **49**, 3459 (1994).
5. H.C. Kim *et al., Phys. Rev.* D **57**, 2859 (1998).
6. T.M. Aliev *et al., Nucl. Phys.* A **678**, 443-454 (2000).
7. L.A. Kontratyuk and L.A. Ponomarev, *Yad. Fiz.* **7**, 11 (1968) [*Sov. J. Nucl. Phys.* **7**, 82 (1968)].
8. A.I. Machavariani *et al., Nucl. Phys.* A **626**, 231 (1999).
9. D. Drechsel *et al., Phys. Lett.* B **484**, 236-242 (2000).
10. R. Beck, H.P.-Krahn *et al., Phys. Rev. Lett.* **78**, 606 (1997).
11. B.M.K. Nefkens *et al., Phys. Rev.* D **18**, 3911 (1978).
12. A. Bosshard *et al., Phys. Rev.* D **44**, 1962 (1991).
13. Th. Walcher, *Prog. Part. Nucl. Phys.* **24**, 189 (1990).
14. J. Ahrens *et al., Nucl. Phys. News* **4**, 5 (1994).
15. I. Anthony *et al., Nucl. Inst. Meth.* A **301**, 230 (1991).
16. R. Novotny, *IEEE Trans. Nucl. Sci.* A **38**, 379 (1991).
17. R. Brun *et al.*, GEANT3.21 CERN program Library, Geneva, 2000.
18. D. Drechsel and M. Vanderhaeghen, hep-ph/0105060.
19. R. Beck, B.M.K. Nefkens *et al.*, Letter of intent MAMI (2001).

Martin Kotulla

Annalisa D'Angelo

MESON PHOTOPRODUCTION AT GRAAL

A. D'ANGELO AND A. d'ANGELO, R. DI SALVO, D. MORICCIANI, C. SCHAERF
INFN Sezione di Roma II and Università di Roma Tor Vergata, 00133 Roma, Italy
E-mail: annalisa.dangelo@roma2.infn.it

O. BARTALINI, P. LEVI SANDRI
INFN Laboratori Nazionali di Frascati, 00044 Italy

F. GHIO, B. GIROLAMI
INFN Sezione di Roma I and Istituto Superiore di Sanitá, 00161 Roma, Italy

M. CASTOLDI, A. ZUCCHIATTI
INFN Sezione di Genova and Università di Genova, 16146 Genova, Italy

V. BELLINI, M.L. SPERDUTO, M.C. SUTERA
INFN Sezione di Catania and Università di Catania, 95100 Catania, Italy

G. GERVINO
INFN Sezione di Torino and Università di Torino, 10125 Torino, Italy

J.P. BOCQUET, A. LLERES, L. NICOLETTI, D. REBREYEND, F. RENARD
IN2P3 Institut des Sciences Nucléaires, 38026 Grenoble, France

J.-P. DIDELEZ, M. GUIDAL, E. HOURANY, R. KUNNE
IN2P3 Institut de Physique Nucléaires, 91406 Orsay, France

V. KOUZNETSOV, A. LAPIK, V. NEDOREZOV
Institute for Nuclear Research, 117312 Moscow, Russia

N. RUDNEV
Institute of Theoretical and Experimental Physics, 117259 Moscow, Russia

A. TURINGE
RRC Kurchatov Institute, 123182 Moscow, Russia

Precise measurements of the differential cross section and the beam polarization asymmetry Σ have been obtained by the GRAAL collaboration for the $\vec{\gamma} + p \rightarrow \eta + p$ and the $\vec{\gamma} + p \rightarrow \pi^0 + p$ reactions, at incoming photon energies up to 1.48 GeV. Preliminary data are presented and compared with different theoretical calculations. First measurements of the Σ observable are also available for the $\vec{\gamma} + p \rightarrow K^+\Lambda$ reaction channel.

1 Introduction

The existence of excited states of the nucleon can be associated with the production of baryonic resonances in $N\pi$, $N\gamma$, $N\eta$ and other N-meson systems, having well defined quantum numbers. Most of the static and dynamical properties of the baryonic resonances, such as masses, widths, photon coupling amplitudes and partial decay widths, have been extracted from the partial-wave analyses of πN scattering and π photo-production data[1]. These properties form a set of stringent constraints for hadronic quark models, which are quite successful in accounting for the masses and widths of several resonances. However the prediction of dynamical properties still represents a challenge[2].

The extraction of resonance parameters from pion data at energies above the $P_{33}(1232)$ is complicated by the fact that several broad baryon states overlap. Moreover a model independent multipole analysis of photo-production data would require experimental results on a *complete* set of polarization measurements. These must include a minimum of eight single and double polarization observables. In the absence of such complete information, the extraction of resonance parameters depends on both the choice of a theoretical model and the quality of the data.

Recently a new generation of precise data on pseudoscalar meson photo-production, including polarization observables, has become available. This is due to the advent of high quality tagged photon and electron beams (MAMI-B, ELSA, LEGS, GRAAL, JLAB-Hall B), coupled with large solid angle detectors and polarized targets or polarized photon beams.

We present here the preliminary results of the GRAAL scientific program on meson photo-production. After the publication of the first data on the beam polarization asymmetry, Σ, for the η photo-production in the energy range from threshold to $E_\gamma = 1056$ MeV[3], the data set has been extended to energies up to $E_\gamma = 1487$ MeV. Precise results on the differential cross section have also been obtained up to $E_\gamma = 1100$ MeV, fully covering the $S_{11}(1535)$ resonance for the first time. Precise results for the differential cross section and the beam polarization asymmetry are also available for the $\vec{\gamma}+p \to \pi^0+p$ reaction in the energy range from $E_\gamma = 580$ MeV to $E_\gamma = 1487$ MeV. Finally, first preliminary results are available for the $\vec{\gamma}+p \to K^+ + \Lambda$ channel.

2 The Graal Experimental Set-up

Our measurements have been obtained using the GRAAL tagged and polarized photon beam, produced by the backscattering of linearly polarized laser light on the 6.04 GeV electrons circulating in the storage ring of the European Synchrotron Radiation Facility in Grenoble. Two sets of data have been obtained using the green and the UV lines of an Argon Laser, which provide maximum energies of 1.1 GeV and 1.48 GeV, respectively. By comparing the results of the two data sets for the same energy bins and different degrees of polarization it was possible to test the internal consistency of the data and to verify our knowledge of the γ-ray beam polarization. A tagging detector provides the photon beam energy with a resolution of 16 MeV (FWHM), limited by the energy spread of the electron beam.

The Lagrange apparatus covers the entire solid angle around a cryogenic H_2 target:

• Polar angles $\theta < 25°$ are covered by two sets of plane chambers, having a track resolution of 0.5°. These are followed by a double wall of plastic scintillators, located three meters away from the target, having an angular resolution of 2°, and providing time of flight (TOF) and ΔE information for charged particles. This scintillator wall is backed by a shower detector made of several layers of lead and plastic scintillators, segmented in 16 vertical modules also having 2° angular resolution. This detector provides TOF information for charged and neutral particles and has good photon efficiency.

• Angles in the $25° < \theta < 125°$ range are covered by an excellent electromagnetic calorimeter, segmented in 480 BGO crystals of 21 radiation length thickness[4]. Charged particles discrimination is obtained by an internal barrel made of 32 plastic scintillators and charged particle identification is possible using ΔE information. Precise tracking is provided by two internal cylindrical wire chambers.

• Backward angles are covered by two disks of plastic scintillators, separated by a layer of 1cm of lead.

Photon flux is monitored by a pair of plastic scintillators, in coincidence with the tagging detector, located after a thin copper photon converter and positioned downstream the Lagrange detector. The photon detection efficiency is measured by the counting ratio at low rates with respect to a total absorption *spaghetti* calorimeter, which also serves as a beam dump. Samples of events from the two detectors are used to measure the TOF with respect to the tagging detector for accidentals corrections.

3 η Photoproduction

The study of the η meson photo-production offers the advantage of a reduced complexity for the resonances involved in the reaction. Since η carries isospin $I = 0$, only $I = 1/2$ N^* resonances may be excited and only those having

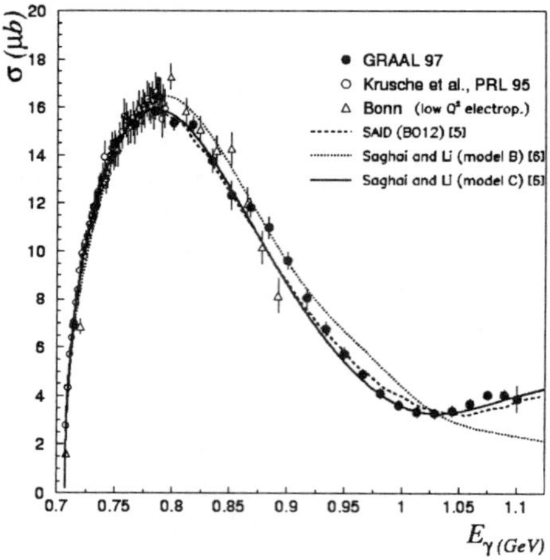

Figure 1. Preliminary results for the total cross section for the reaction: $\vec{\gamma}+p \to \eta+p$. Full circles show the Graal results, covering the full energy range of the $S_{11}(1535)$ resonance for the first time. Open circles and triangles show the results from [7] and [8] respectively. Dashed line shows the latest Partial Wave Analysis of the SAID group [5], including GRAAL data. The solid and the dotted curves show the analysis based on the Li and Saghai quark model [6], with and without the insertion of a third "missing" $S_{11}(1730)$ resonance, which can be interpreted as a ***quasi-bound*** K-Λ state.

significant ηN branching ratio may contribute.

Measurements of the Σ polarization observable add the ability of pinning down small contributions from higher multipole resonances through their interference with the main terms. These small multipoles are hidden under the dominant multipoles in unpolarized measurements. Extraction of the ηN partial widths and photocoupling amplitudes of the corresponding resonances are then possible, even if the ηN branching ratios are very small[9].

The η photo-production data were first to be analyzed. Protons were detected either in the BGO calorimeter or in the forward wall. The η were identified by detecting the photons from both $\eta \to 2\gamma$ and the $\eta \to 3\pi^0 \to 6\gamma$ decay channels in the BGO detector. The η missing mass spectrum was reconstructed from the proton momentum and the η invariant mass was calculated from photon information. With the addition of the energy measurement from the tagger, the kinematics of the reaction is overdetermined and the use of kinematical cuts easily selects the ηp events. Results for the differential cross section have been obtained from the reaction threshold up to $E_\gamma = 1.1$ GeV.

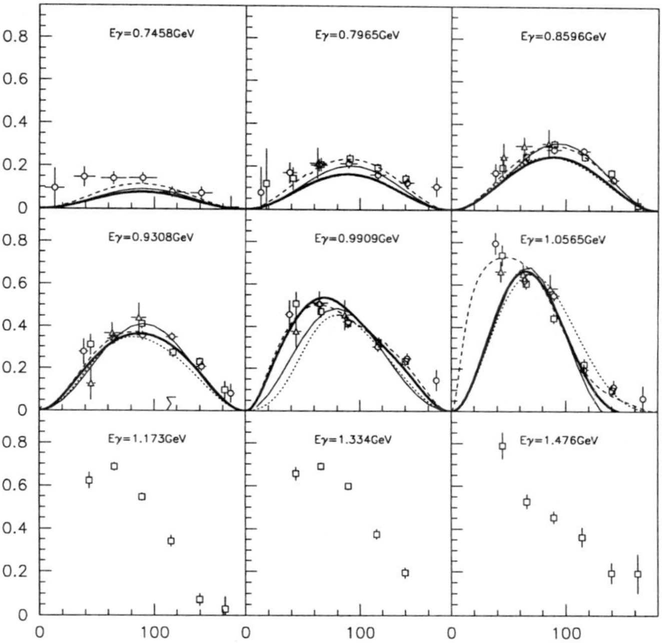

Figure 2. Preliminary results for the beam polarization asymmetry for the reaction: $\vec{\gamma}+p \rightarrow \eta + p$. The present results (open squares) are obtained with the UV Laser line and a maximum γ-ray energy of 1487 MeV. They are compared with the previous results obtained with a maximum γ-ray energy of 1.1 GeV, shown with open circles when the two γ coming from the η decay are detected in the central detectors and triangles when one of the two γ is in the forward direction. Dashed lines show the latest Partial Wave Analysis of the SAID group[5], including GRAAL data. Bold and dotted lines show the analyses based on the Li and Saghai quark model[6]. Solid lines show a global fit [9] of the GRAAL beam asymmetries [3], the Mainz differential cross sections [7] and the Bonn target asymmetries[10].

They cover the full angular range, for a total of 233 data points. The data are in good agreement with existing Mainz[7] data and confirm the nearly isotropic behavior of the angular distribution up to $E_\gamma = 0.9$ GeV, arising from the dominance of the $S_{11}(1535)$ excitation. A multipole expansion up to second order shows that deviations from isotropy at higher energies are mainly due to quadrupole terms, associated with the $D_{13}(1520)$ resonance. However, the onset of the P-wave is clear at energies above $E_\gamma = 1.0$ GeV, confirming similar results from η electro-production[13].

Preliminary results for the total cross section up to $E_\gamma = 1.1$ GeV are

Table 1. Values of the $S_{11}(1535)$ resonance parameters extracted from η photoproduction.

$\gamma + p \to S_{11}(1535) \to \eta N$					
	IB[11]	CC[12]	Chiral[6]	Krusche[7]	PDG[1]
Mass (MeV)	1541	1556	1542	1544	1536
Γ (MeV)	191	252	162	212	150
$A_{1/2}$ ($10^{-3}\sqrt{GeV}$)	118		64	125±25	90±30
$\Gamma_{\eta N}/\Gamma$		0.50	0.62		0.3 - 0.55

reported in Fig. 1. These measurements cover the full energy range of the $S_{11}(1535)$ resonance for the first time. Data are compared with the multipole analysis performed by the GWU group[5] including all GRAAL data (BO12 solution), plotted as a dashed line. Curves from a new analysis based on a chiral constituent quark model[6] that includes all known resonances up to 2 GeV and does not incorporate t-channel exchange terms, are also reported. This model requires the inclusion of a third $S_{11}(1730)$ $K\Lambda$ *quasi-bound* state to reproduce the forward peak in the cross section. Bold and dotted curves in Fig. 1 show the results *with* and *without* the third S_{11} "missing" resonance. Figure 2 shows our published Σ beam asymmetry results[3] together with a sample of preliminary data up to $E_\gamma = 1.48$ GeV. The observable is dominated by the interference of the $D_{13}(1520)$ with the main S-wave. Deviations from the $sin^2\theta$ distribution are due to contributions from other multipoles. Results from the previously described theoretical analyses are also shown. A global fit[9], combining Mainz differential cross section data[7], published GRAAL asymmetry data[3] and Bonn target asymmetry results[10], is plotted as a solid line.

Our results have also been included in the MAID 2000 analysis using an isobar model[11] and in a coupled-channel analysis using an Effective Lagrangian model and Bethe-Salpeter equation in K-matrix approximation[12].

The values for the $S_{11}(1535)$ resonance parameters, extracted using these different approaches, are summarized in Table 1 and are compared with the values quoted by the Particle Data Group. Clear discrepancies still remain among the values of the resonance width and of the photocoupling amplitude.

4 π^0 Photoproduction

Results on the differential cross section and Σ have been obtained for the $\vec{\gamma} + p \to \pi^0 + p$ reaction, using a very similar analysis procedure as in the case of η photo-production. Comparison with existing well established results provided excellent confidence on the quality of the data and the knowledge of the detector efficiency. Preliminary results for Σ are shown in Fig. 3 together

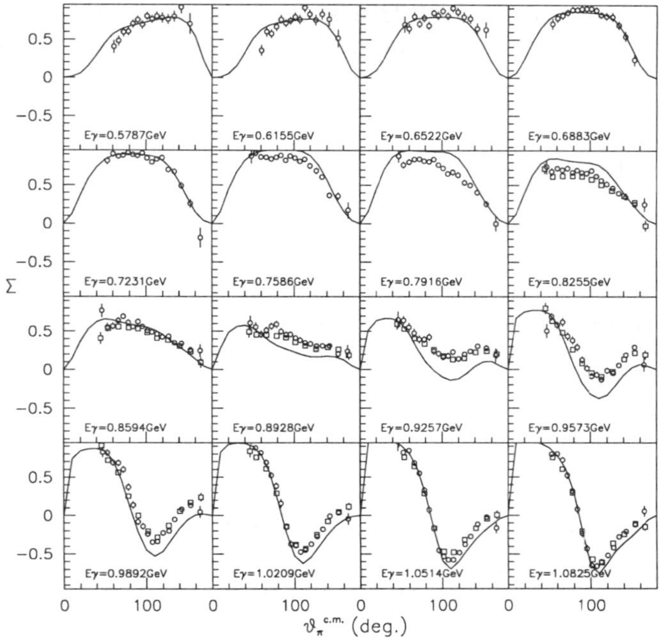

Figure 3. Preliminary results for the beam polarization asymmetry for the reaction: $\vec{\gamma}+p \to \pi^0 + p$. The results have been obtained under two different beam conditions: a maximum γ-ray energy of 1149 (circles) MeV and a maximum γ-ray energy of 1487 MeV (squares). Full lines show the SM00 solution of Partial Wave Analysis of the SAID group [5], obtained by a global fit of all the available points (including the GRAAL green line data).

with the SM00 solution of the GWU Partial Wave Analysis[5]. Important discrepancies have been found in the energy range between $E_\gamma = 0.8$ GeV and $E_\gamma = 1.0$ GeV. These have stimulated an update of the analysis and the inclusion of our results in the database.

5 K^+ Λ Photoproduction

Very preliminary data are also available for the Σ beam asymmetry in the $K^+\Lambda$ reaction channel. Data analysis relies heavily on charged particles track reconstruction from the wire chamber detectors. New data from the SAPHIR collaboration[14] show a structure in the differential cross section that has been reproduced[15] using an isobar model which includes a *missing* $D_{13}(1960)$ resonance. The same model predicts trends of opposite sign for the Σ if the missing resonance is included or not. Our very preliminary results have clearly positive values in the energy range from $E_\gamma = 1050$ MeV to $E_\gamma = 1400$ MeV and confirm the presence of the $D_{13}(1960)$ resonance.

6 Acknowledgments

We are grateful to C. Bennhold, A. Waluyo, Wen-Tai Chiang, B. Saghai and I. Strakovsky for useful discussions and communication of their work prior to publication. We thank the technical staff of INFN and IN2P3 for their essential contribution in the realization and maintenance of the apparatus and the ESRF staff for the stable and reliable operation of the storage ring.

References

1. Rev. of Part. Phys. 2000, *Eur. Phys. J.* C **15**, 1-878 (2000).
2. S. Capstick and W. Roberts, *Phys. Rev.* D **47**, 1994 (1993); *Phys. Rev.* D **49**, 4570 (1994); N. Isgur and Karl, *Phys. Lett.* B **72**, 109 (1977); N. Isgur and R. Koniuk, *Phys. Rev.* D **21**, 1868 (1980); F. E. Close and Z. Li, *Phys. Rev.* D **42**, 2194 (1990); R. Bijker *et al.*, *Phys. Rev.* D **55**, 2862 (1997).
3. J. Ajaka *et al.*, *Phys. Rev. Lett.* **81**, 1797 (1998).
4. F. Ghio *et al.*, *Nucl. Instrum. Methods* A **404**, 71 (1998); M. Castoldi *et al.*, *Nucl. Instrum. Methods* A **403**, 22 (1998).
5. Partial Wave Analysis by SAID at GWU, I. Strakovsky *et al.*, BO12 solution and SM00 solutions.
6. B. Saghai and Z. Li, DAPNIA/SPhN-01-04, subm. to *Eur. Phys. J.* A.
7. B. Krusche *et al.*, *Phys. Rev. Lett.* **74**, 3736 (1995).
8. B. Schoch, *Prog. Part. Nucl. Phys.* **34**, 43 (1995).
9. L. Tiator, D. Drechsel, and G. Knöchlein *Phys. Rev.* C **60**, 035210 (1999).
10. A. Bock *et al.*, *Phys. Rev. Lett.* **81**, 534 (1995).
11. Wen-Tai Chiang *et al.*, *these proceedings.*
12. A. Waluyo and C. Bennhold *these proceedings*; C. Bennhold *et al.*, nucl-th/9901066, nucl-th/0008024; T. Feuster and U. Mosel, *Phys. Rev.* C **59**, 460 (1999).
13. R. Thompson *et al.*, *Phys. Rev. Lett.* **86**, 1702 (2001); J. Mueller, *these proceedings*; C. S. Armstrong *et al.*, *Phys. Rev.* D **60**, 052004 (1999).
14. M.Q. Tran *et al.*, *Phys. Lett.* B **445**, 20 (1998).
15. C. Bennhold *et al.*, nucl-th/0008024.

MAXIMUM LIKELIHOOD TECHNIQUES FOR PWA OF TWO-PION PHOTOPRODUCTION

J. P. CUMMINGS
Department of Physics,
Rensselaer Polytechnic Institute
Troy, NY, 12018 USA
E-mail: cummij@rpi.edu

Experiences from multi-particle meson spectroscopy are drawn upon to aid in the analysis of multi-pion decays of baryon resonances. Analysis of such decays may be necessary to identify missing baryon states.

1 Introduction

A great deal has been learned about the non-strange baryon spectrum by studying final states containing a single pion, produced either with a pion or photon beam. Multi-pion final states such as $p\pi\pi$ have been investigated far less thoroughly, partly due to additional complexities in the analysis not present in single pion production. The acceptance is a more complicated function of much higher dimension (4 instead of 1), for instance. Also the interference of isobars in the final state makes the interpretation of projections such as two body mass plots and angular distributions much more difficult. So why pursue such an analysis?

Quark model calculations predict baryon states that have not been observed experimentally, a situation sometimes referred to as the "Missing Baryon Problem". One possible explanation for not seeing these states is simply that previous searches have been in the wrong place. If these missing states have a small coupling to $p\pi$ they would be found in neither experiments with a pion beam nor experiments looking in the $p\pi$ or $n\pi$ decay channels. We must look in reactions such as $\gamma p \to p\pi\pi$, with no $N\pi$ system in either the initial or final state.

2 Analysis Technique

The conceptually straightforward technique generally used to analyze single pion final states is to correct the data for the acceptance of the measuring apparatus and fit the data to various models of the differential cross section. One advantage of this procedure is the ability to share the data with others who would like to try fitting their model. Since the experimenters have removed any details of their apparatus from the data by performing the acceptance correction, those fitting their model to this data do not need to know specifics of the individual experiment. Indeed, this may be, and often is, collected in a world data base available to the public.[3]

Unfortunately, this becomes more difficult with multi-pion final states. The acceptance function becomes multi-dimensional and, for a modern high rate experiment using many different types of detectors, has many 'edges' leading to a complicated function of many variables.

This problem has been dealt with successfully in multi-particle meson spectroscopy, where 3, 4, and even 5 particle final states have been analyzed using extended maximum likelihood fitting methods. In this technique, the acceptance correction is done at fitting time, and while the partial-wave amplitudes that result from the fit are acceptance corrected, the raw data is never corrected itself. This is related to an inherent drawback of this method, expressed by Burnett and Sharpe[2] in their exotic meson review article:

> The separation into partial waves, isobars, and perhaps incoherent background is accomplished by a complex fitting program involving a maximum likelihood optimization. There are a number of inherent drawbacks. (1) It is difficult to judge the validity of the analysis. One must sometimes rely on the reputation of the program being used!

To study the reputation of our own program which we have used for meson analyses, we investigated its heritage. Interestingly, the roots of our partial wave analysis program can be traced back to a baryon analysis done in the sixties by SLAC/LBL. [4] These authors first used the maximum likelihood method not for the acceptance problem mentioned above, but because of another great strength of this technique: it makes maximal use of limited statistics.

In principle there are many choices for an expansion to describe the cross section, ranging from an almost purely mathematical moment expansion to a description in terms of processes involving sequential decays using the isobar model. A moment expansion, although model independent, involves a complicated mapping from moments to amplitudes in all but the simplest cases. Our program uses the isobar model to evaluate decay amplitudes for states produced either in the s or t channel, and fits the complex production amplitudes associated with them.

The partial wave expansion assumes the cross section or intensity can be written in terms of these amplitudes as

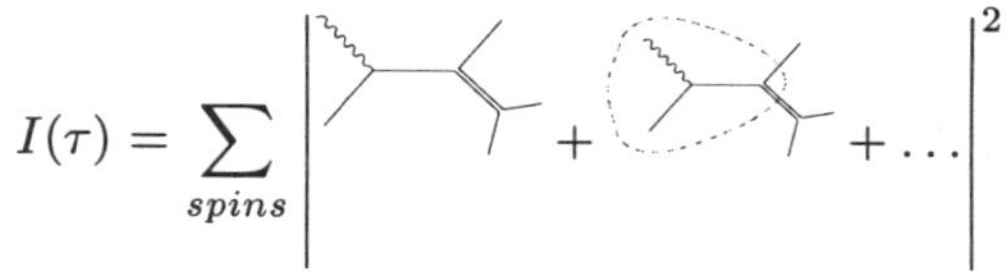

where the dashed loop encloses the parts of the diagram that are fit. The calculation of the decay amplitude includes all that is outside the loop: angular distributions, isobar masses, *etc.* These variables are collectively referred to here as τ. Symbolically, this can be written

$$I(\tau) = \sum_{spins} \left| \sum_{\alpha} V_{\alpha} \psi_{\alpha}(\tau) \right|^2 \tag{1}$$

where the V_{α} are the production amplitudes for the state α, the $\psi_{\alpha}(\tau)$ are the decay amplitudes for a state α as a function of τ, and the amplitudes are summed incoherently over unmeasured external spins. The $\psi_{\alpha}(\tau)$ are calculated using the isobar model in the helicity basis. The V_{α}'s are complex fit parameters assumed constant in bins of W.

Although we have implied only s-channel processes in our expansion above, it is possible to include other diagrams in this expansion. t-channel processes will show up in many s-channel waves, as a matter of fact, a complete description of a t-channel process in terms of s-channel waves is possible only with all s-channel waves, an infinite number. It is this fact which allows mixing of bases: as long as the s-channel basis is truncated, there will be no ambiguity introduced by including t-channel processes.

The fitting technique used to find the production amplitudes is an extended maximum likelihood fit. In a maximum likelihood fit an *unbinned* distribution is fit by maximizing the product of the probability distribution evaluated at each data point. By removing the need for binning as would be needed for a χ^2 fit, maximal use of the statistics is realized and no bias is introduced due to integration over bins.

Figure 1 compares a χ^2 and likelihood fit to a small set of one dimensional data for illustration. Fifty masses were generated according to a Breit-Wigner distribution with a mass of 1.5 GeV/c^2 and a width of 0.250 GeV/c^2. In this example both techniques did well finding the value of the mass, and while both got the correct width within the error, the likelihood fit gives almost a factor of 5 smaller errors.

The likelihood function used for the partial wave analysis is similarly

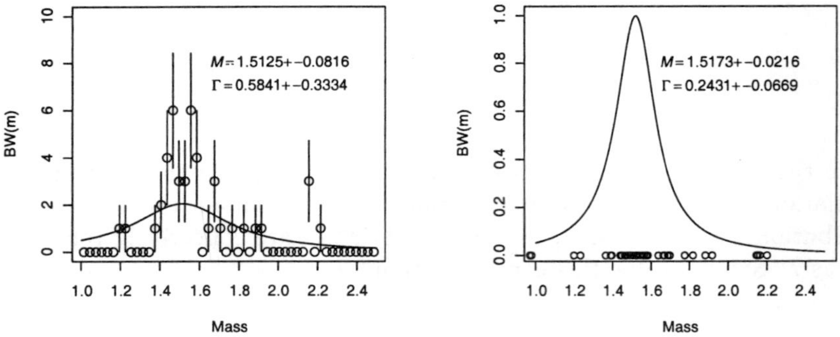

Figure 1. Comparison of statistical strength of likelihood and χ^2 fits. 50 masses generated according to Breit-Wigner with $E_0 = 1.5$ and $\Gamma = 0.250$.

defined as the product of probabilities,

$$\mathcal{L} = \left[\frac{\bar{n}^n}{n!}e^{-\bar{n}}\right]\prod_i\left[\frac{I(\tau_i)}{\int I(\tau)\eta(\tau)dP.S.}\right]$$

where the term outside the product is related to the normalization by accounting for the Poisson probability of measuring $\bar{n}$ events in a particular W bin. $I(\tau)$ is the intensity function and $\eta(\tau)$ is the acceptance function. Since $\bar{n}$ is equal to the integral in the denominator, some simple algebra gives us

$$\mathcal{L} = \exp\left(-\int I(\tau)\eta(\tau)dP.S.\right)\prod_i I(\tau_i).$$

The program actually minimizes

$$-\ln(\mathcal{L}) = -\sum_i I(\tau_i) + \int I(\tau)\eta(\tau)dP.S.$$

where the integral term is referred to as the 'normalization integral',

$$\begin{aligned}\int I(\tau)\eta(\tau)dP.S. &= \int\sum_{\alpha\alpha'} V_\alpha V_\alpha'^{*}\psi_\alpha(\tau)\psi_\alpha'^{*}(\tau)\eta(\tau)dP.S.\\ &= \sum_{\alpha\alpha'} V_\alpha V_\alpha'^{*}\int\psi_\alpha(\tau)\psi_\alpha'^{*}(\tau)\eta(\tau)dP.S.\end{aligned}$$

$$= \sum_{\alpha\alpha'} V_\alpha V_\alpha'^{*} \Psi^x_{\alpha\alpha'}$$

and the accepted normalization integral $\Psi^x_{\alpha\alpha'}$ is evaluated numerically,

$$\Psi^x_{\alpha\alpha'} = \frac{1}{M} \sum_i^{M_x} \psi_\alpha(\tau_i) \psi_\alpha'^{*}(\tau_i) \ ,$$

the sum being over phase space events accepted by a detector simulation program. Notice that the acceptance correction is done at fitting time by the normalization integral which contains $\eta(\tau)$. The acceptance correction is thus done "wave-by-wave" by integrating the acceptance separately for each wave, using the integration measure defined by the angular distribution of this wave. In other words we acceptance correct the basis states that we are using in the fit.

3 Preliminary Tests

Very preliminary tests have been done using the technique described above and a selection of CLAS $\gamma p \to p\pi^+\pi^-$ data. Two positive tracks were detected and identified using time of flight information, and the π^- was identified by missing mass. We start with a very narrow bin of $p\pi\pi$ mass, or W, between 1.440 and 1.445 GeV/c^2. The Dalitz plot for this bin is shown in Fig. 2. The

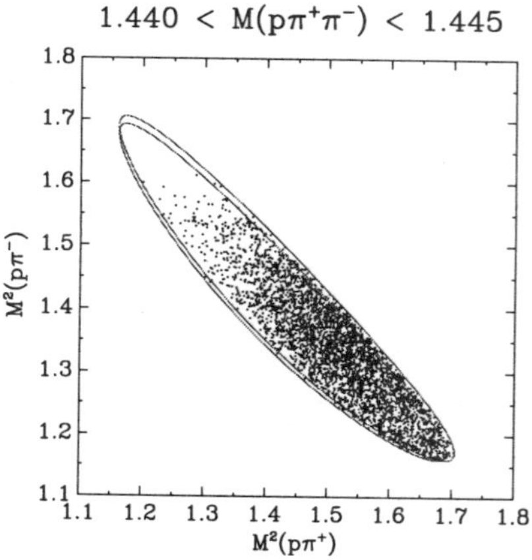

Figure 2. Dalitz plot for $p\pi^+\pi^-$ system produced at CLAS.

lack of events at low $p\pi^+$ mass is due to acceptance.

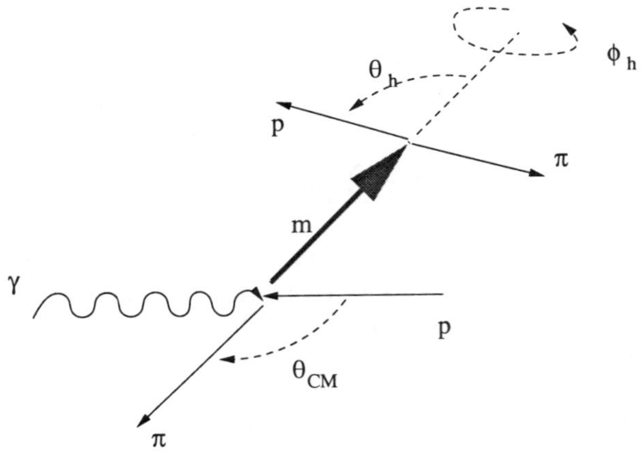

There are many possibilities for a set of kinematic variables which will describe the final state, but we will need 4 numbers: 3 four-vectors - 3 masses - 4 E and p constraints - 1 arbitrary angle = 4 values to fix an event. We choose $\tau = \theta_{CM}$, $M(p\pi^+)$, θ_h, and ϕ_h, where θ_h and ϕ_h are the angles of the proton in the $p\pi^+$ rest frame.

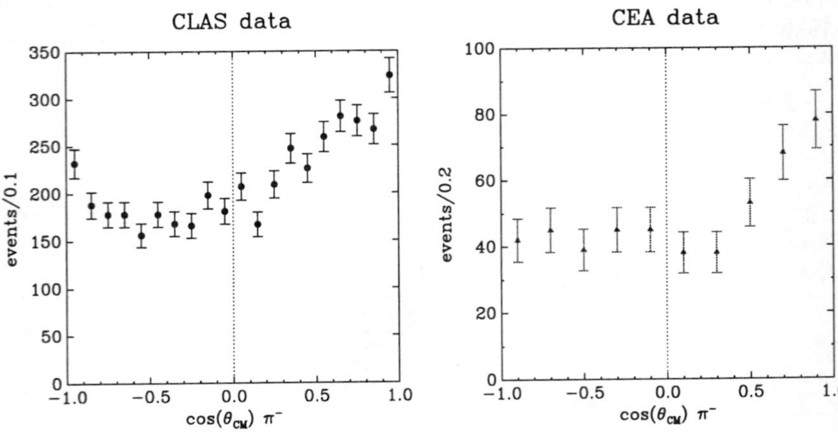

Figure 3. Distributions of $\cos(\theta)$ of the π^- in the CM system for CLAS and CEA data.

Since the π^- was not required to be detected in the data selection, the acceptance is fairly flat in $cos(\theta_{CM})$ of the π^-, and we can compare, in Fig. 3, our distribution with similar data taken using a bubble chamber at the Cambridge Electron Accelerator. [1] The CEA W bin is much wider and includes our bin. Our data compare very well with the 100% acceptance CEA data but with, 4 times the statistics in this sample of CLAS data we can see a rise in the backward direction not seen in the CEA data.

A simple fit was done to these data using only three waves, which we identify by the quantum numbers $I(J^P)$ and the ℓ of the decay into $\Delta\pi$: $\frac{1}{2}(\frac{1}{2}^+)$, $\frac{1}{2}(\frac{3}{2}^-)$ decaying in a S wave, and $\frac{1}{2}(\frac{3}{2}^-)$ decaying in a D wave. Figure 4 shows the comparison of the $\cos(\theta_{CM})$ distribution of the π^- for the data in

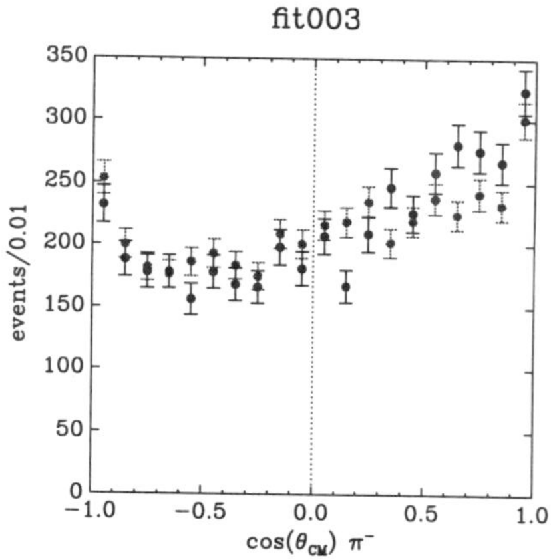

Figure 4. Comparison of data and fit predicted distributions for $\cos(\theta)$ of the π^- in the CM system.

this bin and the prediction of the fit. A very good description of the forward peaking can be obtained using only these three waves, which is due to S-P wave interference.

Similar fits are done in other bins of W scanning the range from 1.35 to 1.65 GeV/c^2. The mass dependence of the amplitudes squared is shown in Fig.5 The intensities are corrected for photon flux, however the photon flux normalization procedure for this run is *very* preliminary. The $\frac{1}{2}(\frac{3}{2}^-)$ S-wave dominates and is probably due to the $\Delta\pi$ contact term, which is an S-wave. The $\frac{1}{2}(\frac{3}{2}^-)$ shows strength above 1.5 GeV/c^2 which may be due to the $D_{13}(1530)$.

The differences in phase of two waves will show 'motion' as a function of mass when at least one of the waves contains a resonance. Figure 6 shows the phase difference of the $\frac{1}{2}(\frac{1}{2}^+)$ against the $\frac{1}{2}(\frac{3}{2}^-)$ S-wave and D-wave separately. The solid line is the expected difference of two Breit-Wigners with

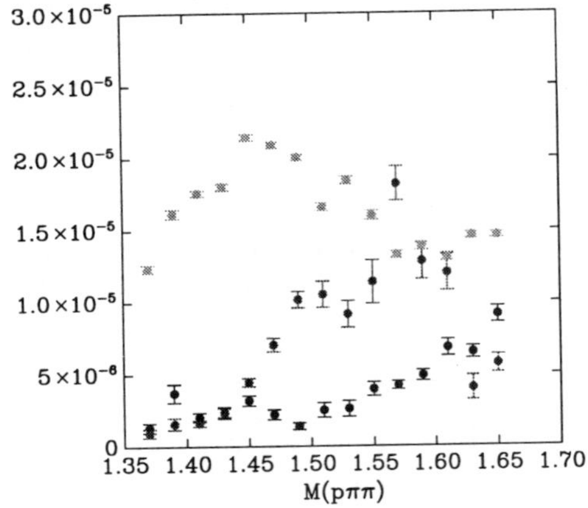

Figure 5. Mass independent analysis of $p\pi\pi$ spectrum. The blue points are $\frac{1}{2}(\frac{1}{2}^+)$, the red ones are $\frac{1}{2}(\frac{3}{2}^-)$ D-wave, and the green ones are $0.2 \times \frac{1}{2}(\frac{3}{2}^-)$ S-wave.

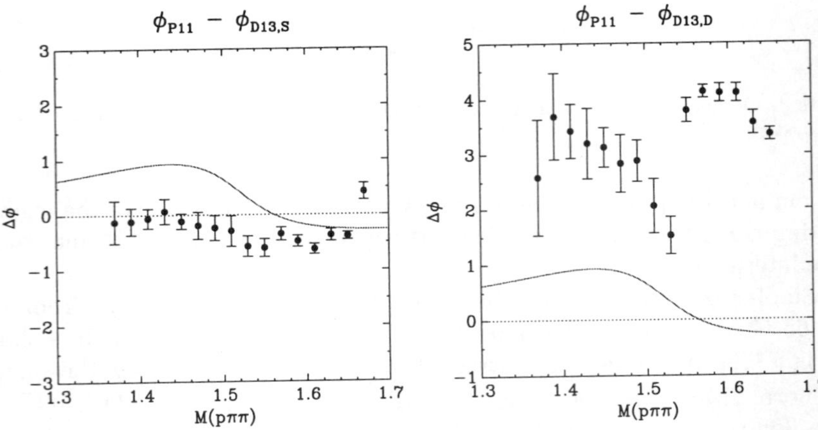

Figure 6. Phase differences from fit compared to PDG values (red line).

PDG values. No backgrounds have been considered.

4 Conclusion

Simple maximum likelihood fits using an isobar model can adequately describe multi-pion photoproduction data with a few waves. At higher energies it may be necessary to add more states in the *s*-channel as well as *t*-channel processes to the description. Complicated acceptance functions such as that of CLAS may be corrected in a very general fashion using this technique.

References

1. H. R. Crouch *et al., Phys. Rev.* **163**, 1510 (1967).
2. T.H. Burnett and S.R. Sharpe, *Ann. Rev. Nucl. Part. Sci.* **40**, 327 (1990).
3. R.A. Arndt, I.I. Strakovsky, and R.L. Workman, *Phys. Rev.* C **53**, 430 (1996).
4. D. Herndon, *et al., Phys. Rev.* D **11**, 3183 (1975).

John Cummings

Frank Lee

James Mueller

η ELECTROPRODUCTION WITH CLAS

JAMES A. MUELLER
Department of Physics and Astronomy,
University of Pittsburgh Pittsburgh, PA 15360, USA
E-mail: mueller+@pitt.edu

1 Introduction

Photo- and electroproduction experiments on the nucleon provide a clean probe of nucleon structure since Quantum Electrodynamics is well understood. The reaction $ep \to ep\eta$ is an especially clean reaction for studying excited states of the proton, since it selects out isospin = 1/2, N^* resonances. The cross section for this reaction appears to be dominated by the $S_{11}(1535)$ resonance, which has quantum numbers $IJ^P = \frac{1}{2}\frac{1}{2}^-$, and is reached from the nucleon through an electric dipole transition. This is the only baryon resonance known to have a branching ratio into ηp larger than a few percent. Past experiments[1] have established that the photocoupling amplitude ($A_{1/2}$) for the $S_{11}(1535)$ has a slower falloff with Q^2 indicating a more compact object than other N^* resonances. In this paper, I describe new measurements of η electroproduction with nearly complete angular coverage. These data are used to look for additional contributions to this reactions beyond the $S_{11}(1535)$ and to extract new values for $A_{1/2}$ of the S_{11}.

2 Results from First Data Run

This experiment was done using the CEBAF Large Acceptance Spectrometer (CLAS)[3] at the Thomas Jefferson National Accelerator Facility (TJNAF), nee CEBAF. In early 1998, the newly commissioned CLAS took its first production data with an electron beam on a hydrogen target. These results were the basis of the thesis of Richard Thompson, and have been published in Physical Review Letters.

2.1 The CLAS detectector and Event Selection

Toroidal magnet coils separate CLAS into 6 largely identical sectors, each covering 54° in ϕ. Tracking chambers in CLAS measure angles and momenta of charged particles for lab polar angles in the range $8° < \theta < 142°$. Outside the DC scintillation counters (SC) provide time-of-flight measurements with which we can separate the charged hadrons into pions, kaons and protons. For lab angles $\theta < 48°$, threshold Cerenkov detectors (CC) and Electromagnetic Calorimeters (EC) distinguish electrons from charged hadrons with high accuracy. A coincidence of the EC (with energy above $\sim$ 300 MeV) and the CC was required in the trigger for the data presented here.

For this analysis, events were selected with an identified electron and proton. A fiducial cut on these particles was applied to avoid the complicated regions near the magnetic coils and the edges of the CC. The momentum of the electron was required to be above 500 MeV in order to be well above the trigger threshold.

2.2 Differential Cross Sections

Events were then binned according Q^2, W, and the center-of-mass angles of the putative η ($\cos\theta^*_\eta$ and ϕ^*_η). The η yield was determined by fitting the distribution of missing mass recoiling against the *e-p* system. The fit is the sum of a peak at the η mass with a radiative tail plus a simple background function modified by the geometric acceptance for this reaction. The rms resolution for the missing mass peak is about 6 MeV. Acceptance for this reaction was calculated using a GEANT-based Monte Carlo simulation. The event generator included radiative effects using the peaking approximation, and the cross sections have been corrected for radiation. The acceptance varied from bin to bin with a high of 54%. Bins with acceptance less than 5% were not used in this analysis.

For each W and Q^2 bin, the differential cross sections were fit to a form that comes from an expansion of the response functions in terms of orthogonal polynomials,

$$\frac{d^2\sigma}{d\Omega^*_\eta} = \frac{|p^*_\eta|}{K}\left[R_T + \epsilon R_L + \sqrt{2\epsilon(1+\epsilon)}R_{TL}\cos\phi^*_\eta + \epsilon R_{TT}\cos 2\phi^*_\eta\right] \tag{1}$$

$$\approx \frac{|p^*_\eta|}{K}\left[A + B\cdot\cos\theta^*_\eta + C\cdot P_2(\cos\theta^*_\eta) + \left(D\cdot\sin\theta^*_\eta + E\cdot\sin\theta^*_\eta\cos\theta^*_\eta\right)\cos\phi^*_\eta + F\cdot\sin^2\theta^*_\eta\cos 2\phi^*_\eta\right] .$$

Under the assumption that the cross section is dominated by the S_{11} partial wave, the A parameter represents the S-wave contribution, B and D come from S/P-wave interference, and C, D, and F are due to S/D-wave interference. Thus, nonzero values for B-F are evidence for nonresonant mechanisms or other N^* resonances. The results of these fits are shown in Fig. 1. The terms representing R_{LT} and R_{TT} are small in agreement with theoretical expectations[5]. A is the largest contribution and has, at low W, the expected

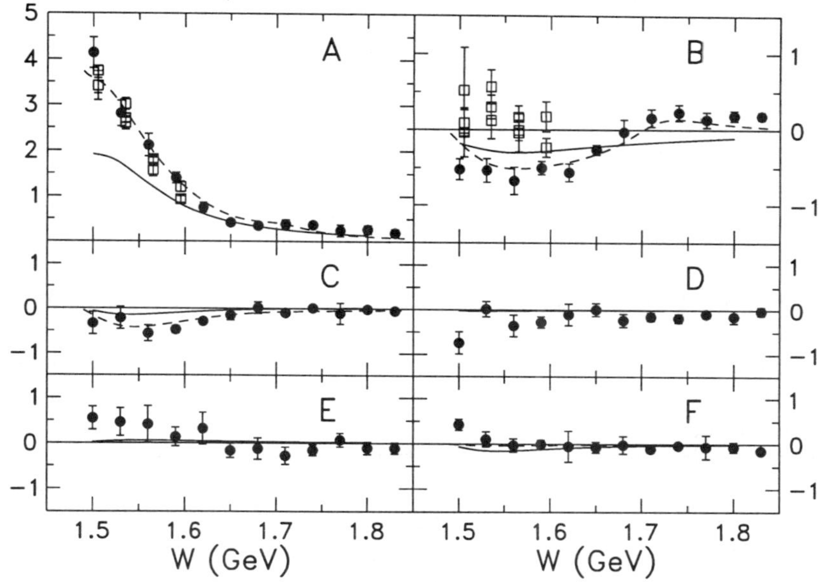

Figure 1. The results of the fit to the differential cross section using Equation 1. The solid curve is from the prediction in reference 5. The dashed curve is the result of a five-resonance fit using relativistic Breit-Wigner amplitudes with energy dependent widths.

Breit-Wigner shape due to an S-wave resonance close to threshold. C is slightly negative for low W, as is also seen in photoproduction, where it is due to the $D_{13}(1520)$. The most striking feature not seen in previous data is that the B parameter starts out negative at low W, but changes sign around 1.7 GeV. This could indicate variation of the relative phase between the S and P wave, or perhaps multiple canceling contributions whose relative magnitudes are changing. This feature is also seen[6] by the GRAAL experiment but was not reproduced by the earlier predictions[5] of the Mainz group. We attempted a simple resonance fit to our data. We included 5 resonances in the fit. The $S_{11}(1535)$, $S_{11}(1650)$, $P_{11}(1710)$, and $D_{13}(1520)$ were required to reproduce our data. We also included the $P_{11}(1440)$ in order to model the observed P-wave strength below the 1710, although this could also be non-resonant. Resonance masses and widths were fixed to their PDG values, with only the amplitudes of the Breit-Wigners allowed to float. The results of this fit is also shown in Fig. 1.

Since this data was published, a new model, based on the MAID formalism has been developed for η electro- and photoproduction. A talk at this conference[7] showed preliminary results. We look forward to comparing this to current and future data from CLAS.

2.3 Integrated Cross Sections

By summing events over $\cos\theta_\eta^*$ and ϕ_η^*, adequate statistics are available to bin the events finer in both W and Q^2. These cross sections are presented in Fig. 2. The prominent peak at $W \sim 1.5$ GeV is the $S_{11}(1535)$, as seen in the isotropic part of the angular distributions. Fits to a relativistic Breit-Wigner with an energy dependent width describe the low W region well, but there are deviations for $W > 1.65$ GeV. It is interesting to note that this is the same energy region where the B parameter derived from the angular distributions shows a strong variation.

For each Q^2 bin, the peak cross section extracted from the fit can be used to extract $A_{1/2}$ with the assumption that $S_{1/2}$ is small. Consistent with PDG and Armstrong et al., a value of the full width of 150 MeV and an $S_{11} \to \eta N$ branching fraction of 0.55 were used. The results of this measurement of $A_{1/2}$ are shown in Fig. 3 along with some previous results converted to be consistent with out choice of Γ and b_η. Although the exact normalization of $A_{1/2}$ depends on the choice of parameters for the contributing resonances, and thus on the results of more detailed theoretical analyses, the *shape* is well determined.

Four theoretical calculations[8] within the Constituent Quark Model are superimposed on the plot. Two of these are nonrelativistic, while the other two include features of relativity in the photon absorption and the 3-quark wave function. These are not the only theoretical predictions for $A_{1/2}$, but they give some indication of the range of models. For comparison, the plot also contains the dependence for a dipole form factor.

3 Prospects and Conclusions

There have been additional runs of this experiment since the data reported on here were collected. The second e1 running period occurred in February and March of 1999. We are now analyzing that data. With this new data we will be able to measure $A_{1/2}$ for $0.13\text{GeV}^2 < Q^2 < 3.0\ \text{GeV}^2$. Higher energy data already taken, and yet to be taken will allow this to be extended to even higher Q^2. As an indication of the improvement expected, the measurements presented in the lowest Q^2 bin are based on only 15 thousand η events taken with a electron beam energy of 1.6 GeV. We are currently examining a sample

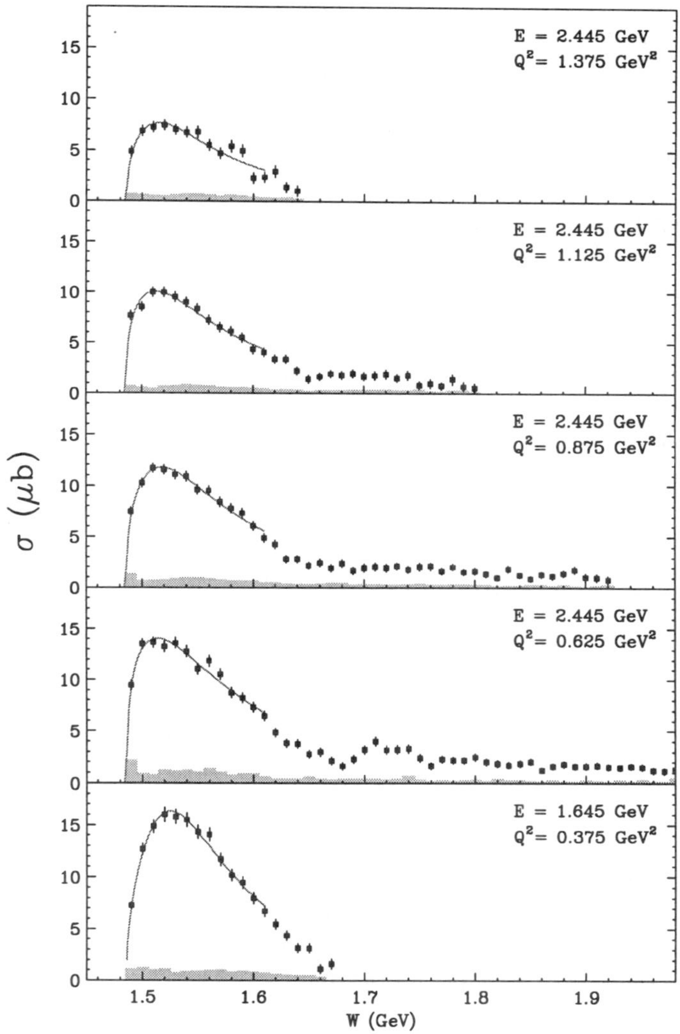

Figure 2. The integrated cross section measured for this experiment. The error bars on the points are statistical only. The size of the systematic uncertainty is indicated by the histogram at bottom of each plot. The curves ae fits to a single Breit-Wigner with an energy dependent width.

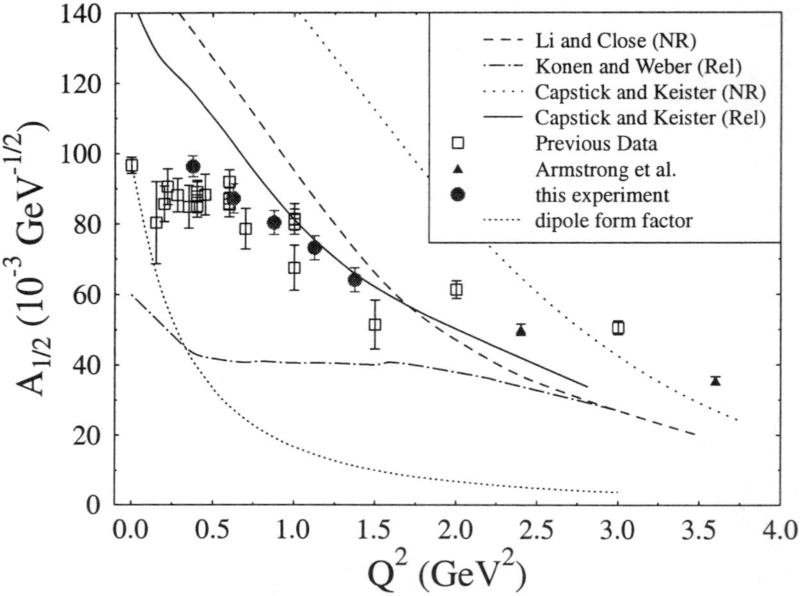

Figure 3. Values of the photon coupling amplitude, $A_{1/2}$ obtained from the integrated cross sections compared to previous experiments and selected calculations. The previous results have been converted to use consistent choice for the full width of the $S_{11}(1535)$ and the partial width into η. Four theoretical predictions are shown as well. As reference, a fifth curve shows the shape of a dipole form factor.

taken in 1999 with a beam energy of 1.5 GeV, containing 220 thousand η's. These data will also allow us higher statistics to further examine the features we have observed above the $S_{11}(1535)$. Polarized electron beams were also used during these runs, opening up the ability to measure additional response functions with different dependences on the underlying physics.

The new η electroproduction measurement shown here is one of the first results of the CLAS at TJNAF. It has broader kinematic coverage than any previous experiment and covers the W region at and above where the $S_{11}(1535)$ is important. The analysis presented here provides new evidence for the unusually slow falloff with Q^2 of the $S_{11}(1535)$ photocoupling amplitude. New structure is seen above the $S_{11}(1535)$ both in the total cross section, and in the term in the partial wave analysis most naturally associated with interference between S and P-waves.

4 Acknowledgements

I would like to thank Wen-Tai Chaing and Annalisa d'Angelo for interesting discussions on this subject during the conference. I would also like to thank the organizers for providing the pleasant environment in which those discussions took place.

References

1. F. Brasse *et al.*, *Z. Phys.* C **22**, 33 (1984). H. Breuker *et al.*, *Phys. Lett* B **74**, 409 (1978). F. Brasse *et al.*, *Nucl. Phys.* B **139**, 37 (1978). P. Kummer *et al.*, *Phys. Rev. Lett.* **30**, 873 (1973). U. Beck *et al.*, *Phys. Lett* B **51**, 103 (1974). J. Alder *et al.*, *Nucl. Phys.* B **91**, 386 (1975). C. S. Armstrong *et al.*, *Phys. Rev.* D **60**, 052004 (1999).
2. B. Krusche *et al.*, *Phys. Rev. Lett.* **74**, 3736 (1995).
3. W. Brooks, Nucl. Phys. A **663-664**, 1077 (2000).
4. R. Thompson *et al.*, Phys. Rev. Lett. **86**, 1702 (2001).
5. G. Knöchlein, D. Drechsel, and L. Tiator, *Z. Phys.* A **352**, 327 (1995); L. Tiator, C. Bennhold, and S.S. Kamalov, *Nucl. Phys.* A **580**, 455 (1994).
6. A. D'Angelo, *these proceedings.*
7. W-T.Chaing, *these proceedings.*
8. Z. Li and F. Close, *Phys. Rev.* D **42**, 2207 (1990). W. Konen and H. J. Weber, *Phys. Rev.* D **41**, 2201 (1990). S .Capstick and B. D. Keister, *Phys. Rev.* D **51**, 3598 (1995).

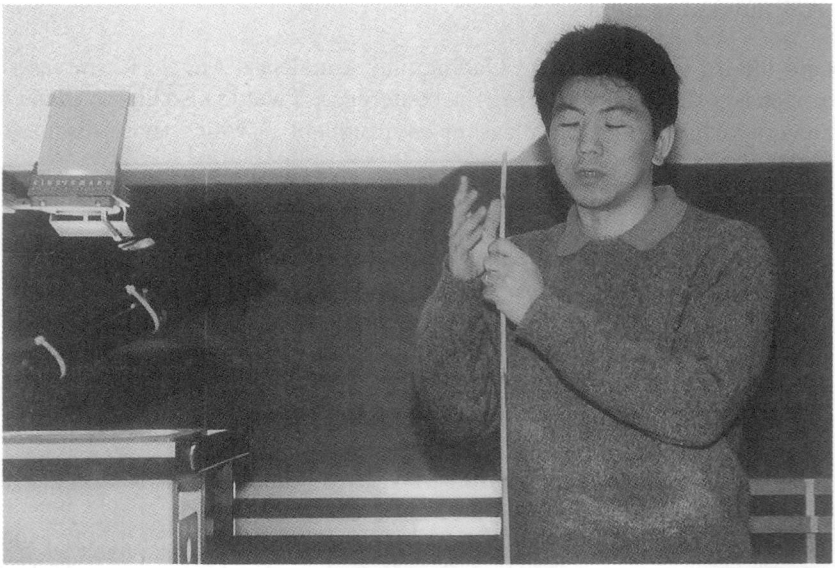

Shoichi Sasaki

Brian Raue

KAON ELECTROPRODUCTION AND Λ POLARIZATION OBSERVABLES MEASURED WITH CLAS

B. A. RAUE
Department of Physics
Florida International University
University Park
Miami, FL, USA 33199
E-mail: baraue@fiu.edu

for the CLAS Collaboration

An extensive program of kaon electroproduction is being conducted using CLAS at Jefferson Lab. The program includes measurements of cross sections, angular distributions, and polarization observables over a broad range of kinematical variables. Preliminary results will be presented for measurements of $\frac{d\sigma}{d\Omega}$ as a function of both $\cos\theta_K^*$ and ϕ^* for $e + p \to e' + K^+\Lambda^0$ and $e + p \to e' + K^+\Sigma^0$ for a beam energy of 4.2 GeV. These results indicate that electroproduction of the lambda final state is dominated by t-channel production whereas the sigma final state has a stronger s-channel component. The large acceptance of CLAS also enables measurement of the induced and transferred polarization of the recoiling lambda. Preliminary results for beam energies of 2.5 and 4.2 GeV indicate a large lambda polarization with a small variation over the covered range of W: ≈ 1.5 to 2.5 GeV. These results are providing tantalizing evidence that s-channel production is not playing the dominant roll in $K - \Lambda$ electroproduction.

1 Introduction

The present state of understanding of the kaon photo- and electroproduction processes is limited by a sparsity of data. Existing cross section measurements cover a limited range of kinematics and suffer from relatively large experimental uncertainties (reviewed in Ref. [1]). Below W of around 2-2.2 GeV, isobaric models utilizing Feynman-graph techniques [2,3,4] can describe the existing kaon photo- and electroproduction data reasonably well. These models have a sizeable number of free parameters (such as coupling constants $g_{K\Lambda N}$ and $g_{K\Sigma N}$, and hyperon and kaon form factors) and include any number of intermediate resonances. As such, these models can generally fit the existing experimental results quite well even though, in some cases, the models rely on different underlying mechanisms to match the data. For example, Mart and Bennhold [5] can explain recent SAPHIR photoproduction cross sections [6] with the inclusion of a previously unseen $D_{13}(1960)$ resonance. Saghai [7] can explain the same data by instead adding in off-shell effects within his model.

The lack of data also prevents any serious constraint of QCD-based models [8,9,10] that predict the formation and decay properties of participating strange baryons. The $D_{13}(1960)$ resonance mentioned above is predicted in the quark model of Capstick and Roberts [8]. However, without a clear agreement as to the need for this state in explaining the data, its status will remain in doubt.

There have been some recent improvements made to the database including the SAPHIR [6] measurements as well as recent Jefferson Lab [11] and GRAAL [12] experiments. However, additional measurements of cross sections and polarization observables over a larger range of kinematics are necessary to test the hadronic-field theories and probe the underlying reaction mechanism. In this paper, we will outline the extensive program of kaon electroproduction being conducted with the CEBAF Large Acceptance Spectrometer (CLAS) in Hall B of the Thomas Jefferson National Accelerator Facility (Jefferson Lab). With the goals of this workshop in mind, data taken with CLAS can be used in two ways to look for N^* resonances: by constraining the hadrodynamic models that include the resonances or as part of a partial-wave analysis that includes all of the existing data.

2 Formalism and Kinematics

The general form for the differential cross section for the exclusive kaon electroproduction reaction is given by the product of the virtual photon flux factor Γ_ν and the $\gamma^* p \to K^+ Y$ virtual photo-absorption differential cross section $d\sigma_\nu/d\Omega_K$. With the momenta and angles of the particles involved in the reaction defined in Fig. 1, the most general form for the $p(e, e'K^+)$ c.m. differential cross section for a polarized beam experiment is

$$\frac{d\sigma_\nu}{d\Omega_K} = \sigma_0(1 + hA_{TL'} + P_{x'}S_{x'} + P_{y'}S_{y'} + P_{z'}S_{z'}). \tag{1}$$

Here we have broken the cross section down into the usual unpolarized cross section,

$$\sigma_0 \equiv \sigma_T + \epsilon\sigma_L + \sqrt{\frac{1}{2}\epsilon(1+\epsilon)}\sigma_{TL}\cos\phi^* + \epsilon\sigma_{TT}\cos 2\phi^*, \tag{2}$$

the cross section asymmetry for polarized beam, $A_{TL'}$, and the recoil hyperon polarization which has induced components $P^0_{j'}$ and transferred components $P'_{j'}$,

$$P_{j'} = P^0_{j'} + hP'_{j'}, \quad j = x, y, z. \tag{3}$$

The various pieces of the cross section are sensitive to different aspects of the underlying physics. For example, extraction of the separated L and T

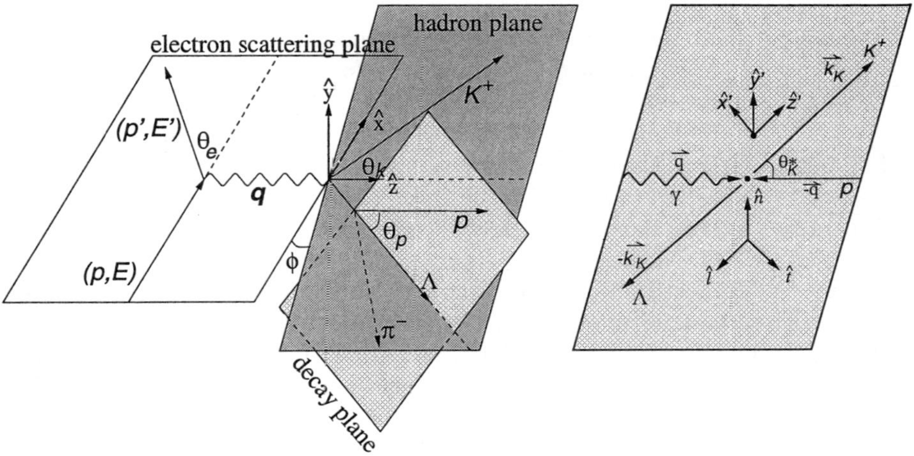

Figure 1. Kinematics for $K\Lambda$ electroproduction showing angles and polarization axes in the laboratory (left) and center-of-mass (right) coordinate systems.

cross sections and the TT and TL interference terms will allow for increased understanding of the underlying reaction mechanism. These separate terms provide sensitivity to the prescription for establishing current conservation in electroproduction models, especially the combination of the non-gauge invariant terms involving the electric proton and kaon form factors. The fifth structure function, $A_{TL'}$, is sensitive to final-state interactions between the kaon and the hyperon. The recoil polarization observables may be particularly sensitive to the reaction channel. In order to fully understand the kaon-electroproduction process, it is necessary to measure as many of these observables as possible over the largest range of kinematics as possible.

3 The CLAS Program

The CLAS kaon electroproduction program includes four approved experiments that will measure several pieces of the cross section. These are shown in the table below. Beam energies for these experiments range from 2.5 to $\sim$ 5.7 GeV. This covers a W range from threshold for Λ production to approximately 3.0 GeV and Q^2 from about 1.0 to 6.0 GeV^2.

Experiment	Spokespersons	Main Goals
E93-030	K. Hicks M.D. Mestayer	Structure function separation: σ_T, σ_L, σ_{TT}, σ_{TL}
E89-043	L. Dennis H. Funsten	Λ and $\Lambda(1520)$ electroproduction study of s, t, and u-channel
E99-006	D.S. Carman, K. Joo L. Kramer, B.A. Raue	polarization observables: P_j and $A_{TL'}$
E00-112	D.S. Carman, K. Joo G. Niculescu, B.A. Raue	Extension of program to ~ 6 GeV

The CLAS electroproduction program will complement other measurements. These include photoproduction measurements done with CLAS [14] and the previously mentioned SAPHIR and GRAAL measurements. In addition there is the electroproduction experiment in Hall A that recently ran [13]. This and the Hall C experiment do high precision $L-T$ separations but cover a significantly smaller range of kinematics.

The large acceptance of CLAS offers a number of advantages over other experiments. The detector system surrounds the target with approximately 50% of 4π sr coverage. The acceptance for $e'K^+$ is about 20% and for $e'K^+p$ is about 5%. With a large acceptance for $\cos\theta_K^*$ (or t), analyses can be performed where the reaction is dominated by the different channels–s, t, or u–thus limiting the influences of intermediate resonances. For the $e'K^+p$ final state, the proton comes about from the weak decay of the Λ. We can therefore use the angular distribution of the proton in the Λ-rest frame to determine the Λ polarization. The angular yield for the "self-analyzing" decay is given by $Y(\theta_p) = 1 - \alpha P \cos\theta_p$, where $\alpha = 0.642 \pm 0.013$ is the weak decay coefficient. The comparative disadvantages of CLAS are that the instrument is limited to low luminosity ($< 10^{34}/\text{cm}^2/\text{s}$), thus, low statistics, has a limited resolution, and requires significant work to understand the acceptance corrections.

4 Results

The data to be presented in this report are from the "E1c" running period of February to April 1999. Only results for beam energies of 2.567 and 4.247 GeV will be presented, however, data at 4.056 and 4.462 GeV were also taken. The targets for this run period were liquid hydrogen targets with lengths of either 3.8 or 5.0 cm. The electron beam typically had a polarization of between 61 and 70% with an average of 70.2±1.7% for the data shown here.

One of the key steps for this analysis is to cleanly identify the kaons and hyperons. The kaons are identified through mass reconstruction using momentum and time-of-flight information. The hyperons are identified through

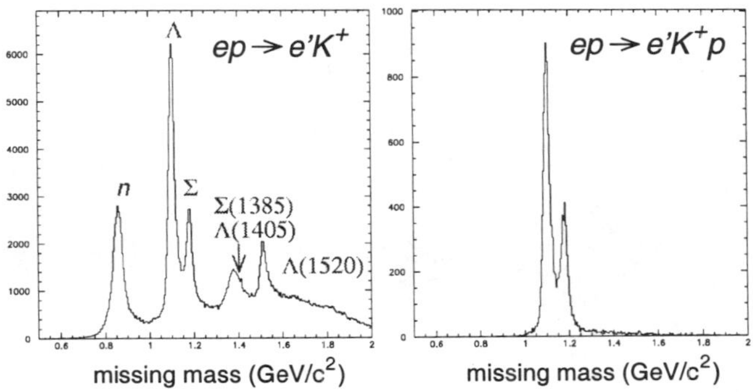

Figure 2. Left: Missing mass spectrum for events in which an electron and kaon have been identified. Pions that have been misidentified as kaons are the primary source of background and lead to the neutron peak. Right: Missing mass spectrum for the same events with the additional requirements that a proton is detected and that a π^0 can be reconstructed.

missing-mass techniques. This is illustrated in Fig. 2. There is a large number of pions that are misidentified as kaons that lead to the background in the hyperon missing-mass spectrum. For cross sections measurements, the results must take this background into account. In the case of the recoil polarization measurements, the additional requirement of detecting the decay proton reduces the background. An additional cut can be placed on the data wherein a π^0 is reconstructed from the $ep \to e'K^+p$ final state.

Fig. 3 shows the preliminary differential cross section for Λ and Σ electroproduction at fixed Q^2 and W as a function of $\cos\theta_K^*$ and ϕ^*. There are clear differences between the shapes of the angular distributions for Λs and Σs with the Λs much more forward peaked. This indicates that Λ electroproduction is dominated by t-channel exchange whereas the peaking of the Σs in the central angle range indicates a larger s-channel component. These distributions can be fit with a function of the form $a + b\cos\phi^* + c\cos 2\phi^*$ with the coefficients b and c proportional to the cross section terms σ_{LT} and σ_{TT}, respectively. These results are also shown in Fig. 3.

It can be shown that the components of the Λ^0 polarization can be determined from the yield asymmetry from different beam-helicity states:

$$A = \frac{N^+ - N^-}{N^+ + N^-} = \frac{\alpha P' P_e \cos\theta_P}{1 + \alpha P^0 \cos\theta_P}, \tag{4}$$

where $N^{+/-}$ are the numbers of events for the two helicity states, P_e is the electron-beam polarization, and θ_P is the angle of the proton decay relative

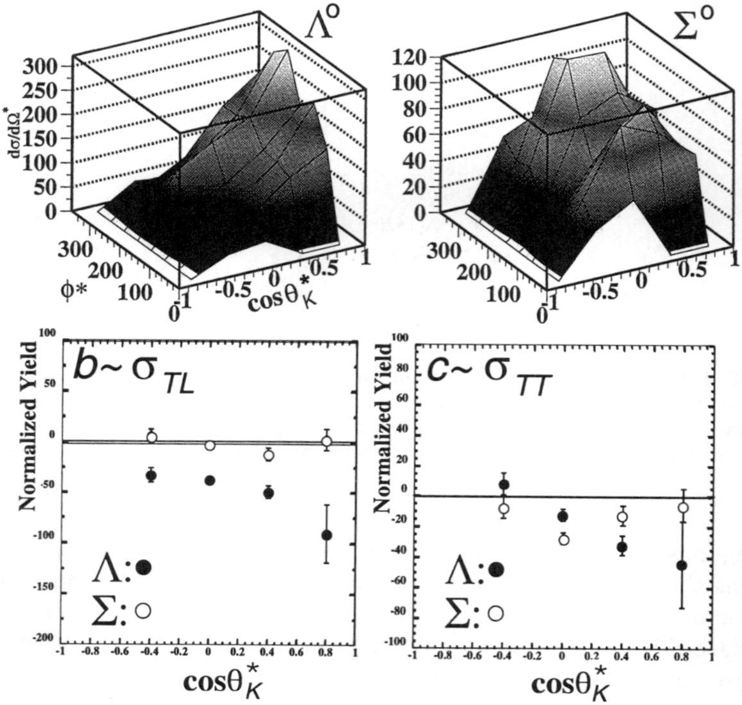

Figure 3. Preliminary differential cross sections for $ep \to e'K^+ + \Lambda^0$ (top left) and $ep \to e'K^+ + \Sigma^0$ (top right) as a function of $\cos\theta^*_K$ and ϕ^* for $Q^2 = 1.25$ GeV^2 and $W = 1.90$ GeV. $\cos\theta^*_K$ dependence of the LT (left) and TT interference terms derived from the differential cross sections as indicated. The results are displayed as normalized yields.

to Λ the momentum vector in the Λ rest frame. The advantage of using this method is that acceptance corrections can be ignored to first order. Because of low statistics, it is necessary to sum over the entire range of ϕ^*, the angle of the hadron plane relative to the electron scattering plane. As a result, three of these integrated polarization terms go to zero: $P_t^0 = P_l^0 = P_n' = 0$. This is seen in our result for P_n' as shown in Fig. 4. Preliminary results indicate that the l and t components of the transferred polarization are large with magnitudes in the neighborhood of 0.5. There is also evidence that the variations with W and Q^2 are small. The relative flatness with W is also seen in the results for the induced polarization which is shown in Fig. 5.

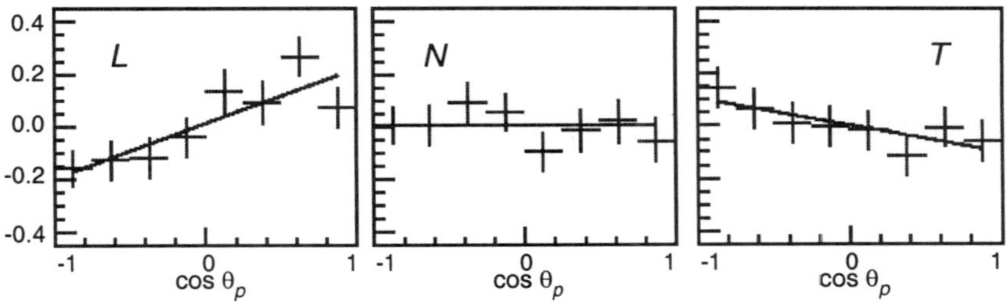

Figure 4. Λ-yield asymmetry as a function of $\cos\theta_P$ for a beam energy of 4.247 GeV, $Q^2 = 1.05$ GeV^2 and $W = 1.75$ GeV. The slope of the fitted line is proportional to the transferred polarization P'.

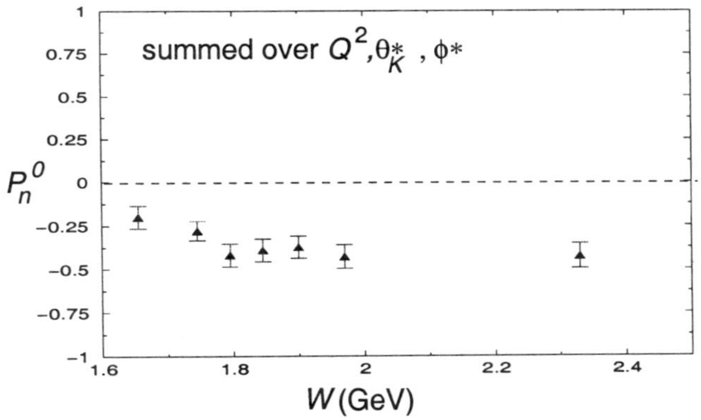

Figure 5. Λ polarization P_n^0. The data have been summed over kinematic variables Q^2, θ_K^*, and ϕ^*.

5 Summary

It is clear from our cross section measurements that the production mechanisms for lambdas and sigmas are different. Specifically, lambdas have a larger t-channel component while sigmas have more s-channel. For the first time, σ_{LT} and σ_{TT} have been separated. More analysis and detailed comparisons to model predictions are necessary before these results reveal anything about the underlying physics. Our preliminary Λ-recoil polarization observables do not exhibit a strong W dependence. Given the suggested dominance of t-channel production, this should not be surprising since there would be relatively little contribution to the cross section from intermediate $N^\star$ resonances. A more complete analysis with higher statistics is forthcoming. It should enable us to look at the reaction at larger values of θ_K^* where s-channel production will be more important. It is possible that one may then observe a more pronounced variation with W that can be used to search for N^* resonances.

Acknowledgments

The results presented in this talk were largely obtained through the efforts of Daniel Carman and Gabriel Niculescu from Ohio University, Si McAleer from Florida State University, and Rob Feuerbach from Carnegie Mellon University. This work was supported by the U.S. Department of Energy and the National Science Foundation.

References

1. R.A. Adelseck and B. Saghai, *Phys. Rev.* C **42**, 108 (1990).
2. J.C. David, C. Fayard, G.H. Lamot, and B. Saghai, *Phys. Rev.* C **53**, 2613 (1996); T. Mizutani, C. Fayard, G.H. Lamot, and B. Saghai, *Phys. Rev.* C **58**, 75 (1998).
3. R.A. Williams, C. Ji, and S.R. Cotanch, *Phys. Rev.* C **46**, 1617 (1992).
4. T. Mart and C. Bennhold, and C.E. Hyde-Wright, *Phys. Rev.* C **51**, 1074 (1995).
5. T. Mart, C. Bennhold, *Phys. Rev.* C **61**, 012201 (2000).
6. M.Q. Tran *et al., Phys. Lett.* B **445**, 20 (1998); and K.-H. Glander, *these proceedings.*
7. B. Saghai, Preprint DAPHIA-SPHN-00-24, Apr 2000.
8. S. Capstick and W. Roberts, *Phys. Rev.* D **58**, 74011 (1998).
9. E.S. Ackleh, T. Barnes, and E.S. Swanson, *Phys. Rev.* D **54**, 6811 (1996).
10. A. LeYaouanc *et al., Phys. Rev.* D **8**, 2223 (1973); *Phys. Rev.* D **9**, 1415 (1974); *Phys. Rev.* D **11**, 1272 (1975).
11. G. Niculescu *et al., Phys. Rev. Lett.* **81**, 1805 (1998).
12. A. D'Angelo, *these proceedings.*
13. F. Garibaldi, S. Frullani, J. LeRose, P. Markowitz, T. Saito *et al.*, JLab Experiment E94-107.
14. R.A. Schumacher *et al.*, JLab Experiment E89-004.

RECENT RESULTS ON KAON PHOTOPRODUCTION AT SAPHIR IN THE REACTIONS $\gamma p \rightarrow K^+\Lambda$ AND $\gamma p \rightarrow K^+\Sigma^0$

K.-H. GLANDER, representing the SAPHIR Collaboration
Physikalisches Institut, Universität Bonn,
Nussallee 12,
53115 Bonn, Germany
E-mail: glander@lisa.physik.uni-bonn.de

The measurement of photoproduction reactions with open strangeness is one of the central issues of the physics program at *SAPHIR*. We report here on the analysis of the reactions $\gamma p \rightarrow K^+\Lambda$ and $\gamma p \rightarrow K^+\Sigma^0$ using data taken after 1997. The measured cross sections suggest the existence of a missing $D_{13}(1895)$ resonance, for which the previous *SAPHIR* data[1] gave a first hint with lower statistical and systematic accuracy. A substantial variation of the differential cross sections as a function of the K^+ production angle (in c.m.s) is observed throughout the measured photon energy range between threshold and 2.6 GeV.

1 Introduction

The *SAPHIR* detector[2] at the electron stretcher facility *ELSA* was built to measure photon induced reactions in the threshold and resonance regions up to photon energies of 2.6 GeV. Of large interest was the measurement of photoproduction processes with open strangeness, especially the reactions $\gamma p \rightarrow K^+\Lambda$ and $\gamma p \rightarrow K^+\Sigma^0$. These reactions were already studied using *SAPHIR* data from the first data taking period from 1992 to 1994[1]. During this first period 30 million triggers were taken for photon energies up to 2.0 GeV. In the second data taking period from 1997 to 1998 180 million triggers were recorded for photon energies up to 2.6 GeV. Equivalently the statistical errors on previous results have been reduced by at least a factor of two and the energy range has been extended into the continuum region. For this latter period the *SAPHIR* setup has been equipped with a new tagging system and a planar forward drift chamber, in addition a modified trigger was used. Major improvements have been achieved in the reconstruction software. In particular the fit for tracks in the drift chamber system, which is the main input for physics analyses, has been improved by including the hits in the forward drift chamber. This led to a five times better momentum resolution on average. The better understanding of the drift chamber performance induced changes in the detector simulation which improved the acceptance calculations and helped to reduce systematic errors.

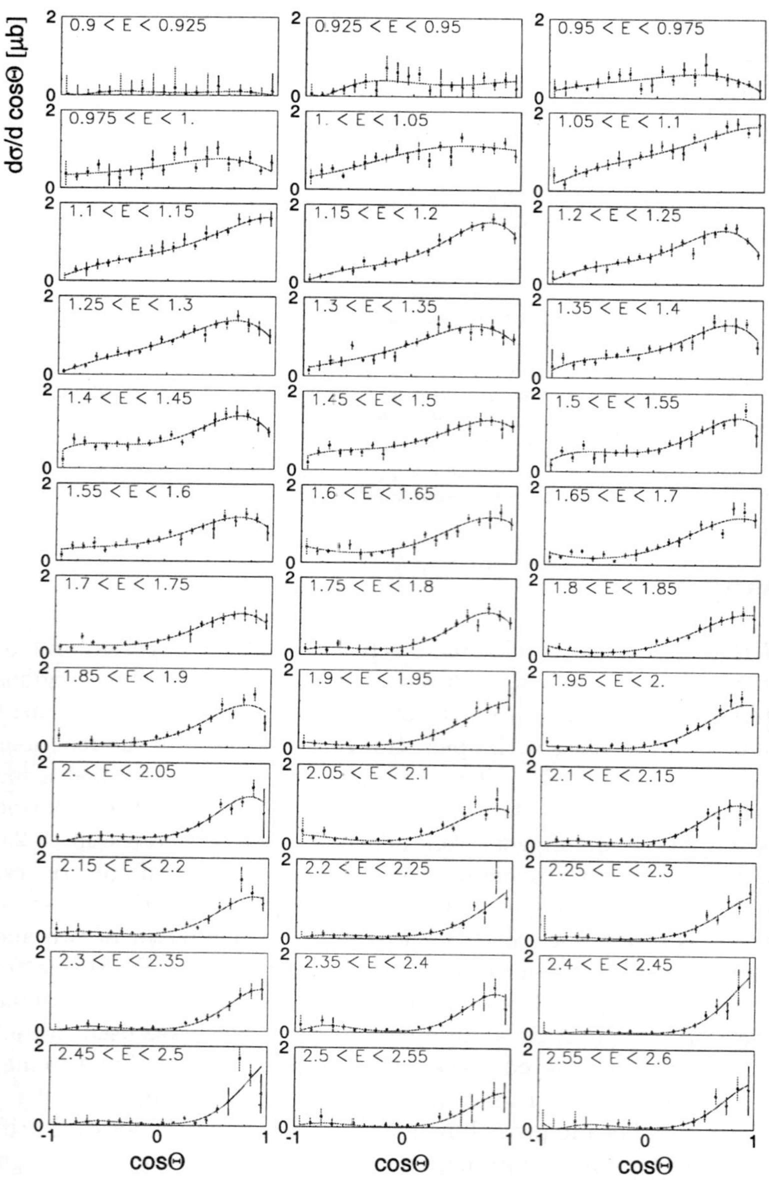

Figure 1. Differential cross sections of the reaction $\gamma p \to K^+\Lambda$ as a function of the production angle θ of the K^+ in the c.m.s. and the photon energy. The curves show Legendre polynomial fits (see text). The data are preliminary.

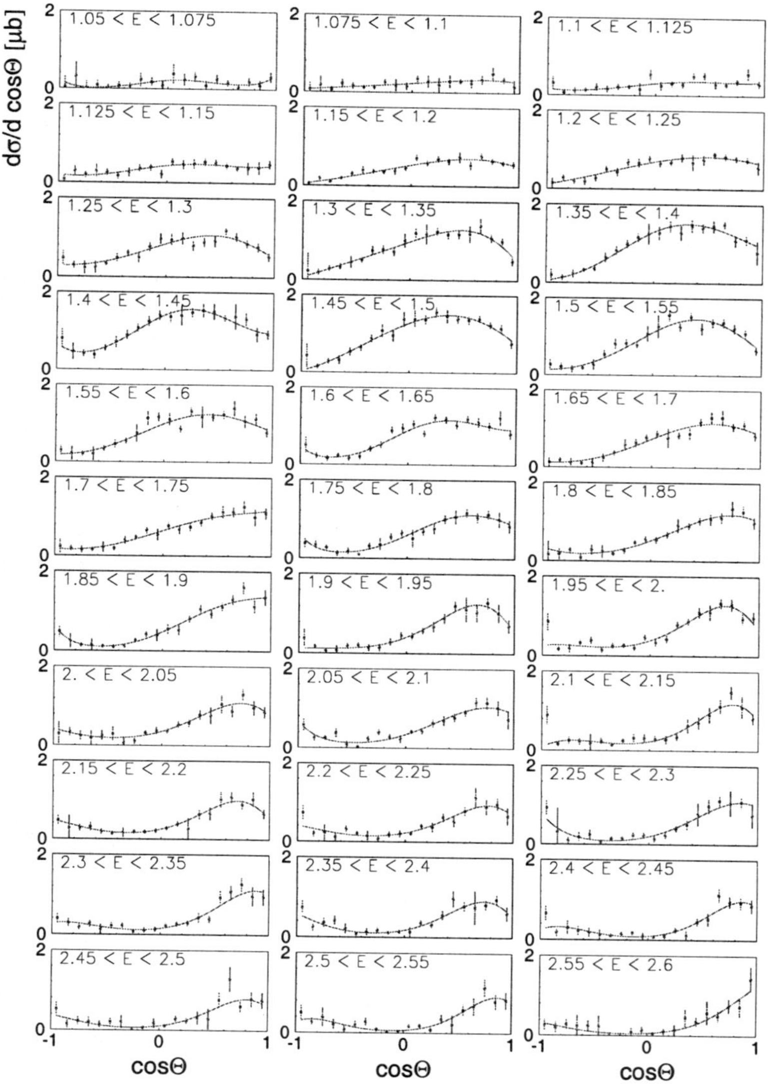

Figure 2. Differential cross sections of the reaction $\gamma p \rightarrow K^+\Sigma^0$ as a function of the production angle θ of the K^+ in the c.m.s. and the photon energy. The curves show Legendre polynomial fits (see text). The data are preliminary.

2 Cross Sections

2.1 Differential cross sections

Fig.1 shows the differential cross sections for the reaction $\gamma p \rightarrow K^+\Lambda$ from threshold up to a photon energy of 2.6 GeV. The kaon production angular range in the c.m.s. is fully covered. In view of the statistical and systematic errors we split the data into 20 angular bins for 36 photon energies. For every data point the error includes the statistical error and a first estimate of systematic errors added in quadrature. The data shown here are selected with restrictive cuts to avoid background contributions. The dependence of the observed cross sections on the selection criteria is presently investigated (typical variation is 10 percent).
For the reaction $\gamma p \rightarrow K^+\Lambda$ a flat angular distribution right at threshold indicates *s*-wave production. Already at the next energy bin (photon energies between 925 and 950 MeV) the differential cross section is no longer flat, indicating the existence of partial waves of higher angular momentum. Coupled channels calculations[3] describe the data in this range if the known resonances $S_{11}(1650)$, $P_{11}(1710)$ and $P_{13}(1720)$ are included, which dominate the cross section close to threshold. At the highest measured photon energies the cross sections show a forward peak. This can be described with exchanges of K mesons in the *t*-channel. The curves to the data are Legendre fits of the form:

$$\frac{d\sigma}{d\,cos(\theta_{K^+}^{CMS})} = \frac{q}{k}(\sum_{l=0}^{4} a_l\, P_l(cos\,\theta_{K^+}^{CMS}))^2 \qquad (1)$$

where q and k are the momenta of the kaon and photon in the c.m.s. In Fig.3 the fit coefficients are shown as a function of the photon energy. The coefficients show clear structures, which give hints for rapid changes of the production mechanisms.
The differential cross sections for $\gamma p \rightarrow K^+\Sigma^0$ are given in Fig.2. The cross section is comparatively flat for the first 100 MeV of photon energy indicating that *p*-wave contributions due to the resonances $P_{11}(1710)$ and $P_{13}(1720)$ are small. The almost symmetrical angular distribution around a photon energy of 1.45 GeV corresponding to a c.m.s. energy of 1900 MeV together with the enhancement in the total cross section (see section 2.2) indicate a substantial resonance contribution. Hitherto no reliable partial wave analysis exists that fixes the quantum numbers of this resonance, there might be even more than one resonance. Isobar model calculations done by C. Bennhold and T. Mart et al.[4] based on our previous data[1] allow for contributions from the $S_{31}(1900)$ and $P_{31}(1910)$ Δ resonances. Our new data will hopefully bring

more light into this situation. The Legendre fit results given in Fig.4 show a

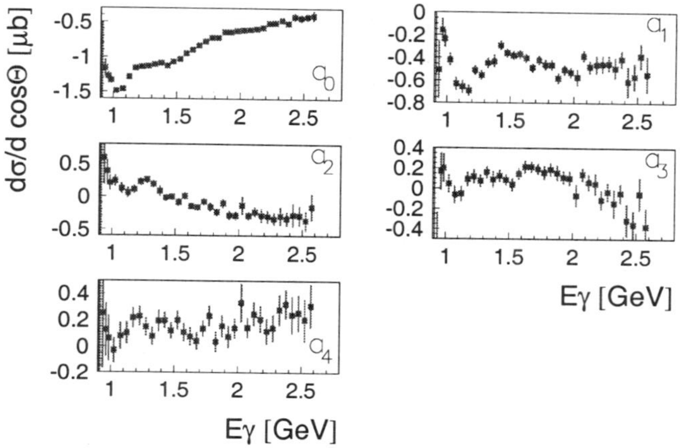

Figure 3. Legendre polynomial fit coefficients from differential cross sections for the reaction $\gamma p \to K^+\Lambda$ as a function of the photon energy.

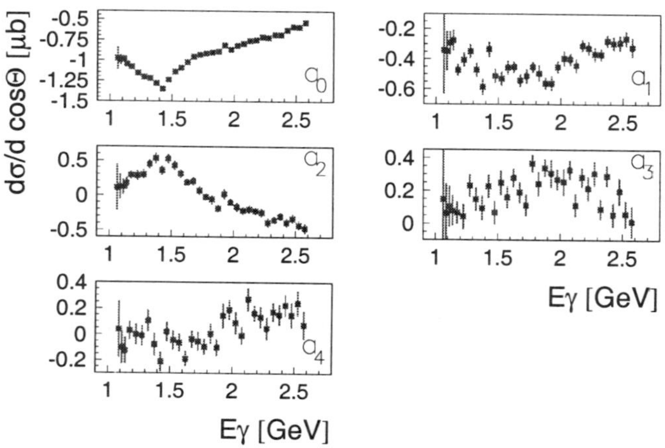

Figure 4. Legendre polynomial fit coefficients from differential cross sections for the reaction $\gamma p \to K^+\Sigma^0$ as a function of the photon energy.

peak structure at a photon energy of 1.425 GeV for the coefficients a_0 and a_2, which could be interpreted as signal of two resonances, one in s-wave and

one in d-wave.Clearly the Legendre fits corresponding to Eq. (1) ignore the spins of the participating particles so that the fit coefficients cannot directly be translated into partial wave amplitudes.
For photon energies higher than 1.8 GeV the differential cross sections for $\gamma p \to K^+\Sigma^0$ show an enhancement in the backward direction (see Fig.2). As for the reaction $\gamma p \to K^+\Lambda$ the differential cross section for $\gamma p \to K^+\Sigma^0$ shows a forward peaking for the highest photon energies.

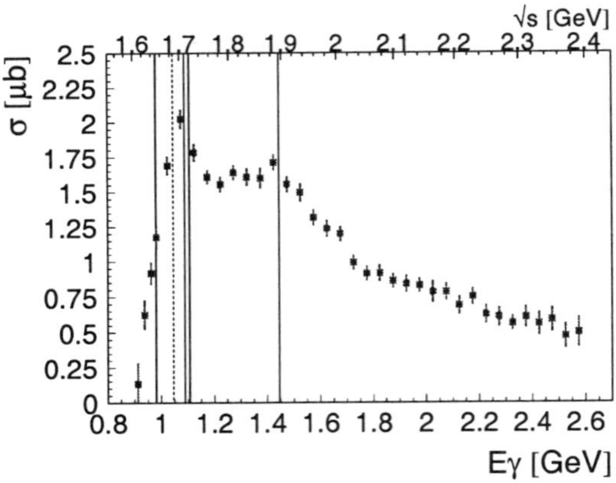

Figure 5. Total cross section for the reaction $\gamma p \to K^+\Lambda$. The vertical lines are explained in the text. The data are preliminary.

2.2 Total cross sections

Integrating the differential cross sections gives the total cross sections shown in Figs. 5 and 6. The total cross section rises for $\gamma p \to K^+\Lambda$ much more steeply than for $\gamma p \to K^+\Sigma^0$. This difference comes presumably from the strong contributions of the resonces $S_{11}(1650)$, $P_{11}(1710)$ and $P_{13}(1720)$ to $\gamma p \to K^+\Lambda$ (the masses being as indicated by vertical full lines). Their contributions to $\gamma p \to K^+\Sigma^0$ seem to be small. The $S_{11}(1650)$ lies below the threshold of 1.046 GeV. The threshold for $\gamma p \to K^+\Sigma^0$ can be seen as the dashed line in Fig.5. The marked maximum at a photon energy of 1.075 GeV may be caused not by resonances alone but also by a cusp of the opening of the Σ^0-channel.
A clear evidence of a peak structure in $\gamma p \to K^+\Lambda$ is seen at a photon energy of 1.45 GeV corresponding to $\sqrt{s} = 1895$ MeV (see Fig.5). Isobar model

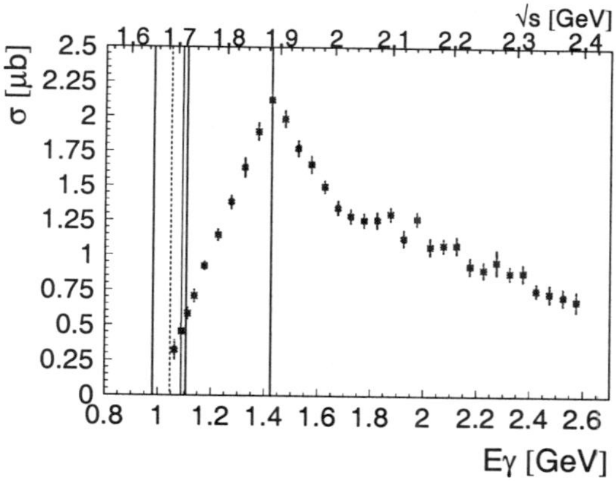

Figure 6. Total cross section for the reaction $\gamma p \to K^+\Sigma^0$. The vertical lines are explained in the text. The data are preliminary.

calculations[7] describe this observation by including a $D_{13}(1895)$, which can be identified with a missing resonance $D_{13}(1960)$ calculated in the relativized quark model of Capstick and Roberts[5]. This resonance is calculated to have a big partial width for a decay into $K^+\Lambda$. In a covariant quark model of H. Petry, B. Metsch et al.[6] two D_{13}-resonances have been predicted at a similar mass around 1900 MeV, but their partial widths have not yet been calculated. In the isobar model calculations the exitation of the D_{13} in the reaction $\gamma p \to K^+\Lambda$ is predicted to have a peak in the photon energy for backward kaon production angles. This peak is significant in our data, as can been seen in Fig.7. A measurement of the beam asymmetry of this reaction by *GRAAL* supports the existence of a D_{13} resonance in this mass region[8] (see the talk and report by A. D'Angelo). Two recent publications[9,10] come to different conclusions concerning the D_{13} case. Whilst B. Saghai's model describes the total cross section without introducing this resonance, S. Janssen, J. Ryckebusch et al. claim the need for this resonance.

In the total cross section for $\gamma p \to K^+\Sigma^0$ a strong bump is seen at a photon energy of 1.425 GeV corresponding to $\sqrt{s} = 1885$ MeV, which could be due to strong resonance contributions, however the differential cross sections (and the Legendre coefficients) are not yet conclusive (see section 2.1).

For photon energies above 1.8 GeV both total cross sections show a similar behaviour accordantly to the observation of the same forward peaking in the differential cross sections. The t-channel exchanges of $K^*(892)$ and $K_1(1210)$ are expected to be dominant.

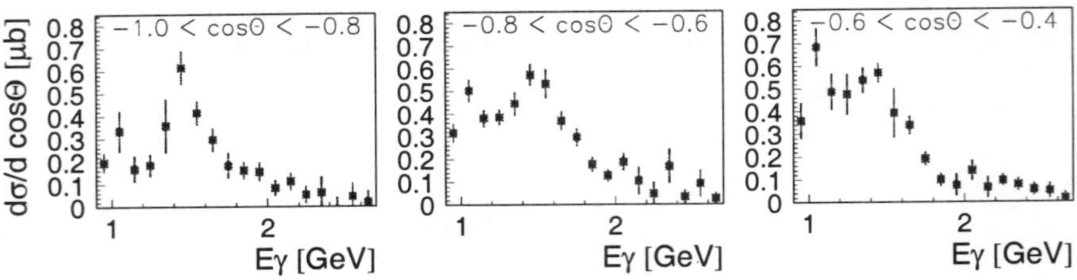

Figure 7. Cross section differential in photon energy and kaon production angle for the reaction $\gamma p \to K^+\Lambda$ indicating a peak structure for backward angles at $E_\gamma = 1.45$ GeV.

2.3 Final remarks

Work on final results is in progress and will be completed during this year. The final analysis will include the measurement of the hyperon polarizations.

References

1. M.Q. Tran *et al.*, "Measurement of $\gamma p \to K^+\Lambda$ and $\gamma p \to K^+\Sigma^0$ at photon energies up to 2 GeV", *Phys. Lett.* B **445**, 20-26 (1998).
2. W.J. Schwille *et al.*, "Design and construction of the SAPHIR detector", *Nuclear Instruments and Methods in Physics Research* A **344**, 470-486 (1994).
3. T. Feuster and U. Mosel, *Phys. Rev.* C **58**, 457 (1998); *Phys. Rev.* C **59**, 460 (1999).
4. C. Bennhold, T. Mart, A. Waluyo, H. Haberzettl, G. Penner, T. Feuster, and U. Mosel, nucl-th/9901066.
5. S. Capstick and W. Roberts, *Phys. Rev.* D **58**, 074011 (1998).
6. U. Löring, B. Ch. Metsch, and H. R. Petry, "The light baryon spectrum in a relativistic quark model with instanton-induced quark forces. The non-strange baryon spectrum and ground-states", hep-ph/0103289, accepted for publication in *Eur. Phys. J.* A.
7. C. Bennhold *et al.*, *Phys. Rev.* C **61**, 012201 (2000).
8. A. D'Angelo, *these proceedings.*
9. B. Saghai, nucl-th/0105001.
10. S. Janssen, J. Ryckebusch *et al.*, nucl-th/0105008.

VECTOR MESON DECAY OF BARYON RESONANCES

U. MOSEL AND M. POST*

Institut für Theoretische Physik, Universität Giessen, D-35392 Giessen

**Email: Marcus.Post@theo.physik.uni-giessen.de*

We investigate the coupling of vector mesons with nucleons to nucleon resonances in an isospin-selective VMD approach and explore the in-medium properties of vector mesons.

1 Introduction

One of the evidences for the discovery of a new state of matter that was quoted by the CERN press release in 2000 was the result of the CERES experiment[1] that showed an excess of dileptons at invariant masses below the vector meson mass. In order to understand this phenomenon it is essential to consider the effect of conventional hadronic interactions on the vector mesons. We emphasize the relevance of the excitation of baryon resonances in this context. To this end we analyze the decay of nucleon resonances into vector mesons within an isospin selective VMD model. In the isovector channel we compare the results to fits to the hadronic $N\rho$ decay width. In the isoscalar channel we then predict the ω coupling strength.

2 VMD Analysis of the Electromagnetic Resonance Decay

Vector Meson Dominance (VMD)[2], a theory which describes photon-hadron interactions exclusively in terms of vector meson-hadron interactions, relates the hadronic coupling strength of resonances to vector mesons $f_{RN\rho(\omega)}$ and the isoscalar and isovector part of the photon-coupling:

$$f_{RN\omega} = g_s \, m_\omega \, \frac{2\, g_\omega}{e} \quad , \quad f_{RN\rho} = g_v \, m_\rho \, \frac{2\, g_\rho}{e} \quad . \tag{1}$$

As values for g_ρ and g_ω – the coupling strengths of ρ and ω meson to the photon – we take $g_\rho = 2.5$ and $g_\omega = 8.7$ (Ref. [2]). The isoscalar and isovector coupling strength of the resonance to the $N\gamma$ system is given by g_s and g_v, respectively, see Eq. (2). Thus VMD gives access to both $f_{RN\omega}$ and $f_{RN\rho}$, if it is possible to obtain g_s and g_v from experimental data. In order to achieve this goal, the coupling has to be decomposed into an isoscalar and an isovector part, which is readily done by constructing suitable linear combinations of proton- and neutron-amplitudes[3].

The isospin part I of the electromagnetic coupling is given by:

$$I = \chi_R^{I\,\dagger}\,(g_s + g_v\,\tau_3)\,\chi_N^I \qquad (2)$$

For simplicity we restrict ourselves here to the case of isospin 1/2; the case of isospin 3/2 contains additional Clebsch-Gordan coefficients which are given in Ref.[3]. The spinors χ_R^I and χ_N^I represent resonance and nucleon isospinors and I_R denotes the isospin of the resonance. τ_3 refers to the Pauli matrix. From the structure of the isospin coupling I it follows that the linear combinations

$$\mathcal{M}_{s/v} = \frac{1}{2}\,(\mathcal{M}_p \pm \mathcal{M}_n) \qquad (3)$$

are proportional to g_s and g_v respectively.

At the pole-mass of the resonance the helicity amplitudes $A_{\frac{1}{2}}^{p/n}$ and $A_{\frac{3}{2}}^{p/n}$ are known from experiment. Therefore also g_s and g_v are determined except for a normalization factor. We calculate this factor by introducing the γ-width $\Gamma_{s/v}^{\gamma}$, defined in terms of the helicity amplitudes $A_{s/v}$ as follows[4]:

$$\Gamma_{s/v}^{\gamma}(m_R) = \frac{\mathbf{q_{cm}}^2}{\pi}\,\frac{2m_N}{(2j_R+1)m_R}\left(|A_{\frac{1}{2}}^{s/v}|^2 + |A_{\frac{3}{2}}^{s/v}|^2\right) \quad , \qquad (4)$$

with j_R and m_R denoting spin and pole-mass of the resonance and $\mathbf{q_{cm}}$ the cm-momentum of the photon. Clearly, $\Gamma_{s/v}^{\gamma}$ can also be expressed using Feynman amplitudes:

$$\Gamma_{s/v}^{\gamma}(k^2) = \frac{1}{(2j_R+1)}\,\frac{\mathbf{q_{cm}}}{8\,\pi\,k^2}\,|\mathcal{M}_{s/v}|^2 \quad , \qquad (5)$$

where $\sqrt{k^2}$ is the invariant mass of the resonance. After summing over the photon polarizations, $|\mathcal{M}_{s/v}|^2$ assumes the following form:

$$|\mathcal{M}_{s/v}|^2 = 4\,m_N\,m_R\,\kappa\,g_{s/v}^2\,q^2\,F(k^2) \quad . \qquad (6)$$

The formfactor $F(k^2)$ at the $RN\gamma$ vertex is taken from Ref.[5]. The numerical factor κ depends on the quantum numbers of the resonance (Ref.[3]). The two expressions Eqs. (4) and (5) can now be equated allowing us to solve for $g_{s/v}$:

$$g_{s/v}^2 = \frac{4}{\kappa}\,\frac{|A_{\frac{1}{2}}^{s/v}|^2 + |A_{\frac{3}{2}}^{s/v}|^2}{\mathbf{q_{cm}}} \qquad (7)$$

In this way it is possible to obtain $g_{s/v}$ from helicity amplitudes. The hadronic couplings $f_{RN\omega(\rho)}$ are then readily deduced from the VMD relation Eq. (1). The corresponding values are listed in Tables 1 and 2.

3 The ρ Meson

3.1 VMD in the Isovector Channel

In this section we investigate the applicability of VMD for the isovector channel of the resonance decay. For the helicity amplitudes we use different parameter sets in order to provide an estimate for the experimental uncertainties entering this analysis. They are taken from Arndt *et al* [6] and Feuster *et al* [5]. The ρ decay widths are taken from the analysis of Manley *et al* [7].

In Table 1 the results for the coupling constants and the corresponding error-bars are given. As a general tendency, VMD works well within a factor of two. This can be seen particularly well in the case of the $D_{13}(1520)$ and the $F_{15}(1680)$, which are the most prominent resonances in photon-nucleon reactions, and whose ρ decay widths are also well under control. Note in particular the large coupling constant for the $D_{13}(1520)$ in the hadronic fit which is possible only because of the large width of the ρ meson[8,9].

For the $P_{13}(1720)$ and the $F_{35}(1905)$ resonances VMD is off by an order of magnitude. We argue that this mismatch does not necessarily indicate a failure of VMD, but can be traced back to the unsatisfactory experimental information on these two resonances. Neither the helicity amplitudes nor the partial $N\rho$ width are well determined from experiment[5,6,7]. Obviously, the extraction of the resonance parameters is very complicated and might be sensitive to the details of the underlying theoretical model, such as the treatment of the non-resonant background. For a conclusive VMD analysis of these resonances it is therefore mandatory to enlarge the data base and to describe hadron- and photoinduced reactions within one and the same

	$f_{RN\rho}$(Arndt)	$f_{RN\rho}$(Feuster)	$f_{RN\rho}$(Manley)
$D_{13}(1520)$	3.44±0.18	2.67	6.67±0.78
$S_{31}(1620)$	0.89±0.42	0.10	2.14±0.30
$S_{11}(1650)$	0.70±0.08	0.59	0.47±0.19
$F_{15}(1680)$	3.48±0.39	—	6.87±1.57
$D_{33}(1700)$	3.96±0.77	3.68	1.962±0.67
$P_{13}(1720)$	0.25±0.42	0.93	13.17±3.35
$F_{35}(1905)$	2.47±0.55	—	17.97±1.14
$P_{33}(1232)$	13.40±0.2	11.96	—

Table 1. Results for $f_{RN\rho}$ from a VMD analysis (1st and 2nd column) and a hadronic fit (3rd column).

analysis.

We conclude that VMD works remarkably well in the isovector channel. Therefore, our approach should yield reasonable predictions of the unknown coupling constants $f_{RN\omega}$.

3.2 The ρ *Spectral Function in Nuclear Matter*

Using the coupling constants obtained from the hadronic $N\rho$ decay width of the resonance, we calculate the spectral function $A_\rho^{T/L}(\omega, \mathbf{q})$ of the ρ meson in nuclear matter[8,9] at density $\rho = \rho_0$. It is defined as:

$$A_\rho^{T/L}(\omega, \mathbf{q}) = \frac{1}{\pi} \frac{\text{Im } \Sigma^{T/L}(\omega, \mathbf{q})}{(\omega^2 - \mathbf{q}^2 - m_\rho^2 + \text{Re } \Sigma^{T/L}(\omega, \mathbf{q}))^2 + \text{Im } \Sigma^{T/L}(\omega, \mathbf{q})^2} \quad . \tag{8}$$

Note that in nuclear matter transverse and longitudinal modes – denoted by T and L, respectively – have to be treated independently. Here ω and $\mathbf{q}$ denote energy and momentum relative to the rest frame of nuclear matter. The selfenergy $\Sigma^{T/L}(\omega, \mathbf{q})$ is a sum of vacuum and in-medium contributions. The vacuum part is given by the 2-pion decay mode and we estimate the in-medium part within the low-density approximation:

$$\Sigma_{med}^{T/L}(\omega, \mathbf{q}) = \frac{1}{8m_N} \rho_N \, T_{tot}^{T/L}(\omega, \mathbf{q}) \quad . \tag{9}$$

The main quantity entering this expression is the ρN forward scattering amplitude $T_{tot}^{T/L}$. In Fig. 1 we show the results for $A_\rho^{T/L}(\omega, \mathbf{q})$. They highlight an important consequence of the strong coupling of the $D_{13}(1520)$ to the $N\rho$ system, namely the strong modification of the ρ spectral function in nuclear matter. In particular at low momenta the mass spectrum is dominated by the excitation of a $D_{13}(1520)$, leading to a substantial shift of spectral strength down to lower invariant masses. This, of course, is of relevance for the interpretation of the CERES data. At large momenta the $P_{13}(1720)$ and the $F_{35}(1905)$ dominate the spectrum. The predominant feature is the different modification of transverse and longitudinal ρ mesons.

4 The ω Meson

4.1 VMD in the Isoscalar Channel

In this section we present our results for the coupling constants $f_{RN\omega}$ and discuss their compatibility with experimental information obtained from pion- and photon-induced ω-production cross sections.

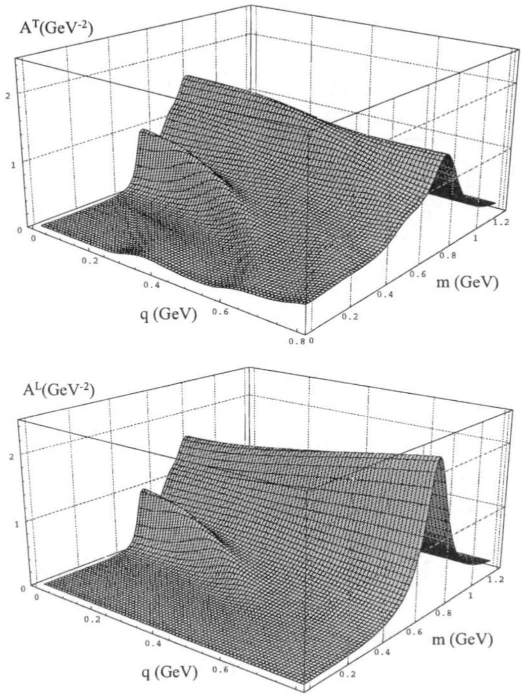

Figure 1. Top: A_ρ^T as a function of invariant mass m and 3-momentum $\mathbf{q}$. Bottom: Same for A_ρ^L.

All nucleon resonances with $j_R < \frac{7}{2}$, for which helicity amplitudes have been extracted, are included. Thus we consider only one resonance above the $N\omega$ threshold in our analysis, namely the $D_{13}(2080)$. We use again the helicity amplitudes from Arndt *et al* [6] and Feuster *et al* [5]and consider also the PDG estimates[4]. The corresponding results for $f_{RN\omega}$ together with the errorbars are given in Table 2. We find a strong coupling to the $N\,\omega$ channel in the S_{11}, D_{13} and F_{15} partial waves; especially the $S_{11}(1650)$, the $D_{13}(1520)$ and the $F_{15}(1680)$ resonances show a sizeable coupling strength to this channel.

It is noteworthy that the resonances with the largest coupling are well below the $N\omega$ threshold. Subthreshold resonances in the $N\omega$ channel are also reported elsewhere[10,11]. The coupled-channel analysis[10] of $\pi\,N$ scattering enforces resonant structures in the $N\omega$ channel, in particular in the S_{11} and D_{13} partial waves. However, the coupling strength extracted in their analysis is

	$f_{RN\omega}$(Arndt)	$f_{RN\omega}$(Feuster)	$f_{RN\omega}$(PDG)
$S_{11}(1535)$	1.27±1.58	1.36	0.76±1.23
$S_{11}(1650)$	1.59±0.29	0.56	1.12±1.09
$D_{13}(1520)$	2.87±0.76	2.28	3.42±0.87
$D_{13}(1700)$	− − −	1.88	0.65±2.76
$D_{13}(2080)$	− − −	− − −	1.13±1.46
$P_{11}(1440)$	0.61±0.68	1.26	0.85±0.48
$P_{11}(1710)$	0.14±0.85	0	0.20±1.02
$P_{13}(1720)$	0.29±1.30	2.18	1.79±3.18
$F_{15}(1680)$	6.89±1.38	− − − −	6.52±1.49

Table 2. VMD predictions for the coupling strength $f_{RN\omega}$.

$f_{RN\omega} \approx 6.5$, nearly twice as large as our value. In the quark model calculation of Ref.[11] a value of about $f_{RN\omega} \approx 2.6$ is found, which is surprisingly close to our result.

The quality of the VMD predictions can further be tested by a comparison with experimental data on the reactions $\pi^- p \rightarrow \omega n$ and $\gamma p \rightarrow \omega p$. Comparison with data also allows to discuss the results for the $D_{13}(2080)$, the only resonance in our analysis above threshold. We find for this resonance an ω decay width of about 70 MeV and argue that its contribution to both reactions is too small to be seen in experiment. As a first approximation, we take the full production amplitude as an incoherent sum of *Breit-Wigner* type amplitudes, describing s-channel contributions.

The results are shown in Fig. 2. The data for the π-induced reaction are taken from Ref.[12] and we use the photoproduction data of Ref.[13]. The cross-section for $\pi^- p \rightarrow \omega n$ is reproduced rather well near threshold. This is in agreement with the findings of Ref.[10], where a satisfactory description of this process around threshold in terms of near or subthreshold resonances is presented. This suggests that the excitation of subthreshold resonances constitutes an essential ingredient to the production mechanism. The contribution coming from the only resonance above threshold – the $D_{13}(2080)$ – is about 0.1 mb, roughly 10% of the total cross-section.

On the other hand, the photoproduction data cannot be saturated within the resonance model; this gives little hope to find the $D_{13}(2080)$ in this reaction. However, adding the contribution from π^0-exchange yields a qualitative explanation of the data over the energy range under consideration.

Overall it seems that the predictions of the resonance model are in reasonable agreement with the data and can be viewed as a confirmation of the

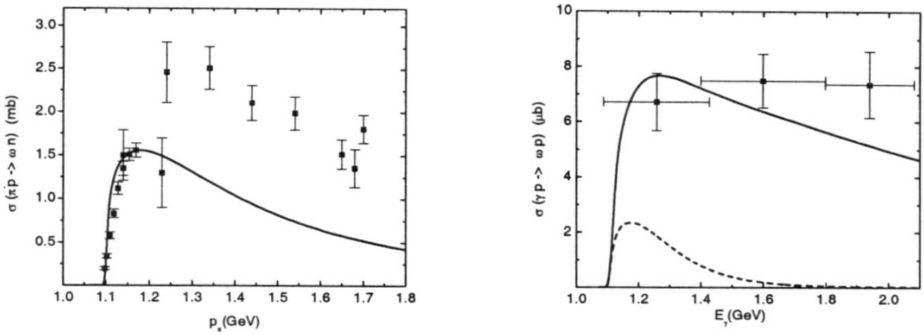

Figure 2. Left: Total cross-section for the reaction $\pi^- p \to \omega n$. Right: Total cross-section for the reaction $\gamma p \to \omega p$ with resonance contribution only (dashed line) and added pion-exchange (solid line).

VMD analysis.

4.2 The ω Spectral Function in Nuclear Matter

Within the same formalism as for the ρ meson we investigate the effects of resonance-excitation on the properties of ω mesons in nuclear matter. Again the calculations are performed at $\rho = \rho_0$. We find a broadening of the ω meson of about 50 MeV and a repulsive mass shift of roughly 20 MeV, see Fig. 3. These findings are in surprising agreement with those of various other groups[10,14,15]. The ω meson thus is much less modified in nuclear matter than the ρ meson, which follows within our model from the much smaller coupling constants. As can be seen in Fig. 3 the in-medium effects are most pronounced at small momenta.

5 Conclusions

We have demonstrated that through the excitation of baryon resonances the in-medium spectral functions of vector mesons receive a substantial shift of spectral strength down to lower invariant masses. These effects play a key role in understanding the in-medium properties of vector mesons.

6 Acknowledgments

This work was supported by DFG and BMBF.

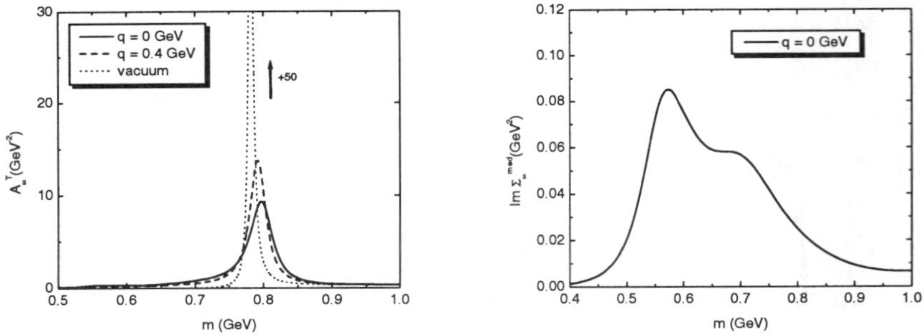

Figure 3. Left: $A_\omega^{T/L}$ at momenta 0 GeV (straight) and 0.4 GeV (dashed). For comparison also the vacuum result is shown (dotted). Right: The imaginary part of the ω selfenergy.

References

1. G. Agakichiev *et al.*, CERES collaboration, *Phys. Rev. Lett.* **75**, 1272 (1995).
2. H. B. O'Connell, B. C. Pearce, A. W. Thomas, and A. G. Williams, *Prog. Part. Nucl. Phys.* **39**, 201 (1997).
3. M. Post and U. Mosel, *Nucl. Phys.* A (2001) in press, nucl-th/0008040.
4. Particle Data Group, *Eur. Phys. J.* C **3**, 1 (1998).
5. T. Feuster and U. Mosel, *Phys. Rev.* C **59**, 460 (1999).
6. R. A. Arndt *et al.*, *Phys. Rev.* C **52**, 2120 (1995).
7. D.M. Manley and E.M. Saleski, *Phys. Rev.* D **45**, 4002 (1992); *Phys. Rev.* D **30**, 904 (1984).
8. W. Peters, M. Post, H. Lenske, S. Leupold, and U. Mosel, *Nucl. Phys.* A **45**, 4002 (1992).
9. M. Post, S. Leupold, and U. Mosel, *Nucl. Phys.* A (2001) in press, nucl-th/0008027.
10. B. Friman, M. Lutz and G. Wolf, nucl-th/0003012.
11. D. O. Riska and G. E. Brown, *Nucl. Phys.* A **679**, 577 (2001).
12. A. Baldini *et al.*, Landolt-Börnstein, vol 1/12 a (Springer, Berlin, 1987).
13. ABBHHM-Colloboration, *Phys. Rev.* **75**, 1669 (1968).
14. F. Klingl, N. Kaiser, and W. Weise, *Nucl. Phys.* A **624**, 527 (1997).
15. M. Effenberger, E.L. Bratkovskaya, W. Cassing, and U. Mosel, *Phys. Rev.* C **60**, 27601 (1999).

HIGHER AND MISSING RESONANCES IN ω PHOTOPRODUCTION

Y. OH
Institute of Physics and Applied Physics, Yonsei University, Seoul 120-749, Korea
E-mail: yoh@phya.yonsei.ac.kr

A. I. TITOV
Bogoliubov Laboratory of Theoretical Physics, JINR, Dubna 141980, Russia
E-mail: atitov@thsun1.jinr.ru

T.-S. H. LEE
Physics Division, Argonne National Laboratory, Argonne, Illinois 60439, U.S.A.
E-mail: lee@theory.phy.anl.gov

We study the role of the nucleon resonances (N^*) in ω photoproduction by using the quark model resonance parameters predicted by Capstick and Roberts. The employed $\gamma N \to N^*$ and $N^* \to \omega N$ amplitudes include the configuration mixing effects due to the residual quark-quark interactions. The contributions from the nucleon resonances are found to be important in the differential cross sections at large scattering angles and various spin observables. In particular, the parity asymmetry and beam-target double asymmetry at forward scattering angles are suggested for a crucial test of our predictions. The dominant contributions are found to be from $N\frac{3}{2}^+(1910)$, a missing resonance, and $N\frac{3}{2}^-(1960)$ which is identified as the $D_{13}(2080)$ of the Particle Data Group.

The nucleon resonances predicted by the constituent quark models have a much richer spectrum than what has been observed in pion-nucleon scattering[1]. The origin of this "missing resonance problem" has been ascribed to the possibility that many predicted nucleon resonances (N^*) could couple weakly to the πN channel[2]. Therefore it would be legitimate to study the nucleon resonances in other reactions, and vector meson electromagnetic production is one of them which under investigation at current experimental facilities such as TJNAF, ELSA-SAPHIR, GRAAL, and LEPS of SPring-8. Theoretically the role of the nucleon resonances was studied by Zhao *et al.*[3,4] based on an effective Lagrangian method within the SU(6) $\times$ O(3) constituent quark model.

Our study on the nucleon resonances in vector meson photoproduction[5] is based on the quark model predictions by Capstick and Roberts[6], where the configuration mixing effects due to the residual quark-quark interactions are included and the hadron decays are calculated by using the $^3P_0$ model. The

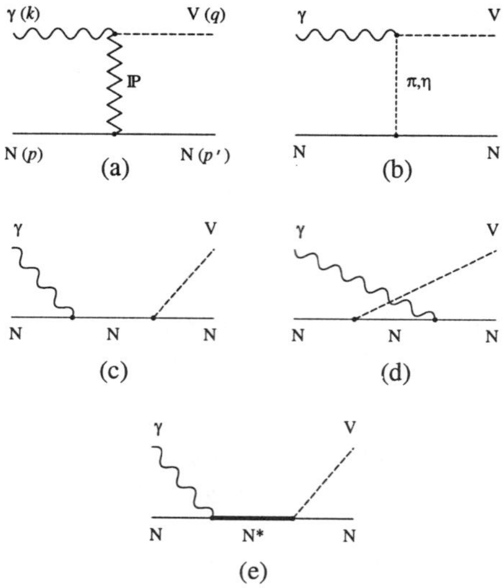

Figure 1. Diagrammatic representation of ω photoproduction mechanisms: (a) Pomeron exchange, (b) (π, η) exchange, (c) direct nucleon term, (d) crossed nucleon term, and (e) s-channel nucleon excitations.

predicted baryon wave functions and the N^* decay amplitudes are considerably different from those of the SU(6) $\times$ O(3) quark model[3]. Thus it would be interesting to find the differences in the model predictions on vector meson photoproduction, which can be tested experimentally.

In this work we focus on ω photoproduction[5] since its non-resonant reaction mechanisms are rather well understood and the isosinglet nature of the ω meson allows the contributions from the isospin-1/2 nucleon resonances only. Earlier studies on ω photoproduction[7] show that the reaction is dominated by diffractive processes at high energies, i.e., via the Pomeron exchange, and by one-pion exchange at low energies. It is therefore reasonable to follow the earlier theoretical analyses and assume that the non-resonant amplitude of ω photoproduction can be calculated from these two well-established mechanisms with some refinements. The resulting model can then be a starting point for investigating the N^* effects.

Our model for ω photoproduction, therefore, can be described by the diagrams shown in Fig. 1. The Pomeron exchange [Fig. 1(a)] is known to govern the total cross sections and differential cross sections at low $|t|$ in the high energy region for electromagnetic production of vector mesons. For this model, we follow the Donnachie-Landshoff model[8]. The details of this model can be found, e.g., in Refs. [9,10] and the resulting amplitude and the parameters are summarized in Refs. [5,11].

The pseudoscalar meson exchange amplitudes [Fig. 1(b)] are calculated from the following effective Lagrangians:

$$\mathcal{L}_{\omega\gamma\varphi} = \frac{eg_{\omega\gamma\varphi}}{M_V}\epsilon^{\mu\nu\alpha\beta}\partial_\mu\omega_\nu\partial_\alpha A_\beta\,\varphi,$$
$$\mathcal{L}_{\varphi NN} = -ig_{\pi NN}\bar{N}\gamma_5\tau_3 N\pi^0 - ig_{\eta NN}\bar{N}\gamma_5 N\eta, \tag{1}$$

where $\varphi = (\pi^0, \eta)$ and A_β is the photon field. We use $g_{\pi NN}^2/4\pi = 14$ and $g_{\eta NN}^2/4\pi = 0.99$ for the πNN and ηNN coupling constants, respectively. The coupling constants $g_{\omega\gamma\varphi}$ can be estimated through the decay widths of $\omega \to \gamma\pi$ and $\omega \to \gamma\eta$ [12] which lead to $g_{\omega\gamma\pi} = 1.823$ and $g_{\omega\gamma\eta} = 0.416$. The higher mass of the η and the associated small coupling constants suppress the η exchange contribution compared with the π exchange. The φNN and $\omega\gamma\varphi$ vertices are dressed by the form factors,

$$F_{\varphi NN}(t) = \frac{\Lambda_\varphi^2 - M_\varphi^2}{\Lambda_\varphi^2 - t}, \qquad F_{\omega\gamma\varphi}(t) = \frac{\Lambda_{\omega\gamma\varphi}^2 - M_\varphi^2}{\Lambda_{\omega\gamma\varphi}^2 - t}, \tag{2}$$

with $\Lambda_\pi = 0.6$ GeV and $\Lambda_{\omega\gamma\pi} = 0.7$ GeV [5], $\Lambda_\eta = 1.0$ GeV and $\Lambda_{\omega\gamma\eta} = 0.9$ GeV [13].

The nucleon pole terms [Fig. 1(c,d)] are calculated from the following interaction Lagrangians:

$$\mathcal{L}_{\gamma NN} = -e\bar{N}\left(\gamma_\mu\frac{1+\tau_3}{2}A^\mu - \frac{\kappa_N}{2M_N}\sigma^{\mu\nu}\partial_\nu A_\mu\right)N,$$
$$\mathcal{L}_{\omega NN} = -g_{\omega NN}\bar{N}\left(\gamma_\mu\omega^\mu - \frac{\kappa_\omega}{2M_N}\sigma^{\mu\nu}\partial_\nu\omega_\mu\right)N, \tag{3}$$

with the anomalous magnetic moment of the nucleon $\kappa_{p(n)} = 1.79\ (-1.91)$. For the coupling constants we use $g_{\omega NN} = 10.35$ and $\kappa_\omega = 0$ [14,15]. In order to dress the vertices, we include the form factor[16],

$$F_N(r) = \frac{\Lambda_N^4}{\Lambda_N^4 - (r - M_N^2)^2}, \tag{4}$$

where $r = s$ or t and $\Lambda_N = 0.5$ GeV.

With the non-resonant amplitudes discussed above we estimate the nucleon resonance contributions by making use of the quark model predictions[6] on the resonance photoexcitation $\gamma N \to N^*$ and the resonance decay $N^* \to N\omega$. In this work we consider the s-channel diagrams shown in Fig. 1(e).

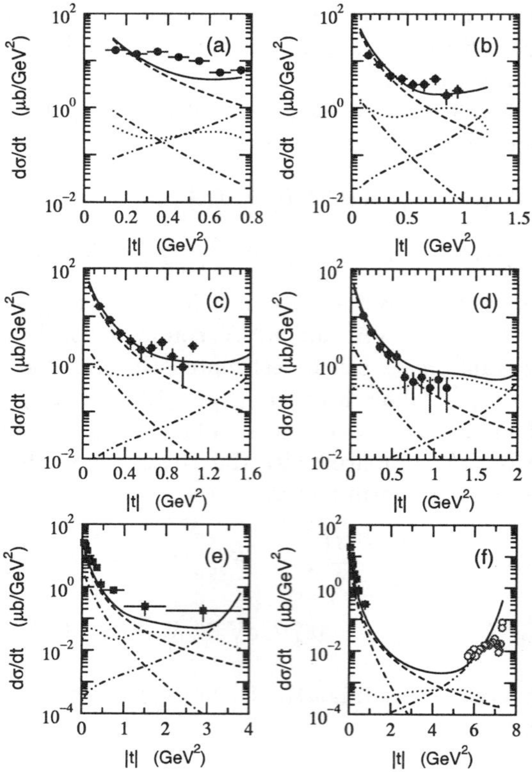

Figure 2. Differential cross sections for $\gamma p \to p\omega$ reaction as a function of $|t|$ at $E_\gamma =$ (a) 1.23, (b) 1.45, (c) 1.68, (d) 1.92, (e) 2.8, and (f) 4.7 GeV. The results are from pseudoscalar-meson exchange (dashed), Pomeron exchange (dot-dashed), direct and crossed nucleon terms (dot-dot-dashed), N^* excitation (dotted), and the full amplitude (solid). Data are taken from Ref. [17] [filled circles in (a,b,c,d)], Ref. [18] [filled squares in (e,f)], and Ref. [19] [open circles in (f)].

The crossed diagrams cannot be calculated from the informations available in Ref. [6]. The resonant amplitude in the center of mass frame is written as

$$I^{N^*}_{m_f,m_\omega,m_i,\lambda_\gamma}(\mathbf{q},\mathbf{k}) = \sum_{J,M_J} \frac{1}{\sqrt{s} - M^J_R + \frac{i}{2}\Gamma^J(s)} \mathcal{M}_{N^* \to N'\omega}(\mathbf{q}; m_f, m_\omega; J, M_J) \times \mathcal{M}_{\gamma N \to N^*}(\mathbf{k}; m_i, \lambda_\gamma; J, M_J), \quad (5)$$

where M^J_R is the N^* mass of spin quantum numbers (J, M_J), and m_i , m_f, λ_γ,

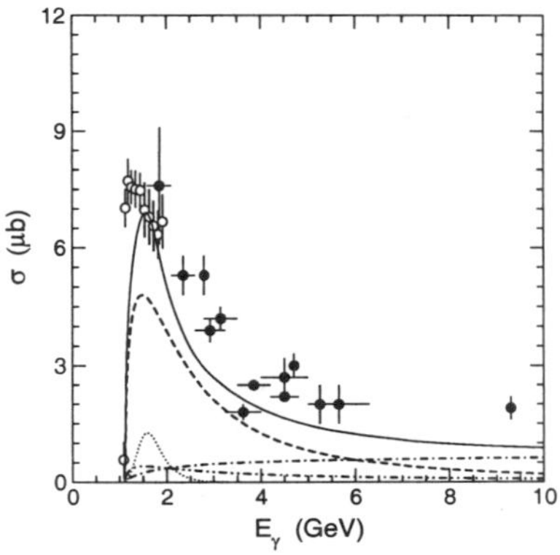

Figure 3. Total cross section of ω photoproduction. The notations are the same as in Fig. 2. Data are taken from Refs. [17,18,20].

and m_ω are the spin projections of the initial nucleon, final nucleon, incoming photon, and outgoing ω meson, respectively. In this study, we consider 12 positive parity and 10 negative parity nucleon resonances up to spin-9/2. The explicit form of the resonant amplitude and the details on the calculations can be found in Ref. [5] as well as the considered nucleon resonances and their parameters.

Our results for the differential cross section are shown in Fig. 2, which shows that the data can be described to a very large extent in the considered energy region, $E_\gamma \leq 5$ GeV. It is clear that the contributions due to the N^* excitations (dotted curves) and the direct and crossed nucleon terms (dot-dot-dashed curves) help bring the agreement with the data at large angles. The forward angle cross sections are mainly due to the interplay between the pseudoscalar-meson exchange (dashed curves) and the Pomeron exchange (dot-dashed curves). The main problem here is in reproducing the data at $E_\gamma = 1.23$ GeV. This perhaps indicates that the off-shell contributions from N^*'s below ωN threshold are important at very low energies.

The contribution from the nucleon resonances to total cross sections is shown in Fig. 3. To have a better understanding of the resonance contributions, we compare the contributions from the considered N^*'s to the differential and total cross sections. We found that the contributions from $N\frac{3}{2}^+(1910)$ and $N\frac{3}{2}^-(1960)$ are the largest at all energies up to $E_\gamma = 3$ GeV. The $N\frac{3}{2}^+(1910)$ is a missing resonance, while $N\frac{3}{2}^-(1960)$ is identified as a two star resonance $D_{13}(2080)$ of the Particle Data Group[12]. This result is significantly different from the quark model calculations of Ref. [3]. The difference between the two calculations is not surprising since the employed quark models are rather different. In particular, our predictions include the configuration mixing effects due to residual quark-quark interactions. The discrepancy of our prediction with the experimental data at very low energy is again expected to be due to the nucleon resonances below ωN threshold, which are missed in our calculation.

Instead of the cross sections, the polarization asymmetries[21] provide more appropriate tools to investigate the role of the nucleon resonances in ω photoproduction[5]. We first examined the single spin asymmetries[5]. Although our predictions are significantly different from those of Ref. [3], we confirm their conclusion that those asymmetries are sensitive to the nucleon resonances but mostly at large $|t|$ region.

More clear signals for the nucleon resonances can be found from the parity asymmetry (or photon polarization asymmetry) and the beam-target double asymmetry. These asymmetries are sensitive to the N^* contributions at forward angles, where precise measurements might be more favorable because the cross sections are peaked at $\theta = 0$. The parity asymmetry is defined as[22]

$$P_\sigma = \frac{d\sigma^N - d\sigma^U}{d\sigma^N + d\sigma^U} = 2\rho^1_{1-1} - \rho^1_{00}, \tag{6}$$

where σ^N and σ^U are the cross sections due to the natural and unnatural parity exchanges respectively, and $\rho^i_{\lambda,\lambda'}$ are the vector-meson spin density matrices. The beam-target double asymmetry is defined as[21]

$$C^{BT}_{zz} = \frac{d\sigma(\uparrow\downarrow) - d\sigma(\uparrow\uparrow)}{d\sigma(\uparrow\downarrow) + d\sigma(\uparrow\uparrow)}, \tag{7}$$

where the arrows represent the helicities of the incoming photon and the target proton. In Fig. 4, we show the results from calculations with (solid curves) and without (dotted curves) including N^* contributions. The difference between them is striking and can be unambiguously tested experimentally. Here we also find that the $N\frac{3}{2}^+(1910)$ and $N\frac{3}{2}^-(1960)$ are dominant. By keeping only these two resonances in calculating the resonant part of the amplitude, we obtain the dashed curves which are not too different from the full calculations (solid curves).

To summarize, we investigated the role of the nucleon resonances in ω photoproduction. We found that the inclusion of the resonance amplitudes leads to a better description of the observed total and differential cross sections. It is also found that the $N\frac{3}{2}^+(1910)$ and $N\frac{3}{2}^-(1960)$ are dominant in the resonance amplitudes. As a further study on the nucleon resonances, we

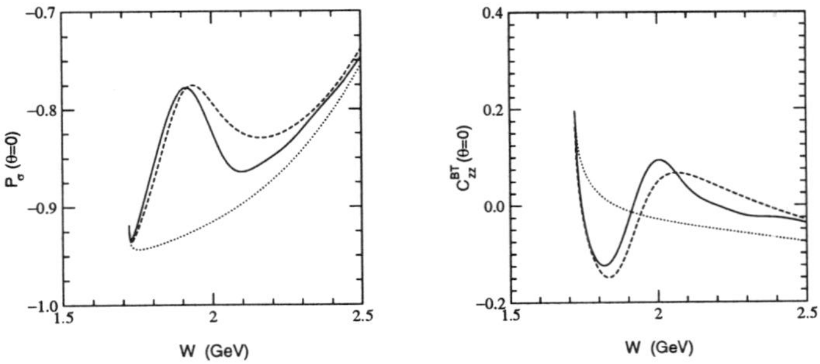

Figure 4. Parity asymmetry P_σ and beam-target double asymmetry C_{zz}^{BT} at $\theta = 0$ as functions of W. The dotted curves are calculated without including N^* effects, the dashed curves include contributions of $N\frac{3}{2}^+(1910)$ and $N\frac{3}{2}^-(1960)$ only, and the solid curves are calculated with all N^*'s up to spin-9/2.

suggest to measure the parity asymmetry and beam-target double asymmetry. Experimental tests of our predictions will be a useful step toward resolving the so-called "missing resonance problem" or distinguishing different quark model predictions and could be done at current electron/photon facilities. Theoretically the predictions should be improved further. For example, the form factor of the vertices including the nucleon resonances should be studied in detail in the given quark models and the crossed N^* terms as well as the nucleon resonances below ωN threshold should be studied. Finally, the effects due to the initial and final state interactions must be also investigated, especially at low energies.

Acknowledgments

This work was supported in part by the Brain Korea 21 project of Korean Ministry of Education, the International Collaboration Program of KOSEF under Grant No. 20006-111-01-2, Russian Foundation for Basic Research under Grant No. 96-15-96426, and U.S. DOE Nuclear Physics Division Contract No. W-31-109-ENG-38.

References

1. N. Isgur and G. Karl, *Phys. Lett.* B **72**, 109 (1977); *Phys. Rev.* D **18**, 4187 (1978); **19**, 2653 (1979), **23**, 817(E) (1981); R. Koniuk and N. Isgur, *ibid.* **21**, 1868 (1980).

2. See, e.g., S. Capstick and W. Roberts, nucl-th/0008028.
3. Q. Zhao, Z. Li, and C. Bennhold, *Phys. Rev.* C **58**, 2393 (1998); Q. Zhao, *Phys. Rev.* C **63**, 025203 (2001).
4. Q. Zhao, *these proceedings.*
5. Y. Oh, A. I. Titov, and T.-S. H. Lee, *Phys. Rev.* C **63**, 025201 (2001); talk at SPIN 2000 Symposium, nucl-th/0012012.
6. S. Capstick, *Phys. Rev.* D **46**, 2864 (1992); S. Capstick and W. Roberts, *ibid.* **49**, 4570 (1994).
7. ABBHHM Collaboration, R. Erbe *et al., Phys. Rev.* **175**, 1669 (1968); K. Schilling and F. Storim, *Nucl. Phys.* B **7**, 559 (1968); H. Fraas, *ibid.* B **36**, 191 (1972); J. Ballam *et al., Phys. Rev.* D **7**, 3150 (1973); P. Joos *et al., Nucl. Phys.* B **122**, 365 (1977); B. Friman and M. Soyeur, *ibid.* A **600**, 477 (1996).
8. A. Donnachie and P. V. Landshoff, *Nucl. Phys.* B **44**, 322 (1984); B **267**, 690 (1986); *Phys. Lett.* B **185**, 403 (1987); **296**, 227 (1992).
9. J.-M. Laget and R. Mendez-Galain, *Nucl. Phys.* A **581**, 397 (1995).
10. M. A. Pichowsky and T.-S. H. Lee, *Phys. Rev.* D **56**, 1644 (1997).
11. Y. Oh, A. I. Titov, and T.-S. H. Lee, talk at NSTAR 2000 Workshop, nucl-th/0004055.
12. Particle Data Group, D. E. Groom *et al., Eur. Phys. Jour.* C **15**, 1 (2000).
13. A. I. Titov, T.-S. H. Lee, H. Toki, and O. Streltsova, *Phys. Rev.* C **60**, 035205 (1999).
14. T. Sato and T.-S. H. Lee, *Phys. Rev.* C **54**, 2660 (1996).
15. Th. A. Rijken, V. G. J. Stoks, and Y. Yamamoto, *Phys. Rev.* C **59**, 21 (1999).
16. H. Haberzettl, C. Bennhold, T. Mart, and T. Feuster, *Phys. Rev.* C **58**, 40 (1998).
17. F. J. Klein, Ph.D. thesis, Bonn Univ. (1996); SAPHIR Collaboration, F. J. Klein *et al., πN Newslett.* **14**, 141 (1998).
18. J. Ballam *et al., Phys. Rev.* D **7**, 3150 (1973).
19. R. W. Clifft *et al., Phys. Lett.* B **72**, 144 (1977).
20. J. Ballam *et al., Phys. Lett.* B **30**, 421 (1969); AHHM Collaboration, W. Struczinski *et al., Nucl. Phys.* B **108**, 45 (1976); D. P. Barber *et al., Z. Phys.* C **26**, 343 (1984).
21. A. I. Titov, Y. Oh, S. N. Yang, and T. Morii, *Phys. Rev.* C **58**, 2429 (1998).
22. K. Schilling, P. Seyboth, and G. Wolf, *Nucl. Phys.* B **15**, 397 (1970).

PHOTOPRODUCTION OF BARYON RESONANCES, FIRST DATA FROM THE CB-ELSA EXPERIMENT

U. THOMA FOR THE CB-ELSA COLLABORATION

Institut für Strahlen- und Kernphysik, Universität Bonn, Nussallee 14-16, D-53115 Bonn, Germany

E-mail: thoma@iskp.uni-bonn.de

Photoproduction data on various final states involving neutral mesons have been taken by the CB-ELSA-experiment at the electron stretcher accelerator ELSA (Bonn). The data show clear structures due to resonance production. Evidence for successive decays of high-mass nucleon resonances via $\Delta(1232)\pi$ and via resonances of higher mass is observed.

1 Introduction

It is widely accepted that QCD is most probably the correct theory of strong interactions. While QCD is in excellent agreement with the high energy data, a major goal of QCD is still unfulfilled: to provide the theory of quark confinement. Instead, constituent quark models have been developed which describe the baryon spectrum with good success. The observed baryon resonances seem to be consistent with being three-quark systems. However there is an interesting controversy in baryon spectroscopy which concerns the "missing resonances". Constituent quark model calculations predict much more resonances than have been observed so far. Two very different explanations have been proposed:

- The "missing" states are not missing. They have not been observed so far because of lack of high quality data in channels different from πN. Most available experimental data stem from πN scattering experiments. If these states decouple from πN they would not have been observed so far. This seems reasonable following quark model predictions [1,2].

- The "missing" states are not missing, they do not exist. As proposed by Lichtenberg, the nucleon-resonances could have a quark-diquark structure. This reduces the number of internal degrees of freedom and therefore the number of existing states [3]. Of course one might also think of other hidden symmetries.

When these "missing" states exist they are expected to couple to channels like, e.g. ηN, $\eta' N$, ωN, $\Delta\pi$ or ρN and to γp [2,4]. So photoproduction experiments investigating one of these channels have a big discovery potential if these resonances really exist. The investigation of different final states are subject of several CB-ELSA proposals [5,6,7].

There is another interesting aspect related to the relevant degrees of freedom which determine the baryon spectrum. It is observed that some resonances show a strong η decay, while other resonances with the same quantum numbers almost decouple from the η channel. The $S_{11}(1535)$ decays, e.g., strongly into ηN while for most other resonances this decay is in the order of a few %. This led to many very different interpretations of the nature of

Table 1. The underlined resonance shows a strong decay into $N\eta$, the others do not. The resonances in the upper line have an intrinsic spin of $s = 3/2$, the states in the lower row of $s = 1/2$.

N(1650)S_{11}	N(1700)D_{13}	N(1675)D_{15}
N(1535)S_{11}	N(1520)D_{13}	

the $S_{11}(1535)$. This resonance is of special interest since it is the lowest mass nucleon resonance which can be reached by an electric dipole transition.

In the model of Isgur and Karl, the $S_{11}(1535)$ and the $S_{11}(1650)$ mix strongly with each other so that the $S_{11}(1650)$ decouples from ηN while the $S_{11}(1535)$ couples strongly to it [8]. Glozman and Riska describe the fine structure interaction between the constituent quarks in terms of Goldstone boson exchange which implies a clusterization of the wave function in a quark and a diquark. The different spin-flavor symmetry of the two states causes the $S_{11}(1535)$ to decay into ηN while the $S_{11}(1650)$ does not [9]. Finally, Kaiser, Siegel and Weise explain the $S_{11}(1535)$ as a quasi-bound $K\Sigma$-$K\Lambda$-state which decays strongly into $N\eta$ [10] because of coupled channel dynamics. These very different explanations clearly show that the $S_{11}(1535)$ is certainly a still debated object deserving further attention.

We believe that maybe the systematics in the η-decays of the different resonances can help to understand the situation. As shown in Table 2 the Λ and the Σ-states show a similar behavior, only one of the states decays strongly into its ground state and an η (see Table 2). For the Λ-resonances the low mass S_{01}-resonance decays into an η; for the Σ-resonances, the situation is reversed. This situation led Glozman and Riska to suggest that not the $\Sigma(1620)S_{11}$ and the $\Sigma(1670)D_{13}$ are the first resonances in these partial waves but the $\Sigma(1750)$ and the $\Sigma(????)D_{13}$. Given the mass of the $\Sigma(1775)D_{15}$ I consider this scenario to be unlikely.

Table 2. Baryons with negative parity and their decays into η-mesons. The underlined resonances show a strong decay into their ground state and an η.

$\Lambda(1800)S_{01}$	$\Lambda(????)D_{03}$	$\Lambda(1830)D_{05}$	$\underline{\mathbf{\Sigma(1750)S_{11}}}$	$\Sigma(????)D_{13}$	$\Sigma(1775)D_{15}$
$\underline{\mathbf{\Lambda(1670)S_{01}}}$	$\Lambda(1690)D_{03}$		$\Sigma(1620)S_{11}$	$\Sigma(1670)D_{13}$	
$\Delta(1900)S_{31}$	$\mathbf{\Delta(1940)D_{33}}$	$\Delta(1930)D_{35}$			
$\Delta(1620)S_{31}$	$\mathbf{\Delta(1700)D_{33}}$				

Investigating the η-decays of the Δ-resonances would allow one to extend this pattern and would possibly give further insight in a possible hidden symmetry [12]. With a very naive picture in mind, assuming that the nucleon resonance decays without a spin flip (and the η takes the parity away), one would expect that the higher mass $\Delta(1940)D_{33}$ decays into $\Delta\eta$. Of course this also assumes a 3-quark nature of the states and that the internal quark spin s is approximately a good quantum number. This would explain the pattern observed for the N^* and the Λ-resonances. Nefkens reported from bubble chamber data some evidence for the $\Delta\eta$-decay of the lower mass $\Delta(1700)D_{33}$ [11]. Following the predictions of Capstick and Roberts [13] the $\Delta\eta$ strength should be distributed over the different resonances.

The investigation of the nucleon resonance spectrum is an important aim of the ongoing CB-ELSA experiment at Bonn. At ELSA, photoproduction of resonances up to a mass of 2.6 GeV can be investigated.
In the following, the present setup of the detector is described and first data are shown.

2 The CB-ELSA Detector

After seven years of successful operation at the low energy antiproton ring (LEAR) at CERN, the Crystal Barrel calorimeter was moved to Bonn to perform photo- and electroproduction experiments at the electron stretcher ring ELSA. The setup of the CB-ELSA experiment used for a first series of photoproduction experiments is shown in Fig. 1. The calorimeter consists of 1380 CsI crystals which cover about 98% of 4π. It is especially suited to measure photons. A scintillating fibre detector was built to identify charged particles leaving the target. It provides an intersection point of the particle trajectory with the detector. This detector consists of 3 layers of scintillating fibres; one straight layer and two layers in $\pm 25°$. It surrounds the liquid hydrogen target which has a length of 5 cm and a diameter of 3 cm.

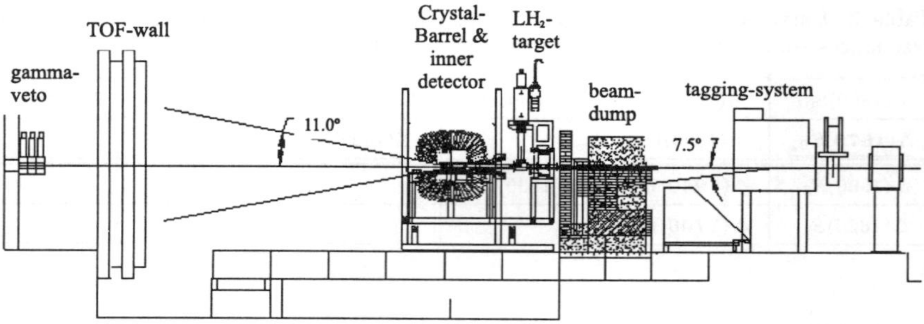

Figure 1. Current setup of the CB-ELSA detector.

This detector system can be extended in forward direction by various other detectors depending on the physics one is interested in. Available are an electromagnetic spectrometer (EMS) and a time of flight detector (TOF) of the former ELAN experiment [14]. In addition, measurements with the TAPS detector [15,16,17,6] built up as a hexagonal wall of 528 BaF_2 crystals are planned. A lead glas wall consisting of about 900 lead glas detectors is also available. The present setup uses TOF as forward detector for protons. The tagger system provides photons with energies from 22 - 93% of the incoming electron energy. The setup allows to reach nucleon resonances up to a mass of 2.6 GeV.

3 First Data

So far, data at four different e^--beam energies have been taken with the CB-ELSA experiment: 800 MeV, 1400 MeV, 2600 MeV and 3200 MeV. For the first calibration of the detector system at ELSA, the 800 MeV data was used. At this energy, the forward boost is small, the cross section for π^0 production is large and each crystal gets enough statistics to perform the calibration. The calibration is done using the known π^0 mass as reference. The spectra (as well as the following ones) are preliminary and not corrected for acceptance. Fig. 2 shows the $\gamma\gamma$ invariant mass of selected proton 2γ events. A clear π^0 and η peak above a small background is observed. The spectra for the two beam-energies look similar. Of course the energy window of the incoming photons relative to the particle thresholds effects the relative strength of the η and π^0 peak. Fig. 3 shows the $p\pi^0$ invariant mass of selected $p\pi^0$ events. The

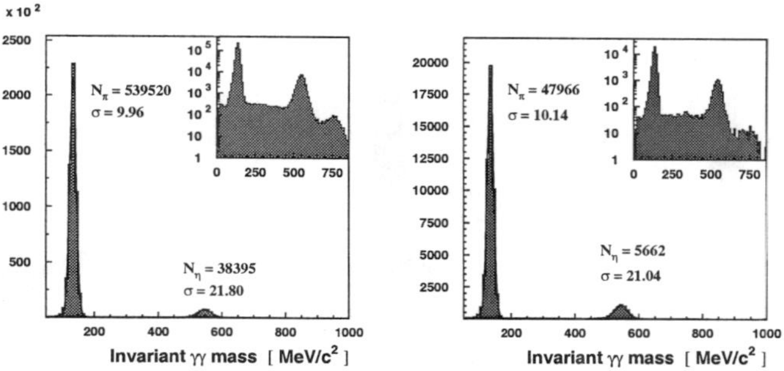

Figure 2. Invariant $\gamma\gamma$-mass of p$\gamma\gamma$-events; beam energies: E_{e^-}=2.6 GeV (left), 3.2 GeV (right).

different resonance regions starting with the $\Delta(1232)$, (1400 MeV data), to the second and third resonance regions are clearly visible. The reaction

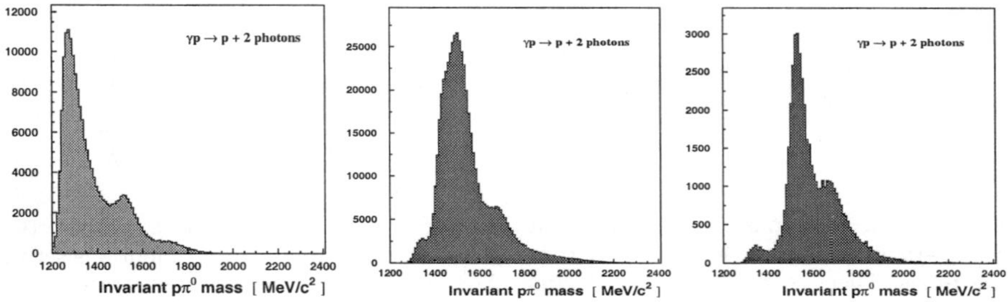

Figure 3. pπ^0 invariant mass for different beam-energies: 1400 MeV, 2600 MeV and 3200 MeV. Clearly observed are the different resonances regions. The small enhancement at low masses is due to accidentals.

$\gamma p \rightarrow p\eta$ was observed in the $\eta \rightarrow 2\gamma$ decay discussed above, but also in the $\eta \rightarrow 3\pi^0$ decay (see Fig. 4). The η-events clearly show up above a small background. The pη invariant mass shows a strong peaking structure due to the $S_{11}(1535)$. Further resonance intensity may be hidden in the spectrum; this has to be investigated further. Photoproduction of $2\pi^0$ is also interesting for the search of missing resonances [7]. Final states with four photons have to be investigated. These data include the reaction $\gamma p \rightarrow p\pi^0\eta$. Fig. 5 shows p$4\gamma$ events. In the left picture a clear peak due to $\pi^0\pi^0$ events is observed. In addition, an enhancement in the region of the $\pi^0\eta$ events is visible; it gets more obvious in the right picture where a horizontal cut on the dominant $2\pi^0$ peak was applied. Fig. 6 shows the p$\pi^0\pi^0$ and the pπ^0 invariant masses as well

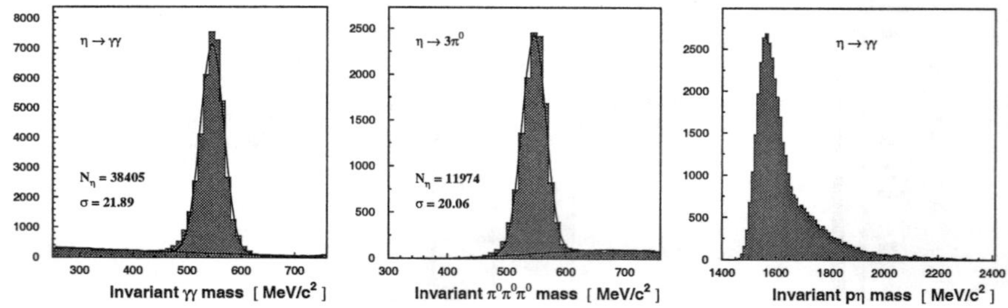

Figure 4. Observed pη-events, left: invariant $\gamma\gamma$ mass, middle: invariant $3\pi^0$-mass. Right: invariant pη-mass. A peak due to the $S_{11}(1535)$ is clearly observed in the data.

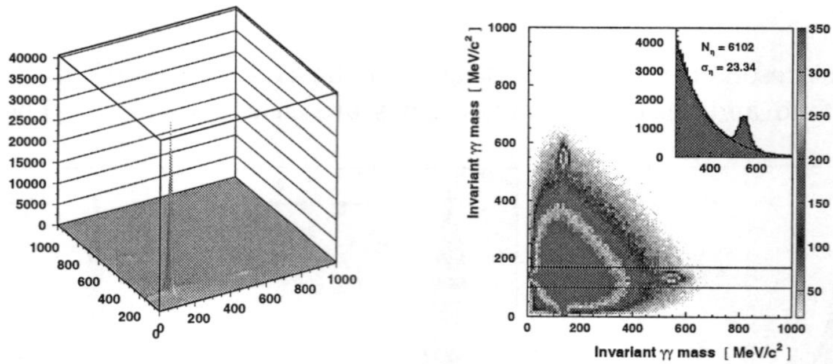

Figure 5. In both plots the $\gamma\gamma$-invariant mass of one $\gamma\gamma$-pair is plotted against the $\gamma\gamma$-invariant mass of the other pair, left: complete spectrum, right: a maximum was set to the z-axis. There is strong $\pi^0\pi^0$ production; the $\pi^0\eta$ system is also clearly observed.

as a two-dimensional plot where one mass is plotted against the other one. Two resonance structures are clearly visible: one around 1500 MeV decaying via $\Delta\pi$, and the other around 1700 MeV decaying via $\Delta\pi$ and possibly into a higher mass resonance (see two-dimensional plot). The resonance region around 1700 MeV gets more clearly visible in the data taken at the higher beam energy. As expected, the second peak in the pπ^0 invariant mass is also enhanced. These plots clearly show that we are observing the structures we are looking for; exited resonances which decay via $\Delta(1232)\pi$, where we expect missing resonances to occur. Also additional decay modes via higher mass resonances are visible in the data.

Fig. 7 shows similar plots for the p$\pi^0\eta$ data set. The $p\pi^0\eta$ invariant mass shows an enhancement at threshold which may be related to resonance production. The inset and the projections show that there is enhanced intensity

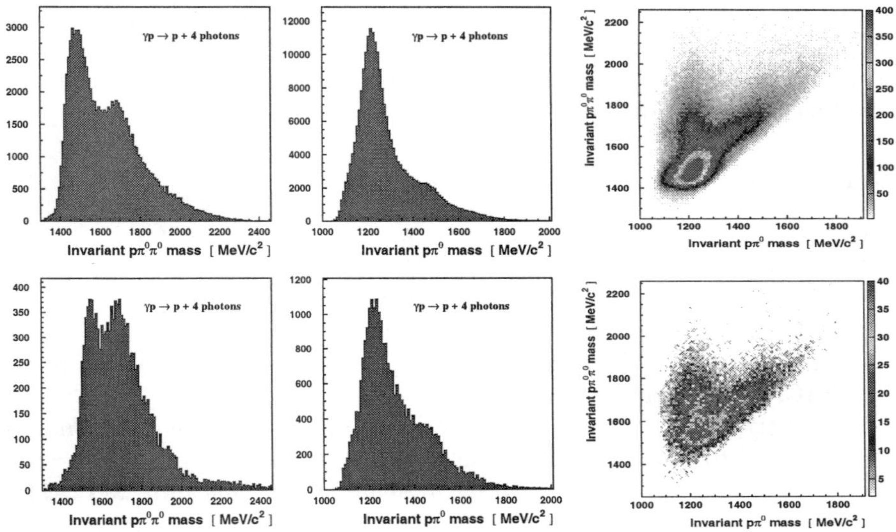

Figure 6. $\pi^0\pi^0$p-data measured at two different beam-energies: E_{e^-}=2.6 GeV (upper row), 3.2 GeV (lower row).

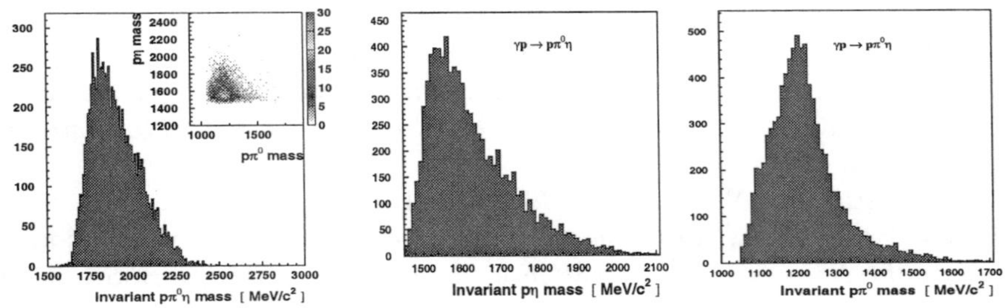

Figure 7. $\pi^0\eta$p-data measured at: E_{e^-}=2.6 GeV.

in the $\Delta(1232)$ and in the $S_{11}(1535)$ mass region which hints to the observation of possible decay modes via $\Delta(1232)\eta$ and $S_{11}(1535)\pi$. As in the $\pi^0\pi^0$p case we find evidence in the data for the decay modes we are searching for.

4 Conclusions

A first look into the data taken so far with the CB-ELSA experiment already shows promising resonance structures and evidence for decay modes via $\Delta\pi$ and higher mass resonances. Further investigation of the data with improved statistics, including an improved reconstruction and a partial wave analysis, should lead to interesting results on nucleon resonances and their properties. This should help to better understand the spectrum of nucleon resonances and finally QCD in the non-perturbative regime.

References

1. S. Capstick and W. Roberts, *Phys. Rev.* D **47**, 1994 (1993).
2. S. Capstick and W. Roberts, *Phys. Rev.* D **49**, 4570 (1994).
3. D.B. Lichtenberg, *Phys. Rev.* **178**, 2197 (1969).
4. S. Capstick, *Phys. Rev.* D **46**, 2864 (1992).
5. A. Fösel *et al.*, *Photoproduction of η and η'-mesons* , Proposal to the PAC, Bonn-Mainz.
6. M.Ostrick *et al.*, *Production of ω-mesons with linearly polarised photons*, Proposal to the PAC, Bonn-Mainz.
7. U.Thoma *et al.*, *Study of baryon resonances decaying into $\mathbf{\Delta(1232)}\pi^{\mathbf{0}}$ in the reaction $\gamma\mathbf{p} \to \mathbf{p}\pi^{\mathbf{0}}\pi^{\mathbf{0}}$ with the Crystal Barrel detector at ELSA*, Proposal to the PAC, Bonn-Mainz.
8. N. Isgur and G. Karl, *Phys. Lett.* B **72**, 109 (1977).
9. L.Y. Glozman and D.O. Riska, *Phys. Lett.* B **366**, 305 (1996).
10. N. Kaiser, P.B. Siegel, and W. Weise, *Phys. Lett.* B **362**, 23 (1995).
11. B.M.K. Nefkens, HADRON'97, AIP (1998) 514.
12. J.Smyrski *et al.*, *A search for new baryonic resonances and for the exotic meson $\hat{\rho}(1380)$ in the reaction $\gamma p \to p\pi^0\eta$ using the CB-ELSA detector at ELSA.*
13. S. Capstick and W. Roberts, *Phys. Rev.* D **57**, 4301 (1998).
14. R.W.Gothe *et al.*, *An optimized experimental setup to measure and analyze the $N \to \Delta(1232)$ transition at the accelerator facility ELSA, to be published in NIM.*
15. S.Schadmand *et al.*, *Experimental study of the in-medium properties of the $S_{11}(1535)$ resonance* , Proposal to the PAC, Bonn-Mainz.
16. J.Messchendorp *et al.*, *Studying the in-medium ω-mass in $\gamma + A \to \pi^0\gamma + X$ reactions* , Proposal to the PAC, Bonn-Mainz.
17. H.Löhner *et al.*, *Nucleon resonance decay by the $K^0\Sigma^+$ channel near threshold*, Proposal to the PAC, Bonn-Mainz.

LASER-ELECTRON PHOTON PROJECT AT SPRING–8

T. NAKANO
RCNP, Osaka University,
10-1 Mihogaoka, Ibaraki, Osaka 567-0047, JAPAN
E-mail: nakano@rcnp.osaka-u.ac.jp

A new photon beam facility (LEPS) has been constructed at SPring–8 in Japan to study N^* physics and the nonperturbative nature of QCD in the GeV energy region. The backward Compton scattering of laser photons from 8 GeV electrons produces linearly polarized photon beam with the maximum energy of 2.4 GeV. We report the status and prospects of the LEPS. Preliminary results from the first physics run are also presented.

1 Physics Motivation

1.1 ϕ photoproduction near threshold

At high energies, diffractive photoproduction of a ϕ meson from a proton target is well described as a pomeron-exchange process in the framework of the Regge theory and of the Vector Dominance Model. However, at low energies other contributions arising from meson(π, η)-exchange [1], a scaler(0^{++} glueball)-exchange [2], and $s\bar{s}$ knock-out [3] may be detectable. Precise measurements of spin observables with linearly polarized photons will be very useful to decompose these contributions [4]. The measurements of the cross sections and the spin observables are underway at the LEPS.

1.2 K^+ photoproduction

Recent measurements for $K^+\Lambda$ photoproduction at SAPHIR indicated a structure around $W = 1.9$ GeV in the total cross section [5]. It attracted theorist's interest to study missing nucleon resonances in this process. Bennhold *et al.* [6] showed that the SAPHIR data can be reproduced by inclusion of a new resonance $D_{13}(1960)$ which has large couplings both to the photon and the $K\Lambda$ channels according to the quark model calculation [7]. Furthermore, they showed that the photon polarization asymmetry is very sensitive to the missing nucleon resonance. The measurement in the energy region below $E_\gamma =$ 1.5 GeV will be carried out by GRAAL collaboration soon [8]. The LEPS collaboration will measure the asymmetry in the energy region of 1.5–2.4 GeV.

1.3 $\Lambda(1405)$ *photoproduction from proton and nuclei*

The mass of the $\Lambda(1405)$ resonance is just below KN threshold. And it is well described by a chiral unitary model [9] where the resonance is generated by the meson-baryon interaction. The model predicts a drastic change of the decay width of the resonance in a nucleus because of its meson-baryon molecular nature [10]. On the other hand, a recent quenched lattice QCD calculation gives the right mass for the $\Lambda(1405)$ [11] and suggests it is a genuine three-quark state. Photoproduction of the $\Lambda(1405)$ from a proton and a nucleus will elucidate the nature of the Λ.

1.4 ω *photoproduction*

Missing nucleon resonances could couple to the π-nucleon channel weakly but couple to the ω-nucleon channel strongly [12]. The LEPS can measure a forward proton from backward ω photoproduction. In this kinematic region u-channel and s-channel contributions dominate and the effects of missing resonances would be large. The old differential cross section data for backward ω photoproduction [13] shows a structure around u = -0.15 GeV, which is very hard to reproduce by model calculations [14]. New precise measurements are awaited to confirm the structure.

2 LEPS Facility

The Spring-8 facility is the most powerful third-generation synchrotron radiation facility with 61 beamlines. We use a beamline, BL33LEP (Fig. 1), for the quark nuclear physics studies. The beamline has a 7.8-m long straight section between two bending magnets. Polarized laser photons with a wavelength of 351 nm are injected from a laser hutch toward the straight section where Backward-Compton scattering (BCS) [15] of the laser photons from the 8 GeV electron beam takes place. The maximum energy is 2.4 GeV, well above the threshold of ϕ-photoproduction from a nucleon (1.57 GeV). The polarization of the photons is the highest at the maximum energy and it drops as the photon energy decreases as shown in Figs. 2 and 3. However, the energy of laser photons is easily changed so that the polarization remains reasonably high in the energy region of interest. The intensity, position, and polarization of the laser lights which do not interact with the electron beam are monitored at the end of the beamline.

The incident photon energy is determined by measuring the energy of a recoil electron with a tagging counter. The tagging counter is located at the

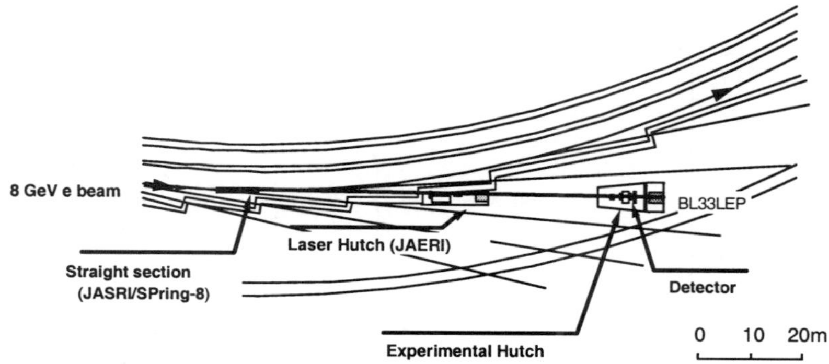

Figure 1. Plan view of the Laser-Electron Photon facility at SPring–8 (LEPS).

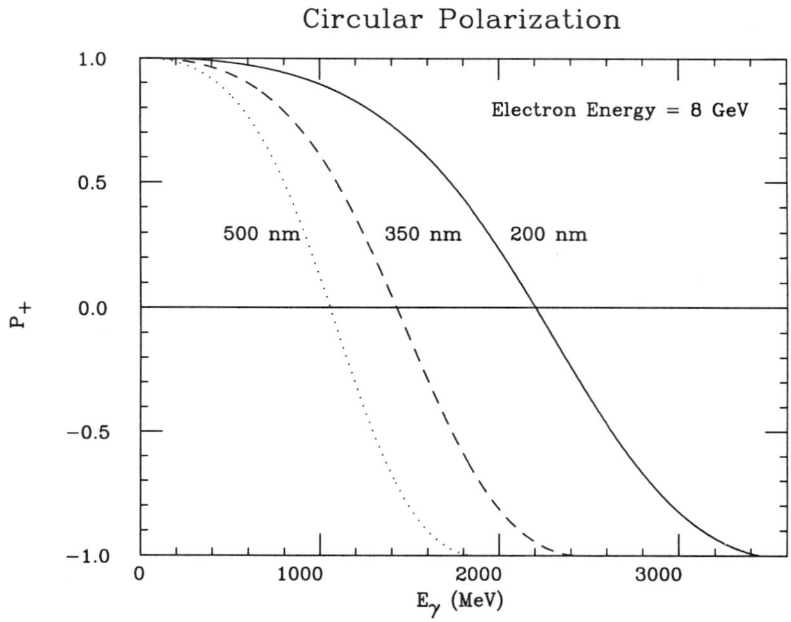

Figure 2. Circular polarization as a function of photon energy. Laser photons at various wavelengths are assumed to have 100 % circular polarization.

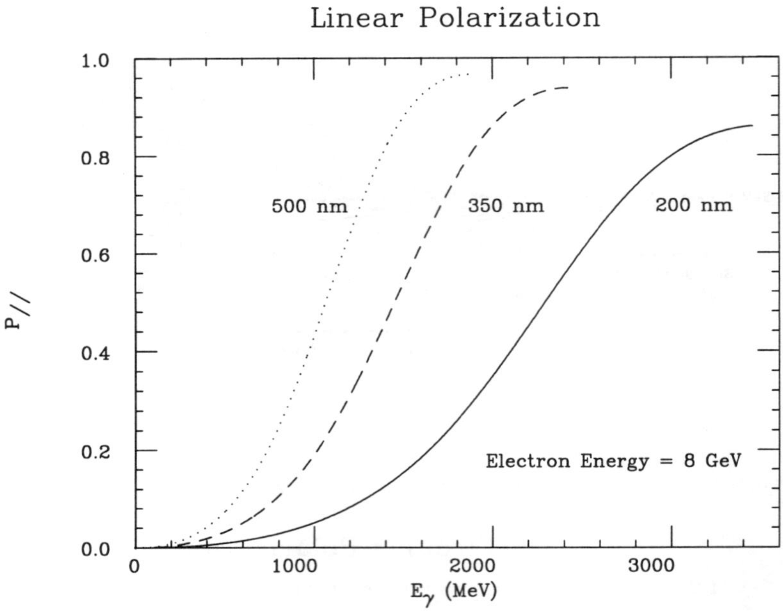

Figure 3. Linear polarization as a function of photon energy. Laser photons at various wavelengths are assumed to have 100 % linear polarization.

exit of the bending magnet after the straight section. It consists of multi-layers of a 0.1 mm pitch silicon strip detector (SSD) and plastic scintillator hodoscopes. Photons with the energy more than 1.5 GeV are tagged with a typical energy resolution of 15 MeV. The intensity of the photon beam is about 2.5×10^6 photons/sec and that of the tagged photon beam is about 0.8×10^6 photons/sec.

3 Detector

The LEPS detector (Fig. 4) consists of a plastic scintillator to detect charged products after a target, an aerogel Cerenkov counter with a refractive index of 1.03, charged-particle tracking counters, a dipole magnet, and a time-of-flight TOF wall. The design of the detector is optimized for ϕ photo-production in forward angles.

The opening of the dipole magnet is 135-cm wide and 55-cm high. The length of the pole is 60 cm, and the field strength at the center is 1 T. The

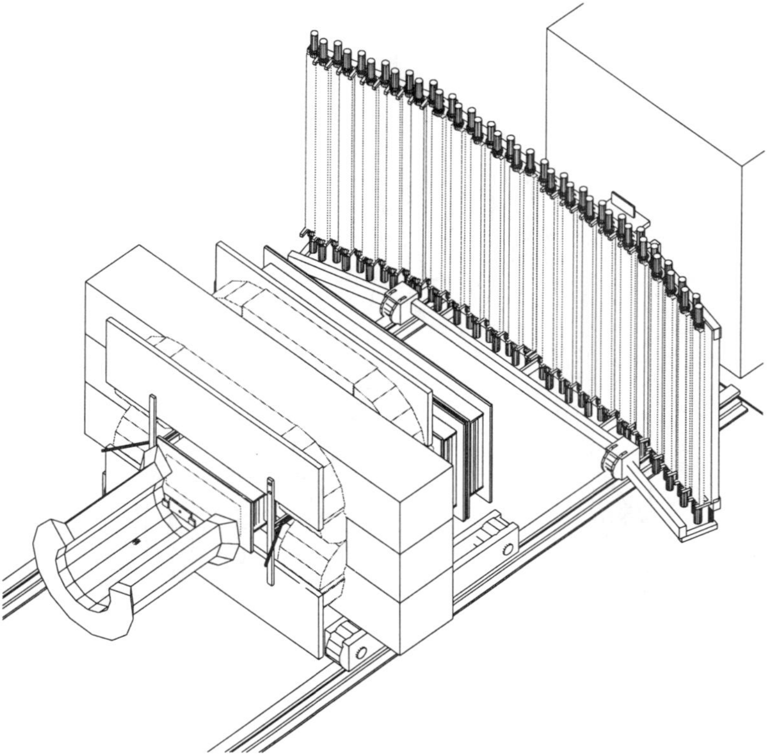

Figure 4. The LEPS detector setup.

vertex detector consists of 2 planes of single-sided SSDs and 6 planes of multi-wire drift chamber (MWDC). The position of a charged track after the magnet are measured by two sets of 5-plane MWDCs.

The identification of momentum analyzed particles is performed by measuring a time of flight from the target to the TOF wall consisting of 40 2m-long plastic scintillation bar with a cross section of 4cm (t) $\times$ 12 cm (w). The resolution of the TOF counter is about 150 psec. The start signal for the TOF measurement is provided by a RF signal from the 8-GeV ring, where electrons are bunched at every 2 nsec (508MHz) with a width (σ) of 12 psec.

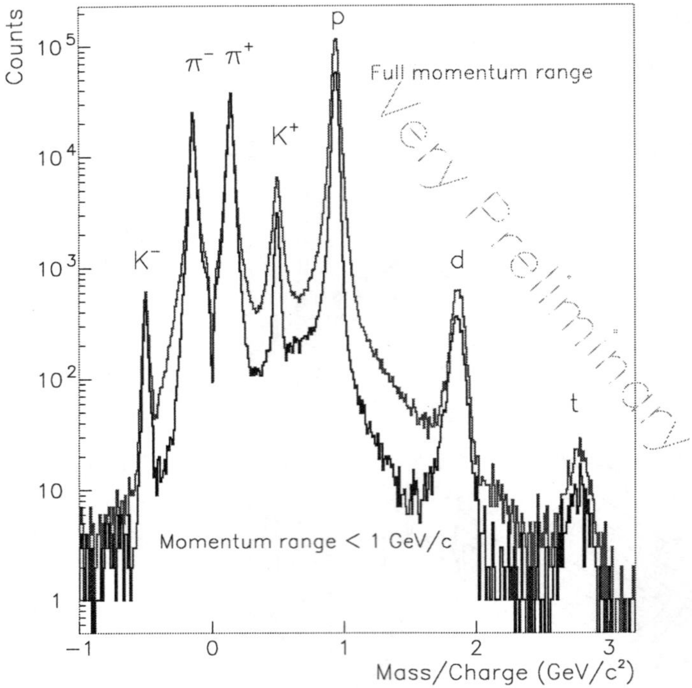

Figure 5. A mass distribution of charged particles reconstructed from momentum and TOF information.

4 Results and Prospects

The physics run with a 5-cm long liquid H_2 target started in December, 2000. The trigger required a tagging counter hit, no charged particle before the target, charged particles after the target, no signal in the aerogel Cerenkov counter, at least one hit on the TOF wall. A typical trigger rate was about 20 counts per second. The run will be continued until the end of June, 2001.

Figure 5 shows a preliminary mass distribution of charged particles reconstructed from momentum and TOF information. ϕ mesons are successfully identified through the reconstruction of the KK invariant mass as shown in Fig. 6. Figure 7 shows a missing mass distribution of the (γ, K^+) reactions. The Λ and Σ peaks are clearly identified, but the $\Lambda(1405)$ is not separated from the $\Sigma(1385)$. We will construct a time-projection chamber in 2002 to decompose the two resonaces by detecting their decay products.

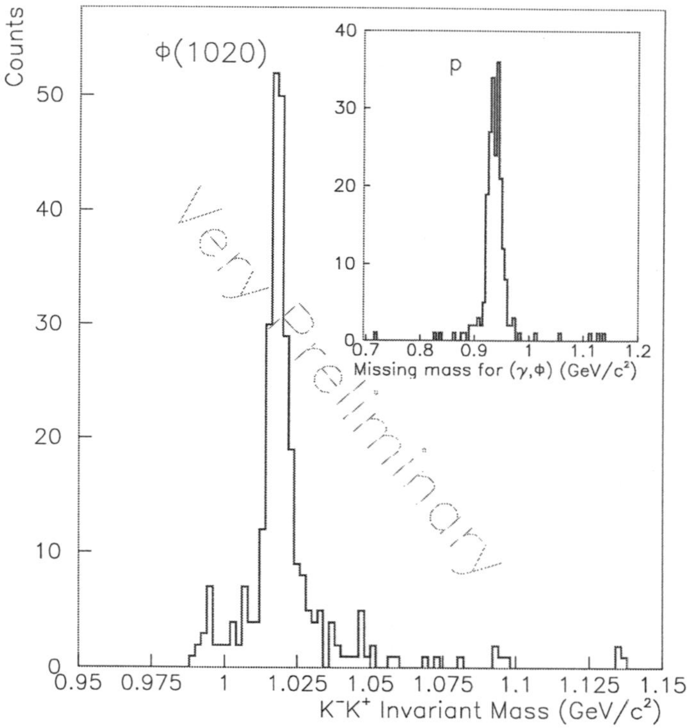

Figure 6. A two-kaon invariant mass distribution. The ϕ peak is clearly identified.

References

1. M.A. Pichowsky and T.-S. H. Lee, *Phys. Rev.* D **56**, 1644 (1997).
2. T. Nakano and H. Toki, in Proc. of Intern. Workshop on Exciting Physics with New Accelerator Facilities, SPring–8, Hyogo, 1997, World Scientific, 1998, p.48.
3. A.I. Titov, Y. Oh, and S.N. Yang, *Phys. Rev. Lett.* **79**, 1634 (1997); A.I. Titov, Y. Oh, and S.N. Yang, *Phys. Rev.* C **58**, 2429 (1998).
4. A.I. Titov, T.-S. H. Lee, and H. Toki, *Phys. Rev.* C **59**, 2993 (1999).
5. K.-H. Glander, *these proceedings.*
6. C. Bennhold, *these proceedings.*
7. S. Capstick and W. Roberts, *Phys. Rev.* D **58**, 074011 (1998).
8. A. D'Angelo, *these proceedings.*

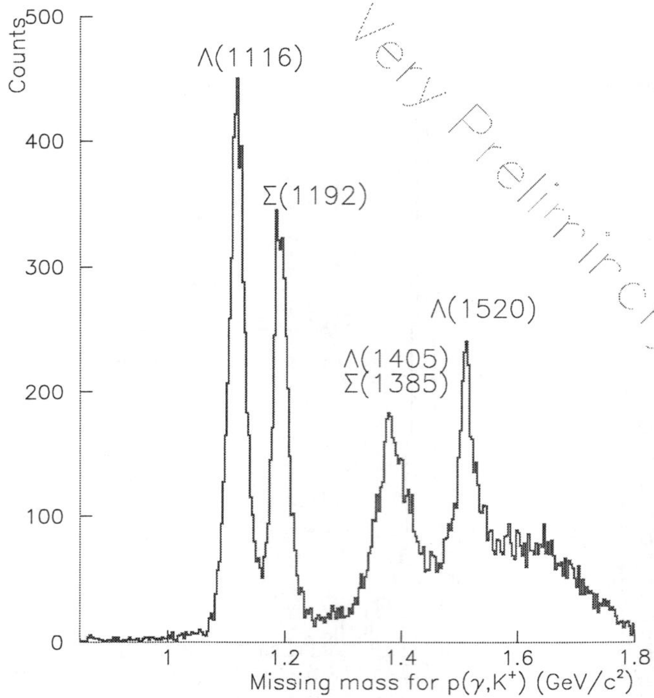

Figure 7. A missing mass distribution of $(\gamma, K+)$ reactions.

9. A. Ramos, *these proceedings.*
10. J.C. Nacher, E. Oset, H.Toki, A. Ramos, *Phys. Lett.* B **455**, 55 (1999).
11. S. Sasaki, private communication.
12. Y. Oh, *these proceedings*; Q. Zhao, *these proceedings.*
13. R.W. Clifft *et al., Phys. Lett.* B **72**, 144 (1977).
14. T.-S. H. Lee, private communication.
15. R.H. Milburn, *Phys. Rev. Lett.* **10**, 75 (1963).

THE BARYON RESONANCE PROGRAM AT BES

B. S. ZOU (BES COLLABORATION)
Institute of High Energy of Physics, Chinese Academy of Science,
Beijing 100039, P.R.China
E-mail: zoubs@mail.ihep.ac.cn

Physics motivation, data status, partial wave analyses and future prospects are presented for the baryon resonance program at BES.

1 Physics motivation

Spectroscopy has long proven to be a powerful tool for exploring internal structures and basic interactions of the microscopic world. Ninety years ago detailed studies of atomic spectroscopy resulted in the great discovery of Niels Bohr's atomic quantum theory[1]. Forty to sixty years later, still detailed studies of nuclear spectroscopy resulted in the Nobel Prize winning discoveries of the nuclear shell model[2] and the collective motion model[3] by Aage Bohr *et al.* Comparing with the atomic and nuclear spectroscopy at those times, our present baryon spectroscopy is still in its infancy[4]. Many fundamental issues in baryon spectroscopy are still not well understood[5]. The possibility of new, as yet unappreciated symmetries could be addressed with accumulation of more data.

Joining the new effort on studying the excited nucleons, N^* baryons, at new facilities such as CEBAF at JLAB, ELSA at Bonn, GRAAL at Grenoble and SPRING8 at KEK, we also started a baryon resonance program at BES[6], at Beijing Electron-Positron Collider (BEPC). The J/ψ and ψ' experiments at BES provide an excellent place for studying excited nucleons and hyperons – N^*, Λ^*, Σ^* and Ξ^* resonances[7]. The corresponding Feynman graph for the production of these excited nucleons and hyperons is shown in Fig. 1 where Ψ represents either J/ψ or ψ'.

Comparing with other facilities, our baryon program has advantages in at least three obvious aspects:

(1) We have pure isospin 1/2 πN and $\pi\pi N$ systems from $J/\psi \to \bar{N}N\pi$ and $\bar{N}N\pi\pi$ processes due to isospin conservation, while πN and $\pi\pi N$ systems from πN and γN experiments are mixtures of isospin 1/2 and 3/2, and suffer from difficulties concerning the isospin decomposition;

(2) ψ mesons decay to baryon-antibaryon pairs through three or more gluons. This is a favorable place for producing hybrid (qqqg) baryons, and for looking for some "missing" N^* resonances which have weak coupling to both

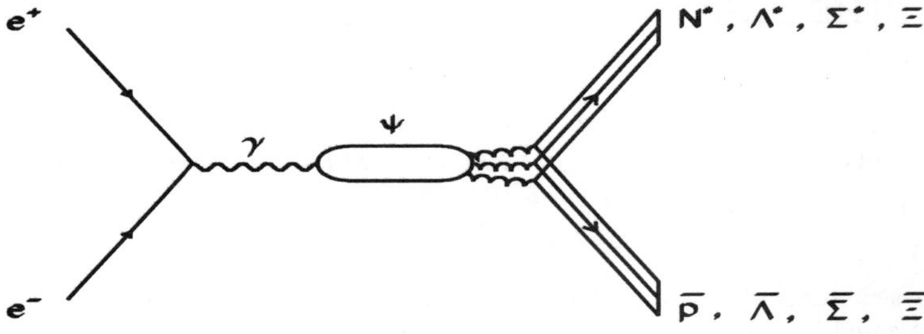

Figure 1. $\bar{p}N^*$, $\bar{\Lambda}\Lambda^*$, $\bar{\Sigma}\Sigma^*$ and $\bar{\Xi}\Xi^*$ production from e^+e^- collision through Ψ meson.

πN and γN, but stronger coupling to g^3N;

(3) Not only N^*, Λ^*, Σ^* baryons, but also Ξ^* baryons with two strange quarks can be studied. Many QCD-inspired models[8,9] are expected to be more reliable for baryons with two strange quarks due to their heavier quark mass. More than thirty Ξ^* resonances are predicted where only two such states are well established by experiments. The theory is not at all challenged due to the lack of data.

2 Data Status

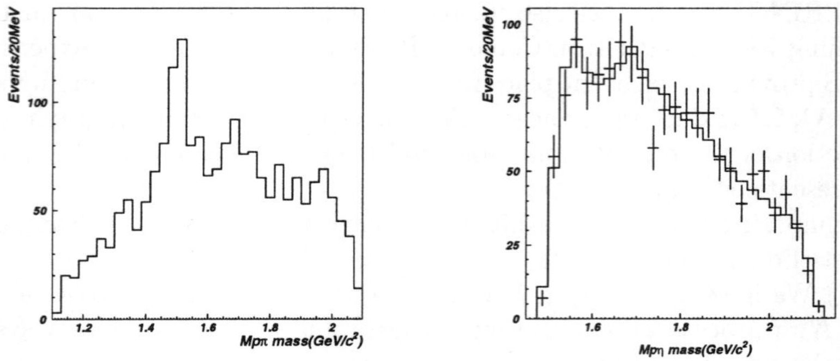

Figure 2. left: $p\pi^0$ invariant mass spectrum for $J/\psi \to \bar{p}p\pi^0$; right: $p\eta$ invariant mass spectrum for $J/\psi \to \bar{p}p\eta$. BESI data.

The BEjing Spectrometer (BES) started data-taking in 1989 and was upgraded in 1998. The upgraded BES is named BESII while the previous one is called BESI. BESI collected 7.8 million J/ψ events and 3.7 million ψ' events. BESII has collected 50 million J/ψ events.

Based on 7.8 million J/ψ events collected at BESI before 1996, the events for $J/\Psi \to \bar{p}p\pi^0$ and $\bar{p}p\eta$ have been selected and reconstructed with π^0 and η detected in their $\gamma\gamma$ decay mode[6]. The corresponding $p\pi^0$ and $p\eta$ invariant mass spectra are shown in Fig. 2 with clear peaks around 1500 and 1670 MeV for $p\pi^0$ and clear enhancement around the $p\eta$ threshold, peaks at 1540 and 1650 MeV for $p\eta$.

With 50 million new J/ψ events collected by BESII of improved detecting efficiency, we expect to have one order of magnitude more reconstructed events for each channel. As an example, we show in Fig. 3 preliminary result of the $N\pi$ invariant mass spectrum for 27,000 $J/\psi \to p\bar{n}\pi^-$ events reconstructed from half BESII data[10]. It looks similar to the $p\pi$ invariant mass spectrum for $J/\psi \to p\bar{p}\pi^0$ as in Fig. 2, but with much higher statistics.

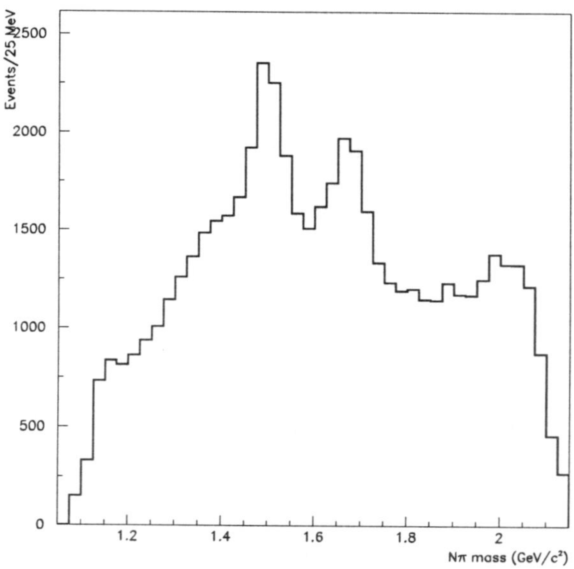

Figure 3. $N\pi$ invariant mass spectrum for $J/\psi \to p\bar{n}\pi^-$. Preliminary result from half BESII data.

We are also reconstructing $J/\psi \to pK^-\bar{\Lambda}$, $\bar{p}K^+\Lambda$, $\bar{p}p\omega$ channels. Clear N^* and Λ^* peaks are observed from preliminary selections.

3 Partial Wave Analyses

In order to get more useful information about properties of the baryon resonances, such as their J^{PC} quantum numbers, mass, width, production and decay rates, etc., partial wave analyses (PWA) are necessary.
The basic procedure for our partial wave analyses is the standard maximum likelihood method:
(1) construct amplitudes A_i for each i-th possible partial waves;
(2) from linear combination of these partial wave amplitudes, get the total transition probability for each event as $w = |\sum_i c_i A_i|^2$ with c_i as free parameters to be determined by fitting the data;
(3) maximize the following likelihood function $\mathcal{L}$ to get the parameters c_i as well as mass and width parameters for the resonances.

$$\mathcal{L} = \prod_{n=1}^{N} \frac{w_{data}}{\int w_{MC}},$$

where N is the number of reconstructed data events and w_{data}, w_{MC} are evaluated for data and Monte Carlo events, respectively.

For the construction of partial wave amplitudes, we assume the effective Lagrangian approach[11,12] with the Rarita-Schwinger formalism[13,14,15]. In this approach, there are three basic elements for constructing amplitudes: particle spin wave functions, propagators and effective vertex couplings; the amplitude can be written out by Feynman rules for tree diagrams.

For example, for $J/\psi \to \bar{N}N^*(3/2+) \to \bar{N}(k_1, s_1)N(k_2, s_2)\eta(k_3)$, the amplitude can be constructed as

$$A_{3/2+} = \bar{u}(k_2, s_2)k_{2\mu}P^{\mu\nu}_{3/2}(c_1 g_{\nu\lambda} + c_2 k_{1\nu}\gamma_\lambda + c_3 k_{1\nu}k_{1\lambda})\gamma_5 v(k_1, s_1)\psi^\lambda$$

where $u(k_2, s_2)$ and $v(k_1, s_1)$ are 1/2-spinor wave functions for N and $\bar{N}$, respectively; ψ^λ is the spin-1 wave function, *i.e.*, polarization vector, for J/ψ. The c_1, c_2 and c_3 terms correspond to three possible couplings for the $J/\psi \to \bar{N}N^*(3/2+)$ vertex. The c_1, c_2 and c_3 can be taken as constant parameters or with some smooth vertex form factors in them if necessary. The spin 3/2 propagator $P^{\mu\nu}_{3/2}$ for $N^*(3/2+)$ is

$$P^{\mu\nu}_{3/2} = \frac{\gamma \cdot p + M_{N*}}{M^2_{N*} - p^2 - iM_{N*}\Gamma_{N*}} \left[g^{\mu\nu} - \frac{1}{3}\gamma^\mu\gamma^\nu - \frac{2p^\mu p^\nu}{3M^2_{N*}} + \frac{p^\mu\gamma^\nu - p^\nu\gamma^\mu}{3M_{N*}} \right]$$

with $p = k_2 + k_3$.

Other partial wave amplitudes can be constructed similarly[16]. Programing these amplitudes and maximum likelihood method to fit the data is straightforward, but very tedious. Fortunately we have a programing expert, Dr. Jian-xiong Wang, who developed an automatic Feynman Diagram Calculation (FDC) package[17] and has now extended it to work for our partial wave analyses of baryon resonance channels. Using the extended FDC package, we have performed a partial wave analysis of the $\bar{p}p\eta$ channel[18] and are now working on more channels.

4 Future Prospects

BESII just finished data-taking for 50 million J/ψ events at the end of last March. We are now working on partial wave analyses of $J/\psi \to p\bar{n}\pi^-$, $p\bar{p}\omega$ channels to study N^* resonances, and $pK^-\bar{\Lambda}$, $\bar{p}K^+\Lambda$ channels to study Λ^* resonances as well as $N^* \to \Lambda K$. As a next step, we are going to investigate $\Lambda\bar{\Sigma}^-\pi^+$, $pK^-\bar{\Sigma}^0$ channels to study Σ^* resonances, and the $K^-\Lambda\bar{\Xi}^+$ channel to study Ξ^* resonances. These channels are relative easy to be reconstructed by BES. For example, for $K^-\Lambda\bar{\Xi}^+$, we can select events containing K^- and Λ with $\Lambda \to p\pi^-$, then from the missing mass spectrum of $K^-\Lambda$ we should easily identify the very narrow $\bar{\Xi}^+$ peak. We will investigate more complicated channels when we get more experienced and more manpower.

A major upgrade of the collider to BEPC2 has been approved by the Chinese central government very recently. A further two order of magnitude more statistics is expected to be achieved. Such statistics will enable us to perform partial wave analyses of plenty important channels not only from J/Ψ but also Ψ' decays which will allow us to study heavier baryon resonances, *e.g.*, for masses up to 2.36 GeV for Ξ^* resonances. We expect BEPC2 to play a very important and unique role in studying excited nucleons and hyperons, i.e., N^*, Λ^*, Σ^* and Ξ^* resonances.

Acknowledgments

We would like to thank H.C.Chiang, X.B.Ji, W.H.Liang, G.X.Peng, P.N.Shen, X.Y.Shen, J.X.Wang H.X.Yang, J.J.Zhu and Y.C.Zhu for the collaboration on the BES Baryon Resonance Program. The work is partly supported by National Science Foundation of China under contract Nos. 19991487 and 19905011.

References

1. N.Bohr, *Phil. Mag.* **26**, 1-25, 471-502, 857-875 (1913).
2. M.Mayer and J.Jensen, *Elementary Theory of Nuclear Shell Structure*, John Wiley and Sons, Inc., New York (1955).
3. A.Bohr and B.Mottelson, *Nuclear Structure Vol.I, Single-Particle Motion* (1969) and *Vol.II, Nuclear Deformations* (1975), W.A.Benjamin, Inc., Reading, Massachusetts.
4. Particle Data Group, *Euro. Phys. J.* C **15**, 1 (2000).
5. S.Capstick *et al.*, hep-ph/0012238.
6. BES Collaboration, H.B.Li *et al., Nucl. Phys.* A **675**, 189c (2000); BES Collaboration, B.S.Zou *et al., Excited Nucleons and Hadronic Structure*, Proc. NSTAR 2000, eds. V.Burkert *et al.*, World Scientific, Singapore (2001) p.155.
7. B.S.Zou, *Nucl. Phys.* A **684**, 330 (2001).
8. S.Capstick and N.Isgur, *Phys. Rev.* D **34**, 2809 (1986).
9. L.Glozman, W.Plessas, K.Varga, and R.Wagenbrunn, *Phys. Rev.* D **58**, 094030 (1998).
10. X.B.Ji (BES Collaboration), Ph.D. thesis of Shandong University, (2001).
11. M.Benmerrouche, N.C.Mukhopadhyay, and J.F.Zhang, *Phys. Rev. Lett.* **77**, 4716 (1996); *Phys. Rev.* D **51**, 3237 (1995).
12. M.G.Olsson and E.T.Osypowski, *Nucl. Phys.* B **87**, 399 (1975); *Phys. Rev.* D **17**, 174 (1978); M.G.Olsson *et al.*, *ibid.* **17**, 2938 (1978).
13. W.Rarita and J.Schwinger, *Phys. Rev.* **60**, 61 (1941).
14. C.Fronsdal, *Nuovo Cimento Suppl.* **9**, 416 (1958); R.E.Behrends and C.Fronsdal, *Phys. Rev.* **106**, 345 (1957).
15. S.U.Chung, *Spin Formalisms*, CERN Yellow Report 71-8 (1971); *Phys. Rev.* D **48**, 1225 (1993); J.J.Zhu and T.N.Ruan, *Comm. Theor. Phys.* **32**, 293, 435 (1999).
16. J.X.Wang, J.J.Zhu, and B.S.Zou, in preparation.
17. J.X.Wang, *Comput. Phys. Commun.* **77**, 263 (1993).
18. BES Collaboration, "Study of N^* Production from $J/\psi \to p\bar{p}\eta$", to be published.

FLAVOR SYMMETRY STUDIES WITH NEW HYPERON DATA FROM THE CRYSTAL BALL

B. M. K. NEFKENS, E. BERGER, S. McDONALD, N. PHAISANGITTISAKUL, S. PRAKHOV, J. W. PRICE, A. STAROSTIN, AND W. B. TIPPENS FOR THE CRYSTAL BALL COLLABORATION* †

UCLA, Los Angeles, CA 90095, USA

New hyperon data produced in the reactions $K^-p \to$ *neutrals* have been measured using the Crystal Ball 4π multiphoton spectrometer. They provide striking evidence for the important role of flavor symmetry in QCD at low energy. We will show that the pronounced features seen in $\pi^-p \to \eta n$ near threshold are also exhibited in $K^-p \to \eta\Lambda$. The unique features of the $\pi^-p \to \pi^0\pi^0 n$ Dalitz plot are replicated in the flavor–symmetric reaction $K^-p \to \pi^0\pi^0\Lambda$, while $K^-p \to \pi^0\pi^0\Sigma^0$, which is not flavor–related, is much different. The measured mass, width, and spin/parity of the Λ^* octet hyperons are found to be the flavor symmetric copy of the N^* states. Flavor symmetry combined with the narrowness of Ξ^* states implies that searching for the missing baryon states may be done more profitably in the Ξ^* family rather than in the N^* or Δ^* family.

1 Introduction

Data on hyperon resonances are needed to evaluate the role of the approximate flavor symmetry of QCD in the non–perturbative regime. The QCD Lagrangian can be divided rather conveniently into two parts,

$$\mathcal{L}_{QCD} = \mathcal{L}_0 + \mathcal{L}_m, \tag{1}$$

$$\mathcal{L}_0 = -\frac{1}{4}F_{\mu\nu}F^{\mu\nu} + i\sum_q \overline{\psi}_q\gamma^\mu D_\mu\psi_q, \tag{2}$$

where the symbols are defined in the literature [1]. $\mathcal{L}_0$ depends on quark and gluon fields, their derivatives, and on one universal coupling constant. $\mathcal{L}_0$ has

*THE CRYSTAL BALL COLLABORATION: E. BERGER, M. CLAJUS, A. MARUŠIĆ, S. MCDONALD, B.M.K. NEFKENS, N. PHAISANGITTISAKUL, S. PRAKHOV, M. PULVER, A. STAROSTIN AND W.B. TIPPENS, *UCLA*, D. ISENHOWER AND M. SADLER, *ACU*, C. ALLGOWER AND H. SPINKA, *ANL*, J. COMFORT, K. CRAIG AND T. RAMIREZ, *ASU*, T. KYCIA, *BNL*, J. PETERSON, *UCO*, W. BRISCOE AND A. SHAFI, *GWU*, H.M. STAUDENMAIER, *UKA*, D.M. MANLEY AND J. OLMSTED, *KSU*, D. PEASLEE, *UMD*, V. BEKRENEV, A. KOULBARDIS, N. KOZLENKO, S. KRUGLOV AND I. LOPATIN, *PNPI*, G.M. HUBER, N. KNECHT, G.J. LOLOS AND Z. PAPANDREOU, *UREG*, I. SLAUS AND I. SUPEK, *RBI*, D.GROSNICK, D. KOETKE, R. MANWEILER AND S. STANISLAUS, *VALU*.

†SUPPORTED BY US DOE, NSF, NSEKC, RMS AND US.

flavor symmetry, which means that the six quark flavors have identical strong interactions. This symmetry of $\mathcal{L}_0$ is broken by the mass term

$$\mathcal{L}_m = -\overline{\psi}_u m_u \psi_u - \overline{\psi}_d m_d \psi_d - \overline{\psi}_s m_s \psi_s - \cdots . \tag{3}$$

The characteristic hadronic energy scale is 1 GeV; this is much larger than the masses of the light quarks, $m_u \simeq 5$ MeV, $m_d \simeq 8$ MeV and $m_s \simeq 170$ MeV. We may therefore expect that flavor symmetry is only broken mildly and can be used profitably in reactions involving the light quarks u, d, and s. The flavor symmetry of the u and d quarks alone is known as isospin invariance. Its validity is well–established, not only in low energy nuclear physics but also in meson and baryon physics at medium and high energy [2]. Flavor symmetry of u, d, and s–quark systems is often explored as SU(3)–flavor symmetry. The breaking due to the mass term is more pronounced than for isospin. Various properties of the three light meson families and the six light baryon families are related by SU(3)–flavor symmetry with a correction for the breaking due to the different actual quark masses. For instance, we have the well–established Gell–Mann baryon–decuplet ground state and the Gell–Mann–Okubo baryon–octet ground state mass relations. There are also the relations among the baryon magnetic dipole moments obtained by naive quark models. We will investigate the use of flavor symmetry as it manifests itself in the dynamics of selected reactions. The unique features of η threshold production seen in $\pi^- p \to \eta n$ are duplicated in $K^- p \to \eta\Lambda$. Also, we have found that the Dalitz–plot characteristics of $\pi^- p \to \pi^0\pi^0 n$ are replicated in $K^- p \to \pi^0\pi^0\Lambda$, but not in $K^- p \to \pi^0\pi^0\Sigma^0$ as anticipated by flavor symmetry.

New results on η, $2\pi^0$, π^0, γ and $\overline{K}^0$ production in $K^- p$ interactions up to $p_K = 760$ MeV/c have been obtained with the Crystal Ball (CB), a 4π multiphoton spectrometer. The CB was built and used originally at SLAC. It is now set up in a separated K^- beam of the AGS at Brookhaven National Laboratory. A 10 cm long liquid H_2 target has been installed in the center of the CB. The properties of the Crystal Ball (CB) multiphoton spectrometer have been discussed in this workshop by W. Briscoe. Details of the CB and data analysis are given in Refs. [3,4,5,6]. We have measured simultaneously all neutral final states in $K^- p$ interactions. We report here on the production of η and $2\pi^0$, and compare with our results from $\pi^- p$. In Sec. 4 we apply SU(3)–flavor symmetry to calculate the masses of the Λ^* resonances from the known N^* masses. In Sec. 5 we apply SU(3)–flavor to the search for the "missing" baryon resonances.

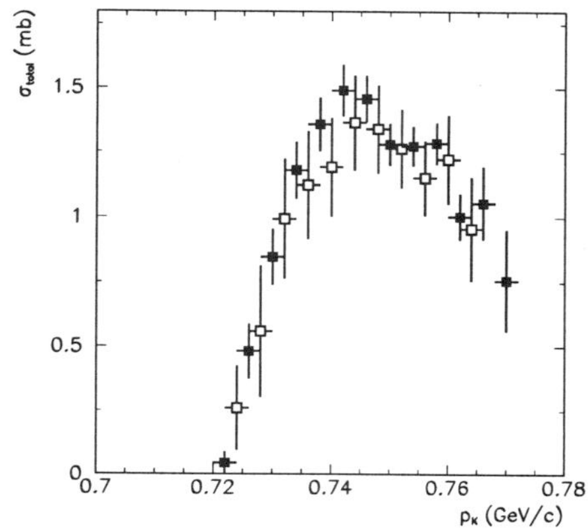

Figure 1. Preliminary CB results for $\sigma_t(K^- p \to \eta\Lambda)$ as function of p_{K^-}. The open squares are for the $\eta \to 3\pi^0$ and the solid squares for the $\eta \to 2\gamma$ decay mode.

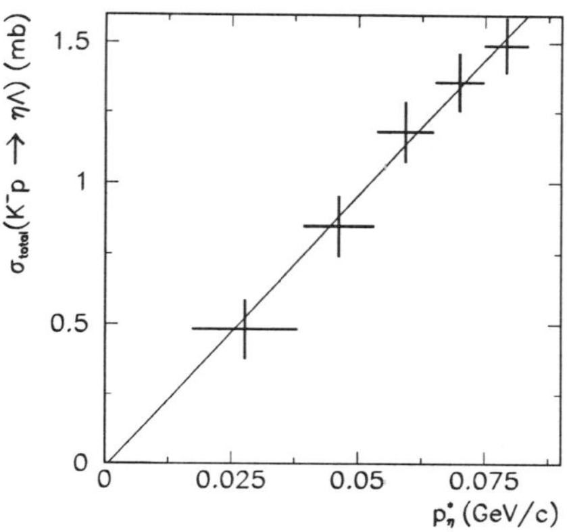

Figure 2. Preliminary $\sigma_t(K^- p \to \eta\Lambda)$ as function of p^*_η.

2 $K^- p \to \eta\Lambda$

The η decays 39% into 2γ and 32% into $3\pi^0 \to 6\gamma$, while 36% of Λ decay is to $\pi^0 n \to 2\gamma n$. We have measured the reaction $K^- p \to \eta\Lambda$ from the four photon cluster event sample when two clusters are due to $\eta \to 2\gamma$ and the other two to $\Lambda \to \pi^0 n \to 2\gamma n$; the neutron is not detected. We have also used eight cluster events detecting $\eta \to 3\pi^0 \to 6\gamma$ as well as $\Lambda \to \pi^0 n \to 2\gamma n$. Both event samples are very clean, and our two separate sets of results are in good agreement.

Shown in Fig. 1 is $\sigma_t(K^- p \to \eta\Lambda)$ as a function of the incident K^- momentum from threshold to 770 MeV/c. Note the steep onset of strong η production directly at threshold. In the interval from threshold at $p_K =$ 722 MeV/c to 740 MeV/c, the rise in σ_t is as expected for s–wave production (see Fig. 2). We find that

$$\sigma_t = C \times p^*_\eta, \tag{4}$$

where C is a constant, characteristic of s–wave production reactions, and p^*_η is the η c.m. momentum. Our data yield $C = (19 \pm 4)\mu\text{b}/(\text{MeV}/c)$.

The maximum of $\sigma_t(K^- p \to \eta\Lambda)$ is 1.5 ± 0.1 mb. It occurs at about $p_K = 740$ MeV/c. This is a large cross section, considering the phase space is small, p^*_η is merely 81 MeV/c. This should be compared to the $\bar{K}N$ final

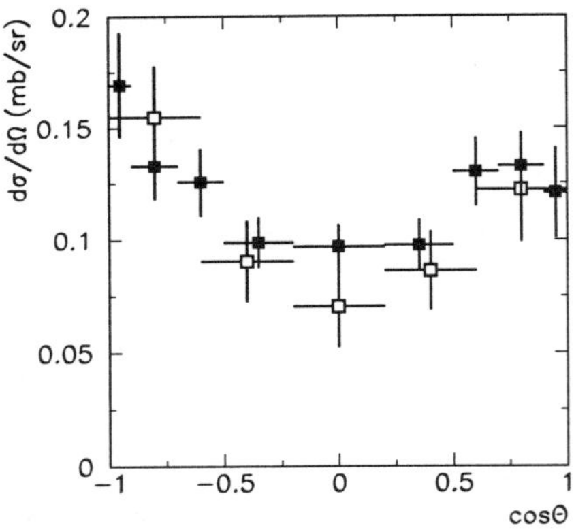

Figure 3. Preliminary $d\sigma/d\Omega(K^-p \to \eta\Lambda)$ at $p_{K^-} = 745\,\mathrm{MeV}/c$.

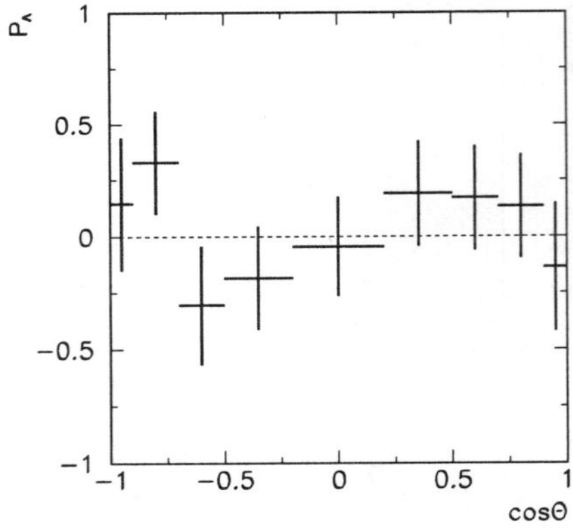

Figure 4. Preliminary Λ polarization in $K^-p \to \eta\Lambda$ at 735 MeV/c.

state, for which p_K^* is 414 MeV/c.

$d\sigma/d\Omega(K^-p \to \eta\Lambda)$ is shown in Fig. 3. The bowl shape is indicative of a *small* d–wave contribution. The polarization of the recoil Λ is displayed in Fig. 4. It is small, as expected when s–wave production is dominant.

The various features of η threshold production in $K^-p \to \eta\Lambda$, as displayed in Figs. 1–4, can be understood when we assume that η production is dominated by the excitation of the $\Lambda(1670)\frac{1}{2}^-$ resonance with a minimal contribution by other Λ^* states. It resembles in its various aspects the reaction $\pi^-p \to \eta n$. This includes:

1. The steep rise in σ_t at the opening of the η channel.

2. s–wave dominance with similar characteristic C–coefficients (see Eq. (4)),

$$\begin{aligned}\sigma_t(K^-p \to \eta\Lambda) &= (19 \pm 4)\,(\mu\mathrm{b}/(\mathrm{MeV}/c)) \times p_\eta^* \\ \sigma_t(\pi^-p \to \eta n) &= (21 \pm 2)\,(\mu\mathrm{b}/(\mathrm{MeV}/c)) \times p_\eta^* \end{aligned} \quad (5)$$

3. Large value for σ_t not far from the opening of the η channel; the maximum value for $\sigma_t(K^-p \to \eta\Lambda)$ is 1.5 ± 0.1 mb with $p_\eta^* = 81$ MeV/c, while for $\sigma_t(\pi^-p \to \eta n)$ the max is 2.6 ± 0.3 mb with $p_\eta^* = 182$ MeV/c [7].

4. The angular distribution somewhat above threshold has a bowl–like shape which may be described by a second order Legendre polynomial with similar coefficients for both $K^-p \to \eta\Lambda$ and $\pi^-p \to \eta n$. This is not a

trivial observation as η production by photons [8] or protons [9] has the opposite shape (it is dome–like).

We summarize as follows: η threshold production by K^- and π^- have the same features. This is a clear manifestation of the important role that approximate flavor symmetry plays in the dynamics of hadron interactions at intermediate energies. Furthermore, the two important intermediate state resonances, the $\Lambda(1670)\frac{1}{2}^-$ and the $N(1535)\frac{1}{2}^-$, belong to the same $SU(3)$ octet. This brings into question the proposal that the $N(1535)\frac{1}{2}^-$ may be mainly a $K\Sigma - K\Lambda$ bound state as proposed in Ref. [10] rather than a simple three–quark state.

3 $\pi^- p \to \pi^0\pi^0 n$, $K^- p \to \pi^0\pi^0\Lambda$ and $K^- p \to \pi^0\pi^0\Sigma^0$

Another case which illustrates the permeating role of SU(3)–flavor symmetry is $2\pi^0$ production. Consider $\pi^- p \to \pi^0\pi^0 n$. The data are a byproduct of our investigations of rare η decay [4]. The Dalitz–plot (D–plot) at $p_\pi = 720$ MeV/c is shown in Fig. 5a. The $\pi^0\pi^0 n$ final state has two identical π^0's; we have taken this into account in Fig. 5a by double plotting each event, once for each π^0. The D–plot is dominated by a high concentration of events on an "island". There is only a faint indication of a vertical band structure which one naively would expect from a N^* or Δ^* intermediate state. There is no uniform horizontal density distribution that would signal the significant production of the f_0 (also called σ) s–wave dipion phenomenon. The projection of the D–plot distribution on the $m^2(\pi^0\pi^0)$ axis is shown in Fig. 5c together with the phase space; $m^2(\pi^0\pi^0)$ is the square of the $\pi^0\pi^0$ invariant mass. The $\pi^0\pi^0$ spectrum increases markedly with increasing $m(\pi^0\pi^0)$. Figure 5b represents the D–plot projection on the $m^2(\pi^0 n)$ axis. It has a towering peak at $m^2(\pi^0 n) = (1.2\ \text{GeV})^2$, which is close to the mass of the $\Delta(1232)\frac{3}{2}^+$ resonance. The width of the peak is about 110 MeV, which is close to the Δ width. Our interpretation of Figs. 5a–5c is that $2\pi^0$ production here is dominated by the sequence $\pi^- p \to N^* \to \pi^0\Delta(1232)\frac{3}{2}^+ \to \pi^0\pi^0 n$. The p–wave decay of the Δ is likely responsible for the "island" feature of the D–plot. There is only a small contribution from $\pi^- p \to N^* \to f_0\, n \to \pi^0\pi^0 n$. At our incident π^- momentum of 720 MeV/c, the intermediate state N^* is a melange of $N(1440)\frac{1}{2}^+$, $N(1535)\frac{1}{2}^-$, and $N(1520)\frac{3}{2}^-$. The flavor analog processes of the ones just discussed are $K^- p \to \Lambda^* \to \pi^0\Sigma^0(1385)\frac{3}{2}^+ \to \pi^0\pi^0\Lambda$ and $K^- p \to \Lambda^* \to f_0\Lambda \to \pi^0\pi^0\Lambda$. Here, Λ^* is a medley of the $\Lambda(1600)\frac{1}{2}^+, \Lambda(1670)\frac{1}{2}^-$ and $\Lambda(1690)\frac{3}{2}^-$. Note that the $N(1440)\frac{1}{2}^+$ and $\Lambda(1660)\frac{1}{2}^+$ belong to the

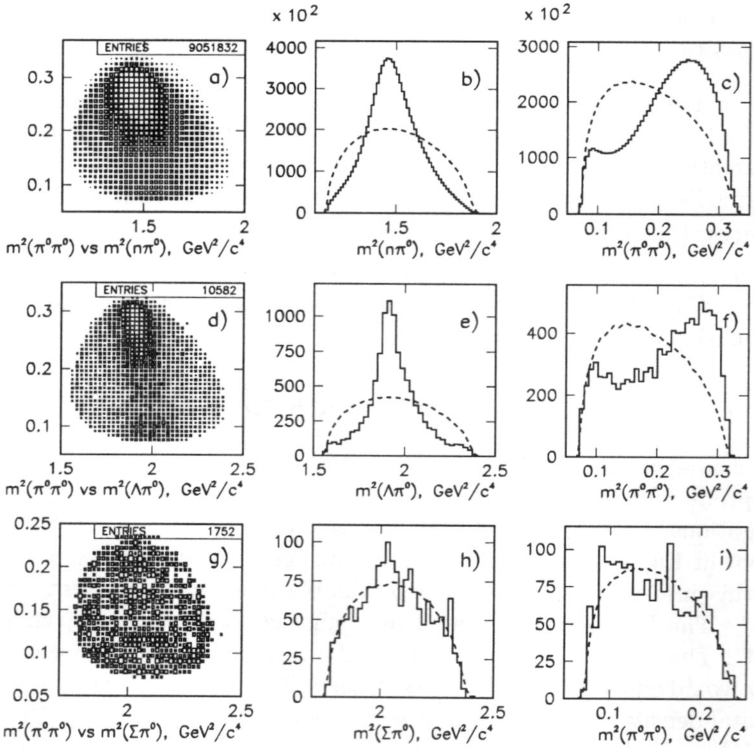

Figure 5. Figures a–c: $\pi^- p \to \pi^0\pi^0 n$ at $p_{\pi^-} = 720$ MeV/c; d–f: $K^- p \to \pi^0\pi^0\Lambda$ at $p_{K^-} = 750$ MeV/c; g–i: $K^- p \to \pi^0\pi^0\Sigma^0$ at $p_{K^-} = 750$ MeV/c. Preliminary data. The dashed lines show the phase space.

same SU(3) octet; similarly, the $N(1535)\frac{1}{2}^-$, and the $\Lambda(1670)\frac{1}{2}^-$; also the $N(1520)\frac{3}{2}^-$ and the $\Lambda(1690)\frac{3}{2}^-$, while the $\Delta(1232)\frac{3}{2}^+$ and $\Sigma(1385)\frac{3}{2}^+$ are part of the same SU(3) decuplet.

The CB results on $K^- p \to \pi^0\pi^0\Lambda$ at $p_{K^-} = 750$ MeV/c are shown in Figs. 5d–5c. They cover about the same $m(2\pi^0)$ region as $\pi^- p \to \pi^0\pi^0 n$ at 720 MeV/c. The resemblance of the outstanding features of Figs. 5a–5c and 5d–5f is a clear illustration of the significant role played by flavor symmetry when dealing with the dynamics of 3–body processes. The minor dissimilarities originate in the difference in the width of the intermediate states; the $\Sigma(1385)\frac{3}{2}^+$ width is about half that of the $\Delta(1232)\frac{3}{2}^+$. The origin of the width difference was discussed earlier at this workshop by D. Riska who suggested it comes mainly from the availability of the u and d quarks, which yields a ratio of the Σ to Δ width of $(2/3)^2 = 4/9$.

We have found that the magnitude of $\sigma_t(K^- p \to \pi^0\pi^0\Lambda)$ is about half that of $\sigma_t(\pi^- p \to \pi^0\pi^0 n)$, which is not too different from the proportionality seen in η production by K^- and π^- as mentioned in the previous section.

We want to contrast the two $2\pi^0$ production reactions just discussed with $K^- p \to \Sigma^* \to \pi^0\pi^0\Sigma^0$. Here Σ^* stands for the $\Sigma(1660)\frac{1}{2}^+$, the $\Sigma(1670)\frac{3}{2}^-$ and possibly one or more states which are listed in the Review of Particle Physics [1] as questionable states and are not given in the Summary table. The decay of the Σ^* could involve an intermediate state, the $\Lambda(1405)\frac{1}{2}^-$, thus $K^- p \to \Sigma^* \to \pi^0\Lambda(1405)\frac{1}{2}^- \to \pi^0\pi^0\Sigma^0$, the $\Lambda(1405)$ is not a flavor analog of $\Sigma(1385)\frac{3}{2}^+$ and $\Delta(1232)\frac{3}{2}^+$ that dominate the $K^- p \to \pi^0\pi^0\Lambda$ and $\pi^- p \to \pi^0\pi^0 n$ reactions. The $\Lambda(1405)$ is the $J^P = \frac{1}{2}^-$ SU(3) singlet, which has a totally antisymmetric flavor state function. In contrast, the $\Sigma(1385)$ and $\Delta(1232)$ states are members of the $J^P = \frac{3}{2}^+$ decuplet, which requires a totally symmetric flavor state function. Flavor symmetry implies that $\Sigma^* \to f_0\Sigma^* \to \pi^0\pi^0\Sigma^0$ and $\Lambda^* \to f_0\Lambda \to \pi^0\pi^0\Lambda$ should have similar features, though there is a small kinematic difference because the Σ^0 is 77 MeV heavier than the Λ.

The CB results on $K^- p \to \pi^0\pi^0\Sigma^0$ at $p_K = 750$ MeV/c are shown in Figs. 5g–5i. Because of the greater Σ^0 mass, the $\pi^0\pi^0$ invariant mass region is smaller than the one of Figs. 5c or 5f. The D–plot of Fig. 5g does not have the "signature" island that characterizes Figs. 5a and 5d. But the near uniform horizontal density distribution seen in Fig. 5d is consistent with a healthy f_0 role in $K^- p \to \pi^0\pi^0\Sigma^0$. There is one unexpected feature, namely $m(\pi^0\pi^0)$ peaks at low m (see Fig. 5i). This needs to be investigated at other p_K. Our preliminary results indicate that the Σ^0 production is considerably smaller than Λ production: $\sigma_t(K^- p \to \pi^0\pi^0\Sigma^0) < \frac{1}{5}\,\sigma_t(K^- p \to \pi^0\pi^0\Lambda)$.

The above results on $2\pi^0$ production have interesting consequences for current efforts in nuclear physics to find f_0 modification in the nuclear medium [5] as suggested by several theorists. Our preliminary results indicate that the preferred case for this is K^- interactions leading to the $f_0\Sigma^0$ final state.

4 Flavor Symmetry and the Masses of the Hyperons

In the limit $\mathcal{L}_m \to 0$ the Λ^* octet states are the same as the N^* states; they have identical masses and spin/parities. When the finite quark masses are

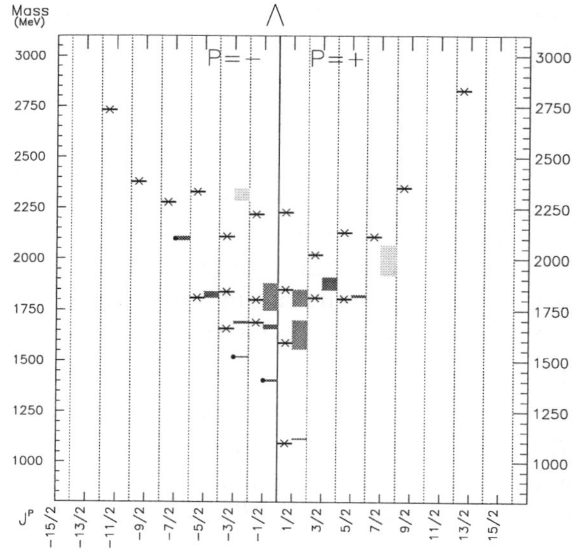

Figure 6. The spectrum of all Λ^* states arranged by spin, parity and mass. The boxes show the value of the mass and its uncertainty. Four–star resonances are shown with the darkest shading, one–star with the lightest. The lines with the crosses are the predictions for all Λ^*–octet states. The Λ^*–singlet states are marked with dots.

taken into consideration, the mass relation for flavor analog states is $m(\Lambda^*) = m(N^*) + \delta m$. In lowest order, $\delta m = m_s - m_d \simeq 150$ MeV. This already accounts for all known octet Λ^* states to a level better than ~ 80 MeV. To obtain the next–order correction, a model is needed such as the relativized constituent quark model [11]. Shown by shaded rectangular boxes in Fig. 6 are the masses of all Λ^*'s [1]. The states have been arranged according to spin and parity. The lines with crosses give our predictions for the masses of all octet Λ^* states, the input is the *experimental* N^* masses and δ_m as calculated from Ref. [11]. The agreement is spectacular.

5 Flavor Symmetry and the Search for the Missing Baryon Resonances

The known baryons in the various quark models are all (qqq) states and they obey approximate SU(3)–flavor symmetry. Quark models give a good account of the gross features of the observed baryons but fail in the details.

Furthermore, most quark models predict the existence of many more states than have actually been observed. Proponents have dubbed the predicted but non-observed N^*'s the "missing" resonances. They allegedly have a very small coupling to the crucial πN-channel. It is also possible that the missing states are non-existing; the quark models could be incomplete because they do lack certain features such as a diquark substructure. There are efforts under way to find the missing states but with little success so far. Shown in Fig. 7 by shaded boxes are the known N^*'s, while the predicted ones by Capstick-Roberts [12] are indicated by a line with a triangle. In the mass region from 1850 to 2150 MeV there are 17 missing N^*'s: $3(\frac{1}{2}^+), 3(\frac{3}{2}^+), 1(\frac{5}{2}^+), 4(\frac{1}{2}^-), 4(\frac{3}{2}^-)$ and $2(\frac{5}{2}^-)$ states. Figure 8 shows the width of all known baryon resonances. In the region 1850–2150 MeV the N^* states are some 300 MeV wide; this fact greatly reduces the prospect of finding them. Note that the width of the known Ξ^* states is about 1/9 of the N^* or Δ^* width as expected from the Riska argu-

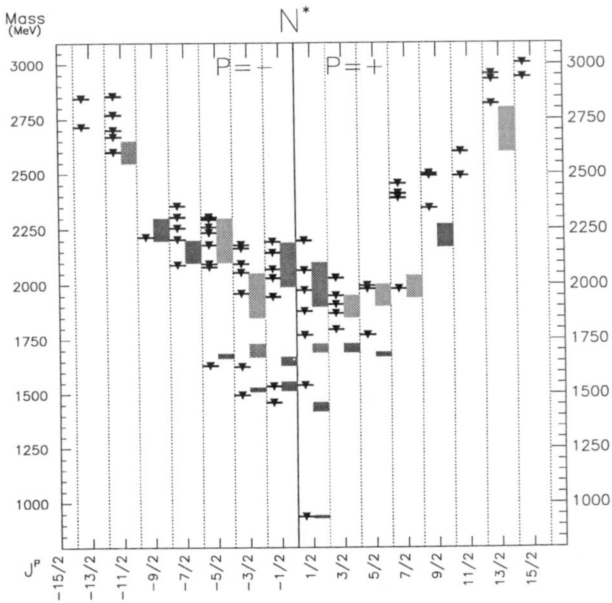

Figure 7. The spectrum of all N^* states arranged by spin, parity and mass. The boxes are the experimental data [1]. Four-star resonances are shown with the darkest shading, one-star with the lightest. The quark-model predictions [12] are marked by the lines with dark triangles.

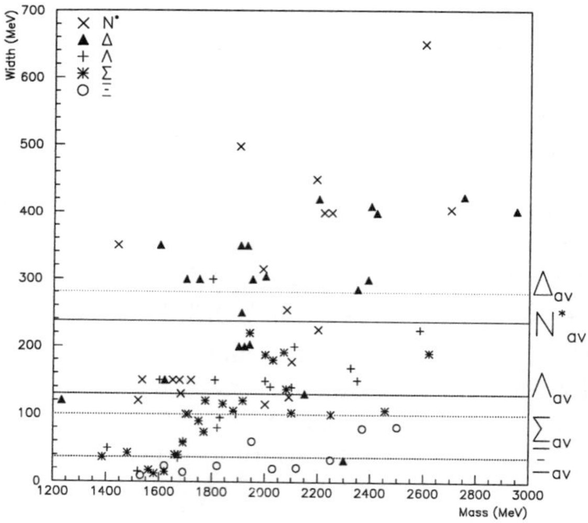

Figure 8. The width of all known baryon states.

ment mentioned earlier. Flavor symmetry requires that for every missing N^* there should be a missing Ξ^* state that is about $3 \times (m_s - m_d) \simeq 450$ MeV heavier but with a narrow width, typically 30 MeV. To search for missing baryons, the Ξ^* family appears to have the edge. Such a search has been initiated at JLab [13].

6 Summary and Conclusion

New data on hyperon production obtained by the Crystal Ball collaboration have provided impressive manifestations of the large role played by approximate SU(3)–flavor symmetry. Specifically, we have shown the similarity in η threshold production by π^- and K^- as well as the similarity in $2\pi^0$ production, especially the sensitive Dalitz plot distribution in $\pi^- p \to \pi^0\pi^0 n$ and $K^- p \to \pi^0\pi^0\Lambda$ contrasting them with the very different $K^- p \to \pi^0\pi^0\Sigma^0$ Dalitz plot. Flavor symmetry explains the similarity of the octet Λ^* and N^* states, directly relating their masses and widths. Flavor symmetry considerations lead us to suggest to shift the search for missing baryons from the N^* and Δ^* to the Ξ^* family.

It is remarkable that baryon spectroscopy is rather successful without using directly a gluon degree of freedom. After all, the gluon plays an important role in the Lagrangian of QCD (see Eq. (1)).

References

1. D.E. Groom *et al., Eur. Phys. J.* C **15** (2000).
2. G. Miller, B. Nefkens, and I. Slaus, *Phys. Rep.* **194**, 1 (1990).
3. `http://bmkn8.physics.ucla.edu/Crystalball/crystalball.html`.
4. S. Prakhov *et al., Phys. Rev. Lett.* **84**, 4802 (2000).
5. A. Starostin *et al., Phys. Rev. Lett.* **85**, 5539 (2000).
6. A. Starostin *et al.*, unpublished.
7. D. M. Binnie *et al., Phys. Rev.* D **8**, 2789 (1973).
8. B. Krusche *et al., Phys. Rev. Lett.* **74**, 3736 (1995).
9. H. Calen *et al., Phys. Lett.* B **458**, 190 (1999).
10. N. Kaiser *et al., Phys. Lett.* B **362**, 23 (1995).
11. S. Capstick and N. Isgur, *Phys. Rev.* D **34**, 2809 (1986).
12. S. Capstick and W. Roberts, *Phys. Rev.* D **49**, 4570 (1992).
13. J.W. Price, unpublished.

Ben Nefkens

Jian-Ping Chen

Marco Ripani

HIGHER RESONANCES AND THE EXAMPLE OF TWO PION ELECTROPRODUCTION WITH THE CLAS DETECTOR AT JEFFERSON LAB

M. RIPANI
FOR THE CLAS COLLABORATION
Istituto Nazionale di Fisica Nucleare,
Via Dodecaneso 33, I-16146 Genova (Italy)

Quark model predictions on the light-quark baryon properties need to be tested with a new generation of precise experimental data. In this report it is shown how multiple pion production can clarify the nature of higher mass nucleon resonances and may produce evidence of "missing" states; in particular, this scientific program is pursued by CLAS collaboration experiment E-93-006 about two pion electroproduction in Hall B at Jefferson Laboratory. Some preliminary data and cross sections from CLAS will be shown, along with results of a phenomenological analysis.

1 Physics Issues

Successful models of baryonic excitations are based on three constituent quarks confined in a potential well and bound by forces whose unperturbed Hamiltonian obeys SU(6)⊗O(3) symmetry[1,2], differing in the specific implementation of important basic ingredients like confining potential, quark-quark spin/flavour interactions, etc. In addition to the baryon spectrum, wave functions of the various states are calculated and it is therefore important to have access to them in experiments; moreover, a well known puzzle of baryon physics is that the number of observed baryon resonances is lower than predicted by theories, originating in the "missing states" issue, although other models such as the Quark Cluster Model[3] predict a smaller number of states based on reduced degrees of freedom. Measurements of electromagnetic transition amplitudes from the nucleon to its excited states are indeed sensitive to the spatial and spin structure of the transition, therefore providing a benchmark for baryon models. A relevant phenomenological aspect is that all existing information on baryon resonances has been obtained in experiments where a pion is present in either the incoming or the outgoing channel ($\pi N \to \pi N$, $\gamma N \to \pi N$, etc.); it is clear that these experiments are not suited to look for states with a weak pion coupling: actually, many of the nucleons excited states in the mass region around and above 1.7 GeV tend to decouple from the single-pion and eta channels, while decaying predominantly in multipion channels, such as $\Delta\pi$ or $N\rho$[4]. A measurement of the transition form factors of these states is very important for testing symmetry properties of the quark model. A similar situation is expected for the "missing states", as QCD mixing effects could decouple many of these states from the pion-nucleon channel[1], while strongly coupling them to channels such as $\Delta\pi$[1,5,6,7]. Search for the states still missing from the experimental evidence is therefore crucial in understanding the basic degrees of freedom in baryon structure. So far, possible evidence of missing states has been reported as a new resonant wave in $\pi N \rightarrow \eta N$ [8], thereby associated with the quark model $P_{11}(1880)$; it has been suggested that a missing state may explain the discrepancy between data and theory in the η photoproduction beam asymmetry from GRAAL[9], while a missing state was advocated in Ref. [10] as a way to better fit the SAPHIR data on $K^+\Lambda$ production: in that case, a $D_{13}(1900)$ is supposed to correspond to the quark model $D_{13}(1960)$.

Studying high lying resonances involves new features and open issues. The presence of many broad, overlapping states makes it necessary to use appropriate filters to enhance particular states or group of states. This is naturally accomplished in a large-acceptance, multipurpose detector like CLAS, where using different decay channels allows to study selected states that manifest through a correspondingly large branching fraction. In particular, an important part of the CLAS experimental program is devoted to the study of multipion channels, like $\Delta\pi$ and ρN, with the goal of extracting information on the electromagnetic excitation of high-lying states weakly visible in single pion production, like the $D_{33}(1700)$, and of discovering "missing" states. An example of this line of research is experiment E-93-006, where double pion electroproduction is chosen as a filtering channel to isolate a particular family of light-quark baryons: in the following I will illustrate some general aspects of double pion electroproduction, as well as experimental aspects and first achievements of CLAS experiment E-93-006.

2 Kinematics and Observables

Two pion production from other experiments[11,12,13] shows the presence of isobar "quasi-two-body" states, $\Delta\pi$ and ρN, on top of an $N\pi\pi$ "phase space" which uniformly populates the Dalitz plot of the invariant masses $M_{p\pi^+}$ and $M_{\pi^+\pi^-}$. The correct description of a three-body collision is based on five independent kinematical variables in the most general case[14] and moreover the isobar quasi-two-body production and subsequent decay involves all of them[15]. In particular, it is necessary to choose a complete set of independent kinematic variables, in order to apply the appropriate acceptance corrections to the data. For instance, a convenient set is made up from the invariant mass of the $p\pi^+$ and the $\pi^+\pi^-$ pairs, then the polar angles θ and ϕ of the π^- and finally the rotation freedom ψ of the $p\pi^+$ pair with respect to the so-called hadronic plane defined by the incoming virtual photon, incoming proton and the outgoing π^-; this set of variables is very appropriate for the analysis of resonance decay into $\Delta^{++}\pi^-$. We chose this method, instead of applying a maximum likelihood procedure on an event-by-event basis, because in the analysis of the data in the resonance region it is not sure *a priori* that models employed in the physics analysis are correctly describing the measured data distributions; on the other hand, a general partial-wave expansion is certainly very complicated and affected by strong ambiguities; instead, we derived experimental cross sections by binning and correcting the data in the full kinematic space: the observables derived can then be compared to any model calculation, knowledge of specific experimental features of the detector being not necessary.

3 Theoretical Aspects

There has been recently an increasing theoretical and phenomenological activity in the field of double pion electromagnetic production; a preliminary survey was done in Ref.[16]; an accurate model for this reaction, with a comprehensive description of all charge channels, has been developed at low energy ($W \leq 1.6$ GeV) by the Valencia group[17] and compared to the recent precise data from MAMI[13]. A similar approach has been discussed in Ref. [18], focusing on specific non-resonant diagrams; ρ meson electroproduction has been calculated by Ref. [19], introducing σ meson exchange as the main contributor in the N^* region. ρ meson photoproduction has been also calculated in the framework of a quark model[20]; two pion contributions to the GDH sum rule[13] have also been calculated[17,21]. We developed a phenomenological model, in cooperation with a Moscow State University group[22], providing an extensive description of two pion cross section and helicity asymmetry (GDH) for both real and virtual photons; a specific goal of our work was the initial interpretation of data from E-93-006 at JLAB. Remarkable aspects of double pion electromagnetic production are the presence of many non-resonant mechanisms, characterized by a large number of partial waves, and the mathematical ambiguities that make the model-independent extraction of partial waves not unique as experiments are limited to a restricted set of unpolarised observables; therefore only a model can guide to the extraction of the resonant amplitudes, but needs first to be tested on cross sections derived in a model-independent way.

4 Data Collection with CLAS

The CLAS scientific program in Hall B at JLAB contains a class of electron scattering experiments focused on N* physics[23]. Important features of the CLAS detector[24] are large kinematic coverage for charged particles and good momentum resolution of the order of 1%. In the three running periods completed between 1998 and 2000 many exclusive hadronic final states were measured simultaneously. Data reconstruction has been completed for both the 1998 and 1999 running periods, while it is under way for the year 2000 data. Different channels were separated through subsequent particle identification using Time of Flight information and kinematic cuts. Typical beam currents of a few nA were delivered on a liquid hydrogen target, corresponding to luminosities of almost 10^{34}. Data were taken with several beam energies ranging from 1.5 to 5.5 GeV and different magnetic fields. A total of about half a billion inclusive electrons were collected in the 1999 run. I will present two pion electroproduction cross sections derived from the 1999 run at 2.567 GeV beam energy. Reaction $ep \to ep\pi^+\pi^-$ was identified using the missing mass technique, measuring in CLAS the final state $ep \to ep\pi^+$.

5 Preliminary Data, Cross Sections and Physical Analysis

Using the full statistics collected in the 1999 run at 2.567 beam energy (at two magnetic field settings), two pion event candidates were filtered, then further reduced after applying fiducial and missing mass cuts. Using a GEANT-based simulation acceptance and efficiency were evaluated in the reaction kinematic space and preliminary cross sections were derived. In Fig. 1 we report the total virtual photon cross section for Q^2 between 0.5 and 0.8 GeV^2 and between 0.8 and 1.1 GeV^2, for the full accessible W range; CLAS is providing very high statistics and accuracy and moreover the data points exhibit structures not visible in previous data. We can notice that the relative weight of the bumps slightly changes at the two different momentum trasnfers, a sign of the expected different Q^2 evolution of different resonant states.

To perform a first physical analysis of the data, we started from a phenomenological calculation[22] that describes the two relevant intermediate isobar production mechanisms, $\gamma p \to \Delta\pi \to p\pi^+\pi^-$ and $\gamma p \to \rho^0 p \to p\pi^+\pi^-$; a third mechanism, the direct $p\pi^+\pi^-$, was included as a simple phase space amplitude and fitted from the data. In this calculation non-resonant terms for $\Delta\pi$ and ρN were introduced via Regge theory and diffractive production, respectively; reggeon coupling and diffractive amplitude were derived from the comparison to our data; resonance electromagnetic excitation was described

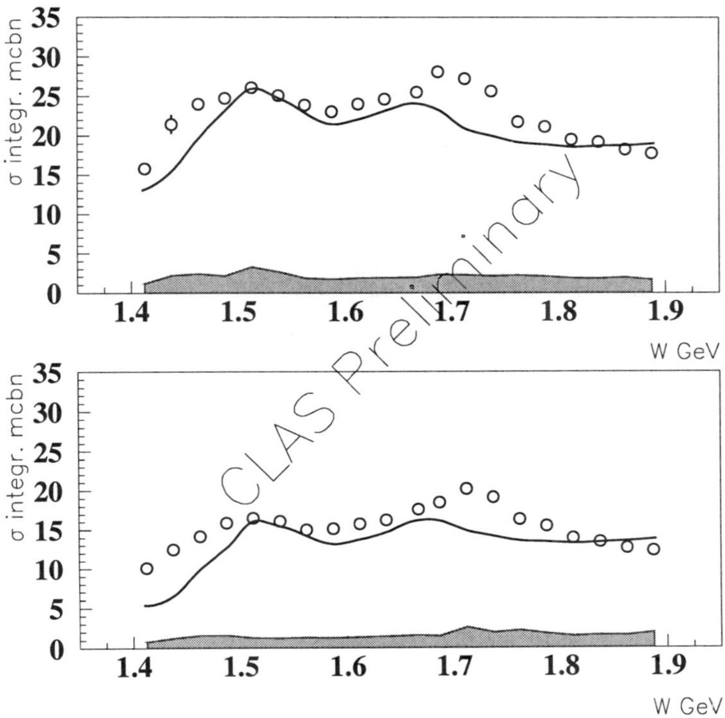

Figure 1. Total virtual photon cross section as a function of the CMS energy W for $ep \to ep\pi^+\pi^-$ at Q^2 between 0.5 and 0.8 GeV^2 (top) and Q^2 between 0.8 and 1.1 GeV^2 (bottom). Data from CLAS (grey bands show systematic errors); curves represent the Genova-Moscow phenomenological calculation described in the text.

through a Single Quark Transition Model fit[25], while partial decay widths were taken from a previous analysis of hadronic data[26] and renormalised in order for the total width to be consistent with PDG. Since the derived experimental cross section is 5 times differential as discussed in section 2, it is convenient to first analyse reduced single differential observables. For instance, we report in Fig. 2 the differential cross section with respect to the $p\pi^+$, the $\pi^+\pi^-$ invariant mass and the angle θ_{π^-}, for W=1.71 GeV: in the first invariant mass we can notice the bump corresponding to Δ^{++} intermediate

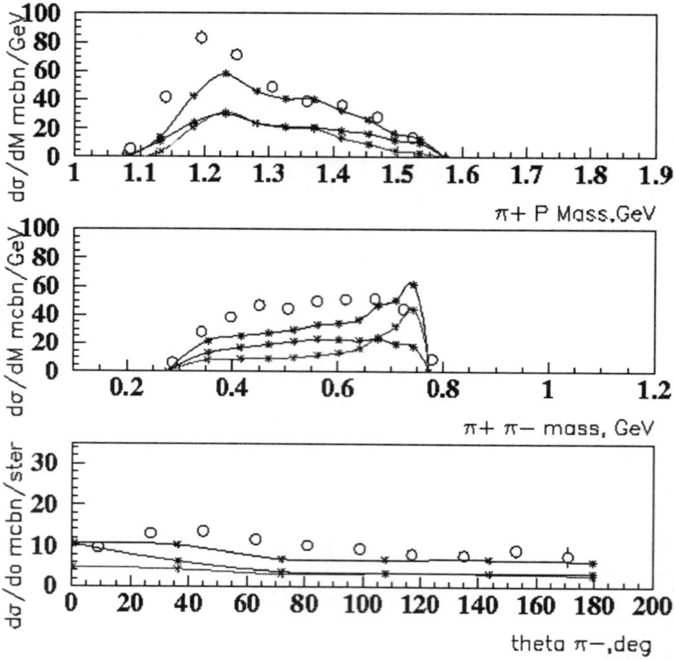

Figure 2. Some preliminary differential cross sections from CLAS, for W=1.71 GeV, at Q^2=0.5-0.8 GeV^2. Top curve is our calculation including non-resonant and resonant contributions, while the two bottom curves represent each contribution separately.

production; the various curves represent our calculation as described above. In this case, we notice that there is a clear excess of ρ meson production in the pion-pion mass distribution: in our analysis, this is directly connected to the strength of the $P_{13}(1720)$ state, whose photocouplings should therefore be reduced with respect to the SU(6) fit we used as input value[25]. Fig. 1 shows the result of our calculation for the total cross section at the two Q^2 points mentioned. We found that the structure observed at W around 1.5 GeV is indeed strongly resonant, as expected in particular from the $D_{13}(1520)$, and well described from our calculation; there is instead a discrepancy around 1.4 GeV that is still to be understood in a global analysis involving also real photon data; however, that may be an indication of the existence of a strong longitudinal resonance coupling via a second "Roper", as reported in Ref. [27]. Also the structure observed at W around 1.7 GeV contains sizeable resonance excitations, but our calculation underestimates the strength: it turns out that the $\Delta\pi$ contribution needs to be pumped up in this energy region to reproduce the data, while the calculated ρ meson production exceeds what seen in the mass distributions, as mentioned above illustrating Fig. 2 . This may indicate that some poorly known states like the $S_{31}(1620)$, $D_{33}(1700)$ and the $P_{13}(1720)$ may have a different Q^2 behavior than assumed in our calculation. Moreover, this discrepancy in the observed resonance strength is currently being investigated in terms of missing states.

6 Summary and Conclusions

High lying resonances pose new challenges and opportunities to the experimental searches: broad overlapping states are more difficult to identify and isolate, but measuring several exclusive decay branches in electron scattering allows one to separate them and to obtain a great deal of additional information on hadronic properties calculated in many recent and advanced quark models. As a promising channel for extending our knowledge of light-quark baryons, two-pion production is one of the main subjects of investigation in Hall B at Jefferson Lab, being related to poorly known and "missing" baryon resonances. A sizeable statistics for the two-pion production at different momentum transfers has already been collected with CLAS: these data allow us to achieve an outstanding accuracy in the study of the isobar formation in the final state, which is clearly evident in the data. Preliminary cross sections have been derived, manifesting so far unobserved structures as a function of the CMS energy; comparison with a phenomenological calculation containing a detailed description of the relevant physics shows clear discrepancies with the observed strength, with a corresponding enhancement or depletion in the two intermediate isobar states. Further analysis focusing on poorly known states and possible contributions from "missing" resonances is under way and will soon provide definite results.

References

1. R. Koniuk and N. Isgur, *Phys. Rev. Lett.* **44**, 845 (1980); *Phys. Rev.* D **21**, 1868 (1980).
2. M.M. Giannini, *Rep. Prog. Phys.*, **54**, 453 (1990).
3. K.F. Liu and C.W. Wong, *Phys. Rev.* D **28**, 170 (1983).
4. Particle Data Group, 1998.
5. R. Koniuk, *Nucl. Phys.*, B **195**, 452 (1982).
6. S. Capstick and W. Roberts, *Phys. Rev.* D **49**, 4570 (1994).
7. F. Stancu and P Stassart, *Phys. Rev.* D **47**, 2140 (1993).
8. S. Capstick *et al., Phys. Rev.* C **59**, 3002 (1999).
9. A. D'Angelo, *these proceedings.*
10. T. Mart and C. Bennhold, *Phys. Rev.* C **61**, 12201 (2000).
11. D. Lüke and P. Söding, *Springer Tracts in Mod. Phys.* **59** (1971).
12. V. Eckart *et al., Nucl. Phys.* B **55**, 45 (1973); P. Joos *et al., Phys. Lett.* B **52**, 481 (1974); K Wacker *et al., Nucl. Phys.* B **144**, 269 (1978).
13. Andreas Thomas, *these proceedings.*
14. Byckling and Kajantie, Particle Kinematics, Wiley and Sons (1973).
15. H.M. Pilkuhn, Relativistic Particle Physics, Springer Verlag (1979).
16. L.Y. Murphy, J.M. Laget, DAPNIA-SPHN-96-10, Mar. 1996.
17. J.C. Nacher, *these proceedings.*
18. K. Ochi, M. Hirata, T. Takaki, *Phys. Rev.*, C **56**, 1472 (1997).
19. Y. Oh, *these proceedings.*
20. Q. Zhao, *these proceedings.*
21. M. Vanderhaeghen *et al.*, unpublished.
22. V. Mokeev *et al., these proceedings* and references therein.
23. V. Burkert, in Proceedings of PANIC '99, Uppsala, Sweden, 10-16 June 1999, published in *Nucl. Phys.* A **663&664** and CLAS collaboration contributed papers therein.
24. W. Brooks, in Proceedings of PANIC99, Uppsala, Sweden, 10-16 June 1999, published in *Nucl. Phys.* A **663&664**.
25. V. D. Burkert, *Czech. Journ. of Phys.*, **46**, 627 (1996).
26. D.M.Manley and E.M.Salesky, *Phys. Rev.* D **45**, 4002 (1992).
27. H.P. Morsch and P. Zupranski, *Phys. Rev.* C **61**, 024002 (1999).

NUCLEON RESONANCES AND MESONS IN NUCLEI

V. METAG
II. Physikalisches Institut, Universität Giessen, Heinrich Buff Ring 16
D-35392 Giessen, Germany
E-mail:volker.metag@exp2.physik.uni-giessen.de

The photoexcitation of nucleon resonances and their subsequent meson decay allows one to probe the structure of free and bound nucleons. A series of photonuclear experiments has been performed at the Mainz Microtron MAMI using the photon spectrometer TAPS. Differences in the photoproduction of mesons on the free proton and on nuclei are attributed to changes in the properties of hadrons in the nuclear medium. The results are discussed in the context of related experiments at other accelerators.

1 Introduction

The origin of mass of composite hadronic systems is one of the central questions in hadron physics. For atoms and nuclei the masses of the constituents sum up to the mass of the composite system, apart from some small binding energy corrections. In contrast the bare masses of quarks, the constituents of the nucleon, only amount to a few percent of the total nucleon mass [1]. The main physics goal is to understand the origin of hadron masses in the context of spontaneous chiral symmetry breaking in QCD and their modification in a hadronic environment due to chiral dynamics and partial restoration of chiral symmetry.

In this paper, medium effects in photonuclear and pion-induced reactions will be presented and discussed. The photoproduction of mesons on free and bound nucleons is used as a tool for studying medium modifications of nucleon resonances which decay into mesons and the nucleon ground state. These decays may be affected by in-medium modifications of mesons which feed back on the properties of the decaying states. In this context, the $D_{13}(1520)$ resonance plays a special role as it is the only nucleon resonance up to the second resonance regime which has a vector meson decay branch; vector mesons are predicted to undergo dramatic medium modifications. A model-independent signature for chiral symmetry restoration in the nuclear medium would be the degeneracy in mass of chiral partners like the pion ($J^\pi = 0^-$) and the "σ" -meson ($J^\pi = 0^+$), a system of two pions coupled to the I=J=0 channel. Modification of two-pion correlations in nuclei would thus provide a testing ground for possible chiral symmetry restoration effects.

2 Photoabsorption on Free and Bound Nucleons

The absorption of photons on free and bound nucleons in the 0.1 - 2.0 GeV energy regime has been studied by several groups. Fig. 1 shows the results ob-

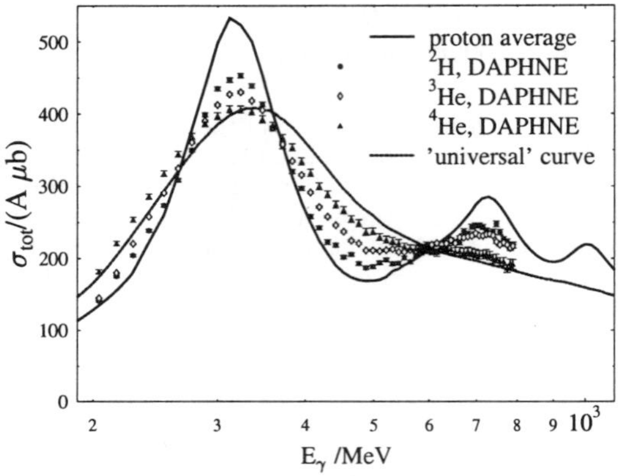

Figure 1. Comparison of photoabsorption cross sections on free and bound nucleons [2]. The figure has been prepared by J. Ahrens.

tained with the DAPHNE detector at MAMI [2]. The distinct structures in the excitation function on the free proton can be associated with excited states of the nucleon. These are the $\Delta(1232)$ - resonance, the lowest non-strange excitation of the nucleon, overlapping resonances in the second resonance regime ($P_{11}(1440)$, $D_{13}(1520)$, $S_{11}(1535)$) (see Fig.2) and still higher lying states in the third resonance regime. Due to meson decay via the strong interaction the resonances are very short lived ($\tau \approx 3 \cdot 10^{-24}$s) and correspondingly have widths of the order of 100-150 MeV.

The cross section per nucleon for photoabsorption on nuclei differs dramatically from the cross section on the free nucleon. Whereas the structure assigned to the Δ - excitation of a bound nucleon is still clearly pronounced, the bump associated with the second resonance regime is washed out. This behaviour points to a strong medium effect, a modification of hadronic properties in nuclear matter. To trace the medium modification to individual resonances, it is not sufficient to study the inclusive photoabsorption process. Instead, the excitation of the individual resonances in exclusive reactions has to be investigated. Here, the characteristic decay properties of the different resonances can be exploited. The $\Delta(1232)$ - resonance decays via single pion emission and the resonance structure is most pronounced in the neutral pion channel where non-resonant contributions (Born-terms) are suppressed. The

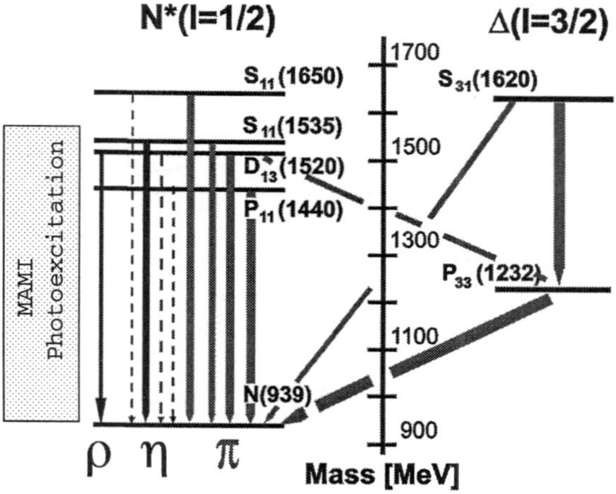

Figure 2. Excitation energy spectrum of the nucleon. The states are separated according to their isospin. Main decay modes and the excitation energy range accessible at MAMI are indicated.

observation of an η meson is characteristic for the excitation of the $S_{11}(1535)$ resonance since none of the other nucleon resonances has a comparable η branching ratio (≈ 50 %) [3]. Similarly, the $D_{13}(1520)$ resonance can be identified via its predominant decay mode into two pions. Most decays occur via sequential emission of the two pions with the $\Delta(1232)$ as intermediate state [4,5], but some pions are correlated, e.g., in the decay via a ρ meson, as recently observed [6]. In the following sections, these decays are used to establish differences between free and in-medium properties of the individual resonances.

3 Properties of Bound Nucleon Resonances

The excitation of a bound nucleon to the Δ resonance has been studied in the coherent photoproduction of π^0 mesons on ^{4}He as shown in Fig. 3. Details of the experiment are described in [7]. In the theoretical analysis of Drechsel et al. [8], the interaction of the Δ resonance with the nuclear environment is

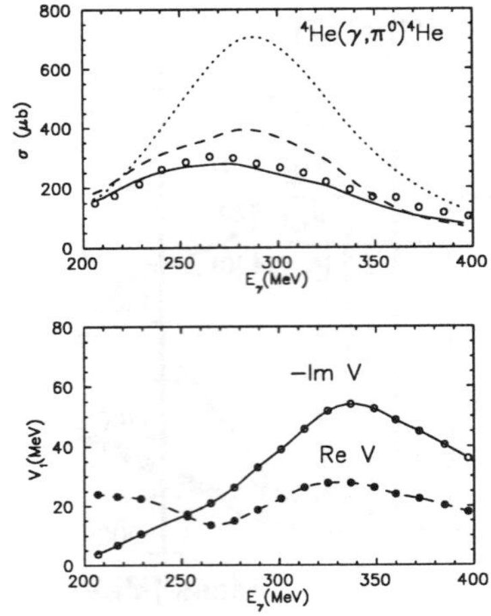

Figure 3. Energy dependence of the total cross section for the $^4He(\gamma,\pi^0)^4He$ reaction. The experimental data are from [7]; the curves represent PWIA and DWIA calculations and calculations including the Δ self energy term [8]. The corresponding parameters of the potential are given in the lower figure.

taken into account by adding a self energy term Σ_Δ to the free Δ-propagator:

$$\frac{1}{W - M_\Delta + i\Gamma_\Delta(W)/2} \Longrightarrow \frac{1}{W - M_\Delta + i\Gamma_\Delta(W)/2 - \Sigma_\Delta} . \tag{1}$$

The self energy is parametrized as

$$\Sigma_\Delta(E_\gamma, q^2) = V(E_\gamma)F(q^2), \qquad F(q^2) = e^{-\beta q^2} \tag{2}$$

where V is a complex potential which reflects the medium modification of the Δ resonance and is obtained from a fit to the data. The real and imaginary part of V represent additional contributions to the resonance mass and width, respectively. They are shown for ^4He in the lower part of Fig. 3. A mass shift of the order of 20 MeV is a small correction ($\approx 2\%$) while the increase in width is of the order of 30%. Remarkably, the same potential parameters also reproduce the cross section for coherent π^0 production on heavier nuclei [9], indicating a saturation of the Δ-nucleus interaction. Similar values for the mass shift and the broadening of the $\Delta(1232)$ resonance have recently been deduced from pion electroproduction off ^3He in the reaction $^3He(e,e'\pi^+)^3H$ [10].

While the $\Delta(1232)$ resonance preserves its identity in nuclei with some modification of mass and width, the disappearance of the second bump in the photoabsorption excitation function on nuclei occurs in the second resonance

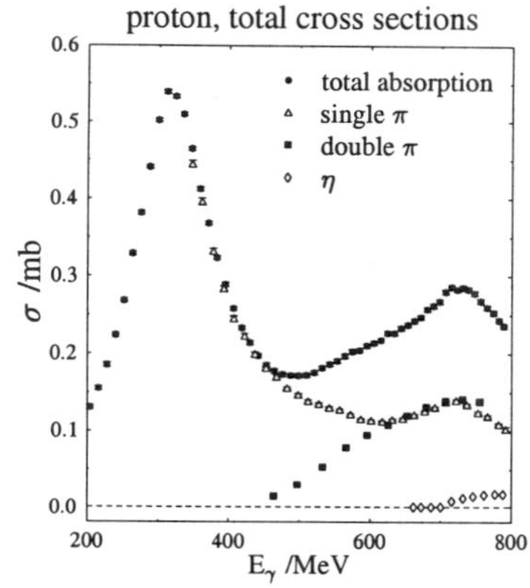

Figure 4. Decomposition of the photoabsorption cross section on the proton into different meson production channels. The figure has been prepared by J. Ahrens.

regime. Fig. 4 shows a decomposition of the inclusive photoabsorption cross section into individual meson production channels. Here, the D_{13} and S_{11} resonances are important. It should be pointed out, however, that the rise in the photoabsorption cross section above 500 MeV is due to the onset of the two-pion channel which - at least for charged pion production - is dominated by non-resonant π-pole and Δ-Kroll-Rudermann terms. Therefore the disappearance of the second bump may not only be related to a possible modification of resonance properties but may also be associated with in-medium changes in the delicate interference of resonant and non-resonant terms as recently discussed by Hirata et al. [11].

The in-medium properties of the S_{11} resonance have been studied by measuring the photoproduction of η mesons on nuclei [12,13]. According to [13], the observed broadening of the S_{11} resonance width can be largely attributed to collisional broadening through the $S_{11}N \to NN$ reaction while the resonance mass is hardly affected. For a more quantitative conclusion, data of highly improved quality will be needed.

The decay of the D_{13} resonance occurs predominantely via single and double pion emission. Recent studies [6,14,15] have, however, revealed an appreciable ρ-strength as illustrated in Fig. 5. The total photoproduction cross section for $\pi^+\pi^0$ pairs and the corresponding invariant mass distributions, recently measured by the TAPS collaboration [6], can only be reproduced by including a $D_{13} \to N\rho$ channel with a branching ratio of about 20%. There are various

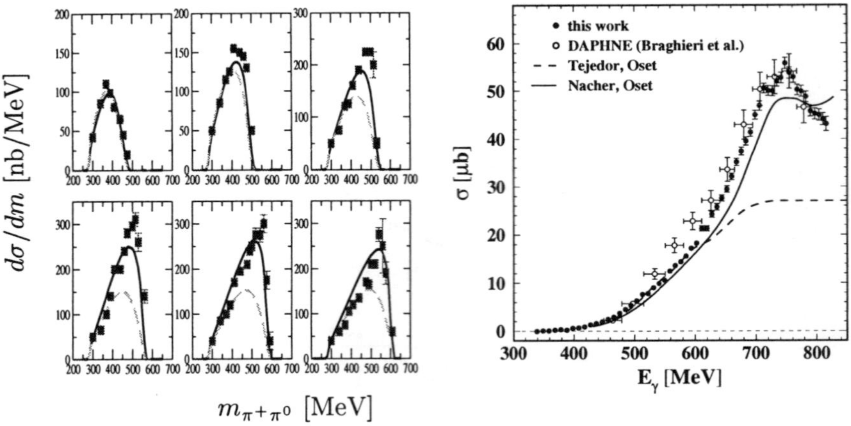

Figure 5. Excitation function and invariant mass distribution for the $p(\gamma,\pi^+\pi^0)p$ reaction in comparison to model calculations [15] with (solid curve) and without (dashed curve) the inclusion of the $D_{13} \to N\rho^+$ decay branch.

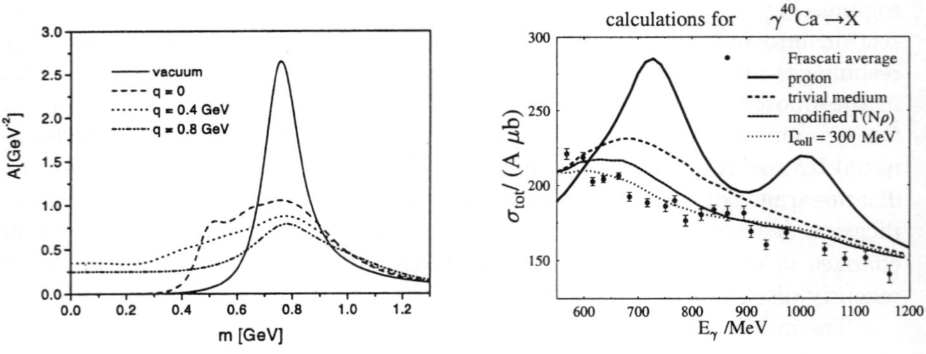

Figure 6. Left: Spectral function of the ρ meson at normal nuclear matter density for different momenta [16]. Right: Photoabsorption cross section in the second resonance regime on the proton and on ^{40}Ca in comparison to theoretical calculations [19] taking different in-medium effects into account.

theoretical predictions for a strong spreading of the ρ meson strength in the nuclear medium because of its couplings to N*- hole excitations [16,17,18]; an example is shown in Fig. 6. As a consequence, the opening of phase space for ρ decay of the D_{13} resonance may provide sufficient resonance broadening to explain the disappearance of structures in the photoabsorption on nuclei. The result of a corresponding transport model calculation [19] is shown in Fig. 6. In this scenario, the disappearance of structures in the photoabsorption on nuclei is traced to medium modifications of a vector meson. In a self-consistent treatment medium modifications of the decay products must have an impact on the in-medium properties of the decaying state, the D_{13} resonance. Recent studies of single π^0 -production on nuclei show, however, a width of the D_{13} resonance which is consistent with the free width folded with Fermi motion [20]. This apparent discrepancy will have to be studied further.

4 Scalar Mesons in the Nuclear Medium

The modification of vector meson properties in the medium has been discussed in the context of a possible partial restoration of chiral symmetry with increasing baryon density and temperature. A model independent signature for chiral symmetry restoration would be the degeneracy in mass of chiral partners. The chiral partner of the pion ($J^\pi = 0^-$) is the σ meson ($J^\pi = 0^+$) whose existence still is a matter of debate. The Particle Data Group lists a particle $f_0(400 - 1200)$ in the most recent edition [1].

Being a Goldstone boson, the mass of the pion is expected to be rather stable as a function of baryon density. In fact, changes in the π mass by some 20 MeV have been reported for neutron rich nuclei (^{207}Pb) at normal nuclear matter density [21]. Partial chiral symmetry restoration in the nuclear medium would thus imply a downward shift of the σ - mass [22], as schematically shown in Fig. 7, leading to an enhancement in the spectral function in the J=I=0 channel close to the 2π threshold [23,24,25]. First experimental signals for such a concentration of strength have been reported in $A(\pi, 2\pi)$ reactions on various nuclei [26,27]. As an example, the data taken with the Crystal Ball at BNL are shown in Fig. 7. The σ strength function is studied in these reactions at densities below normal nuclear matter density since the incoming pions are preferentially absorbed already in the nuclear surface. In photonuclear reactions $A(\gamma, 2\pi)$, as studied with TAPS at MAMI, the strength function is probed over the bulk of the nucleus. An accumulation of $2\pi^0$ strength near $m_{\pi\pi} = 270 MeV$ is also observed in these reactions [28]. As in the other 2π production reactions on nuclei, the rescattering of pions will also contribute to such an enhancement which, however, should be calculable exploiting the knowledge of pion rescattering and -absorption in nuclei. A clean experimental signal for a concentration of the σ spectral function near the 2π threshold would be an important step forward in our understanding of chiral symmetry restoration in the nuclear medium.

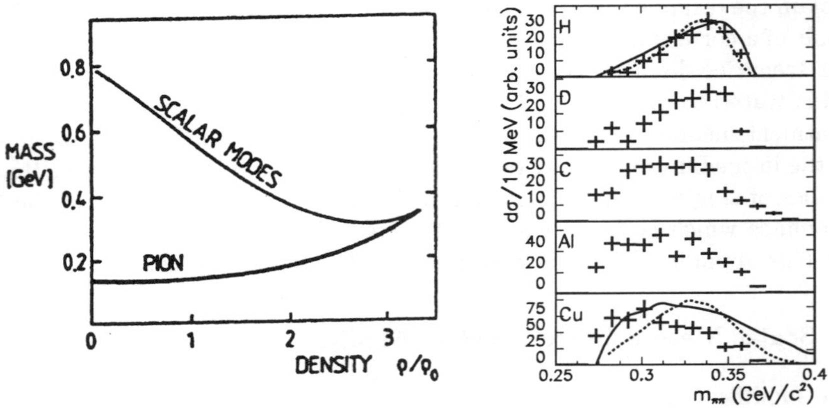

Figure 7. Theoretically expected variation of the σ and π mass for chiral symmetry restoration with increasing nuclear density [22]. Right: $2\pi^0$ invariant mass distributions measured in the reactions $A(\pi^-, \pi^0\pi^0)$ at $p_\pi = 400 MeV/c$ for various nuclei [27].

5 Summary

The effect of the nuclear medium on hadronic properties has been studied by measuring the photoproduction of mesons and by comparing the properties of free and bound nucleon resonances. An intriguing suggestion for the disappearance of structures in the photoabsorption cross section in the second resonance regime may be a broadening of the D_{13} resonance caused by a shift of the ρ strength to lower masses in the nuclear medium. It has to be clarified, however, whether this scenario can be reconciled with the recent observation that, apart from effects due to Fermi motion, the D_{13} resonance maintains its mass and width in the surface region of nuclei. Intriguing results on a downward shift of the two pion invariant mass distribution in nuclei will have to be critically analyzed as a possible signal for a partial restoration of chiral symmetry at normal nuclear matter density.

The results presented in this article are largely based on the work of the TAPS and A2 collaborations at MAMI. I would like to thank the scientists, technicians and in particular the many Ph.D. students who have carried the major load in data taking and analysis. It is a pleasure to acknowledge illuminating discussions with U. Mosel, J.C. Nacher and E. Oset. I am grateful to J. Ritman and S. Schadmand for critical comments on the manuscript.

References

1. D.E. Groom *et al., Eur. Phys. J.* C **15**, 1 (2000).
2. M. MacCormick *et al., Phys. Rev.* C **55**, 1033 (1997); V. Muccifora *et al., Phys. Rev.* C **60**, 064616 (1999).
3. B. Krusche *et al., Phys. Rev. Lett.* **74**, 3736 (1995).
4. F. Härter *et al., Phys. Lett.* B **401**, 229 (1997).
5. M. Wolf *et al., Eur. Phys. J.* A **9**, 5 (2000).
6. W. Langgärtner *et al.*, submitted to *Phys. Rev. Lett.*
7. F. Rambo *et al., Nucl. Phys.* A **660**, 69 (1999).
8. D. Drechsel *et al., Nucl. Phys.* A **660**, 423 (1999).
9. B. Krusche *et al.*, to be published.
10. M. Kohl *et al.*, Proceedings 16th. Int. Conf. on Few-Body Problems in Physics, March 2000, Taipei, Taiwan.
11. M. Hirata *et al., Phys. Rev. Lett.* **80**, 5068 (1998).
12. M. Röbig-Landau *et al., Phys. Lett.* B **373**, 45 (1996).
13. T. Yorita *et al., Phys. Lett.* B **476**, 226 (2000).
14. A. Zabrodin *et al., Phys. Rev.* C **60**, 055201 (1999).
15. J.C. Nacher *et al.*, nucl-th/0012065 and private communication.
16. W. Peters *et al., Nucl. Phys.* A **632**, 109 (1998).
17. F. Klingl *et al., Nucl. Phys.* A **624**, 527 (1997).
18. R. Rapp *et al., Nucl. Phys.* A **617**, 472 (1997).
19. U. Mosel, *Prog. Part. Nucl. Phys.* **42**, 163 (1999).
20. B. Krusche *et al., Phys. Rev. Lett.*, in print.
21. T. Yamazaki *et al., Phys. Lett.* B **418**, 246 (1998).
22. M. Lutz *et al., Nucl. Phys.* A **542**, 521 (1992).
23. M.J. Vicente Vacas and E. Oset, *Phys. Rev.* C **60**, 064621 (1999).
24. R. Rapp *et al., Phys. Rev.* C **59**, 1237 (1999).
25. T. Hatsuda *et al., Phys. Rev. Lett.* **82**, 2840 (1999).
26. F. Bonutti *et al., Phys. Rev. Lett.* **77**, 603 (1996).
27. A. Starostin *et al., Phys. Rev. Lett.* **85**, 5539 (2000).
28. S. Janssen, Ph.D. thesis, University of Giessen, in preparation.

Ralph Minehart

Volker Metag

EXCITATION OF NUCLEON RESONANCES

V. D. BURKERT
Thomas Jefferson National Accelerator Facility,
12000 Jefferson Avenue, Newport News, VA 23606, USA

I discuss developments in the area of nucleon resonance excitation, both necessary and feasible, that would put our understanding of nucleon structure in the regime of strong QCD on a qualitatively new level. They involve the collection of high quality data in various channels, a more rigorous approach in the search for "missing" resonances, an effort to compute some critical quantities in nucleon resonance excitations from first principles, i.e. QCD , and a proposal focussed to obtain an understanding of a fundamental quantity in nucleon structure.

1 Introduction

It is not easy to give an "OUTLOOK" talk after we have heard so many interesting new results, and most speakers talked already about plans for the future. So, I will be just speaking about a few selected aspects of nucleon resonance physics where progress may be possible in the next few years. I will point out a few areas where there have been significant advances recently, and where we may expect important progress soon. Finally, I want to present to the community a proposal for a focussed effort to study the Q^2 evolution of the generalized Bjorken integral from small to large distances. If successful, this would be an *important milestone of nucleon structure studies.*

Electromagnetic production of mesons, which is what we are doing when studying resonances, may be crudely characterized by 3 regions representing different distance scales. At large distances, say 1 fm, we study nucleon properties near the surface. Nucleons and pions are the relevant degrees-of-freedom. Chiral perturbation theory describes many phenomena, and is linked to QCD via chiral symmetry and chiral symmetry breaking. At the other end, at very small distances, we probe the parton structure of the nucleon. Elementary quark and gluon fields are the relevant degrees of freedom used in perturbative QCD. This workshop has been mostly about the regime of intermediate distances, where the excitation of baryon resonances is prominent. Quarks and gluons are relevant; however, they interact more like constituent quarks and glue. This domain is currently addressed theoretically by quark models, flux tube models, QCD sum rules, instanton models, etc., with some success. The relationship of these approaches to QCD is often not fully developed, making it difficult to assess the accuracy of the model prediction.

The regions of different distance scales are likely not strictly separated from each other. This should provide areas of overlap where different theoretical approaches can be used to compute the same observables, thus allowing important checks of the range of validity of a specific approach. What I see as an important task for the community is to anchor more firmly these descriptions to the fundamentals of QCD, and finally come to an understanding of resonance phenomena and nucleon structure from the largest to the smallest distances within fundamental theory.

Nucleon structure studies are often associated with deep inelastic scattering where the interpretation of data in terms of the underlying degrees-of-freedom is usually more straightforward. However, quark structure function measurements and the test of asymptotic sum rules are only one small area of the nucleon structure to be explored, and certainly not the one most strongly related to QCD. Strong interaction plays no role in asymptotic sum rules, and the determination of quark structure functions is related to QCD only via secondary effects such as gluonic corrections. It is understanding the measured parton distributions that is the real challenge for QCD.

Information on nucleon structure from formfactors or nucleon resonance excitations is much richer, albeit more difficult to interpret. Nevertheless these are quantities closer to the "real" world, and they need to be described and understood in terms of the underlying degrees-of-freedom if we want to make progress.

2 The $\gamma N\Delta(1232)$ Transition - From Precision Experiments to Precision Calculations.

This is the region where we aim for precision. The $\Delta(1232)$ is the only resonance that is well separated from all the higher mass states. At low Q^2 the $\Delta(1232)$ has the largest cross section. There has been considerable progress in experiments and analyses over the past 5 years or so. The uncertainties in ratios of multipoles $R_{EM} = E_{1+}/M_{1+}$ and $R_{SM} = S1+/M_{1+}$ have been reduced by an order of magnitude since the early studies. This now allows sensible comparisons with model predictions. One of the most noteworthy results is the quantitative realization that a description of the $N\Delta(1232)$ transition requires the inclusion of pions as effective degrees of freedom. A simultaneous description of both ratios is achieved only by models that include pion d.o.f [1]. Also the total absorption cross sections for the $N\Delta(1232)$ transition is well described by models that include pion cloud effects.

A more precise description of quantities such as R_{EM} in lattice QCD (LQCD) is long overdue. The only LQCD "prediction" of R_{EM} is nearly a decade old[3] and "predicts" a value of $3 \pm 8\%$ at the photon point, while the experimental value is $(-2.75 \pm 0.50)\%$, where the experimental error is estimated generously. An order of magnitude smaller LQCD error is needed to have any impact here. A simple extrapolation of the computer performance using Moore's law, gives precisely the factor ten needed for progress. The next step would be to evolve this quantity in Q^2, and compute the magnitude of the magnetic transition multipole vs. Q^2. This would mark real progress on the lattice!

We also would like to know whether the apparent trend in the R_{em} data really indicates[2] that there will be a zero crossing at $Q^2 \approx 4\text{GeV}^2$. This may give us a clue where leading order pQCD contributions may have some relevance. Since the signal/background at high Q^2 will be a lot smaller we also need to refine our analysis techniques and collect more data that give us more direct information on background contributions. Beam spin asymmetries as well as other polarization observables are needed to reduce the model-dependence of the analysis techniques at high Q^2.

3 The 2nd Resonance Region

The so called 2nd resonance region, comprising the mass range from 1.4 to 1.6 GeV, is of particular interest for nucleon structure studies. It contains 3 states, the $N(1440)P_{11}$ "Roper", the $N^*(1535)S_{11}$, and $N^*(1520)D_{13}$ states, all of which are highly interesting for the study of nucleon structure properties, and for the testing of basic symmetry properties.

3.1 Mysteries of the Roper resonance $N(1440)P_{11}$

A natural candidate for detailed studies beyond the $\Delta(1232)$ would be the Roper resonance $N(1440)P_{11}$. However, more than 35 years after its discovery its structure is basically still unknown. The non-relativistic constituent quark model (nrCQM) puts its mass above 1600MeV, the photocoupling amplitudes are not described well, and the transition formfactors, although poorly determined, are far off. Relativized variations of the nrCQM improved the situation only modestly. To obtain a better description of the data a number of alternative models have been proposed. Does the Roper have a large gluonic component[4]? Is it a small quark core with a large pion cloud[5]? Is it a nucleon-sigma molecule[6]? Or, is it not a single resonance but two appearing in different reactions differently[7]? These questions will be discussed at future workshops, however, it is crucial to get more precise electroproduction data, as it is the Q^2 dependence where the models differ strongly. From the model builders we must require that their models make predictions for the electromagnetic couplings and formfactors.

There is also some good news: lattice QCD calculations are beginning to produce results for the mass of the Roper which may soon be accurate enough to have a real impact.

3.2 The $N^*(1535)S_{11}$ and $N^*(1520)D_{13}$, and the $[70, 1^-]$ supermultiplet

There is some good news from the constituent quark model. The slow fall-off of the transverse $N^*(1535)S_{11}$ transition formfactor, which has been a problem for model builders for a long time, is now quite well described by the CQM using a potential containing a Coulomb form and a linear term[8]. At the same time the $A_{1/2}$ amplitude of the D_{13} is described as well, while there remains a large discrepancy for the $A_{3/2}$ amplitude at small Q^2. Could this be explained by pionic contributions which then would have to contribute to the helicity nonconserving (nonleading) term but not to the helicity conserving (leading) amplitude? Calculations that include pionic contributions explicitly are needed to answer this question.

Another piece of good news comes from LQCD. As already discussed at the previous workshop[9], mass predictions for the lowest N^* state with negative parity agree well with the experimental values. The obvious next step would be to compute the $A_{1/2}$ amplitude for that state at the photon point in LQCD.

For a better understanding of the Roper as well as the $N^*(1520)D_{13}$, data in the $n\pi^+$ channel are crucial to obtain more complete isospin information. Also, beam spin asymmetry measurements will give information about the background amplitudes which are especially important in that mass region. Such data have been taken and are currently being analyzed[11].

The ordering of excited states according to the $SU(6) \otimes O(3)$ symmetry group and the assumption that excitations are due to a single quark transition (SQT), allows predictions for a large number of states belonging to the same supermultiplet based on only three known amplitudes. In the case of the $[70, 1^-]$, the $N^*(1535)S_{11}$ and the $N^*(1520)D_{13}$ may serve that purpose. These are the only states in this multiplet whose transition amplitudes have been measured with some accuracy. This allows tests of the SQT assumption, and how the symmetry will break down as a function of the distance scale. While the predicted photocoupling amplitudes are in quite good agreement with the data, there are not enough data at finite Q^2 to test this simple model at shorter distances. The lack of data for two of the prominent states, the $\Delta(1620)S_{31}$ and $\Delta(1700)D_{33}$, is largely due to the complete lack of data in the $N\pi\pi$ channel. Also, amplitudes for neutron resonances are absent for all states. This situation will hopefully change soon with new data from CLAS[10].

4 Missing Baryon Resonances

Understanding the fundamental structure of baryons remains the main focus of the N^* progam. There is now a significant effort underway to search for some of the states predicted by the symmetric quark model[12] that have not been seen in πN scattering. The importance of this effort lies in the fact that these states can tell us much about internal baryon structure. For example, models that do not have approximate SU(6) symmetry may not predict some or even many of these states to exist[15]. Some of these states are predicted to couple to $\Delta\pi$, $N\omega$, Y^*K, and other hadronic channels, as well as to photons[13]. Photo- or electroproduction may therefore be the only way to search for some of these states. Experiments at GRAAL, JLab, ELSA (Crystal Barrel), Spring-8, and BEPC have begun a vigorous search employing large acceptance detectors[16]. This effort is accompanied by a theoretical effort to understand how these resonances might show up in experimental observables[17,18].

There are indications for one or even two of such states which have been discussed at this workshop. Some of this evidence is, however, due to improvements that model curves show in comparison to data in case such states are included. Clearly, this is not sufficient. Other partial wave contributions need to be tested and excluded. For the evidence to be fully convincing, partial wave analyses must be done that seek to analyse such states in the energy-dependence of partial-wave amplitudes and their phase motion.

The strangeness sector offers excellent prospects in the search for missing states. Hyperon resonances are more narrow than states made of u and d quarks only, and they can be separated more easily from other overlapping resonances[19].

Another kind of "missing" baryons are the gluonic excitations or "hybrid" states where the "glue" or flux tubes are excited and produce a $|q^3G>$ state. They have been predicted in bag models and flux tube models. Lattice QCD predicts such states in the meson sector. They are likely expected in the baryon sector as well, although no LQCD calculations have been performed. In distinction to the meson sector no exotic quantum numbers are expected in the baryon sector. This will make it experimentally more difficult to identify gluonic excitations. QCD sum rules[20], flux tube[21], and bag models[22] predict the lowest gluonic states to be P_{11} or P_{13} with masses between 1.5 GeV for bag models and QCD sum rules, and 1.8 - 1.9 GeV for the flux tube model[21]. Possible signatures could be the overabundance of states, unusual decay channels, form factors which are different from the 3-quark sector due to the larger sizes of $|q^3G>$ states, and different threshold behavior due to different $SU(6) \otimes O(3)$ assignment. Another possibility is the production of hybrid states in the gluon rich environment of J/ψ decays[23].

None of these signatures alone will be convincing. It will take various pieces of evidence, and a good understanding of these states within models that treat 3-quark states and gluonic states on an equal footing, to have sufficient confidence in any discovery in this area.

5 The Nucleon Spin Integral from Small to Large Distances - A Proposal for the Next 5 Years

Coming back to the goals outlined in the introduction one may ask what quantities are most directly accessible to a description within fundamental theory. As "fundamental" I would characterize exact sum rules, such as the GDH and Bjorken sum rules, QCD, pQCD, and chiral perturbation theory. I will argue that $\Delta\Gamma_1^{pn}(Q^2) = \int [g_1^p(x,Q^2) - g_1^n(x,Q^2)]dx$ is such a quantity.

What would be the significance of such a project? Why is it important, and why should the N^* community be involved in this? Clearly, from a physics perspective such a project, if successful, would be a milestone, as it would *mark the first time a fundamental quantity of nucleon structure is described by fundamental theory from small to large distances*, a worthwhile goal of nucleon structure physics, and worth a serious effort by the community. First, the expertise of the N^* community is important as nucleon resonances make significant contributions to the spin integral at medium and large distances[24,25]. Second, such a project provides a focus for the community to solve a fundamental problem. Third, the description of the resonance contributions to the first moment in LQCD may be the biggest effort, and there are proposals from within the community to have significant computing resources available for nucleon structure studies in the next five years, that can be brought to bear on such a project.

5.1 *What is the experimental and theoretical situation?*

The experiments to measure the polarized structure function $g_1(x,Q^2)$ in a large Q^2 range are far along as has been reported at this conference[26,27]. The deep inelastic regime has been studied for decades, and good data are available for $Q^2 > 1.5 GeV^2$ mostly for the proton but also for the neutron. Experiments at JLab in CLAS and in Hall A are near to final results for the range in $Q^2 = 0.1 - 1.0$ GeV^2. These data currently require an extrapolation

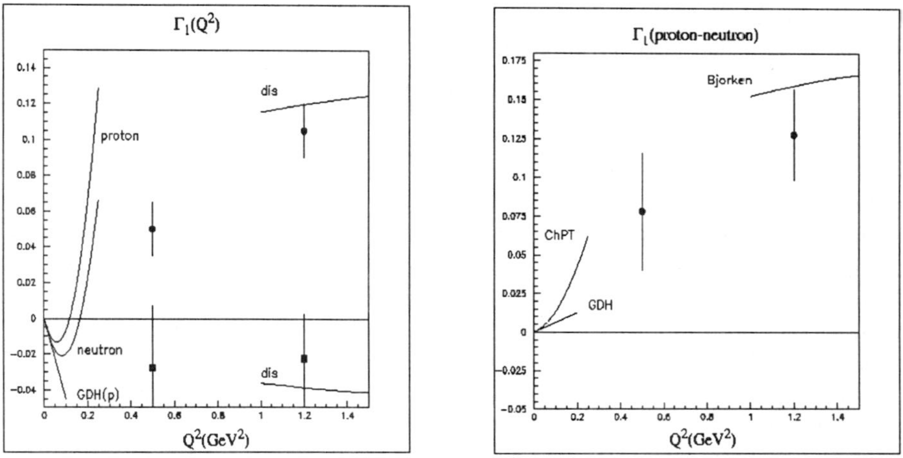

Figure 1. First moment of the spin structure function $g_1(x, Q^2)$ for the proton and neutron (left), and for the proton-neutron difference (right). The curves above $Q^2 = 1\text{GeV}^2$ are pQCD evolutions of the measured Γ_1 for proton and neutron, and the pQCD evolution for the Bjorken sum rule, respectively. The straight lines near $Q^2 = 0$ are the slope given by the GDH sum rule. The curves at small Q^2 represent the NLO HBChPT results.

at small x which adds a small systematic error in the low Q^2 range, however, a significant uncertainty at $Q^2 > 1$ GeV^2. This situation is changing with the new data taken with CLAS in the energy range from 1.6 - 5.75 GeV, and in Hall A with an upcoming experiment at very small Q^2. Also, uncertainties in the extraction of the neutron contribution from measurements on ^3He require an improved treatment of the nuclear effects at small Q^2 where uncertainties are significant.

Within this year the first complete information on $\Gamma_1^p(Q^2)-\Gamma_1^n(Q^2)$ should be available in a Q^2 range from 0.1 - 1 GeV^2 from JLab experiments. At the same time, information on the proton and the neutron separately will be available as well. The current theoretical situation is illustrated in Fig. 1. The left hand panel is for the proton and for the neutron separately. The high Q^2 behavior has been measured, and is known to approach a constant value. The asymptotic behavior has been evolved to lower Q^2 in perturbative QCD to order α_s^3. This is shown by the lines labelled "dis". At the low Q^2 end we have the GDH sum rule believed to be valid at the photon point. It also defines the slope of $\Gamma_1(Q^2 \to 0)$. The slope is negative for both proton and neutron. Heavy Baryon Chiral Perturbation Theory (HBChPT) has been used[28] to evolve the GDH integral to finite Q^2. The curves are from NLO calculations. Unfortunately, for the proton and neutron this expansion appears to break down already at very small Q^2. A potential problem in these calculations is the treatment of the $\Delta(1232)$. To avoid this problem we take the proton-neutron difference, where this contribution is not present. The result is a dramatic improvement in the low Q^2 description of the apparent trend of the data[29]. Q^2 values up to 0.25 GeV^2 or higher might be reachable in this quantity. Taking the proton-neutron difference $\Delta\Gamma_1^{pn}$ is also suggested from the behavior in the deep inelastic regime where the Bjorken sum rule[30] establishes an important constraint for the absolute normalization of the first moment, which has been verified experimentally within 5-10%

The combination of two fundamental sum rules at the opposite sides of the distance scale, with the pQCD evolution at small distances and the ChPT evolution at large distances, provides powerful constraints for the Q^2 evolution of that quantity throughout the entire distance scale. This provides a unique opportunity to describe $\Delta\Gamma_1^{pn}$ within fundamental theory. This may require going to higher order in ChPT, and to lower Q^2 in the Operator Product Expansion of pQCD. In addition, it may be nessecary to employ lattice QCD to cover the intermediate distance scale and provide an overlap with both the higher and the lower Q^2 domains. These calculations must be confronted with precise measurement of resonance contributions to the spin integral.

Closing Remarks

As the last speaker of this workshop I have the honor, and the pleasant obligation and opportunity, to express the gratitude of the participants at this workshop to the organizing committee and its chairman, Dieter Drechsel, for an excellent scientific program, organized in a most friendly atmosphere, for the superb food, and for providing the opportunity for in-depths discussions with colleagues and friends. For this we say

D a n k e !

In Memoriam

One of our best, who is no longer is with us, Nimai Mukhopadhyay, a friend to many of us, and a champion of nucleon structure studies and of baryon resonances, was sorely missed at this workshop. Nimai made many important contributions to this field, and organized workshops like this one. We can honor his name by making baryon resonances an even more visible part of nucleon structure studies in the years to come.

References

1. L.C. Smith, *these proceedings.*
2. L. Tiator, *et al., Nucl. Phys.* A **689**, 205-214 (2001).
3. D. Leinweber, T. Dreper, and R. Wolyshyn, *Phys. Rev.* D **48**, 2230 (1993).
4. Z.P. Li, V. Burkert, and Zh. Li, *Phys. Rev.* D **46**, 70 (1992).
5. F. Cano and P. Gonzales, *Phys. Lett.* B **431**, 270-276 (1998).
6. S. Krewald *et al.*, Excited Nucleons and Hadronic Structure, eds. V. Burkert, L. Elouadrhiri, J. Kelly, and R. Minehart, World Scientific, Singapoe (2001).
7. H.P. Morsch, *these proceedings.*
8. M.M. Giannini and E. Santopinto, *Few Body Syst. Suppl.* **11**, 37-42 (1999).
9. S. Sasaki, Excited Nucleons and Hadronic Structure, eds. V. Burkert, L. Elouadrhiri, J. Kelly, and R. Minehart, World Scientific, Singapoe (2001).
10. M. Ripani, *these proceedings.*
11. K. Joo, talk at this conference.
12. N. Isgur, G. Karl, *Phys. Rev.* D **23**, 817 (1981).
13. S. Capstick and W. Roberts, *Phys. Rev.* D **49**, 4570 (1994).
14. K. Abe *et al., Phys. Rev.* D **58**, 2003 (1998).
15. M. Kirchbach, *Mod. Phys. Lett.* A **12**, 3177 (1997).
16. A. D'Angelo, M. Ripani, S. Nakano, U.Thoma, and B.S. Zou, *these proceedings.*
17. C. Bennhold, *these proceedings.*
18. Y. Oh, *these proceedings.*
19. B. Nefkens, *these proceedings.*
20. L.S. Kisslinger and Z.P. Li, *Phys. Rev.* D **51**, 5986-5989 (1995).
21. P. Page, Excited Nucleons and Hadronic Structure, eds. V. Burkert, L. Elouadrhiri, J. Kelly, and R. Minehart, World Scientific, Singapoe (2001).
22. E. Golowich, E. Haqq, and G. Karl, *Phys. Rev.* D **28**, 160 (1983).
23. B.S. Zou, *these proceedings.*
24. V. Burkert and Zh. Li, *Phys. Rev.* D **47**, 46 (1993).
25. V. Burkert and B. Ioffe, *Phys. Lett.* B **296**, 223 (1992); *J. Exp. Theo. Phys.* **78**, 619 (1994).
26. R. Minehart, *these proceedings.*
27. J.P. Chen, *these proceedings.*
28. X. Ji, C.W. Kao, and J. Osborne, *Phys. Lett.* B **472**. 1-4 (2000).
29. V. Burkert, *Phys. Rev.* D **63**, 97904 (2001).
30. J.D. Bjorken, *Phys. Rev.* **179**, 1547 (1969).

Bing-Song Zou

Volker Burkert

MULTIPOLE ANALYSIS OF A BENCHMARK DATA SET FOR PION PHOTOPRODUCTION*

R.A. ARNDT[1], I. AZNAURYAN[2], R.M. DAVIDSON[3], D. DRECHSEL[4], O. HANSTEIN[4], S.S. KAMALOV[5], A.S. OMELAENKO[6], I. STRAKOVSKY[1], L. TIATOR[4], R.L. WORKMAN[1], S.N. YANG[7]

[1] *Department of Physics, The George Washington University, Washington, D.C. 20052, U.S.A*

[2] *Yerevan Physics Institute, Alikhanian Brothers St. 2, Yerevan, 375036 Armenia*

[3] *Department of Physics, Applied Physics and Astronomy, Rensselaer Polytechnic Institute, Troy, NY 12180, U.S.A*

[4] *Institut für Kernphysik, Universität Mainz, 55099 Mainz, Germany*

[5] *Laboratory of Theoretical Physics, JINR, 141980 Dubna, Russia*

[6] *National Science Center, Kharkov Institute of Physics and Technology, Akademicheskaya St., 1, Kharkov, 61108, Ukraine*

[7] *Department of Physics, National Taiwan University, Taipei, Taiwan*

We have fitted low- and medium-energy benchmark datasets employing methods used in the MAID/SAID and dynamical model analyses. Independent fits from the Mainz, RPI, Yerevan, and Kharkov groups have also been performed over the low-energy region. Results for the multipole amplitudes are compared in order to gauge the model-dependence of such fits, given identical data and a single method for error handling.

1 Overview

The following report summarizes results from a series of fits to selected datasets for pion photoproduction, a project initiated by the Partial-Wave Analysis Working Group of BRAG (the Baryon Resonance Analysis Group). The goal of this work was an evaluation of the model-dependence inherent in multipole analyses of photoproduction data. In the past, groups have constructed their own databases, and the resulting differences have been shown to significantly effect some multipoles, in particular the E2/M1 ratio for the $\Delta(1232)$. The handling of systematic errors has also differed, the common choices being to either ignore them entirely, to combine them in quadrature with statistical errors, or to use them in 'floating' the angular distributions.

The construction of low-energy (180-450 MeV) and medium-energy (180-1200 MeV) datasets was carried out mainly for practical reasons. Many groups have studied the region below 450 MeV, which is dominated by the $\Delta(1232)$.

*SUMMARY OF THE PARTIAL WAVE ANALYSIS GROUP OF BRAG

Only a few groups have extended their analyses over the full resonance region. As the recent MAID and dynamical model fits (see the first contribution) extend to 1 GeV, an upper limit of 1.2 GeV was chosen to include as many independent fits as possible.

The fitted data are listed on the BRAG website[1] and include differential cross section, target asymmetry (T), and photon asymmetry (Σ) data from proton targets only. The low- and medium-energy datasets were constructed to contain mainly recent measurements, with the goal of minimizing redundancies and simplifying the fitting procedure. In order to further simplify the exercise, systematic errors were not included in the fits. This should be kept in mind when χ^2 values are quoted.

Below we have compiled the reports of individual groups, giving details of the methods used, comments on the fit quality, and ways these results could be interpreted. Finally, in the last section, we summarize the findings of this study and suggest ways it could be improved or extended.

2 Multipole Analysis with MAID and a Dynamical Model

[S.S. Kamalov, D. Drechsel, L. Tiator, and S.N. Yang]

During the last few years we have developed and extended two models for the analysis of pion photo and electroproduction, the Dynamical Model [2] (hereafter called Dubna-Mainz-Taipei (DMT) model) and the Unitary Isobar Model [3] (hereafter called MAID). The final aim of such an analysis is to shed more light on the dynamics involved in nucleon resonance excitations and to extract N^* resonance properties in an unambiguous way. For this purpose as a testing ground we will use benchmark data bases recently created and distributed among different theoretical groups.

The crucial point in a study of N^* resonance properties is the separation of the total amplitude (in partial channel $\alpha = \{l, j\}$)

$$t_{\gamma\pi}^{\alpha} = t_{\gamma\pi}^{B,\alpha} + t_{\gamma\pi}^{R,\alpha} \tag{1}$$

in background $t_{\gamma\pi}^{B,\alpha}$ and resonance $t_{\gamma\pi}^{R,\alpha}$ contributions. In different theoretical approaches this procedure is different, and consequently this could lead to different treatment of the dynamics of N^* resonance excitation. As an example we will consider the two different models: DMT and MAID.

In accordance with Ref.[2], in the DMT model the $t_{\gamma\pi}^{B,\alpha}$ amplitude is defined as

$$t_{\gamma\pi}^{B,\alpha}(DMT) = e^{i\delta_\alpha}\cos\delta_\alpha \left[v_{\gamma\pi}^{B,\alpha} + P\int_0^\infty dq' \frac{q'^2\, R_{\pi N}^{(\alpha)}(q_E, q')\, v_{\gamma\pi}^{B,\alpha}(q')}{E - E_{\pi N}(q')} \right], \quad (2)$$

where $\delta_\alpha(q_E)$ and $R_{\pi N}^{(\alpha)}$ are the πN scattering phase shift and full πN scattering reaction matrix, in channel α, respectively, q_E is the pion on-shell momentum. The pion photoproduction potential $v_{\gamma\pi}^{B,\alpha}$ is constructed in the same way as in Ref.[3] and contains contributions from the Born terms with an energy dependent mixing of pseudovector-pseudoscalar (PV-PS) πNN coupling and t-channel vector meson exchanges. In the DMT model $v_{\gamma\pi}^{B,\alpha}$ depends on 7 parameters: The PV-PS mixing parameter Λ_m (see Eq.(12) of Ref.[3]), 4 coupling constants and 2 cut-off parameters for the vector mesons exchange contributions.

In the extended version of MAID, the s, p, d and f waves of the background amplitudes $t_{\gamma\pi}^{B,\alpha}$ are defined in accordance with the K-matrix approximation

$$t_{\gamma\pi}^{B,\alpha}(\mathrm{MAID}) = \exp(i\delta_\alpha)\,\cos\delta_\alpha v_{\gamma\pi}^{B,\alpha}(q, W, Q^2), \quad (3)$$

where $W \equiv E$ is the total πN c.m. energy and $Q^2 = -k^2 > 0$ is the square of the virtual photon 4-momentum. Note that in actual calculations, in order to take account of inelastic effects, the factor $\exp(i\delta_\alpha)\cos\delta_\alpha$ in Eqs.(2-3) is replaced by $\frac{1}{2}[\eta_\alpha \exp(2i\delta_\alpha) + 1]$, where both the πN phase shifts δ_α and inelasticity parameters η_α are taken from the analysis of the SAID group[4].

From Eqs. (2) and (3), one finds that the difference between the background terms of MAID and of the DMT model is that pion off-shell rescattering contributions (principal value integral) are not included in the background of MAID. From our previous studies of the p-wave multipoles in the (3,3) channel [2] it follows that they are effectively included in the resonance sector leading to the dressing of the $\gamma N\Delta$ vertex. However, in the case of s waves the DMT results show that off-shell rescattering contributions are very important for the E_{0+} multipole in the $\pi^0 p$ channel. In this case they have to be taken into account explicitly. Therefore, in the extended version of MAID we have introduced a new phenomenological term in order to improve the description of the π^0 photoproduction at low energies,

$$E_{corr}(MAID) = \frac{\Delta E}{(1 + B^2 q_E^2)^2} F_D(Q^2), \quad (4)$$

where F_D is the standard nucleon dipole form factor, $B = 0.71$ fm and ΔE is a free parameter which can be fixed by fitting the low energy π^0 photoproduction data. Thus the background contribution in MAID finally depends on

8 parameters. Below $\pi^+ n$ threshold for both models we also take into account the cusp effect due to unitarity, as it was described in Ref.[5], i.e.

$$E_{cusp} = -a_{\pi N}\,\omega_c\,\mathrm{Re}E_{0+}^{\gamma\pi^+}\sqrt{1-\frac{\omega^2}{\omega_c^2}}\,, \tag{5}$$

where ω and $\omega_c = 140$ MeV are the π^+ c.m. energies corresponding to $W = E_p + E_\gamma$ and $W_c = m_n + m_{\pi^+}$, respectively, and $a_{\pi N} = 0.124/m_{\pi^+}$ is the pion charge exchange amplitude.

For the resonance contributions, following Ref.[3], in both models the Breit-Wigner form is assumed, i.e.

$$t_{\gamma\pi}^{R,\alpha}(W,Q^2) = \bar{\mathcal{A}}_\alpha^R(Q^2)\,\frac{f_{\gamma R}(W)\Gamma_R\,M_R\,f_{\pi R}(W)}{M_R^2 - W^2 - iM_R\Gamma_R}\,e^{i\phi_R}\,, \tag{6}$$

where $f_{\pi R}$ is the usual Breit-Wigner factor describing the decay of a resonance R with total width $\Gamma_R(W)$ and physical mass M_R. The phase $\phi_R(W)$ in Eq. (6) is introduced to adjust the phase of the total multipole to equal the corresponding πN phase shift δ_α.

The main subject of our study in the resonance sector is the determination of the strengths of the electromagnetic transitions described by the amplitudes $\bar{\mathcal{A}}_\alpha^R(Q^2)$. In general, they are considered as free parameters which have to be extracted from the analysis of the experimental data. In our two models we have included contributions from the 8 most important resonances, listed in the Table 1. The total number of $\bar{\mathcal{A}}_\alpha^R$ amplitudes is 12 which can be also expressed in terms of the 12 standard helicity elements $A_{1/2}$ and $A_{3/2}$. Thus, to analyze experimental data we have a total of 19 parameters in DMT and 20 parameters in MAID. The final results obtained after the fitting of the high-energy (HE) benchmark data base with 3270 data points in the photon energy range $180 < E_\gamma < 1200$ MeV are given in Table 1.

For the analysis of the low-energy (LE) data base with 1287 data points in the photon energy range $180 < E_\gamma < 450$ MeV in the DMT model we used only 4 parameters: The PV-PS mixing parameter and 3 parameters for the $P_{33}(1232)$ and $P_{11}(1440)$ resonances. In MAID we have one more parameter due to the low energy correction given by Eq. (4). The final results for the helicity elements and the E2/M1 ratio (REM) are given in Table 2. In Table 3 we summarize our results and show the χ^2 obtained for different channels and different observables after fitting the LE and HE data bases. Note that the largest χ^2 in the LE fit we get for differential cross sections and target asymmetries in $p(\gamma,\pi^0)p$. Similar results were obtained practically in all other analyses. A detailed comparison with the results of different theoretical groups is given on the website

N^*		MAID current	MAID HE fit	DMT HE fit	PDG2000
$P_{33}(1232)$	$A_{1/2}$	-138	-143	—	-135± 6
	$A_{3/2}$	-256	-264	—	-255± 8
$P_{11}(1440)$	$A_{1/2}$	-71	-81	-77	-65± 4
$D_{13}(1520)$	$A_{1/2}$	-17	-6	-7	-24± 9
	$A_{3/2}$	164	160	165	166± 5
$S_{11}(1535)$	$A_{1/2}$	67	81	102	90± 30
$S_{31}(1620)$	$A_{1/2}$	0	86	37	27± 11
$S_{11}(1650)$	$A_{1/2}$	39	32	34	53± 16
$F_{15}(1680)$	$A_{1/2}$	-10	5	10	-15± 6
	$A_{3/2}$	138	137	132	133 ± 12
$D_{33}(1700)$	$A_{1/2}$	86	119	107	104± 15
	$A_{3/2}$	85	82	74	85± 22
PV-PS mixing:	Λ_m	450	406	302	
	ΔE	2.01	1.73	—	
	χ^2/d.o.f.	11.5	6.10	6.10	

Table 1. Proton helicity amplitudes (in $10^{-3}\, GeV^{-1/2}$), values of the PV-PS mixing parameter Λ_m (in MeV) and low-energy correction parameter ΔE (in $10^{-3}/m_{\pi^+}$) obtained after the high-energy (HE) fit.

N^*		MAID current	MAID LE fit	DMT LE fit	PDG2000
$P_{33}(1232)$	$A_{1/2}$	-138	-142	—	-135± 6
	$A_{3/2}$	-256	-265	—	-255± 8
$P_{11}(1440)$	$A_{1/2}$	-71	-81	-93	-65± 4
	REM(%)	-2.2	-1.9	-2.1	-2.5± 0.5
	χ^2/d.o.f.	4.76	4.56	3.59	

Table 2. Proton helicity elements (in $10^{-3}\, GeV^{-1/2}$) and REM=E2/M1 ratio (in %) obtained from the LE fit.

http://gwdac.phys.gwu.edu/analysis/pr_benchmark.html. Below, in Fig. 1 we show only one interesting example, the E_{0+} multipole in the channel with total isospin 1/2. In this channel contributions from the $S_{11}(1535)$ and $S_{11}(1620)$ resonances are very important. At $E_\gamma > 750$ MeV our values for the real part of the ${}_pE_{0+}^{1/2}$ amplitude are mostly negative and lower than the

Observables	LE N	LE MAID	LE DMT	HE N	HE MAID	HE DMT
$\frac{d\sigma}{d\Omega}(\gamma,\pi^+)$	317	4.68	3.32	871	6.36	5.95
$\frac{d\sigma}{d\Omega}(\gamma,\pi^0)$	354	7.22	5.74	859	6.87	5.85
$\Sigma(\gamma,\pi^+)$	245	2.79	2.57	546	4.57	6.49
$\Sigma(\gamma,\pi^0)$	192	2.22	1.58	488	7.65	7.65
$T(\gamma,\pi^+)$	107	3.28	2.94	265	3.75	4.17
$T(\gamma,\pi^0)$	72	5.18	4.84	241	5.31	5.65
Total	1287	4.56	3.64	3270	6.10	6.10

Table 3. χ^2/N for the cross sections ($\frac{d\sigma}{d\Omega}$), photon (Σ) and target (T) asymmetries in (γ,π^+) and (γ,π^0) channels obtained after LE and HE fit. N is the number of data points.

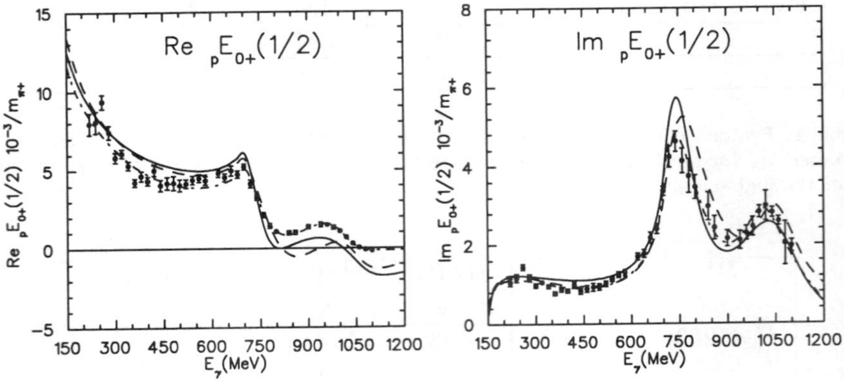

Figure 1. $_pE_{0+}^{1/2}$ multipole obtained after the HE fit using MAID (solid curves) and DMT (dashed curves). The dash-dotted curves and data points are the results of the global and single-energy fits obtained by the SAID group.

results of the SAID multipole analysis. The only possibility to remove such a discrepancy in our two models would be to introduce a third S_{11} resonance. Another interesting result is related to the imaginary part of the $_pE_{0+}^{1/2}$ amplitude and, consequently, to the value of the helicity elements given in Table 1. Within the DMT model for the $S_{11}(1535)$ we obtain $A_{1/2} = 102$ for a total width of 120 MeV, which is more consistent with the results obtained in η photoproduction, than with previous pion photoproduction results obtained by the SAID and MAID groups.

3 Multipole Analysis of Pion Photoproduction with Constraints from Fixed-t Dispersion Relations and Unitarity

[O. Hanstein, D. Drechsel and L. Tiator]

3.1 Outline of the Analysis

The method presented in Ref. [6] has been used to analyze the benchmark data set.

The starting point of our analysis is the fixed-t dispersion relation for the invariant amplitudes[7],

$$\mathrm{Re}A_k^I(s,t) = A_k^{I,\mathrm{pole}}(s,t) + \frac{1}{\pi}\mathcal{P}\int_{s_{\mathrm{thr}}}^{\infty} ds' \left(\frac{1}{s'-s} + \frac{\varepsilon^I \xi_k}{s'-u}\right) \mathrm{Im}A_k^I(s',t), \tag{7}$$

with the Mandelstam variables s, u and t, the isospin index $I = 0, \pm$, $s_{\mathrm{thr}} = (m_N + m_\pi)^2$, $\varepsilon^I = \pm 1$ and $\xi_k = \pm 1$. The pole terms $A_k^{I,\mathrm{pole}}(s,t)$ are obtained by evaluating the Born approximation in pseudoscalar coupling.

The multipole projection of the dispersion relations (7) leads to a system of coupled integral equations of the form

$$\mathrm{Re}\mathcal{M}_l^I(W) = \mathcal{M}_l^{I,\mathrm{pole}}(W) + \frac{1}{\pi}\mathcal{P}\int_{W_{\mathrm{thr}}}^{\infty} dW' \sum_{l'=0}^{\infty} \mathcal{K}_{ll'}^I(W,W')\mathrm{Im}\mathcal{M}_l^I(W'), \tag{8}$$

where $\mathcal{M}_l^I(W)$ denotes any of the multipoles $E_{l\pm}^I$ or $M_{l\pm}^I$. The integral kernels $\mathcal{K}_{ll'}^I(W,W')$ are regular kinematical functions except for the diagonal kernels $\mathcal{K}_{ll}^I$, which contain a term $\propto 1/(W-W')$.

After appropriate simplifications, namely neglect of weakly coupling integral kernels and restriction to the partial waves of low angular momentum, the integral equations (8) can be solved in different ways. We based our analysis on the method of Omnès[8] because this method leads to a parametrization of the multipoles in a natural way and thus allows for analyzing experimental data. As an additional input, the solution of the integral equations requires the phases $\phi_l^I(W)$ of the multipoles on the whole range of integration. According to the Fermi-Watson theorem, these phases are equal to the corresponding πN scattering phase shifts below 2π threshold, $\phi_l^I(W) = \delta_l^I(W)$. Above 2π threshold, we use the ansatzes

$$\phi_l^I(W) = \arctan\left(\frac{1-\eta_l^I(W)\cos 2\delta_l^I(W)}{\eta_l^I(W)\sin 2\delta_l^I(W)}\right) \tag{9}$$

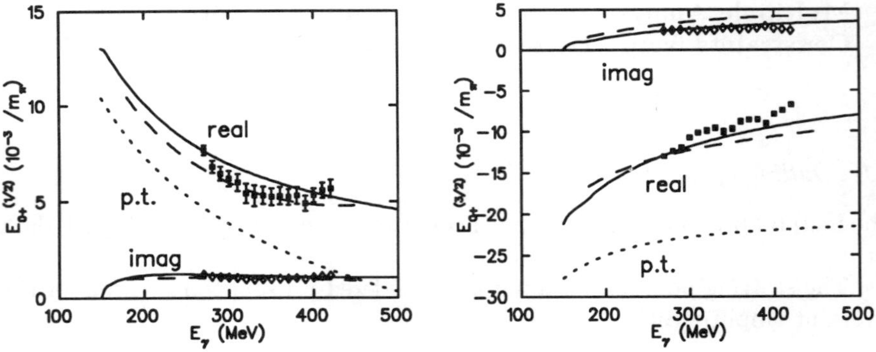

Figure 2. The real and imaginary parts and the pole term (p.t.) contributions of the amplitudes $E_{0+}(1/2)$ and $E_{0+}(3/2)$. The results of our fit (solid lines) are compared with those from Ref. [10] (dashed lines). The data points are the result of our energy independent fit.

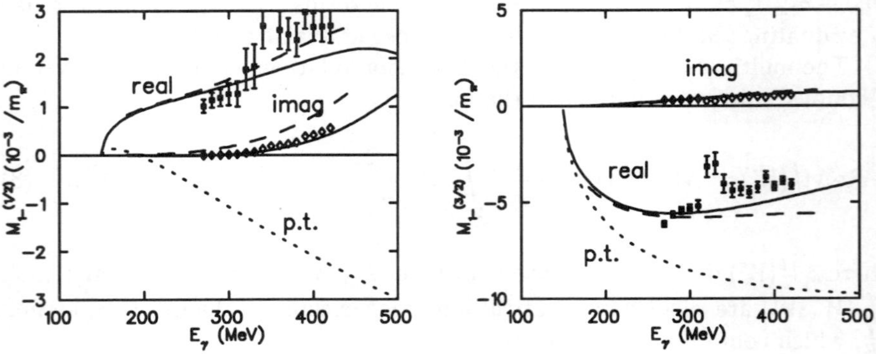

Figure 3. The real and imaginary parts and the pole term (p.t.) contributions of the amplitudes $M_{1-}(1/2)$ and $M_{1-}(3/2)$. Symbols as in Fig. 2.

for the S_{11}, P_{11}, P_{33}, and D_{13} waves, and

$$\phi_l^I(W) = \arctan\left(\frac{\eta_l^I(W)\sin 2\delta_l^I(W)}{1+\eta_l^I(W)\cos 2\delta_l^I(W)}\right) \tag{10}$$

for the S_{31}, P_{13}, and P_{31} waves. These ansatzes are each based on unitarity[9] and an additional assumption. They contain the scattering phases shifts $\delta_l^I(W)$ and the inelasticity parameters $\eta_l^I(W)$ of πN scattering. Both ansatzes give $\phi_l^I(W) = \delta_l^I(W)$ below 2π threshold. Since partial wave analyses of πN

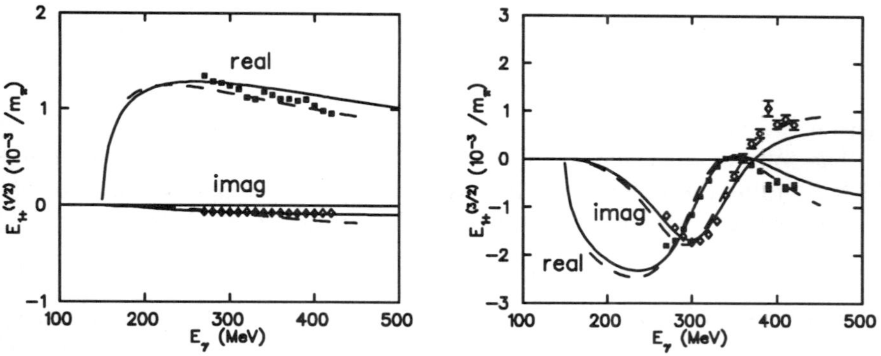

Figure 4. The real and imaginary parts of the amplitudes $E_{1+}(1/2)$ and $E_{1+}(3/2)$. Symbols as in Fig. 2.

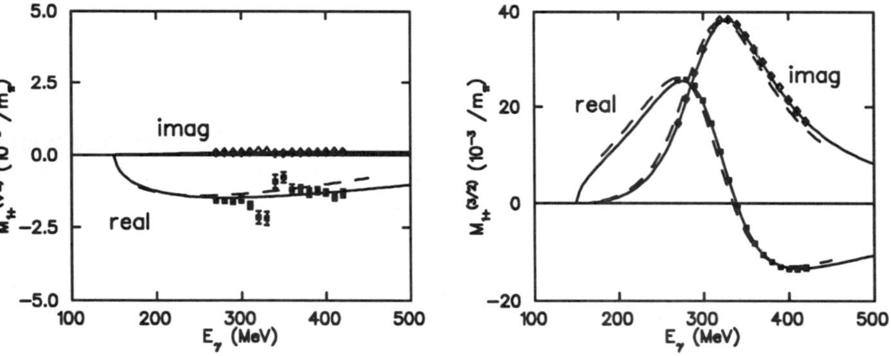

Figure 5. The real and imaginary parts of the amplitudes $M_{1+}(1/2)$ and $M_{1+}(3/2)$. Symbols as in Fig. 2.

scattering are only available up to about $W = 2$ GeV, we cut the integrals off at this energy to avoid the integration of unknown functions. Instead we represented the contributions of the imaginary parts at higher energies by t-channel exchange of vector mesons. The integral equations then take the form

$$\mathrm{M}_l^I(W) = \mathrm{M}_l^{I,\mathrm{pole}}(W) + \frac{1}{\pi} \int\limits_{W_{\mathrm{thr}}}^{\Lambda} \frac{h_l^{I*}(W')\mathrm{M}_l^I(W')dW'}{W' - W - i\varepsilon} \tag{11}$$

$$+\frac{1}{\pi}\sum_{l',I'}\int_{W_{\rm thr}}^{\Lambda} \mathrm{K}_{ll'}^{II'}(W,W')h_{l'}^{I'*}(W')\mathrm{M}_{l'}^{I'}(W')dW' + \mathrm{M}_l^{I,\mathrm{V}}(W),$$

where $\Lambda = 2$ GeV.

The solutions relevant for our case are the sum of a particular solution, which contains the inhomogeneities as driving terms, and a solution to the homogeneous equation multiplied by an arbitrary real coefficients c_l^I:

$$\mathcal{M}_l^I(W) = \mathcal{M}_l^{I,\mathrm{part}}(W) + c_l^I \mathcal{M}_l^{I,\mathrm{hom}}(W). \tag{12}$$

The coefficients c_l^I are the fitting parameters in our procedure. In addition, we varied the coupling constants of the ω and the ρ meson. Since the addition of the homogeneous solution is only allowed for multipoles for which the phase is different from zero in the asymptotic limit, we decided to fit the parameters c_l^I only for the following multipoles: $E_{0+}(0)$, $E_{0+}(1/2)$, $M_{1-}(0)$, $M_{1-}(1/2)$, $E_{1+}(3/2)$ and $M_{1+}(3/2)$. So we end up with a 10 parameter fit, the results of which are discussed in the next section.

3.2 Results

The fit to the benchmark data set leads to an overall χ^2 of 3.7 per data point. This high value is mainly due to the differential cross sections and target asymmetries of π^0 production (see Table 4).

Table 4. The χ^2 per data point in our fit for the individual observables. It is seen that, except for the beam asymmetry Σ, the description of π^0 production data is much poorer than that of π^+ production data.

	$\pi^0 p$	$\pi^+ n$
$d\sigma/d\Omega$	7.4	2.8
Σ	1.2	1.6
T	5.2	2.9

The magnitudes of the vector meson coupling constants as determined by our fit differ somewhat from the values quoted in the literature. This can be attributed to the fact that the dispersion integrals up to 2 GeV already contain a certain fraction of the vector meson contributions (see Table 5).

Our results for the s- and p-wave multipoles are shown in Figs. 2 to 5. As the procedure presented by Inna Aznauryan[10] at this workshop is closely related to ours, her results are also shown for comparison. The real

Table 5. Comparison between the vector meson coupling constants resulting from our fit and values quoted in the literature.

	g_ρ^V	g_ρ^T	g_ω^V	g_ω^T
this work	4.85	15.69	6.78	-1.67
Ref. [11]	3.24	19.81	15.85	0
Ref. [12]	1.99	12.42	20.86	-3.41

parts of the E_{0+} and M_{1-} multipoles differ significantly from the pole term contributions. In our approach, this difference, which in other approaches has to be provided by mechanisms like rescattering, is due to contributions from dispersion integrals.

4 Analysis of Low-Energy Benchmark Data Using Fixed-t Dispersion Relations

[I.G.Aznauryan]

In this analysis the low-energy benchmark data are analyzed using fixed-t dispersion relations within the approach which is close to the approach developed in Refs. [13,14,15,16]. The real parts of the amplitudes are constructed through real parts of invariant amplitudes $A_i^{(\pm,0)}(s,t)$, $i = 1,4$ (Ref. [17]), which are obtained using fixed-t dispersion relations:

$$\mathrm{Re}A_i^{(\pm,0)}(s,t) = A_i^{Pole}(s,t) + \frac{P}{\pi}\int_{s_{thr}}^{s_{max}} \mathrm{Im}A_i^{(\pm,0)}(s',t)\left(\frac{1}{s'-s} + \frac{\eta_i\eta^{(\pm,0)}}{s'-u}\right) ds', \tag{13}$$

with

$$A_i^{Pole}(s,t) = A_i^{Born}(s,t) + A_i^{\omega}(s,t) + A_i^{\rho}(s,t) \tag{14}$$

where $\eta^{(+,0)} = -\eta^{(-)} = 1$, $\eta_1 = \eta_2 = -\eta_3 = \eta_4 = 1$. $A_i^{Born}(s,t)$ are the contributions of the nucleon poles in the s- and u-channels and of the pion pole in the t-channel. In the dispersion relations[13], the dispersion integrals are taken over the resonance energy region up to $s_{max} = (2\ GeV)^2$, and it is supposed that the integrals over higher energies can be approximated by the t-channel ω and ρ-contributions: $A_i^{\omega}(s,t)$, $A_i^{\rho}(s,t)$. These contributions are taken in the form presented in Ref. [6]. The integrals over the resonance

$\mathrm{Im}M_{1+}^{3/2}$	$52.458 \pm 0.002\ 10^{-3} fm$
$\mathrm{Im}E_{1+}^{3/2}$	$-1.19 \pm 0.01\ 10^{-3} fm$
$E2/M1$	-0.023

Table 6. The amplitudes $M_{1+}^{3/2}, E_{1+}^{3/2}$ at the resonance position.

energy region are saturated by the resonances used in the VPI analysis of pion photoproduction data [4]. The coupling constants for all resonances, except $P_{33}(1232)$, are taken from this analysis. The resonance contributions are parametrized in the Breit-Wigner form according to Ref.[18].

The multipole amplitudes $M_{1+}^{3/2}$ and $E_{1+}^{3/2}$ corresponding to the $P_{33}(1232)$ resonance are parametrized using the approach developed in Refs. [9,19]. According to this approach the amplitudes $M_{1+}^{3/2}$, $E_{1+}^{3/2}$ are the solutions of the singular integral equations which follow from dispersion relations for these amplitudes, and have the form:

$$M(s) = M_{part}^{Born}(s) + M_{part}^{\omega}(s) + c_M M^{hom}(s), \tag{15}$$

where $M(s)$ denotes any of the amplitudes $M_{1+}^{3/2}$, $E_{1+}^{3/2}$. $M_{part}^{Born}(s)$ and $M_{part}^{\omega}(s)$ are the particular solutions of the integral equations generated by the Born and ω contributions. Particular solutions have definite magnitudes fixed by these contributions. $M^{hom}(s)$ is the solution of the homogeneous part of the integral equation; it has a certain energy dependence fixed by the integral equation and an arbitrary weight which was found by fitting the data.

In the $P_{33}(1232)$ resonance region we have taken into account also the contributions of the non-resonant multipoles $E_{0+}^{1/2}$, $E_{0+}^{(0)}$, $E_{0+}^{3/2}$, $M_{1-}^{1/2}$, $M_{1-}^{(0)}$ into $\mathrm{Im}A_i^{(\pm,0)}(s,t)$. These multipole amplitudes were found by calculating their real parts from the dispersion relations[13]; then the imaginary parts of the multipole amplitudes at $W < 1.3\ GeV$ were found using the Watson theorem. At higher energies a smooth cutoff of these contributions was assumed.

Our fitting parameters were:

(1) the constants c_M and c_E which correspond to the magnitudes of homogeneous solutions for $M_{1+}^{3/2}$, $E_{1+}^{3/2}$ in Eq. (15) at the resonance position;

(2) the coupling constants g_ω^V, g_ω^T, g_ρ^V, g_ρ^T which describe the ω and ρ contributions in Eq. (14).

(3) we have also included into the fitting procedure the coupling constants for the resonances $S_{11}(1535)$, $P_{11}(1430)$ and $D_{13}(1520)$, namely, the cou-

	Present analysis	Ref.[6]	Ref.[3]	Ref.[20]
g_ω^V	4	6.8	21	8-14
g_ω^T	-24	-1.7	-12	0-(-14)
g_ϱ^V	2.5	4.9	2	1.8-3.2
g_ϱ^T	21	16	13	8-21

Table 7. The ωNN and ϱNN coupling constants in comparison with results from other sources.

	$\sigma(\pi^+)$	$A(\pi^+)$	$T(\pi^+)$	$\sigma(\pi^0)$	$A(\pi^0)$	$T(\pi^0)$
$\chi 2/N$	922/317	440/253	245/107	2568/347	276/192	394/71

Table 8. The values of $\chi 2$.

pling constants for the multipoles $E_{0+}^{1/2}$, $E_{0+}^{(0)}$ of $S_{11}(1535)$, for the multipoles $M_{1-}^{1/2}$, $M_{1-}^{(0)}$ of $P_{11}(1430)$, and for the multipoles $E_{2-}^{1/2}$, $E_{2-}^{(0)}$, $M_{2-}^{1/2}$, $M_{2-}^{(0)}$ of $D_{13}(1520)$. A small variation of these coupling constants around the values obtained in the analysis of Ref. [4] was allowed.

The obtained results are presented in Tables 6-8.

5 The Effective Lagrangian Analysis of the Benchmark Dataset

[R. M. Davidson]

5.1 The Model

Details of the effective Lagrangian approach(ELA) to pion photoproduction may be found in Ref.[21]. Here I briefly summarize the main features of the model. The effective Lagrangian consists of the pseudovector (PV) nucleon Born terms, t-channel ω and ρ exchange, and s- and u-channel $\Delta(1232)$ exchanges. At the tree-level, the amplitude is gauge invariant, Lorentz invariant, crossing symmetric, and satisfies the LET's for these reactions to order m_π/M. However, the tree-level amplitude violates unitarity. To unitarize this amplitude, the tree-level multipoles, $\mathcal{M}_l^T$, are projected out and unitarized via a K-matrix approach;

$$\mathcal{M}_l = \mathcal{M}_l^T \cos\delta_l e^{i\delta_l} , \tag{16}$$

where δ_l is the appropriate πN elastic scattering phase shifts. In practice, this is done only for the s- and p-wave multipoles. In order to keep multipoles of all l values, the unitarized multipoles are added to the tree-level CGLN $\mathcal{F}$'s[22] and the tree-level multipoles are subtracted. To give a simple example, if only E_{0+} is unitarized, then $\mathcal{F}_1$ in this model is

$$\mathcal{F}_1 = \mathcal{F}_1^T + E_{0+}^T(\cos\delta_0 e^{i\delta_0} - 1) , \tag{17}$$

where $\mathcal{F}_1^T$ is the tree-level approximation to $\mathcal{F}_1$.

After unitarization, one is now ready to fix the parameters of the model. The Δ mass, M_Δ, and width Γ are determined by a fit to the P_{33} phase shift and are not varied in the photoproduction fit. The pion and nucleon masses are fixed at 139.6 and 938.9 MeV, respectively. Although one could look at the sensitivity of the photoproduction data to the pion-nucleon coupling constant, I keep the PV coupling constant, f, fixed at a value of 1.0. The ρ and ω are taken to be degenerate with a mass of 770 MeV, and the $V\pi\gamma$ ($V = \rho$ or ω) coupling constants are taken from the known radiative decays $V \to \pi\gamma$. The Dirac and Pauli-like VNN coupling constants were allowed to vary in the fit. The remaining parameters of the model are related to the Δ interactions. Of most interest are the two $\gamma N\Delta$ coupling constants, g_1 and g_2, which are related to M1 and E2 by

$$M1 = \frac{e}{12M}\left(\frac{k_\Delta}{M_\Delta M}\right)^{1/2}\left\{g_1(3M_\Delta + M) - g_2\frac{M_\Delta}{2M}(M_\Delta - M)\right\} ,$$

$$E2 = -\frac{e}{6M}\frac{k_\Delta}{(M_\Delta + M)}\left(\frac{k_\Delta M_\Delta}{M}\right)^{1/2}\left\{g_1 - g_2\frac{M_\Delta}{2M}\right\} ,$$

where $k_\Delta = (M_\Delta^2 - M^2)/(2M_\Delta)$. The remaining parameters are the off-shell parameters, X,Y,Z, associated with the Δ transition vertices. For example, the $\pi N\Delta$ Lagrangian is of the form

$$\mathcal{L} \sim \bar{\Delta}^\mu(g_{\mu\nu} + a\gamma_\mu\gamma_\nu)N\partial^\nu\pi + \text{h.c.} , \tag{18}$$

where a depends linearly on Z. Thus, the off-shell parameters essentially control the relative strength of the $\gamma_\mu\gamma_\nu$ term compared to the $g_{\mu\nu}$ term. It should be noted that when calculating matrix elements involving the off-shell parameters the pole in the Δ propagator is canceled. Thus, the contributions involving the off-shell parameters appear as contact terms. As the off-shell parameters are fitted to the data, these contact terms can partially compensate for the lack of strong form factors which might arise, for example, from pionic dressing of the vertices.

5.2 *Results and Discussion*

The nine parameters of the model were fitted to the low-energy benchmark dataset consisting of 1287 data points. The total χ^2 was 5203 giving a χ^2/df of 4.07. The breakdown of the χ^2 according to observable is given in Table 9. It is seen that the highest χ^2/n are for $d\sigma/d\Omega(\pi^0)$ and $T(\pi^0)$, which was true for the other analyses also. The best agreement was with the photon asymmetry, Σ, for both π^0 and π^+.

As the χ^2 in this analysis is slightly larger than in the other analyses, it is useful to look at the χ^2 breakdown in energy bins. This is shown in Table 10, where $[x, y)$ and $[x, y]$ have their usual mathematical meaning. It is seen that the fit to the lowest energy bin is extremely poor. Most of the data in this energy interval are π^0 differential cross section data, and a comparison of the fit with these data shows that the fit has too large of a forward-backward asymmetry as compared to the data. At these energies, this asymmetry is determined by the interference between the E_{0+} and the p-wave multipoles. A comparison of this multipole obtained in this fit with the same multipole obtained from fits that reproduce these low-energy data shows a significant difference.

Table 9. χ^2 for each observable fitted.

OBS	n	χ^2	χ^2/n
$d\sigma/d\Omega(\pi^+)$	317	988.4	3.1
$d\sigma/d\Omega(\pi^0)$	354	2804.0	7.9
$\Sigma(\pi^+)$	245	388.0	1.6
$\Sigma(\pi^0)$	192	258.1	1.3
$T(\pi^+)$	107	274.9	2.6
$T(\pi^0)$	72	489.6	6.8

At the BRAG workshop, we were able to understand this difference. As L. Tiator pointed out, at low-energies for π^0 production, the coupling to the π^+n channel can be important and one must somehow account for processes like

$$\gamma p \to n\pi^+ \to p\pi^0 \, . \tag{19}$$

In dispersion relation models[6,23] and dynamical models[2], this channel coupling is dynamically taken into account. In partial wave analyses[4], the parametrization is flexible enough to account for this physics. In the isobar model[3], this important physics was included in a semi-phenomenological manner. In the

Table 10. χ^2 as a function of energy bin fitted.

Interval	n	χ^2	χ^2/n
[180, 210)	64	747.7	11.7
[210, 240)	137	767.6	5.6
[240, 270)	208	803.6	3.9
[270, 300)	181	578.3	3.2
[300, 330)	191	516.8	2.7
[330, 390)	157	598.6	3.8
[360, 390)	120	265.5	2.2
[390, 420)	132	307.6	2.3
[420, 450)	97	615.4	6.3

ELA, this channel coupling is only partially taken into account via the unitarization. However, dispersive corrections are not explicitly accounted for. It is implicitly assumed that the main effect of the dispersive corrections is to renormalize the parameters to their physical values. It is further assumed (or hoped) that any additional dispersive corrections can be accounted for by the contact terms coming from the off-shell parameters. Evidently, this is not the case over the entire fitted energy range. Thus, to improve the low-energy fit, without destroying the fit near the peak of the resonance, this additional physics would need to be added by hand to the ELA.

The parameters from the fit are shown in Table 11 along with representative parameters from Ref.[21]. The resulting resonance couplings are

$$\begin{aligned} M1 &= 286.2 \times 10^{-3}\text{GeV}^{-1/2} \\ E2 &= -7.21 \times 10^{-3}\text{GeV}^{-1/2} \\ \frac{E2}{M1} &= -2.55\% \,. \end{aligned} \tag{20}$$

The g_1 coupling, which is mostly responsible for the $M1$ strength, has not changed much compared to the earlier work. g_2, which determines the strength of $E2$ has changed within the error bar. As a rule of thumb, the smaller g_2, the larger in magnitude $E2$. The change in g_2, and hence $E2$, is probably due to the new high-precision Σ data [24,25]. The off-shell parameter Z has not changed much, but the other parameters have. This is partly due to the fact that the vector meson couplings were allowed to vary in this fit, but not in Ref.[21]. Since I did not do an error analysis for the BM fit, it is hard to say if the changes are significant. I do know that the vector meson couplings and off-shell parameters are highly correlated fit parameters. I should point

Table 11. Parameters obtained in this fit (B.M.) compared with those obtained in [21].

Par.	B.M.	DMW
g_1	5.06	5.01 ± 0.22
g_2	4.72	5.71 ± 0.43
Z	-0.25	-0.30 ± 0.12
Y	1.65	-0.38 ± 0.66
X	-5.74	1.94 ± 2.28
$g_{\omega 1}$	-0.09	8
$g_{\omega 2}$	4.46	-8
$g_{\rho 1}$	1.51	2.66
$g_{\rho 2}$	3.66	16.2

out that the vector mesons are put in as point particles, i.e., no form factors. Thus, when I fit the vector meson coupling 'constants', I would expect them to decrease compared to their on-shell values. However, the results for the ω seem unusual. The general conclusion at the BRAG workshop is that the Δ region is not a good place to fit the vector meson couplings.

From a glance through the various multipole solutions, it seems to me that the model dependence is under good control in the Δ region. One multipole that remains to be understood is the $M_{1-}^{1/2}$. Though all analyses pretty much agree on its numerical value, its physical interpretation is quite different in the various models. Is it a crossed-Δ effect or the tail of the Roper? In principle, the amplitude should be crossing symmetric, so if there is an s-channel Δ exchange, there must be a u-channel Δ exchange. In Ref.[21], it was found that the $M_{1-}^{1/2}$ multipole is largely insensitive to the off-shell parameters. Thus, these contributions cannot suppress the u-channel contribution in this multipole, and there is no room for a Roper contribution as large as in the isobar model. What is even more mysterious is that in Ref.[21] it was found that the Roper contribution enters with the opposite sign than the u-channel contribution (see Fig. 8 in [21]). Regarding the Roper contribution to this multipole, there are two clear possibilities. Either someone has made a mistake, or two different things are being compared. Certainly the latter is true to some extent. In Ref.[21], both the s- and the u-channel Roper exchanges were included, whereas in the isobar model[3] only the s-channel contribution was included. Thus, it is possible that the s- and u-channel contributions destructively interfere in this multipole to produce the small effect found in [21]. I do not know the solution to this puzzle, but if dispersion relations are telling us there is a large crossing contribution to this multipole from the $M_{1+}^{3/2}$, then we must necessarily investigate this problem within the framework of a crossing symmetric model.

6 Multipole Analysis of the Benchmark Set Covering the First Resonance Region

[A.S. Omelaenko]

6.1 Introduction

A great deal of experimental data on the photoproduction of single pions is stored in compilations. There are problems connected with the normalization of some systematic measurements of the differential cross section. As a result, in multipole analyses, one has to reject or renormalize up to 10% of experimental points. This can throw some doubt on the extraction of delicate values, such as the E2/M1 ratio for the $\Delta^+(1232)$. For this reason it it interesting to compare the analyses based on different approaches, using the same test database for the best investigated $\gamma p \to p\pi^0(n\pi^+)$ reaction, with a strong accent on the modern data. Here we report on a fit to the low-energy dataset based on (a) the resonance model with polynomial parametrization of the background and (b) energy-independent fitting.

6.2 Formalism in the First Resonance Region

Resonance Model. The description of the P_{33} πN scattering amplitude is simple if it is taken to be purely elastic in the first resonance region. Similar to the form used in Ref.[18], our multi-parameter model for single pion photoproduction is written as a sum of a background plus the $\Delta^+(1232)$ resonance contribution. The real part of the background is given by the electric Born approximation which is completed in s-, p- and d-waves by a cubic polynomial of the form

$$\mathrm{Re}M^I_{l\pm}(E_\gamma) = \sum_{i=1}^{4} \mathrm{Re}M^I_{l\pm}(E^{(i)}_\gamma) \prod_{j=1}^{4\;(j\neq i)} (E_\gamma - E^{(j)}_\gamma)/(E^{(i)}_\gamma - E^{(j)}_\gamma). \qquad (21)$$

Here, depending on the laboratory photon energy E_γ, multipoles $M^I_{l\pm} \equiv A^I_{l\pm}, B^I_{l\pm}$ are defined according to Ref.[18] with the isospin structure

$$A^{1/2} = 1/3A_{\pi^0 p} + \sqrt{2}/3A_{\pi^+ n},$$

$$A^{3/2} = A_{\pi^0 p} - \sqrt{1/2} A_{\pi^+ n}.$$

The index l is the orbital angular momentum, while + and − correspond to the total angular momentum $j = l \pm 1/2$. The imaginary parts of the background multipoles are calculated according to Watson's theorem, using phase shifts from πN elastic scattering analyses:

$$\mathrm{Im} M^I_{l\pm} = \mathrm{Re} M^I_{l\pm} \tan(\delta_{2I,2(l\pm)}). \tag{22}$$

The resonant multipole amplitudes $A^{3/2}_{1+}$ and $B^{3/2}_{1+}$ have the form

$$M^{3/2}_{1+}(E_\gamma) = M^{3/2,R}_{1+}(E_\gamma) + B_M(E_\gamma) \cos\delta_{33} e^{i\delta_{33}} , \tag{23}$$

with the first term describing excitation of the resonance

$$M^{3/2,R}_{1+}(E_\gamma) = C_M \sqrt{\frac{k_0 q_0}{kq}} \frac{W_0 \sqrt{\Gamma \Gamma_\gamma}}{W_0^2 - W^2 - iW_0\Gamma}, \tag{24}$$

with W and W_0 the total c.m. energy and its value at resonance, respectively, C_M being the resonance constant. The energy-dependent widths were parametrized by

$$\Gamma_\gamma(W) = \Gamma_0 \left(\frac{k}{k_0}\right)^2 \left(\frac{k_0^2 - X^2}{k^2 - X^2}\right)^2, \tag{25}$$

$$\Gamma(W) = \Gamma_0 \left(\frac{q}{q_0}\right)^2 \left(\frac{q_0^2 - X^2}{q^2 - X^2}\right)^2 , \tag{26}$$

where k and q are the c.m. momenta of the photon and pion, respectively, and k_0, q_0 are the corresponding values at W_0. The second term in Eq. (23) corresponds to Noelle's unitary treatment of the background[26]. In our approach, the resonance term has a simplified form, not containing the elastic background phase shift δ_B. However, this has practically no influence on the E2/M1 ratio, which is quite independent of δ_B[27]. The real background functions $B_A(E_\gamma)$ and $B_B(E_\gamma)$ are parametrized according to Eq. (21), each in terms of 4 knot values. The phase shift δ_{33} is calculated as the phase of the Breit-Wigner form in Eq. (24):

$$\tan\delta_{33} = \frac{W_0 \Gamma(W)}{W_0^2 - W^2} . \tag{27}$$

The resonance quantities C_A, C_B, W_0, Γ_0 were determined in the fit, along with the knot values of B_A and B_B and the real parts of the background multipoles: A_{0+}, A_{1+}, A_{1-}, A_{2-}, B_{1+}, B_{2-} with $I = 1/2$ and A_{0+}, A_{1-}

with $I = 3/2$. Inclusion of the d-wave multipoles is motivated by a possible influence of the second resonance region.

Energy-Independent Version. In order to fit narrow energy bins, we take the real parts of the above mentioned non-resonant multipoles as independent parameters. Eq. (22) is used to calculate the imaginary parts, taking into account the energy dependence of the phase shifts. To avoid calculational problems at the point where δ_{33} passes through $\pi/2$, the resonant multipoles have been parametrized as

$$A_{1+}^{(3/2)} = A\exp(i\delta_{33}), \tag{28}$$

$$B_{1+}^{(3/2)} = B\exp(i\delta_{33}), \tag{29}$$

with A and B to be determined for each energy bin along with the real parts of all background multipoles.

6.3 Results from the Fits

Resonance Model. The whole energy interval was split into 3 subintervals, with knot values corresponding to the central values for energy-independent fits.

Fit 1	$180 \leq E_\gamma \leq 350$ MeV	$E_\gamma^{(1,2,3,4)}$=200,250,300,350 MeV
Fit 2	$250 \leq E_\gamma \leq 400$ MeV	$E_\gamma^{(1,2,3,4)}$=250,300,350,400 MeV
Fit 3	$300 \leq E_\gamma \leq 450$ MeV	$E_\gamma^{(1,2,3,4)}$=300,350,400,450 MeV

Fit	Data	χ^2	N	χ^2/N
Fit 1	Total	2026.5	930	2.18
	$\gamma p \to p\pi^0$	1332.4	460	2.90
	$\gamma p \to n\pi^+$	694.1	470	1.48
Fit 2	Total	1759.2	871	2.02
	$\gamma p \to p\pi^0$	1115.7	388	2.88
	$\gamma p \to n\pi^+$	643.5	483	1.33
Fit 3	Total	1259.7	697	1.81
	$\gamma p \to p\pi^0$	821.6	306	2.69
	$\gamma p \to n\pi^+$	438.1	391	1.12

Table 12. Fitting in the resonance model framework.

Reaction	Observable	Label	N	χ^2/N
$\gamma p \to p\pi^0$	$d\sigma/d\Omega$	FU96MA	45	2.2
	$d\sigma/d\Omega$	HA97MA	52	10.5
	$d\sigma/d\Omega$	BE97MA	77	1.4
	Σ	BL92LE	16	3.3
	Σ	BE97MA	77	0.8
	T	BO98BO	28	3.1
	T	BE83KH	26	3.6
$\gamma p \to n\pi^+$	$d\sigma/d\Omega$	FI72BO	32	0.9
	$d\sigma/d\Omega$	BR00MA	39	1.9
	$d\sigma/d\Omega$	BE00MA	140	1.7
	Σ	BL00LE	57	1.7
	Σ	BE00MA	140	0.9
	T	DU96BO	75	1.1

Table 13. Fits with the resonance model in the 250-400 MeV interval. See the database summary webpage to match abbreviations with full references.

Interval (MeV)	Mass (MeV)	Width (MeV)	E2/M1 (%)
180-350	1232.6 ± 1.0	121.1 ± 6.0	-1.83 ± 0.31
250-400	1231.7 ± 0.7	118.4 ± 3.9	-2.34 ± 0.13
300-450	1230.6 ± 0.8	116.5 ± 3.9	-2.26 ± 0.14

Table 14. Resonance parameters for the $\Delta(1232)$ from different energy intervals.

For each fit a total of 44 free parameters were searched (the parameter X turned out to be difficult to determine, and was fixed at Walker's value of 185 MeV). The statistical characteristics for all fits are given in Table 12, separately for the neutral and charged particle production. In Table 13, the contribution to χ^2 from individual experiments is given, as calculated from Fit 2. Finally, in Table 14, values for the resonance mass and width are given along with the corresponding value for the E2/M1 ratio.

Energy-Independent Analysis. Results from the resonance model fits were used as starting values for 10 parameters, which were determined by χ^2 minimization from the data contained in each 10 MeV bin. For all formally successful fits, the corresponding central values for photon energy, number of points (N) and χ^2 per degree of freedom is given in Table 15.

E_γ, MeV	χ^2	N	χ^2/df	E_γ, MeV	χ^2	N	χ^2/df
210	10.1	18	1.27	330	100.6	51	2.45
220	18.6	43	0.56	340	20.0	49	0.51
230	23.8	22	1.99	350	118.3	74	1.85
240	52.9	58	1.10	360	6.3	32	0.29
260	112.3	71	1.84	370	28.9	42	0.90
270	33.3	36	1.28	380	93.1	46	2.59
280	171.5	69	2.91	390	16.7	42	0.52
290	72.9	51	1.78	400	86.8	56	1.89
300	81.3	80	1.16	410	16.7	34	0.70
310	121.8	59	2.49	420	67.2	50	1.68
320	96.8	69	1.64				

Table 15. Statistics of the energy-independent fits.

6.4 Concluding Remarks

In the low-energy region, dominated by the first resonance, we find a rather good determination of the main s- and p-wave partial-wave amplitudes for pion photoproduction (on proton targets), taking into account d-wave corrections. Differences between the Fits 1-3 on overlapping energy intervals can be considered as some measure of the model error. In our analysis, a suspicious deviation in the energy dependence of the $M_{1-}^{1/2}$ and $M_{1+}^{1/2}$ was evident over the 380-420 MeV interval. Encouraging was the rather stable determination of the mass and width of the $\Delta^+(1232)$ resonance and E2/M1 ratio within the framework of this model (Fits 2 and 3).

A comment of the chosen database is in order. Of the cross section data for neutral pion production, the set of HA97MA gives a χ^2 which is definitely too large. This is true in all of the fits. Problems seem concentrated mainly at the small and large angles. Unfortunately, the modern polarization data do not represent the complete set of single polarization observables. As a result, in some bins of the energy-independent fit, the experimental data were not sufficient for a successful minimization procedure, or yielded unreasonably low (less than 1) values for χ^2/df. Further examination has shown that the above mentioned deviations in the energy behavior of the d-wave multipoles are due to the lack of reliable data as well. Precise new experiments would be extremely desirable.

7 SAID Multipole Analysis of the Benchmark Dataset

[R.A. Arndt, I.I. Strakovsky, R.L. Workman]

SAID fits were performed on both the low- and medium-energy datasets, using a phenomenological method described in Ref.[4]. Multipoles were parametrized in the form

$$M = (A_{\text{phen}} + Born)\,(1 + iT_{\pi N}) + B_{\text{phen}} T_{\pi N} \tag{30}$$

where A_{phen} and B_{phen} were constructed from polynomials in energy having the correct threshold behavior. An additional overall phase proportional to $\text{Im}T_{\pi N} - T_{\pi N}^2$ was also allowed in order to allow deviations from the above form for waves strongly coupled to channels other than πN. This form clearly satisfies Watson's theorem, where valid, and allows a smooth departure as new channels become important. A real Born contribution is assumed for unsearched high partial-waves.

The overall fit to data is summarized in the Table below. While the fit to most quantities is about 2 per data point, there are clear exceptions. Our fit to the neutral-pion differential cross section and target polarization is significantly worse in the low-energy region. If χ^2 is calculated for the region between 450-1200 MeV, a fairly uniform fit emerges, with all data types fitted equally well.

Observable	Low-Energy Fit (180-450)MeV	Medium-Energy Fit (180-1200)MeV	Medium-Energy Fit (450-1200) MeV
$\sigma(\theta)_{\pi^0 p}$	4.69	3.14	1.80
$\Sigma(\theta)_{\pi^0 p}$	1.24	2.05	2.32
$T(\theta)_{\pi^0 p}$	4.49	2.84	1.89
$\sigma(\theta)_{\pi^+ n}$	2.12	2.55	2.08
$\Sigma(\theta)_{\pi^+ n}$	1.46	1.81	2.36
$T(\theta)_{\pi^+ n}$	2.92	2.25	1.83

Table 16. Comparison of the low- and medium-energy fits in terms of χ^2.

The poor fit to neutral-pion differential cross sections can be traced to the Haerter data from Mainz (Ph.D. Thesis, unpublished - see the database description on the BRAG webpage). This set contributed a χ^2/datum of 10, the remaining data again having an overall χ^2 per data point of about 2.

Why the fit to target polarization should be poor is harder to determine. In the low-energy region, the benchmark dataset has taken target-polarization data from Kharkov and Bonn. Of these, the fit to Kharkov data is particularly bad. Our fit to the Bonn data is reasonable apart from the lowest-energy (and perhaps also the highest-energy) set.

8 Conclusions and Future Projects

8.1 Database Issues

As this exercise was motivated by an earlier study of the E2/M1 ratio, and its database dependence, we should (at least) expect to verify the relative model-independence of this quantity (when a standard dataset is selected). This point was discussed by Davidson[28] in his contribution to NSTAR2001. From the combined set of benchmark fits, the E2/M1 ratio was found to be $-2.38 \pm 0.27\%$. Here the important quantity is the error, since a change in the fitted dataset could shift the central value.

Qualitative features of the fit to data were reasonably similar in the low-energy region. There was general agreement that the differential cross section and target polarization were badly fitted in the neutral-pion channel. In all cases, the thesis data of Haerter were identified as problematic for the differential cross section. [These data can be abandoned, as a new Mainz measurement[29] of the cross section and photon asymmetry for neutral-pion photoproduction has been analyzed and is nearing publication. This new set covers the full angular range and a wide range of energies.]

Unfortunately, such a simple solution was not evident for the target polarization. The fit to both the Bonn and Kharkov data is poor, with some sets suggesting a different shape. In this case, the general *disagreement* should motivate a re-measurement of this quantity, hopefully with better precision. A confirmation of the shape suggested by these data would be very difficult to accommodate.

8.2 Model Dependence

One ingredient common to all analyses is the Born contribution, usually augmented with vector-meson exchange diagrams. While it is tempting to determine an optimal set of vector meson couplings, over the fitted energy range, our study has shown this to be a very unstable procedure, when fits are restricted to the low-energy database.

The $M_{1-}^{1/2}$ multipole, and its low-energy behavior, provides an interesting example of the model dependence encountered when one tries to decompose a partial-wave into its underlying ingredients. Hanstein has noted that, in his dispersion integral, the coupling $M_{1+}^{3/2} \to M_{1-}^{1/2}$ exceeds $M_{1-}^{1/2} \to M_{1-}^{1/2}$. In the study of Davidson, this is accounted for by a u-channel Δ exchange, a contribution absent in many isobar-model approaches. As a result, the Roper resonance contribution depends strongly on the set of approximations defining a fit.

8.3 Further Comparisons

One unique feature of this study was the construction of a site[1] containing, in one place, both the database and the multipole fits. This facility was built into the SAID site[4], thus allowing a wide range of comparisons. Users can compare fits to observables (in both the benchmark and full database), the resulting multipoles, and search for regions of maximal agreement or disagreement.

8.4 Extensions of this Study

Having the present set of fits as a guide, it would be useful to repeat this study over a more carefully chosen database - containing, in some way, the effects of systematic errors. A second project of interest would be the extension to electroproduction, with the goal of understanding the model dependence seen in the extracted E_{1+}/M_{1+} and S_{1+}/M_{1+} ratios as a function of Q^2.

References

1. The BRAG website (http://cnr2.kent.edu/~manley/benchmark.html) contains a detailed list of included data and the resulting fits.
2. S.S. Kamalov and S.N. Yang, Phys. Rev. Lett. **83**, 4494 (1999); S.N. Yang, *J. Phys.* G **11**, L205 (1985).
3. D. Drechsel, O. Hanstein, S.S. Kamalov, and L. Tiator, *Nucl. Phys.* A **645**, 145 (1999).
4. R.A. Arndt, I.I. Strakovsky, and R.L. Workman, *Phys. Rev.* C **53**, 430 (1996); most recent results are available at http://gwdac.phys.gwu.edu.
5. J. M. Laget, *Phys. Rep.* **69**, 1 (1981).
6. O. Hanstein, D. Drechsel, and L. Tiator, *Nucl. Phys.* A **632**, 561 (1998).
7. J. S. Ball, *Phys. Rev.* **124**, 2014 (1961).
8. R. Omnès, *Nuovo Cimento* A **8**, 316 (1958).
9. D. Schwela and R. Weizel, *Z. Phys.* **221**, 71 (1969).
10. I. Aznauryan, see Section 4 of this contribution.
11. R. Machleidt *et al., Phys. Repts.* **149**, 1 (1987).

12. P. Mergell, diploma thesis, Mainz (1995), and
P. Mergell, U.-G. Meissner, and D. Drechsel, NPA **596**, 367 (1996).
13. R.C.E. Devenish and D.H. Lith, *Phys. Rev.* D **5**, 47 (1972).
14. R.C.E. Devenish and D.H. Lith, *Nucl. Phys.* B **43**, 228 (1972).
15. R.C.E. Devenish and D.H. Lith, *Nucl. Phys.* B **93**, 109 (1975).
16. I.G. Aznauryan and S.G. Stepanyan, *Phys. Rev.* D **59**, 1(54009) (1999).
17. G.F. Chew *et al, Phys. Rev.* **106**, 1345 (1957).
18. R.L. Walker, *Phys. Rev.* **182**, 1729 (1969).
19. D. Schwela, H. Rolnik, R. Weizel, and W. Korth, *Z. Phys.* **202**, 452 (1967).
20. O. Dumbrajs, R. Koch, and P. Pilkuhn *et al., Nucl. Phys.* B **216**, 277 (1983).
21. R.M. Davidson, N.C. Mukhopadhyay, and R.S. Wittman, *Phys. Rev.* D **43**, 71 (1991).
22. G.F. Chew, M.L. Goldberger, F.E. Low, and Y. Nambu, *Phys. Rev.* **106**, 1345 (1957).
23. I.G. Aznauryan, *Phys. Rev.* D **57**, 2727 (1998).
24. G. Blanpied *et al., Phys. Rev. Lett.* **69**, 1880 (1992); *ibid.* **79**, 4337 (1997).
25. R. Beck *et al., Phys. Rev. Lett.* **78**, 606 (1997); R. Beck and H. P. Krahn, *ibid*, **79**, 4510 (1997).
26. P. Noelle, *Prog. Theor. Phys.* **60**, 778 (1978).
27. A.S. Omelaenko and P.V. Sorokin, *Sov. J. Nucl. Phys.* **38**, 398 (1983); P. Christillin and G. Dillon, *J. Phys.* G **15**, 967 (1989).
28. R.M. Davidson, *these proceedings.*
29. R. Leukel, Ph.D. thesis, Mainz 2001.

Author Index

Gerhard Höhler, Dan-Olof Riska

Jürg Jourdan, Anthony Thomas, Wally Melnitchouk, Jian-Ping Chen

NSTAR 2001

WORKSHOP ON THE PHYSICS OF EXCITED NUCLEONS

Wednesday, March 7

Welcome

9:00	Th. Walcher (Mainz)
9:10	J. Reiter (President, Johannes Gutenberg-Universität Mainz)
Chair	**T. Nakano**
9:30	S. Capstick (FSU) "Review of nucleon resonances in the quark model"
10:15	C. Papanicolas (Athens) "Search for quadrupole strength in the Delta region"
10:45	R. Gothe (Bonn) "Pion electroproduction at ELSA"
11:15	*-Coffee -*
Chair	**R. Van de Vyver**
11:45	H. Schmieden (Mainz) "Neutral pion electroproduction in the Delta region at MAMI"
12:15	L.C. Smith (UVA) "Pion electroproduction using CLAS"
12:45	A. Sarty (St. Mary's University, Halifax) "Recoil polarization measurements in π^0 electroproduction at the peak of the $\Delta(1232)$"
13:15	*- Lunch -*
Chair	**E. Oset**
15:00	T.R. Hemmert (TU München) "The Delta resonance in chiral effective theories"
15:30	A. Ramos (Barcelona) "Dynamical resonances with chiral Lagrangians"
16:00	J. Oller (Jülich) "Chiral Lagrangians with resonances"
16:30	*- Coffee -*
Chair	**S. Boffi**
17:00	T. Sato (Osaka) "Pion photo- and electroproduction in dynamical models"
17:30	S.-N. Yang (Taipei) "Dynamical and isobar models"
18:00	S. Krewald (Jülich) "The Jülich model"
18:30	*- End of Session -*

Thursday, March 8

Chair	**L. Elouadrhiri**
8:30	R. Workman (GWU) "Update on partial wave analysis"
9:00	C. Bennhold (GWU) "Higher resonances and coupled channels analysis"
9:30	A.W. Thomas (Adelaide) "Nucleon resonances in chiral quark models and on the lattice"
10:00	- *Coffee* -
Chair	**H. Haberzettl**
10:30	D.-O. Riska (Helsinki) "The role of the pion in nucleon resonance structure"
11:00	S. Simula (Roma) "Relativistic quark models"
11:30	S. Scherer (Mainz) "Generalized polarizabilities in a constituent quark model"
12:00	V. Lyubovitskij (Tübingen) "Nucleon properties in the perturbative chiral quark model"
12:30	- *Lunch* -

Parallel Session A "Analysis"

Chair	**L. Tiator**
14:00	R. Crawford (Glasgow) "The Glasgow pion photoproduction partial wave analysis, 2001"
14:20	I. Aznauryan (Yerevan) "Residues at the T-matrix poles of the $M_{1+}^{(3/2)}$ and $E_{1+}^{(3/2)}$ multipoles in the framework of dispersion relations"
14:40	W.-T. Chiang (NTU) "An isobar model for eta photo- and electroproduction on the nucleon"
15:00	A. Svarc (Zagreb) "Detailed analysis of eta production in proton-proton collisions"
15:20	V. Mokeev (Moscow) "Phenomenological analysis of on N* excitation in double-charged pion production"
15:40	- *Coffee* -
Chair	**U. Mosel**
16:00	G. Höhler (Karlsruhe) "Results on πN partial wave analysis and resonance parameters"
16:20	J. Nacher (Valencia) "The role of $\Delta(1700)$ excitation and ρ production in double-pion photoproduction"
16:40	G. Penner (Giessen) "The Giessen model - Vector meson production on the nucleon in a coupled channel K-matrix approach"
17:00	S. Kamalov (Dubna) "Partial wave analysis for pion photo- and electro-production with MAID and a dynamical model"
17:20	R. Davidson (RPI) "Model dependence of the Delta E/M ratio"
17:40	A. Waluyo (GWU) "Nucleon resonances in a coupled-channels K matrix formalism"
18:00	- *End of session* -

Parallel Session B "Models"

Chair	**D. O. Riska**
14:00	S. Boffi (Pavia) "Electromagnetic properties of baryons in a covariant quark model"
14:20	Ch. Helminen (Helsinki) "Low lying $qqqq\bar{q}$ states in the baryon spectrum"
14:40	B. Metsch (Bonn) "Parity doublets in a relativistic quark model"
15:00	B. Pasquini (ECT* Trento) "Dispersion analysis of virtual Compton scattering
15:20	A. Buchmann (Tübingen) "The neutron charge form factor and the $N \to \Delta$ quadrupole transition"
15:40	- *Coffee* -
Chair	**S. Scherer**
16:00	S. Gerasimov (Dubna) "Radial excitations of nucleons in a relativistic 'hybrid'-type quasipotential bag model"
16:20	Q. Zhao (Surrey) "Vector meson photoproduction in the quark model"
16:40	L. Fil'kov (Moscow) "Status of nucleon resonances with masses $m < m_N + m_\pi$"
17:00	B.-S. Zou (Beijing) "Partial wave analysis of J/Ψ decays to N* resonances"
17:20	A. L'vov (Moscow) "Spin structure of the Δ (1232) and inelastic Compton scattering
17:40	- *End of session* -

Parallel Session C "Experimental Results"

Chair	**J. Ahrens**
14:00	H.-P. Morsch (Jülich) "Structure of the Roper resonance in αp and πN scattering"
14:20	M. Schumacher (Göttingen) "Structure of the nucleon investigated by Compton scattering"
14:40	S. Jaminion (Clermont-Ferrand) "Virtual Compton scattering at JLab: Preliminary results in the polarizability domain at $Q^2 = 1$ and 1.9 GeV"
15:00	H. Hakobyan (Yerewan) "Cross section asymmetry measurements in π^0 photoproduction in the energy range of 0.5-1.1 GeV"
15:20	V. Kouznetsov (Moscow) "Beam asymmetries in positive pion photoproduction and Compton scattering at GRAAL"
15:40	- *Coffee* -
Chair	**J. Jourdan**
16:00	G. Laveissiere (Clermont-Ferrand) "Preliminary results of $p(e, e'p)X$ experiments in the nucleon resonance region
16:20	K. Joo (JLab) "Electron asymmetry in π^0 production near the $\Delta(1232)$"
16:40	W. Briscoe (GWU) "Non-strange baryon resonance experiments with the Crystal Ball"
17:00	M. G. Sokol (Moscow) "Observation of eta-mesic nuclei and a study of the $S_{11}(1535)$ nucleon resonance inside nuclei"
17:20	I. Lopatin (Petersburg) "Baryon spectroscopy: Experimental program at PNPI"
17:40	A. Sirunyan (Yerevan) "Measurement of the cross section asymmetry in deuteron photodisintegration with linearly polarized photons in the energy range of 0.8-1.6 GeV"
18:00	- *End of session* -

Friday, March 9

Chair	**G. Rosner**
8:30	F. Lee (GWU) "Nucleon resonances in lattice QCD"
9:00	S. Sasaki (Tokyo) "Lattice study of nucleon properties with domain wall fermions"
9:30	W. Melnitchouk (Adelaide) "Quark-hadron duality: Resonances and the onset of scaling"
10:00	*- Coffee -*
Chair	**P. Pedroni**
10:30	J.-P. Chen (JLab) "Spin-dependent electroproduction in the resonance region and the extended GDH sum rule"
11:00	R. Minehart (UVA) "Helicity dependent electroproduction"
11:30	A. Thomas (Mainz) "The helicity structure of the nucleon excitation spectrum and the GDH sum rule"
12:00	M. Kotulla (Giessen) "The static magnetic moment of the $\Delta(1232)$"
12:30	*- Lunch -*
Chair	**B. Krusche**
14:00	A. d'Angelo (Roma) "Meson photoproduction at GRAAL"
14:30	J. Cummings (RPI) "Maximum likelihood techniques for PWA of two-pion photoproduction"
15:00	J. Mueller (Pittsburgh) "Photo- and electroproduction of eta mesons"
15:30	B. Raue (FIU) "Kaon electroproduction and Λ polarization observables measured with CLAS"
16:00	*- Coffee -*
Chair	**F. Klein**
16:30	K.-H. Glander (Bonn) "Kaon and vector meson production at SAPHIR"
17:00	U. Mosel (Giessen) "Vector meson decay of nucleon resonances"
17:30	Y. Oh (Yonsei) "Higher and missing resonances in vector meson production"
18:00	*- End of Session -*

Saturday, March 10

Chair	**E. Klempt**
8:30	U. Thoma (Bonn) "Photoproduction of baryon resonances, first data from the CB-ELSA experiment"
9:00	T. Nakano (Osaka) "SPring-8/LEPS collaboration"
9:30	B.-S. Zou (Beijing) "The baryon resonance program at BES"
10:00	B. Nefkens (UCLA) "Hyperon resonances with the Crystal Ball"
10:30	*- Coffee -*
Chair	**J.-P. Chen**
11:00	S. Dytman (Pittsburgh) "Report on the BRAG workshop"
11:30	M. Ripani (Genova) "Higher resonances"
12:00	V. Metag (Gießen) "Nucleon resonances and mesons in nuclei"
12:30	V. Burkert (JLab) "Outlook"
13:00	*- End of Workshop -*

Participants

Jürgen Ahrens
Universität Mainz
Institut für Kernphysik
Becherweg 45
55099 Mainz
GERMANY
ahrens@kph.uni-mainz.de
Tel.: +49 6131 392 5195
Fax: +49 6131 392 2964

Hans-Jürgen Arends
Universität Mainz
Institut für Kernphysik
Becherweg 45
55099 Mainz
GERMANY
arends@kph.uni-mainz.de
Tel.: +49 6131 392 5194
Fax: +49 6131 392 2964

Inna G. Aznauryan
Yerevan Physics Institute
Alikhanian Brothers St. 2
Yerevan 375036
ARMENIA
aznaur@jerewan1.yerphi.am
Tel.: +374 1 35 00 30
Fax: +374 1 34 44 39

Reinhard Beck
Universität Mainz
Institut für Kernphysik
Becherweg 45
55099 Mainz
GERMANY
rbeck@kph.uni-mainz.de
Tel.: +49 6131 392 2933
Fax: +49 6131 392 2964

Cornelius Bennhold
George Washington University
Center for Nuclear Studies
Department of Physics
Washington, DC 20052
USA
bennhold@gwu.edu
Tel.: +1 202 994 6274
Fax: +1 202 994 3001

Frederic Bloch
Universität Basel
Institut für Physik
Klingelbergstr. 82
4056 Basel
SWITZERLAND
frederic.bloch@unibas.ch
Tel.: +41 61 267 3731

Jean-Paul Bocquet
Institut des Sciences Nucl.
53, Avenue des Martyrs
38026 Grenoble
FRANCE
bocquet@isn.in2p3.fr
Tel.: +33 4 7628 4185

Sigfrido Boffi
INFN Pavia
Via Bassi 6
27100 Pavia
ITALY
sigfrido.boffi@pv.infn.it
Tel.: +39 0382 507 434
Fax: +39 0382 507 752

Alessandro Braghieri
INFN Pavia
Via Bassi 6
27100 Pavia
ITALY
braghieri@pv.infn.it
Tel.: +39 0382 507 628

William J. Briscoe
George Washington University
Center for Nuclear Studies
Department of Physics
Washington, DC 20052
USA
briscoe@gwu.edu
Tel.: +1 202 994 6788
Fax: +1 202 994 3001

Alfons J. Buchmann
Universität Tübingen
Theoretische Physik
Auf der Morgenstelle 14
72076 Tübingen
GERMANY
alfons.buchmann@uni-tuebingen.de
Tel.: +49 721 608 2250

Volker D. Burkert
Jefferson Lab
12000 Jefferson Ave., MS 12 H
Newport News, VA 23606
USA
burkert@jlab.org
Tel.: +1 757 269 7540

Simon C. Capstick
Florida State University
Department of Physics
Tallahassee, FL 32306-4350
USA
capstick@csit.fsu.edu
Tel.: +1 850 644 1724
Fax: +1 850 644 4478

Ralph Castelijns
KVI Groningen
Zernikelaan 25
9747 AA Groningen
NETHERLANDS
castelijns@kvi.nl
Tel.: +31 50 363 6192
Fax: +31 50 363 4003

Jian-Ping Chen
Jefferson Lab
12000 Jefferson Ave., MS 12H
Newport News, VA 23606
USA
jpchen@jlab.org
Tel.: +1 757 269 7413
Fax: +1 757 269 5703

Wen-Tai Chiang
National Taiwan University
Department of Physics
Taipei 10617, Taiwan
R.O.C.
wtchiang@phys.ntu.edu.tw
Tel.: +886 918 58 3152
Fax: +886 2 2363 9984

Ronald L. Crawford
University of Glasgow
Physics & Astronomy
University Avenue
Glasgow G12 8QQ, Scotland
UNITED KINGDOM
r.crawford@physics.gla.ac.uk
Tel.: +44 141 330 4716
Fax: +44 141 330 6436

Volker Crede
Univerität Bonn
ISKP
Nussallee 14-16
53115 Bonn
GERMANY
crede@iskp.uni-bonn.de
Tel.: +49 228 73 9036
Fax: +49 228 73 2505

John P. Cummings
Rensselaer Polytechnic Inst.
Department of Physics
110 8th Street
Troy, NY 12180-3590
USA
cummij@rpi.edu
Tel.: +1 518 276 2542
Fax: +1 518 276 6680

Annalisa D'Angelo
INFN Roma II
Via della Ricerca Scientifica 1
00133 Roma
ITALY
annalisa.dangelo@roma2.infn.it
Tel.: +39 06 7259 4562
Fax: +39 06 204 0309

Eed M. Darwish
Universität Mainz
Institut für Kernphysik
Becherweg 45
55099 Mainz
GERMANY
darwish@kph.uni-mainz.de
Tel.: +49 6131 392 5850
Fax: +49 6131 392 2964

Richard M. Davidson
Rensselaer Polytechnic Inst.
Department of Physics
110 8th Street
Troy, NY 12180
USA
davidr@rpi.edu
Tel.: +1 518 276 8418
Fax: +1 518 276 6680

Dieter Drechsel
Universität Mainz
Institut für Kernphysik
Becherweg 45
55099 Mainz
GERMANY
drechsel@kph.uni-mainz.de
Tel.: +49 6131 392 3695
Fax: +49 6131 392 5474

Steven Dytman
University of Pittsburgh
Physics & Astronomy
Pittsburgh, PA 15260
USA
dytman@pitt.edu
Tel.: +1 412 624 9244
Fax: +1 412 624 9136

Latifa Elouadrhiri
Jefferson Lab
12000 Jefferson Ave.
Newport News, VA 23606
USA
latifa@jlab.org
Tel.: +1 757 269 7303
Fax: +1 757 269 5800

Lev V. Fil'kov
Lebedev Physical Institute
Leninsky Prospect 53
117924 Moscow
RUSSIA
filkov@x4u.lpi.ruhep.ru
Tel.: +7 095 13 58739
Fax: +7 095 13 26567

Jörg Friedrich
Universität Mainz
Institut für Kernphysik
Becherweg 45
55099 Mainz
GERMANY
friedrch@kph.uni-mainz.de
Tel.: +49 6131 392 5829
Fax: +49 6131 392 2964

Khalaf H. Gad
Universität Tübingen
Theoretische Physik
Auf der Morgenstelle 14
72076 Tübingen
GERMANY
kha92@yahoo.com
Tel.: +49 7071 29 74131
Fax: +49 7071 29 5388

Sergo Gerasimov
JINR Dubna
Laboratory of Theor. Physics
141980 Dubna (Moscow)
RUSSIA
gerasb@thsun1.jinr.ru
Tel.: +7 096 21 62749
Fax: +7 096 21 65084

Karl-Heinz Glander
Universität Bonn
Physikalisches Institut
Nussallee 12
53115 Bonn
GERMANY
glander@lisa.physik.uni-bonn.de

Ralf W. Gothe
Universität Bonn
Physikalisches Institut
Nussalle 12
53115 Bonn
GERMANY
gothe@physik.uni-bonn.de
Tel.: +49 228 73 2235
Fax: +49 228 73 7869

Helmut Haberzettl
George Washington University
Center for Nuclear Studies
Deptartment of Physics
Washington, DC 20052
USA
helmut@gwu.edu
Tel.: +1 202 994 0886
Fax: +1 202 994 3001

Hrachya Hakobian
Yerevan Physics Insitute
Alikhanian Brothers St. 2
Yerevan 375036
ARMENIA
hakopian@atlas.yerphi.am
Tel.: +374 1 35 22 81
Fax: +374 1 35 00 30

Christina L. Helminen
Department of Physics
P.O.Box 9
00014 University of Helsinki
FINLAND
chelmine@pcu.helsinki.fi
Tel.: +358 9 191 8374
Fax: +358 9 191 8378

Thomas R. Hemmert
TU München
Physik Department T39
James-Franck-Straße
85747 Garching, GERMANY
themmert@physik.tu-muenchen.de
Tel.: +49 89 2891 2886
Fax: +49 89 2891 2325

Gerhard Höhler
Universität Karlsruhe
Theoretische Teilchenphysik
Engesserstr. 7
76128 Karlsruhe
GERMANY
gerhard.hoehler@physik.uni-karlsruhe.de
Tel.: +49 721 6083 3373

Eid Hourany
IPN Orsay
91406 Orsay
FRANCE
hourany@ipno.in2p3.fr
Tel.: +33 1 6915 5096
Fax: +33 1 6915 6470

Stephanie Jaminion
LPC Universite Blaise Pascal
CNRS-IN2P3
24 Avenue des Landais
63177 Aubiere
FRANCE
jaminion@clermont.in2p3.fr
Tel.: +33 4 73 405 123
Fax: +33 4 73 264 598

Silke Janssen
Universität Gießen
II. Physikalisches Institut
Heinrich-Buff-Ring 16
35392 Gießen, GERMANY
silke.janssen@physik.uni-giessen.de
Tel.: +49 641 993 3227
Fax: +49 641 993 3209

Stijn Janssen
University of Gent
Subatomic & Radiation Physics
Proeftuinstraat 86
9000 Gent, BELGIUM
stijn.janssen@rug.ac.be
Tel.: +32 9 264 6541
Fax: +32 9 264 6697

Kyungseon Joo
Jefferson Lab
12000 Jefferson Ave.
Newport News, VA 23606
USA
kjoo@jlab.org
Tel.: +1 757 269 7764
Fax: +1 757 269 5800

Jürg Jourdan
Universität Basel
Institut für Physik
Klingelbergstr. 82
4056 Basel
SWITZERLAND
juerg.jourdan@unibas.ch
Tel.: +41 61 267 3689
Fax: +31 61 267 3784

Jörg Junkersfeld
Universität Bonn
Physikalisches Institut
Nussallee 14-16
53115 Bonn
GERMANY
junkersf@iskp.uni-bonn.de
Tel.: +49 228 73 9035
Fax: +49 228 73 2505

Sabit Kamalov
JINR Dubna
Laboratory of Theor. Physics
141980 Dubna (Moscow)
RUSSIA
kamalov@thsun.jinr.ru
Tel.: +7 09621 63615
Fax: +7 09621 65084

Chung-wen Kao
Universität Frankfurt
Theoretische Physik
Ditmarstr. 4, apt. 8
60054 Frankfurt/Main
GERMANY
kaochung@th.physik.uni-frankfurt.de
Tel.: +49 69 7982 2628

James J. Kelly
University of Maryland
Department of Physics
College Park, MD 20742
USA
jjkelly@physics.umd.edu
Tel.: +1 301 405 6110
Fax: +1 301 405 8558

Fritz Klein
Universität Bonn
Physikalisches Institut
Nussallee 12
53115 Bonn
GERMANY
klein@physik.uni-bonn.de
Tel.: +49 228 73 2340
Fax: +49 228 73 3518

Eberhard Klempt
Universität Bonn
ISKP
Nussallee 14-16
53115 Bonn
GERMANY
klempt@iskp.uni-bonn.de
Tel.: +49 228 73 2202
Fax: +49 228 73 2505

Willem Kloet
Rutgers University
Department of Physics
136 Frelinghuysen Road
Piscataway, NJ 08854
USA
kloet@physics.rutgers.edu
Tel.: +1 732 445 2517
Fax: +1 732 445 4343

Martin Kotulla
Universität Gießen
II. Physikalisches Institut
Heinrich-Buff-Ring 16
35392 Gießen
GERMANY
martin.j.kotulla@exp2.physik.uni-giessen.de
Tel.: +49 641 993 3223
Fax: +49 641 993 3209

Viatcheslav Kouznetsov
Institute for Nuclear Research
60th Anniversary of October, 7a
117312 Moscow
RUSSIA
slava@cpc.inr.ac.ru
Tel.: +7 095 135 2146
Fax: +7 095 135 2268

Siegfried F. Krewald
Forschungszentrum Jülich
Institut für Kernphysik
52425 Jülich
GERMANY
s.krewald@fz-juelich.de
Tel.: +49 2461 61 4370

Jochen Krimmer
Universität Tübingen
Physikalisches Institut
Auf der Morgenstelle 14
72076 Tübingen, GERMANY
krimmer@pit.physik.uni-tuebingen.de
Tel.: +49 7071 29 78618
Fax: +49 7071 29 5373

Bernd Krusche
Universität Basel
Institut für Physik
Klingelbergstr. 82
4056 Basel
SWITZERLAND
bernd.krusche@unibas.ch
Tel.: +41 61 267 3696
Fax: +41 61 267 3784

Timo A. Lahde
Helsinki Institute of Physics
POB 64
00014 University of Helsinki
FINLAND
talahde@pcu.helsinki.fi

Michael Lang
Universität Mainz
Institut für Kernphysik
Becherweg 45
55099 Mainz
GERMANY
langa2@kph.uni-mainz.de
Tel.: +49 6131 392 5899
Fax: +49 6131 392 2964

Geraud Laveissiere
LPC Université Blaise Pascal
IN2P3-CNRS
24 Avenue des Landais
63177 Aubière
FRANCE
geraud@clermont.in2p2.fr
Tel.: +33 473 405 123
Fax: +33 473 264 598

Frank X. Lee
George Washington University
Center for Nuclear Studies
Department of Physics
Washington, DC 20052
USA
fxlee@gwu.edu
Tel.: +1 202 994 6485
Fax: +1 202 994 3001

Herbert Loehner
KVI Groningen
Zernikelaan 25
9747 AA Groningen
NETHERLANDS
loehner@kvi.nl
Tel.: +31 50 363 3600
Fax: +31 50 363 4003

Igor Lopatin
Petersburg Nucl. Physics. Inst.
Gatchina
Leningrad district, 188350
RUSSIA
lopatin@mail.pnpi.spb.ru
Tel.: +7 812 714 6744
Fax: +7 812 713 7196

Anatoly I. L'vov
Lebedev Physical Institute
Leninsky Prospect 53
117924 Moscow
Russia
lvov@x4u.lpi.ruhep.ru
Fax: +7 095 938 2251

Valery E. Lyubovitskij
Universität Tübingen
Theoretische Physik
Auf der Morgenstelle 14
72076 Tübingen, GERMANY
valeri.lyubovitskij@uni-tuebingen.de
Tel.: +49 7071 29 78637
Fax: +49 7071 29 5388

Wally Melnitchouk
Adelaide University/JLab
CSSM Adelaide
Adelaide, 5005 South Australia
AUSTRALIA
wmelnitc@jlab.org
Tel.: +61 8 8303 3544
Fax: +61 8 8303 3551

Johan Messchendorp
Universität Gießen
II. Physikalisches Institut
Heinrich-Buff-Ring 16
35392 Gießen, GERMANY
j.messchendorp@exp2.physik.uni-giessen.de
Tel.: +49 641 993 3272
Fax: +49 641 993 3209

Volker Metag
Universität Gießen
II. Physikalisches Institut
Heinrich-Buff-Ring 16
35392 Gießen, GERMANY
volker.metag@exp2.physik.uni-giessen.de
Tel.: +49 641 993 3260
Fax: +49 641 993 3209

Bernard Metsch
Universität Bonn
Theoretische Kernphysik
Nussallee 14-16
53115 Bonn
GERMANY
metsch@itkp.uni-bonn.de
Tel.: +49 228 73 2378
Fax: +49 228 23 3728

Werner Meyer
Universität Bochum
Universitätsstr. 150
44780 Bochum
GERMANY
meyer@tau.ep1.ruhr-uni-bochum.de
Tel.: +49 234 322 8558

Ralph Minehart
University of Virginia
Department of Physics
382 McCormick Road
Charlottesville, VA 22904
USA
minehart@virginia.edu
Tel.: +1 804 924 6789
Fax: +1 804 924 4576

Victor I. Mokeev
Moscow State University
Nuclear Physics
Gory MSU INP OEPVAYa
119899 Vorob'evy
RUSSIA
mokeev@depni.npi.msu.su
Tel.: +7 095 030 2558
Fax: +7 095 939 0896

Hans-Peter Morsch
Forschungszentrum Jülich
Institut für Kernphysik
52425 Jülich
GERMANY
morsch@fz-juelich.de
Tel.: +49 2461 61 4403
Fax: +49 2461 61 3930

Ulrich Mosel
Universität Gießen
Theoretische Physik
Heinrich-Buff-Ring 16
35392 Gießen
GERMANY
mosel@physik.uni-giessen.de
Tel.: +49 641 993 3300
Fax: +49 641 993 3309

James A. Mueller
University of Pittsburgh
Physics and Astronomy
100 Allen Hall, 3941 O'Hara St.
Pittsburgh, PA 15360
USA
mueller+@pitt.edu
Tel.: +1 412 624 1566
Fax: +1 412 624 9163

Jose C. Nacher
Inst. Investigación de Paterna
Física Teórica &IFIC
Apdo. Correos 22085
46071 Valencia
SPAIN
nacher@condor4.ific.uv.es
Tel.: +34 96 398 3525
Fax: +34 96 398 3488

Takashi Nakano
University of Osaka
RCNP
10-1 Mihogaoka
Ibaraki, Osaka 567-0047
JAPAN
nakano@rcnp.osaka-u.ac.jp
Tel.: +81 6 6879 8938
Fax: +81 6 6879 8899

Ben M. Nefkens
University of California, UCLA
Physics & Astronomy
5-136 Knudsen Hall
Los Angeles, CA 90095
USA
nefkens@physics.ucla.edu
Tel.: +1 310 825 4970
Fax: +1 310 206 4387

Reiner Neuhausen
Universität Mainz
Institut für Kernphysik
Becherweg 45
55099 Mainz
GERMANY
neuhaus@kph.uni-mainz.de
Tel.: +49 6131 392 5827

Tobias P. Oed
Old Dominion University
Physics Department
4911 Newport ave apt. 3
Norfolk, VA 23508
USA
tobias@physics.odu.edu
Tel.: +1 7757 683 4613

Yongseok Oh
Yonsei University
Physics & Applied Physics
Shin-Chon-Dong,
Seodaemun-Gu
Seoul, 120-749, KOREA
yoh@phya.yonsei.ac.kr
Tel.: +82 2 2123 4558
Fax: +82 2 392 1592

Jose A. Oller
Forschungszentrum Jülich
Institut für Kernphysik
52425 Jülich
GERMANY
u.a.oller@fz-juelich.de
Tel.: +49 2641 61 6307
Fax: +49 2641 61 3930

Eulogio Oset
Inst. Investigación de Paterna
Física Teórica &IFIC
Apdo. Correos 22085
46071 Valencia
SPAIN
oset@condor1.ific.uv.es
Tel.: +34 96 398 3523
Fax: +34 96 398 3488

Michael Ostrick
Universität Bonn
Physikalisches Institut
Nussallee 12
53115 Bonn
GERMANY
ostrick@physik.uni-bonn.de
Tel.: +49 228 73 2447
Fax: +49 228 73 3518

Costas N. Papanicolas
University of Athens
Accelerating Systems &
Applications
IASA, P.O.Box 17214
10024 Athens, GREECE
cnp@cc.uoa.gr
Tel.: +30 1 72 47181
Fax: +30 1 72 95069

Barbara Pasquini
ECT*
Strada delle Tabarelle 286
38050 Villazzano (Trento)
ITALY
pasquini@ect.it
Tel.: +39 0461 314 729
Fax: +39 0461 395 007

Paolo Pedroni
INFN Pavia
Via Bassi 6
27100 Pavia
ITALY
pedroni@pv.infn.it
Tel.: +39 0382 507 635
Fax: +39 0382 423 241

Gregor Penner
Universität Gießen
Theoretische Physik
Heinrich-Buff-Ring 16
35392 Gießen
GERMANY
gregor.penner@theo.physik.uni-giessen.de
Tel.: +49 641 993 3323
Fax: +49 641 993 3309

Marco Pfeiffer
Universität Gießen
II. Physikalisches Institut
Heinrich-Buff-Ring 16
35392 Gießen
GERMANY
marco.post@physik.uni-giessen.de
Tel.: +49 641 993 3227
Fax: +49 641 993 3209

Josef G. Pochodzalla
Universität Mainz
Institut für Kernphysik
Becherweg 45
55099 Mainz
GERMANY
pochodza@kph.uni-mainz.de
Tel.: +49 6131 392 5832

Markus Post
Universität Gießen
Theoretische Physik
Heinrich-Buff-Ring 16
35392 Gießen, GERMANY
marcus.post@theo.physik.uni-giessen.de
Tel.: +49 641 993 3326
Fax: +49 641 993 3309

Milan Potokar
Jozef Stefan Institute
Jamova 39
1000 Ljubljana
SLOVENIA
milan.potokar@ijs.si
Tel.: +386 1 477 3209

Angels Ramos
Universitat de Barcelona
Department ECM
Diagonal 647
08028 Barcelona
SPAIN
ramos@ecm.ub.es
Tel.: +34 93 402 1185
Fax: +34 93 402 1198

Brian A. Raue
Florida International University
Physics Department
University Park
Miami, FL 33199
USA
baraue@sarah.fiu.edu
Tel.: +1 305 348 3958
Fax: +1 305 348 6700

Dominique Rebreyend
ISN Grenoble
53, Avenue des Martyrs
38026 Grenoble
FRANCE
rebreyend@isn.in2p3.fr
Tel.: +33 4 7528 4170
Fax: +33 4 7528 4004

Jörg Reinnarth
Universität Bonn
ISKP
Nussallee 14-16
53118 Bonn
GERMANY
reinnart@iskp.uni-bonn.de
Tel.: +49 228 73 2944

Marco Ripani
INFN Genova
Via Dodecanesco 33
16146 Genova
ITALY
ripani@ge.infn.it
Tel.: +39 010 353 6458
Fax: +39 010 313 358

Dan-Olof Riska
Helsinki Institute of Physics
POB 9
00014 University of Helsinki
FINNLAND
riska@pcu.helsinki.fi
Tel.: +358 0 101 8506
Fax: +358 9 191 8458

Günther Rosner
University of Glasgow
Physics & Astronomy
Univ. Ave., Kelvin Bldg.
Glasgow G12 8QQ, Scotland
UNITED KINGDOM
g.rosner@physics.gla.ac.uk
Tel.: +44 141 330 2774
Fax: +44 141 330 2630

Michael E. Sadler
Abilene Christian University
Physics Department
320 B Foster Science Bldg.
Abilene, TX 79699
USA
sadler@physics.acu.edu
Tel.: +1 915 674 2189
Fax: +1 915 674 2146

Agus Salam
Universität Mainz
Institut für Kernphysik
Becherweg 45
55099 Mainz
GERMANY
salam@kph.uni-mainz.de
Tel.: +49 6131 392 5741

Adam J. Sarty
Saint Mary's University
Astronomy & Physics
923 Robie Street
Halifax, NS, B3H 3C3
CANADA
sarty@ap.stmarys.ca
Tel.: +1 902 420 5664
Fax: +1 902 420 5141

Shoichi Sasaki
University of Tokyo
Physics Department
7-3-1 Hongo
Bunkyo-ku, Tokyo, 113-0033
JAPAN
ssasaki@phys.s.u-tokyo.ac.jp
Fax: +81 3 5841 4224

Toru Sato
Osaka University
Physics Department
Graduate School of Science
Toyonaka, Osaka, 560-0043
JAPAN
tsato@phys.sci.osaka-u.ac.jp
Tel.: +81 6 6850 5345
Fax: +81 6 6850 5529

Susan Schadmand
Universität Gießen
II. Physikalisches Insitut
Heinrich-Buff-Ring 16
35392 Gießen, GERMANY
s.schadmand@exp2.physik.uni-giessen.de
Tel.: +49 641 993 3276
Fax: +49 641 993 3209

Stefan Scherer
Universität Mainz
Institut für Kernphysik
Becherweg 45
55099 Mainz
GERMANY
scherer@kph.uni-mainz.de
Tel.: +49 6131 392 3289
Fax: +49 6131 392 5474

Hartmut Schmieden
Universität Mainz
Institut für Kernphysik
Becherweg 45
55099 Mainz
GERMANY
hs@kph.uni-mainz.de
Tel.: +49 6131 392 2933
Fax: +49 6131 392 2964

Berthold H. Schoch
Universität Bonn
Physikalisches Institut
Nussallee 12
53115 Bonn
GERMANY
schoch@physik.uni-bonn.de
Tel.: +49 228 73 2344
Fax: +49 228 73 3518

Martin Schumacher
Universität Göttingen
II. Physikalisches Institut
Bunsenstr. 7-9
37073 Göttingen, GERMANY
schumacher@physik2.uni-goettingen.de
Tel.: +49 551 39 7644
Fax: +49 551 39 4493

Michael Seimetz
Universität Mainz
Institut für Kernphysik
Becherweg 45
55099 Mainz
GERMANY
seimetz@kph.uni-mainz.de
Tel.: +49 6131 392 2935
Fax: +49 6131 392 2964

Silvano Simula
INFN Sezione Roma III
Dipartimento di Fisica
Via della Vasca Navale, 84
00146 Roma
ITALY
simula@roma3.infn.it
Tel.: +39 06 5517 7053
Fax: +39 06 5517 5059

Albert M. Sirunian
Yerevan Physics Insitute
Alikhanyan Brothers St. 2
Yerevan 375036
ARMENIA
sirunian@uniphi.yerphi.am
Tel.: +374 1 34 27 47
Fax: +374 1 35 00 15

Lee C. Smith
University of Virginia
Physics Department
382 Mc Cormick Road
Charlottesville, VA 22904-4714
USA
cole@nstar.phys.virginia.edu
Tel.: +1 804 924 3781
Fax: +1 804 924 4576

Garry A. Sokol
Lebedev Physical Institute
Leninsky Prospect 53
117924 Moscow
RUSSIA
gsokol@x4u.lebedev.su
Tel.:+7 095 334 0119
Fax: +7 095 938 2251

Igor I. Strakowski
Center for Nuclear Studies
George Washington University
Physics Department
Washington, CD 20052
USA
igor@gwu.edu
Tel.: +1 703 726 8344
Fax: +1 703 726 8248

Alfred Švarc
Rudjer Boškovic Institute
Bijenička C. 54
10000 Zagreb
CROATIA
svarc@rudjer.irb.hr
Tel.: +385 1 456 1090
Fax: +385 1 468 0237

Ulrike Thoma
Universität Bonn
ISKP
Nussallee 14-16
53115 Bonn
GERMANY
thoma@iskp.uni-bonn.de
Tel.: +49 228 73 3275
Fax: +49 228 73 2505

Andreas Thomas
Universität Mainz
Institut für Kernphysik
Becherweg 45
55099 Mainz
GERMANY
thomas@kph.uni-mainz.de
Tel.: +49 6131 392 2948
Fax: +49 6131 392 2964

Anthony W. Thomas
Adelaide University
CSSM
Adelaide 5005 South Australia
AUSTRALIA
athomas@physics.adelaide.edu.au
Tel.: +61 8 8303 5113
Fax: +61 8 8303 3551

Lothar Tiator
Universität Mainz
Institut für Kernphysik
Becherweg 45
55099 Mainz
GERMANY
tiator@kph.uni-mainz.de
Tel.: +49 6131 392 3697
Fax: +49 6131 392 5474

Tim Van Cauteren
Universiteit Gent
Subatomic and Radiation Physics
Proeftunistraat 86
9000 Gent, BELGIUM
tim@talvin5.rug.ac.be
Tel.: +32 9 264 6555
Fax: +32 9 264 6697

Robert E. Van de Vyver
Universiteit Gent
Subatomic and Radiation Physics
Proeftunistraat 86
9000 Gent, BELGIUM
robert@inwfsun1.rug.ac.be
Tel.: +32 9 264 6544
Fax: +32 9 264 6697

Luc Van Hoorebeke
Universiteit Gent
Subatomic and Radiation Physics
Proeftunistraat 86
9000 Gent, BELGIUM
luc@inwfsun1.rug.ac.be
Tel.: +32 9 264 6543
Fax: +32 9 264 6697

Marc Vanderhaeghen
Universität Mainz
Institut für Kernphysik
Becherweg 45
55099 Mainz
GERMANY
marcvdh@kph.uni-mainz.de
Tel.: +49 6131 392 4277
Fax: +49 6131 392 5474

Thomas Walcher
Universität Mainz
Institut für Kernphysik
Becherweg 45
55099 Mainz
GERMANY
walcher@kph.uni-mainz.de
Tel.: +49 6131 392 5196
Fax: +49 6131 392 3825

Agung B. Waluyo
George Washington University
Center for Nuclear Studies
Department of Physics
Washington, DC 20052
USA
waluyoab@gwu.edu
Tel.: +1 202 994 6278
Fax: +1 202 994 3001

Ron Workman
George Washington University
Center for Nuclear Studies
Department of Physics
Washington, DC 20052
USA
workman@gwu.edu

Shin-Nan Yang
National Taiwan University
Physics Department
Taipei 10617, TAIWAN
P.R.C.
snyang@phys.ntu.edu.tw
Tel.: +886 2 2362 6937
Fax: +886 2 2363 9984

Qiang Zhao
University of Surrey
Physics Department
Guildford, GU2 7XH
United KINGDOM
qiang.zhao@surrey.ac.uk
Tel.: +44 1483 87 2730
Fax: +44 1483 87 6781

Bing-Song Zou
Chinese Academy of Science
High Energy Physics
P.O.Box 918(4), Yuquanlu Road
Beijing, 100039
P.R.C.
zoubs@mail.ihep.ac.cn
Tel.: +86 10 6823 6162
Fax: +86 10 6821 8318